MW00526256

Advances in
Atmospheric Chemistry

Volume 1

Advances in Atmospheric Chemistry **ISSN 2425-0015**

Series Editors: J. R. Barker *(University of Michigan, USA)*
A. L. Steiner *(University of Michigan, USA)*
T. J. Wallington *(Ford Motor Company, USA)*

Vol. 1 Advances in Atmospheric Chemistry
edited by J. R. Barker, A. L. Steiner and T. J. Wallington

Advances in Atmospheric Chemistry

Volume 1

Editors

J. R. Barker
University of Michigan, Ann Arbor, USA

A. L. Steiner
University of Michigan, Ann Arbor, USA

T. J. Wallington
Ford Motor Company, USA

NEW JERSEY • LONDON • SINGAPORE • BEIJING • SHANGHAI • HONG KONG • TAIPEI • CHENNAI • TOKYO

Published by

World Scientific Publishing Co. Pte. Ltd.
5 Toh Tuck Link, Singapore 596224
USA office: 27 Warren Street, Suite 401-402, Hackensack, NJ 07601
UK office: 57 Shelton Street, Covent Garden, London WC2H 9HE

Library of Congress Cataloging-in-Publication Data
Names: Barker, John Roger, editor. | Steiner, A. L. (Allison L.), 1972– editor. | Wallington, Timothy J., editor.
Title: Advances in atmospheric chemistry / editors, J.R. Barker, University of Michigan, Ann Arbor, USA, A.L. Steiner, University of Michigan, Ann Arbor, USA, T.J. Wallington, Ford Motor Company, USA.
Description: New Jersey : World Scientific, 2017– | Includes bibliographical references.
Identifiers: LCCN 2016053759 | ISBN 9789813147348 (hardcover : alk. paper : v. 1)
Subjects: LCSH: Atmospheric chemistry.
Classification: LCC QC879.6 .A334 2017 | DDC 551.51/1--dc23
LC record available at https://lccn.loc.gov/2016053759

British Library Cataloguing-in-Publication Data
A catalogue record for this book is available from the British Library.

Typeset by Stallion Press
Email: enquiries@stallionpress.com

Printed in Singapore

Volume 1. Editorial Advisory Board

Introduction to the Series

Scientific study of the atmosphere began more than 200 years ago, when the gas laws were discovered and when physics and chemistry were emerging as distinct scientific fields. The atmosphere provided a convenient medium for research and, as a result, its chemical properties gradually became known. Progress in understanding the atmosphere came as the result of asking questions and developing experimental and theoretical tools.

The same is true today. Advances in atmospheric chemistry are the result of developing new tools, new theories, and new ways of using the old. Advances also come from asking new questions and exploring undiscovered aspects of our atmosphere. Today, the experimental tools at hand are very wide-ranging, including both the oldest (trapping gases and measuring pressures) and the newest (remote sensing from satellite platforms). Today, theory is highly developed and is used in all phases of research, including analyzing experimental and observational data, obtaining predictions from sophisticated climate models, and predicting atmospheric reactivity by solving the Schrodinger equation.

We live in the era of global warming and "carbon pollution" of the atmosphere, which are among the greatest challenges faced by modern society. The atmosphere is the central arena for climate change. Greenhouse gases and short-lived climate forcing agents such as ozone and aerosols are emitted directly into the atmosphere or formed *via* chemical reactions in the atmosphere. The interaction of these atmospheric gases and particles with radiation is the principal source of climate forcing and, as voiced by the Intergovernmental

Panel on Climate Change (IPCC), human activities are largely responsible. To mitigate the effects of anthropogenic influences requires understanding the natural atmosphere, anthropogenic pollutants, and the interactions among all sources of emissions. Moreover, the atmosphere is just one component of the complex Earth system, requiring an understanding of the feedback mechanisms linking the atmosphere to the other components of the climate system.

The aim of this series is to present informed, invited summaries of research on atmospheric chemistry in a changing world. The summaries range from comprehensive reviews of major subject areas to focused accounts of specific topics by individual research groups. Atmospheric chemistry is a multidisciplinary field drawing together elements of physics, chemistry, meteorology, biology, oceanography and computer science. Because the field is so extensive, reviews are important resources for both students and professionals. It is our goal that this series will be a key resource for those who wish to learn about the latest developments directly from the researchers who are leading the way. By gathering these new advances in one place, we aim to catalyze productive communication among the many researchers studying our changing and fascinating atmosphere.

John R. Barker
Allison L. Steiner
Timothy J. Wallington
Editors

Contents

List of Contributors

J. P. D. Abbatt
Department of Chemistry, University of Toronto,
80 St. George Street, Toronto, ON, Canada, M5S 3H6

Mads P. Sulbaek Andersen
Department of Chemistry and Biochemistry,
California State University, Northridge, California 91330, USA

Copenhagen Center for Atmospheric Research,
Department of Chemistry, University of Copenhagen,
Universitetsparken 5, DK-2100 Copenhagen Ø, Denmark

John R. Barker
Climate and Space Sciences and Engineering,
University of Michigan, Ann Arbor, MI 48109-2143, USA

Jiajue Chai
Chemistry Department,
State University of New York-Environmental Science and Forestry,
1 Forestry Drive, Syracuse,
NY USA 13210

Matthew M. Coggon
National Oceanic and Atmospheric Administration, Boulder, CO, USA

Theodore S. Dibble
Chemistry Department,
State University of New York-Environmental Science and Forestry,
1 Forestry Drive, Syracuse, NY USA 13210

Sebastien Dusanter
Sciences de l'Atmosphére et Génie de l'Environnement,
Mines Douai, France

Richard C. Flagan
California Institute of Technology, Pasadena, CA 91125, USA

A. K. Y. Lee
Department of Chemistry, University of Toronto,
80 St. George Street, Toronto, ON, Canada, M5S 3H6

Department of Civil and Environmental Engineering,
National University of Singapore, 1 Engineering Drive 2,
Singapore 117576, Singapore

Hanna Lignell
California Institute of Technology, Pasadena, CA 91125, USA

Renee C. McVay
California Institute of Technology, Pasadena, CA 91125, USA

Thanh Lam Nguyen
Department of Chemistry, The University of Texas Austin,
Texas 78712-0165, USA

Ole John Nielsen
Copenhagen Center for Atmospheric Research,
Department of Chemistry, University of Copenhagen,
Universitetsparken 5, DK-2100 Copenhagen Ø, Denmark

Rebecca H. Schwantes
California Institute of Technology, Pasadena, CA 91125, USA

John H. Seinfeld
California Institute of Technology, Pasadena, CA 91125, USA

John F. Stanton
Department of Chemistry, The University of Texas Austin,
Texas 78712-0165, USA
jfstanton@mail.utexas.edu

Philip S. Stevens
School of Public and Environmental Affairs
and Department of Chemistry, Indiana University,
Bloomington, Indiana, USA

Timothy J. Wallington
Research and Advanced Engineering, Ford Motor Company,
Dearborn, Michigan 48121-2053, USA

C. Wang
Department of Chemistry, University of Toronto,
80 St. George Street, Toronto, ON, Canada, M5S 3H6

Department of Physical and Environmental Sciences,
University of Toronto Scarborough, 1095 Military Trail,
Toronto, ON, Canada, MIC 1A4

F. Wania
Department of Chemistry, University of Toronto,
80 St. George Street, Toronto, ON, Canada, M5S 3H6

Department of Physical and Environmental Sciences,
University of Toronto Scarborough, 1095 Military Trail,
Toronto, ON, Canada, MIC 1A4

Paul O. Wennberg
California Institute of Technology, Pasadena, CA 91125, USA

J. P. S. Wong
Department of Chemistry, University of Toronto,
80 St. George Street, Toronto, ON, Canada, M5S 3H6

School of Earth and Atmospheric Sciences,
Georgia Institute of Technology, 311 Ferst Dr. Atlanta,
GA, USA, 30332

Xuan Zhang
Aerodyne Research Inc., Billerica, MA, USA

R. Zhao
Department of Chemistry, University of Toronto,
80 St. George Street, Toronto, ON, Canada, M5S 3H6

Division of Chemistry and Chemical Engineering,
California Institute of Technology, 1200 E California Blvd,
Pasadena, CA, USA, 91125

S. Zhou
Department of Chemistry, University of Toronto,
80 St. George Street, Toronto, ON, Canada, M5S 3H6

Xianliang Zhou
New York State Department of Health, Wadsworth Center,
State University of New York,
Department of Environmental Health Sciences,
Albany, NY 12201, USA

Lei Zhu
New York State Department of Health, Wadsworth Center,
State University of New York,
Department of Environmental Health Sciences,
Albany, NY 12201, USA

Chapter 1

Science of the Environmental Chamber

Rebecca H. Schwantes*, Renee C. McVay*, Xuan Zhang†, Matthew M. Coggon‡, Hanna Lignell*, Richard C. Flagan*, Paul O. Wennberg* and John H. Seinfeld*§

**California Institute of Technology, Pasadena, CA 91125, USA*

†*Aerodyne Research Inc., Billerica, MA, USA*

‡*National Oceanic and Atmospheric Administration, Boulder, CO, USA*

§*seinfeld@caltech.edu*

Atmospheric chemistry is simulated in the laboratory using several types of environmental chambers; these include the batch chamber, the continuously mixed flow reactor, and the flow tube reactor. These reactors are used to study gas-phase oxidation of volatile organic compounds (VOCs) as well as the formation of secondary organic aerosol (SOA), the process by which VOCs undergo oxidation to form low-volatility products that condense onto particles. This chapter focuses on the design and characterization of environmental chambers, including: (1) radiation conditions; (2) chamber mixing state; (3) chemical blank experiments; (4) free radical generation (principally the hydroxyl (OH) radical); (5) high-*versus*-low-NO conditions that govern the nature of VOC oxidation chemistry; (6) deposition of particles onto chamber walls; (7) deposition of organic vapors onto chamber walls; and (8) determination of the yield of SOA. Comparison of the design and behavior of the different types of reactor is addressed in detail. The performance of the differential mobility analyzer (DMA), the prime instrument for measuring aerosol size distributions in chambers, is addressed.

1.1. Introduction

The environmental chamber is used to isolate atmospheric chemistry under well-controlled conditions. Both gas-phase chemistry and secondary organic aerosol (SOA) formation and growth are studied in such chambers. SOA is formed when volatile organic compounds (VOCs) undergo oxidation to form low volatility products that subsequently partition into the particle phase. Numerous environmental chambers have been constructed and are in use worldwide. The science underlying the environmental chamber can be divided into four parts: (1) design of the chamber, (2) characterization of the chamber, (3) execution of experiments, and (4) interpretation of the data. The purpose of this chapter is to discuss each of these aspects so as to elucidate the considerations in the use of an environmental chamber to perform studies of atmospheric chemistry and aerosol formation. A critical aspect of environmental chamber experiments is the suite of instrumentation used to characterize the gas and particle phases in the chamber. We will address the measurement of particle size distributions in chambers; a number of reviews of gas- and particle-phase chemical composition measurements exist, so we do not address these here.

1.2. Reactor Type

Two broad types of environmental reactors are in common use. The first is simply a batch reactor, the contents of which are well mixed. This type of reactor is referred to as an *environmental chamber.* The second type is a tubular flow reactor into which reactants are introduced at one end and products are withdrawn at the other end, with reactions proceeding as the material flows down the tube. This type of reactor is referred to as a *flow tube reactor.*

There are two modes of operation of the environmental chamber: *batch* and *continuous.* In batch mode, the chamber is initially filled with specified gases (and perhaps particles). After time is allowed for reactants to become sufficiently mixed, reactions are initiated. The chamber walls can be flexible or fixed. Over the duration of an experiment, samples of chamber air containing both gases (and particles) are withdrawn for analysis. The duration of a batch mode experiment is ultimately limited by the volume of chamber air withdrawn for analysis over the course of an experiment and by the magnitude of the accumulated deposition of particles and gases to the walls of the chamber. In a flexible walled chamber, the chamber volume decreases as material is withdrawn for measurement. In a fixed wall chamber, air must be continually replenished to maintain atmospheric pressure in the chamber. Chambers may or may not incorporate active mixing within the chamber. Mixing occurs as a result of the flows used to introduce reactants and withdraw chamber samples, and perhaps as a result of natural movements of the flexible walls of the chamber.

An alternative to the batch-mode (time-dependent) chamber operation is the steady-state continuous flow mode, in which case the reactor is referred to as a continuously mixed flow reactor (*CMFR*). The CMFR is analogous to the batch chamber in physical configuration, but differs in that throughout the course of an experiment there is a continuous flow of reactants into the chamber and a continuous flow (unreacted reactants and reaction products) out of the chamber. After an initial transient start-up period in a CMFR, the gases (and particles) in the reactor eventually achieve steady state. The contents of the CMFR are usually well mixed, whether actively or not, so that the concentrations in the outflow are essentially identical to those in the bulk of the chamber. Continuous sampling of the effluent permits accumulation of arbitrarily large quantities of gases and particles for analysis. The characteristic time scale of a CMFR is the mean residence time in the reactor, the ratio of the volume of the reactor to the volumetric flow rate of air through the chamber.

In the Flow Tube Reactor, reactants are introduced at one end of a tubular reactor and reaction products are withdrawn from the other end. The flow rate through the tube is adjusted to achieve a desired residence time in the reactor. The flow rate of air through the tube and the diameter of the tube itself determine the state of the flow (either laminar or turbulent). Like the CMFR, the Flow Tube Reactor affords continuous sampling of the effluent.

A common material used for flexible-walled batch environmental chambers is fluorinated ethylene propylene (FEP) Teflon film, usually of thickness 0.05 mm (2 mil). Irradiation of the chamber with actual or artificial sunlight is usually required, and Teflon film has the attribute that it is essentially transparent to ultraviolet and visible radiation. While Teflon is a relatively inert material, there is evidence that particles and certain organic molecules can deposit and adhere to the Teflon walls of a chamber. These "wall effects" play an important role in the interpretation of data, and they will be addressed later in this chapter in some detail. Permanent materials for fixed-wall chamber construction are stainless steel and glass. These materials offer the advantage that they can be cleaned between experiments and that, if desired, experiments can be carried out at pressures other than atmospheric. The flow tube reactor is commonly constructed of glass, quartz, or steel, with appropriate attention to transmission of radiation.

1.3. Characterization of the Environmental Chamber

Characterization of the environmental chamber is essential prior to its use in studying atmospheric chemistry. In this section, we outline a number of characterization tests and illustrate the results of such tests carried out on the Caltech chambers, summarized in Table 1.1. Earlier reports of chamber characterizations are those of Cocker *et al.*[1] and Carter *et al.*[2,3] and Wang *et al.*[4]

1.3.1. *Photolytic Environment*

Because the ultimate driving force for atmospheric chemistry is solar radiation, an environmental chamber requires a source of radiation that is either the Sun itself or a source that approximates the solar spectrum. The earliest environmental chambers were constructed outdoors to take advantage of sunlight as the source of radiation. There exist today two major European outdoor chamber facilities, EUPHORE in Valencia, Spain (www.eurochamp.org/chambers/euphore/) and SAPHIR in Juelich, Germany (www.eurochamp.org/chambers/saphir/). For indoor chambers, two types

Table 1.1 Environmental chamber characterization tests

Characterization test	Procedure
Leak rate	Evaluate the extent to which ambient air outside of the chamber leaks into the chamber.
Aerosol background	Use differential mobility analyzer (DMA) coupled to condensation particle counter (CPC) to determine the background particle concentration in the chamber when the chamber is filled with purified air and irradiated.
Mixing time	Evaluate the characteristic mixing time in the chamber: Add an inert gas (e.g., CO), take measurements at different ports and determine the amount of time it takes for the gas to become well mixed.
Irradiation spectrum	Use a portable spectroradiometer to measure the irradiance spectrum of the light source and compare this irradiance spectrum to the solar spectrum.
Light homogeneity	Evaluate the extent to which the radiant flux in the chamber enclosure is uniform: Place photodiodes at varying locations in the chamber enclosure. Record the signal hitting the top (direct) and bottom (reflected) of the photodiode.
Photolysis frequencies of NO_2 and O_3	Calculate photolysis rates by converting the irradiance spectrum determined by the spectroradiometer into actinic flux, and then using the absorption spectrum for the molecule.
Cleaning protocol	Flush chamber with ~10 chamber volumes of clean air over 24 h. During this, periodically bake under full blacklights at 50°C.
NO_x-air photolysis	Photolyze varying amounts of initial NO_2 and monitor the concentrations of NO_2, NO, O_3, and HONO over time.
Vapor wall-deposition	Calculate the first-order wall deposition rate from the decrease in the concentration of a spectrum of compounds under dark conditions. Test vapor-wall deposition under different relative humidity levels (e.g., for glyoxal in Caltech chamber: k_w (dry) = $9.6 \times 10^{-7}\,s^{-1}$, but k_w (RH = 61%) = $4.7 \times 10^{-5}\,s^{-1}$)
Propene-NO_x photooxidation	Add initial amounts of propene and NO_x to the chamber and monitor the concentration of NO_2, NO, O_3, propene, and secondary organic compounds (e.g., HCHO, CH_3CHO, HCOOH, and PAN) during photooxidation. Compare measured concentrations to

(*Continued*)

Table 1.1 (*Continued*)

Characterization test	Procedure
	those predicted by the Master Chemical Mechanism (MCM) for propene. Evaluate HO_x sources and sinks.
Species off-gassing from walls	During the experiments above (e.g., photolysis of NO_x alone and photolysis of propene in the presence of NO_x), use CIMS to evaluate the extent to which species might be off-gassing from the walls.
Photolysis of NO_x in the presence of ammonium sulfate seed aerosol	Add ammonium sulfate seed aerosol and NO_x to the chamber and irradiate this mixture over ~5 h to determine the background SOA production from any residual organic gases in the chamber.
Aerosol generation	Test the aerosol generation equipment by atomizing various solutions of $(NH_4)_2SO_4$ and H_2SO_4 into the chamber and monitoring the aerosol size distribution and aerosol mass concentration resulting.
SOA formation	Carry out dark ozonolysis of α-pinene under dry conditions in the absence of an OH scavenger. α-pinene ozonolysis leads to SOA that is only weakly dependent on the presence or absence of seed particles and chemical properties of the seed. Repeat for dry seed, dry conditions; dry seed, elevated relative humidity; wet seed, elevated relative humidity.
Particle wall deposition	Model particle dynamics as a result of wall deposition by $$dN(D_p,t)/dt = -\beta(D_p)N(D_p,t),$$ where $\beta(D_p)$ = particle wall deposition coefficient. Determine β experimentally by atomizing monodisperse $(NH_4)_2SO_4$ aerosol of different sizes into the chamber and measuring decay of each size class over time. For longer experiments, owing to extraction of air for sampling, the volume of the chamber decreases with time, increasing the surface-to-volume ratio. To simulate this effect, run the same tests with $(NH_4)_2SO_4$ seed particles, but at the beginning of the experiment remove ≈ one-third of the air from the chamber.

of artificial lights are used: (1) radiation sources like argon or xenon arc lamps with specially designed UV filters which give a UV and visible spectrum similar to that of sunlight, and (2) blacklights. For example, xenon lamps, which mimic the solar spectrum, are used in the CESAM (Experimental Multiphasic Atmospheric Simulation Chamber) chamber in the Laboratoire Inter-universitaire des Systemes Atmospheriques (LISA) at Universite Paris VII[4] and in the chamber at the Paul Scherrer Institute in Switzerland (www.psi.ch).

As shown in Fig. 1.1, the blacklight spectral actinic flux differs from that of the Sun. The blacklight spectrum is intended primarily to drive the photolysis of ozone and NO_2. For example, j_{NO2}, the first-order photolysis rate constant of NO_2, was calculated from the spectral actinic flux determined for the Caltech chamber at full irradiation to be $6.63 \times 10^{-3}\,s^{-1}$. The actual j_{NO2} value on Earth depends on many factors, such as latitude, time of day, extent of cloudiness, etc. For comparison, the j_{NO2} value was

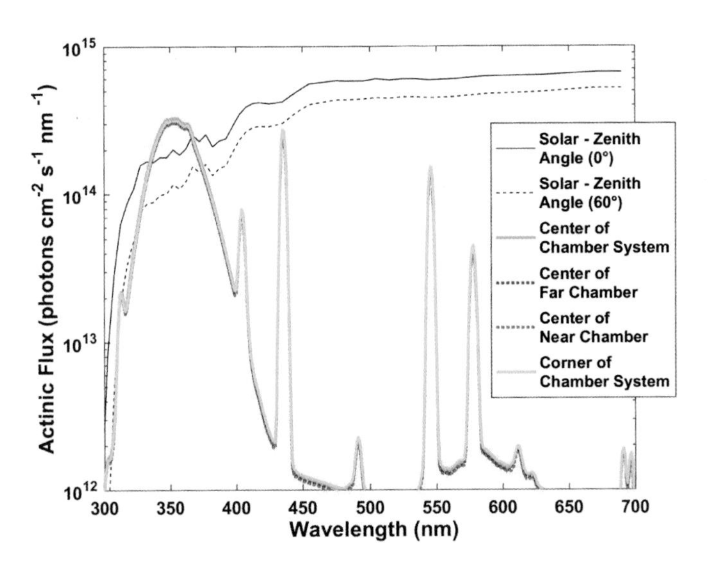

Fig. 1.1 Spectral actinic flux versus wavelength for blacklights in the Caltech chamber facility. Measurements made in the center of the chamber and close to the wall in a corner. "Near" and "far" is the designation for the two otherwise identical chambers. Global solar spectral actinic flux is shown for comparison at zenith angles 0° and 60°.

Table 1.2 Photolysis rate constants in Caltech environmental chamber

Photolysis reactions	Value of j (s^{-1})[a]
$NO_2 + h\nu \rightarrow NO + O$	6.63×10^{-3}
$NO_3 + h\nu \rightarrow NO_2 + O$; $\rightarrow NO + O_2$	3.34×10^{-3}
$O_3 + h\nu \rightarrow O(^1D) + O_2$	7.48×10^{-6}
$CH_3ONO + h\nu \rightarrow CH_3O + NO$	1.67×10^{-3}
$HONO + h\nu \rightarrow OH + NO$	1.62×10^{-3}
$H_2O_2 + h\nu \rightarrow 2OH$	4.85×10^{-6}

[a]Photolysis rate constants calculated using the irradiance spectrum measured for 100% UV lights in the chamber and absorption cross-sections and quantum yields from Ref. [5].

calculated using the global solar spectral actinic flux shown in Fig. 1.1 to be $9.6 \times 10^{-3}\,s^{-1}$ and $5.8 \times 10^{-3}\,s^{-1}$ for zenith angles 0° and 60°, respectively. The j_{NO2} value was confirmed by sending NO_2 gas through a quartz flow cell and measuring the resulting concentrations of NO, O_3, and NO_2 under blacklight irradiation. j_{NO2} values calculated from spectroradiometer measurements fall within the range of the j_{NO2} values calculated by this method. Table 1.2 summarizes photolysis rate constants in the Caltech environmental chamber.

To assess the degree of homogeneity of radiative flux in the chamber enclosure, a spectroradiometer can be used to measure the irradiance (for example from 300 to 850 nm). The irradiance is then converted into spectral actinic flux, as shown in Fig. 1.1. The spectroradiometer can be placed in various locations of the chamber enclosure, to assess the extent of homogeneity of the radiative flux. The irradiances detected in each direction can be averaged since natural air movements in the chamber will mix the fluid elements. Figure 1.1 shows that for the Caltech chamber, the spectra are essentially uniform at all locations in the chamber. If the radiation field is found to be inhomogeneous, then adjustments in the chamber configuration may need to be made.

1.3.2. *Chamber Mixing State*

An inert tracer compound that is conveniently measured, such as CO or hydrogen peroxide (H_2O_2), can be injected into the chamber in order to estimate mixing time in the chamber (i.e., the amount of time it takes for the concentration of an injected compound to become essentially well mixed in the chamber). In the case in which H_2O_2 is used as the tracer, chemical ionization mass spectrometry (CIMS) can be used to monitor the relative concentration of H_2O_2 over time. (H_2O_2 can be heated in a closed glass bulb for $\sim$10 min and then the amount of H_2O_2 vaporized during this period can be injected into the chamber for $\sim$2 min.) The chamber mixing time is then established by the time it takes for the measurement signal to stabilize following injection. Mixing time in the presence and absence of additional inflow air (mixing air) can also be tested. Two ports, 1 and 2, are used in the Caltech chamber for injection of gases. As an example, the mixing time for injection at port 1 with and without mixing air is shown in Fig. 1.2. The mixing time is shorter if extra mixing air is flowing into the chamber. Injection of tracer at different ports led to similar mixing times in both cases (not shown). For example, the average mixing time for injection at port 1 in the chamber with and without air flowing into the chamber was 2 and 8 min, respectively, while the average mixing time for injection at port 2 in the chamber with and without air flowing into the chamber was 2 and 11 min, respectively.

1.3.3. *Wall Penetration*

For Teflon chambers, transport of molecules may occur through the Teflon polymer matrix or in minute leaks at seams. The purpose of the penetration testing is to determine the extent to which transport of material into or out of the chamber is occurring. Isoprene as a molecular tracer, for example, can be injected into the chamber and monitored simultaneously with gas chromatography (GC) and CIMS. Isoprene is an effective tracer for the presence of leaks because it exhibits virtually no wall deposition.

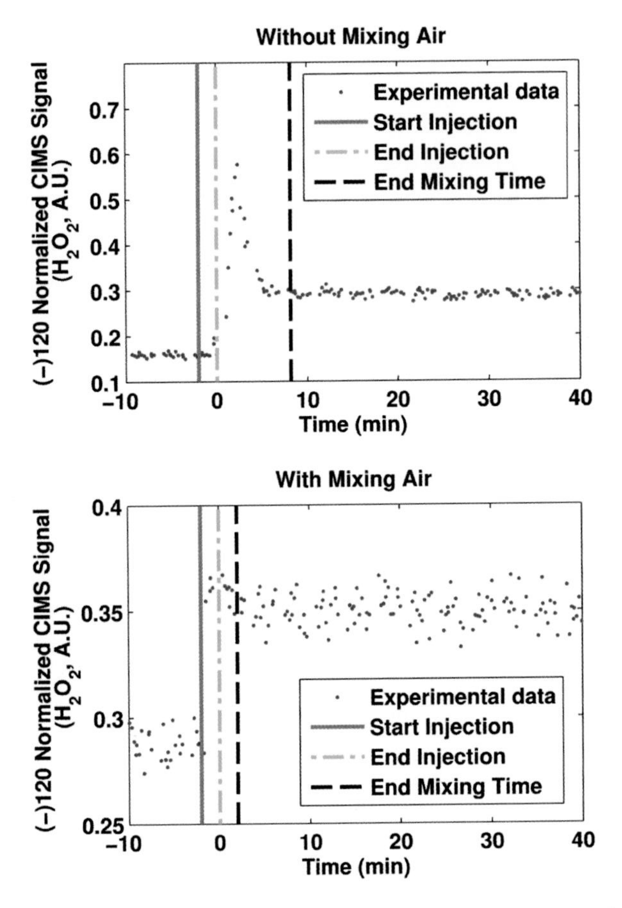

Fig. 1.2 Chamber mixing time for Caltech chamber with and without mixing air (flowing additional air into the chamber other than that carrying the injected hydrogen peroxide).

Water vapor may also leak into the chamber through minute gaps in the seams. In a test on the Caltech chamber, the chamber relative humidity increased by ~1% over 14 h. Based on the temperature and relative humidity readings, the change in the mixing ratio of water increased from 1.9 to 2.1 parts per thousand for one chamber and 2.1 to 2.3 parts per thousand for the other chamber. The CIMS signals (−)103($CF_3O^- \cdot H_2O$) and (−)121 ($CF_3O^- \cdot (H_2O)_2$) are highly water-dependent as they represent the reagent ion-water complex. Infusion of this small amount of water vapor was deemed acceptable in terms of its potential effect on chemistry occurring in the chamber. The change in water vapor should be monitored for all experiments, so that appreciable leaks in the chamber can be detected and repaired.

1.3.4. *Flushing/Cleaning Procedure*

After completion of an experiment, the chamber is routinely flushed with clean air. Results of flushing experiments on the Caltech chamber are shown in Fig. 1.3. Ozone and ammonium sulfate seed particles were injected into both chambers (these chambers, with identical setup, are termed the near and far chambers). Ozone and particle number concentrations were monitored over time as purified

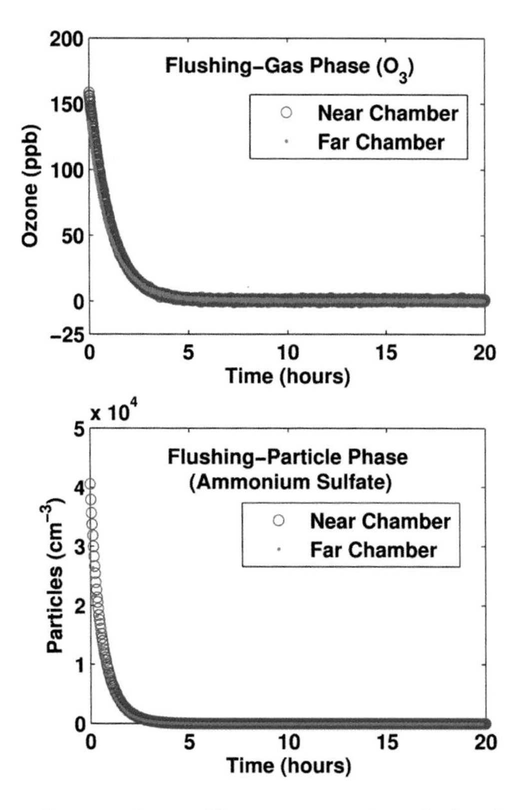

Fig. 1.3 Ozone and ammonium sulfate concentrations during flushing of the two Caltech chambers (near and far).

air was injected into the chamber at the same rate as air was flushed out. As expected, the concentration of species decays exponentially during flushing (concentration $\sim\exp(-t/\tau)$, where τ is the mean residence time of air in the chamber at the particular flow rate of air). Using ozone as a tracer and assuming constant volume and a well-mixed system, it was estimated that the flow during these dry flushing experiments was $370\,L\,min^{-1}$. The mean residence time τ for both chambers for this flow was estimated as 1.1 h. Ozone and particles were flushed sufficiently within five residence times, ~5 h. A typical protocol is to flush the chamber for 24 h between the experiments.

1.3.5. *Chemical Blank Experiments*

The purpose of so-called *blank experiments* is to assess the chemical reactivity of the chamber under conditions when it is filled only with purified air or an oxidant-generating species. Such experiments can reveal chemistry involving residual species present either in the purified air system or from material degassing from the walls of the chamber.

1.3.5.1. *Blank Experiments with Injection Air Only*

The first test is to feed only purified injection air into the chamber. A typical protocol for such a test is as follows. After filling with injection air (~1 h), the lights are turned on. After 5 h with lights on at 20°C, the temperature is increased at a constant rate over 2 h to 40°C. After another ~13 h, lights are turned off, and the experiment terminates. Results of such a test for one (far) of the Caltech chambers are shown in Fig. 1.4. One indication of chemistry occurring during this test is generation of aerosol from gas-phase reactions involving residual species. The volume concentration of aerosol formed in the blank experiment was found to be greater in the near chamber ($\sim1.1\,\mu m^3\,cm^{-3}$) than in the far chamber ($\sim0.2\,\mu m^3\,cm^{-3}$). Prior to this test, the far chamber had been flushing for considerably longer, and this difference is reflected in the amount of aerosol volume generated. As shown in Fig. 1.4, O_3

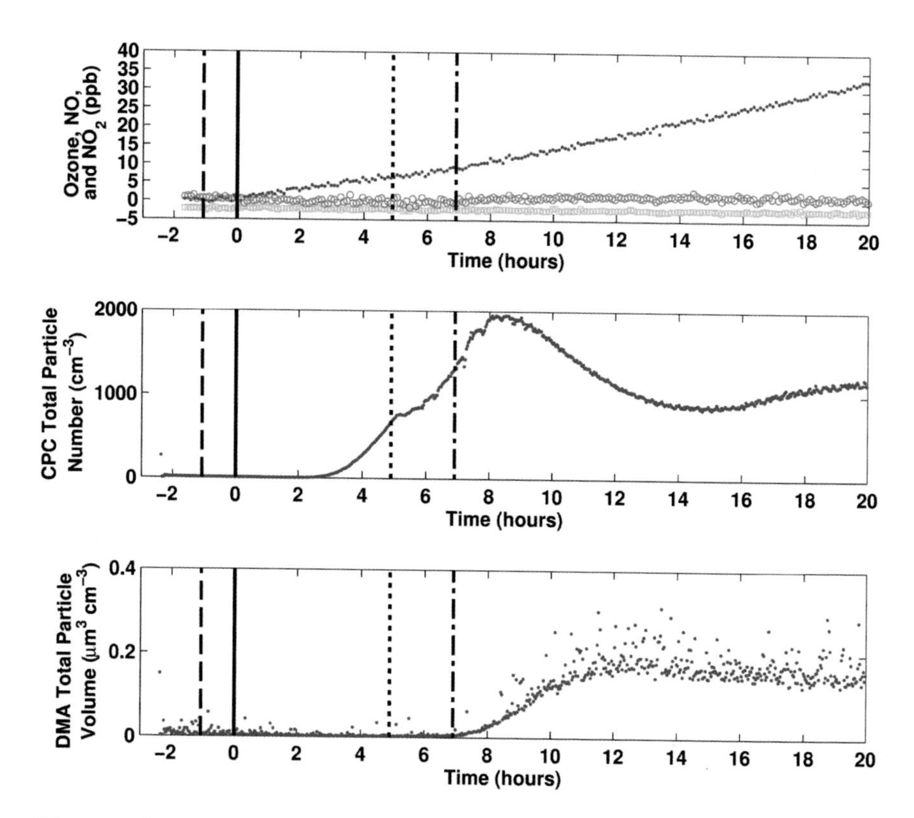

Fig. 1.4 Ozone, NO, and NO_2 mixing ratios, particle number concentration, and particle volume formation during a blank photooxidation experiment in the far chamber with clean injection air.

also formed during the blank experiment, likely from photolysis of trace amounts (below detection limit) of NO_2.

1.3.5.2. *Blank Experiments with OH Generation*

A second blank experiment is one in which OH radicals are intentionally generated in the chamber in the presence of injection air only. In a test of this nature, 210 ppb H_2O_2 was injected into the Caltech chamber, and the same experiment as the blank with injection air only was carried out. Results for this test in the far chamber are shown in Fig. 1.5. A similar amount of ozone and particle

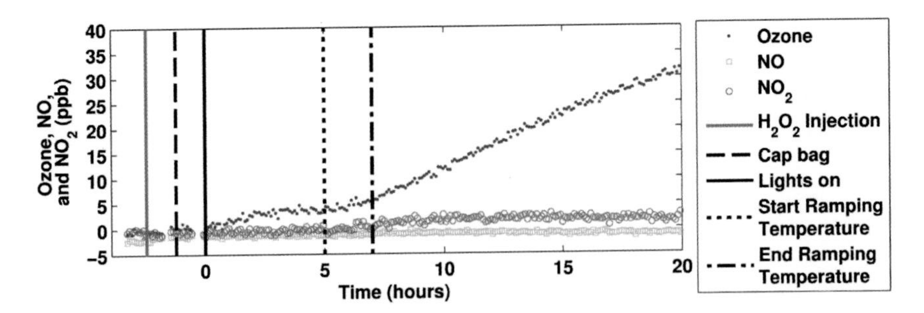

Fig. 1.5 Formation of ozone, NO, and NO_2 during blank photooxidation experiment with added hydrogen peroxide, which photolyzes to produce OH.

volume was generated as in the blank experiment in the absence of H_2O_2. This observation supports the argument that NO_x is the likely agent leading to ozone formation. Adding more oxidant does not increase the ozone concentration significantly, so it is likely that the formation of NO_x is the limiting factor. The aerosol mass and volume concentrations generated for the two chambers were $<0.4\,\mu g\,m^{-3}$ and $0.5\,\mu m^3\,cm^{-3}$ (not shown). For aerosol-generation experiments in which a low aerosol concentration is anticipated, a background aerosol level of this magnitude may be unacceptable. These results emphasize the importance of running blank experiments prior to each new set of the experiments to establish background levels in the chamber.

Aerosol mass spectrometer (AMS) measurements of particle composition for the near chamber initial blank experiment with H_2O_2 are shown in Fig. 1.6. From the normalized carbon spectrum, the m/z 44 signal is indicative of the presence of oxidized organic species in the particles. The aerosol also contains reduced carbon species, as indicated by the signals for m/z 55, 57, 69, and 71. Overall, the AMS data show that the aerosol is carbonaceous in nature, suggesting that the particles result from the oxidation of trace organic species. Additional tests are warranted to determine whether the source of the primary organic material is the injection air or desorption from the chamber walls.

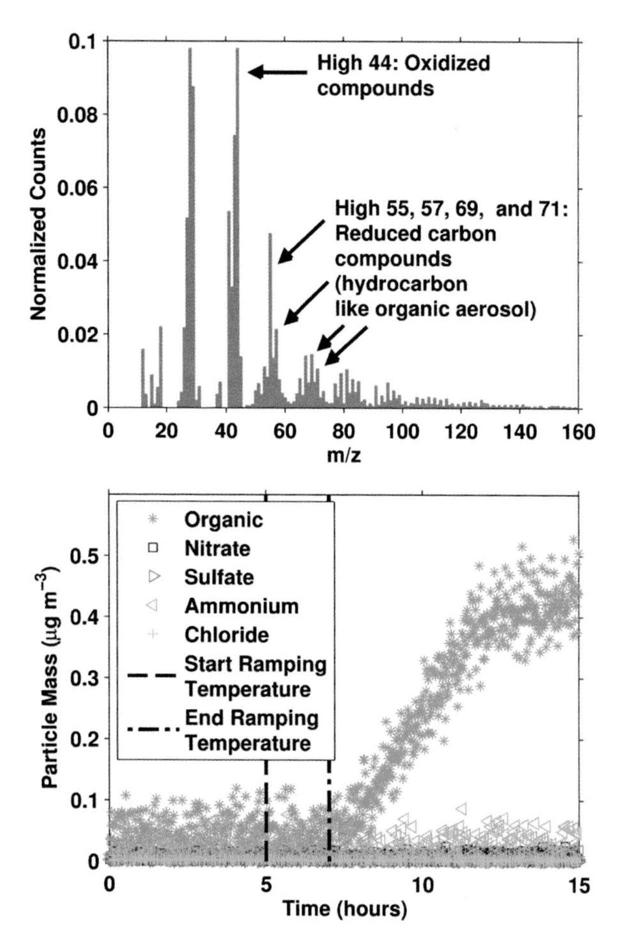

Fig. 1.6 AMS measurements during a blank photooxidation experiment with clean air. The top panel shows the normalized average organic spectrum at time of peak growth. The bottom panel shows the chemical composition of the aerosol formed.

1.4. Chemical Environment in the Chamber

1.4.1. *Hydroxyl Radical Generation*

In the atmosphere, OH radicals are generated via the reaction of H_2O with singlet (O (^1D)) oxygen atoms produced from O_3

photolysis at wavelengths <319 nm. In chamber facilities, the spectrum and wavelength-dependent intensities of radiation sources govern the choice of the OH precursor. For example, blacklights with a spectrum that peaks at 350 nm do not provide sufficient photon intensity to sustain the O_3 photolysis chemistry. OH sources that have been widely used in chambers include H_2O_2, HONO, and CH_3ONO.

1.4.1.1. *H_2O_2 Radical Source*

Photolysis of H_2O_2 yields two OH radicals, which subsequently react with H_2O_2, producing HO_2 radicals:

$$H_2O_2 + hv \rightarrow 2OH$$

$$OH + H_2O_2 \rightarrow HO_2 + H_2O$$

The relatively slow H_2O_2 photolysis rate, together with the suppression of OH propagation by reaction with H_2O_2 itself, makes it possible to sustain a steady OH concentration over a long timescale.

1.4.1.2. *HONO Radical Source*

HONO photolyzes to produce one OH radical and one NO radical. Because NO is produced even if no additional NO_x is added to the system, the RO_2 + NO reaction will always dominate under typical chamber conditions. Thus, for experiments in which HONO is used as a radical source, the peroxy radical will always react preferentially with NO, regardless of the $[VOC]_o/[NO_x]_o$ ratio or initial HONO concentration:

$$HONO + hv \rightarrow OH + NO$$

$$OH + NO + M \rightarrow HONO + M$$

$$OH + HONO \rightarrow NO_2 + H_2O$$

Due to the substantial photolysis rate of HONO, OH radicals generated usually peak at the beginning of the experiment and then rapidly decay to zero. Thus, multiple HONO injections are required for extended duration experiments.

1.4.1.3. *CH_3ONO Radical Source*

CH_3ONO photolysis produces the methoxy radical, CH_3O, and NO. The methoxy radical rapidly reacts with oxygen to yield formaldehyde and HO_2. HO_2 and NO then react to yield OH and NO_2. Thus, in contrast to HONO, which yields OH and NO, the ultimate yield of CH_3ONO photolysis is OH and NO_2:

$$CH_3ONO + hv \rightarrow CH_3O + NO$$

$$CH_3O + O_2 \rightarrow HCHO + HO_2$$

$$HO_2 + NO \rightarrow OH + NO_2$$

NO_2 can then photolyze to generate NO and O_3. One needs also to consider the effect on the system from subsequent reactions of HCHO.

1.4.2. *High- versus Low-NO Conditions*

In atmospheric chemistry the terms "high-NO" and "low-NO" are used to classify photooxidation conditions of VOCs. (The common terminology relating to these conditions has been low-NO_x and high-NO_x, but it is the NO concentration that governs the fate of the peroxy radicals in the system, so low- and high-NO are more appropriate descriptions.) In VOC oxidation, these terms are intended to delineate the gas-phase fate of the peroxy radicals (RO_2) generated by VOC reaction with OH. The peroxy radicals can react via one of three main pathways, as illustrated for n-hexane (C_6H_{14}) in the top panel of Fig. 1.7: reaction with NO, reaction with HO_2, and reaction with other RO_2 radicals. Reaction with NO generates either an alkoxy radical or an alkyl nitrate. Reaction with HO_2 generates a hydrogen peroxide. Reaction with another RO_2 radical generally yields two alkoxy radicals or the combination of an alcohol and a carbonyl.[6] For some VOCs, such as isoprene, a fraction of peroxy radicals may undergo isomerization (autoxidation) under conditions when the lifetime of the peroxy radical against reaction is sufficiently long.[7] In a so-called high-NO regime, reaction with NO dominates the fate of the RO_2 radicals. In a low-NO regime, the RO_2 radical may

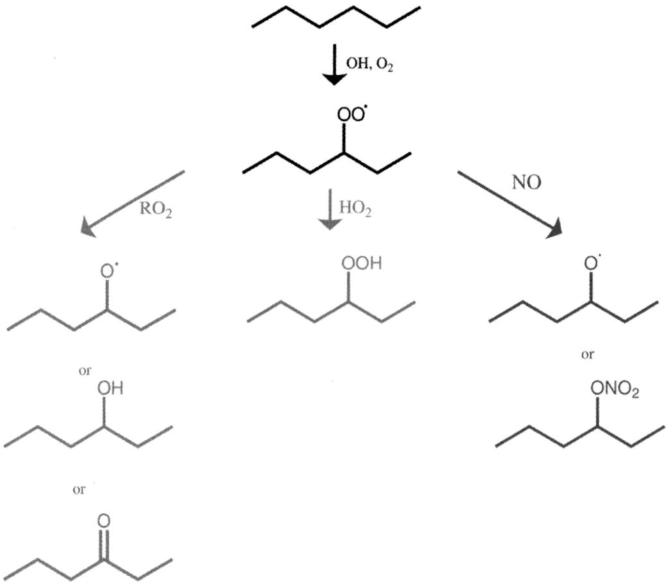

Fig. 1.7 Reactions of the peroxy radicals generated by OH oxidation of n-hexane.

react with HO_2, with RO_2 or undergo intra-molecular isomerization. Kroll *et al.*[8] demonstrated that H_2O_2 may be used as a radical source to isolate the $RO_2 + HO_2$ reaction because the slow photolysis rate of H_2O_2 produces a continuous supply of OH and HO_2 (from OH + H_2O_2). However, the terms high- and low-NO alone are inadequate to characterize the conditions of a VOC photooxidation. A more precise description of the regime is based on the specific fate of the peroxy radicals.

Using hexane as the illustrative VOC, we quantify the conditions needed to isolate different peroxy radical pathways. For specified initial concentrations of hexane and H_2O_2 (50 ppb and 1000 ppb, respectively), we present simulations of the subsequent chemistry at different initial NO levels (assuming a 1:1 initial ratio of NO:NO_2) ranging from 0 to 20 ppb. The fate of the RO_2 radicals at any instant of time is determined from the fractional contribution of each pathway, RO_2+NO, RO_2+HO_2, and RO_2+RO_2. The mechanism for OH oxidation of hexane is derived from the Master Chemical Mechanism

(MCM) v3.2 (http://mcm.leeds.ac.uk/MCM).[9] Photolysis rates for all species are calculated using the UV spectrum of the Caltech laboratory chamber ($j_{NO_2} = 6.63 \times 10^{-3}\,s^{-1}$). We examine H_2O_2, HONO, and CH_3ONO as OH sources.

The relative amounts of RO_2, NO, and HO_2 can vary substantially over the course of a photooxidation. For example, NO tends to be at its maximum concentration at the onset of oxidation prior to consumption by reaction with RO_2 and other species. Consequently, the RO_2 + NO reaction can be disproportionally dominant at the beginning of a photooxidation. This trend is illustrated in Fig. 1.8, where the fate of the RO_2 radical is shown at the start of a hypothetical experiment (solid lines), after 4 h (dashed lines), and after 8 h (dotted lines). As evident in Fig. 1.8, the RO_2 + NO reaction dominates at the start of the experiment, even for what may be considered "low" values of $[NO_x]_0$, i.e., 5 ppb. After 4 h,

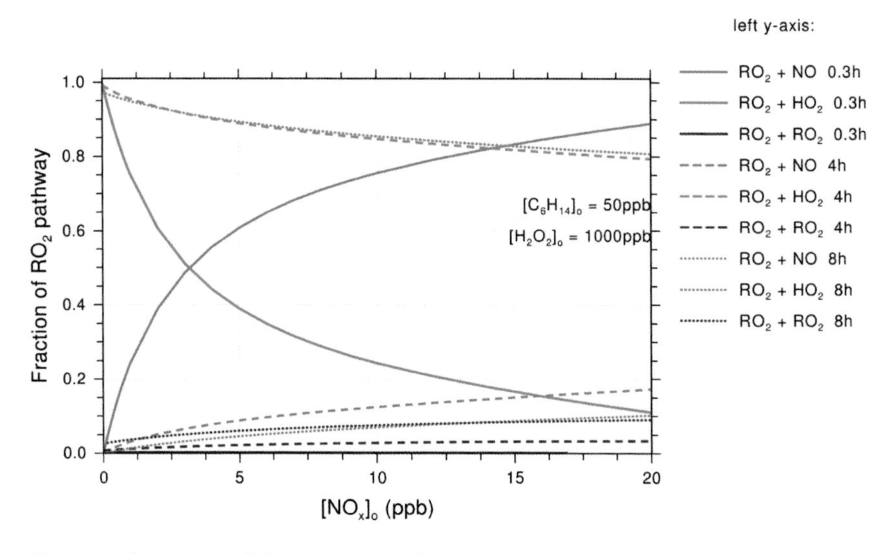

Fig. 1.8 Fate of the RO_2 radical as a function of the initial mixing ratio of NO_x at different times in *n*-hexane photooxidation for a system of 50 ppb initial *n*-hexane and 1000 ppb initial hydrogen peroxide. The sum of all RO_2 + NO reactions is shown in blue, all $RO_2 + HO_2$ reactions in green, and all $RO_2 + RO_2$ reactions in black. The fractions at the start of the experiment are shown by solid lines, the fractions after 4 h are shown by dashed lines, and the fractions after 8 h are shown by dotted lines.

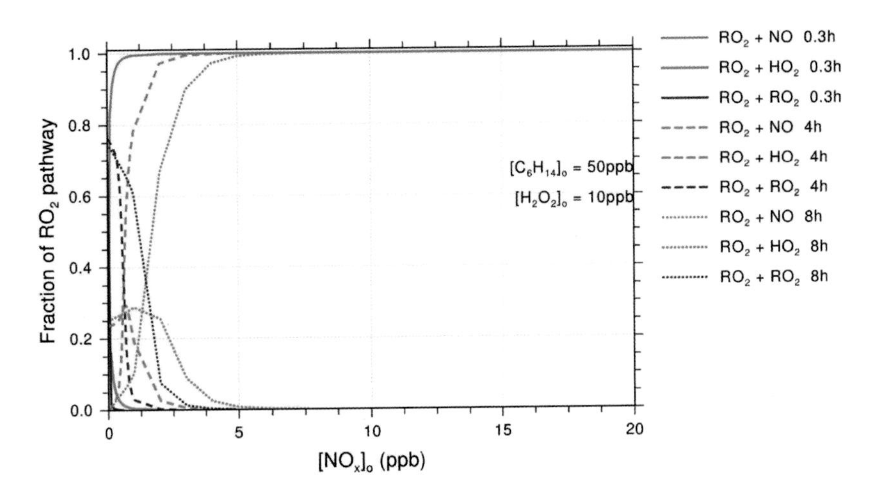

Fig. 1.9 Fate of the RO_2 radical as a function of the initial mixing ratio of NO_x at different times in n-hexane photooxidation with 50 ppb initial n-hexane and 10 ppb initial hydrogen peroxide. The sum of all $RO_2 + NO$ reactions is shown in blue, all RO_2+HO_2 reactions in green, and all RO_2+RO_2 reactions in black. The fractions at the start of the experiment are shown by solid lines, the fractions after 4 h are shown by dashed lines, and the fractions after 8 h are shown by dotted lines.

the peroxy radical reacts predominantly with HO_2 for practically all values of $[NO]_0$ because the substantial initial concentration of H_2O_2, 1000 ppb, eventually generates an excess of HO_2. The behavior of the system at 4 and 8 h is essentially identical. The evolution of the peroxy radical fate over the course of the oxidation is noteworthy, as products from the $RO_2 + NO$ reaction are generated even in the presence of relatively low values of $[NO_x]_0$ until sufficient levels of HO_2 are attained. Figure 1.9 shows the same case as Fig. 1.8 but with an initial H_2O_2 mixing ratio of 10 ppb. In this case, the level of OH generation is considerably smaller and peroxy radical reaction with NO dominates.

The ratio of initial VOC to NO_x concentrations alone does not adequately define the behavior of the peroxy radical when applied to a wide range of initial VOC concentrations and initial H_2O_2 concentrations, as demonstrated in Figs. 1.10 and 1.11. Figure 1.10 shows the fate of the RO_2 radical as a function of $[VOC]_0/[NO_x]_0$ for three different initial hexane mixing ratios and a range of $[NO_x]_0$,

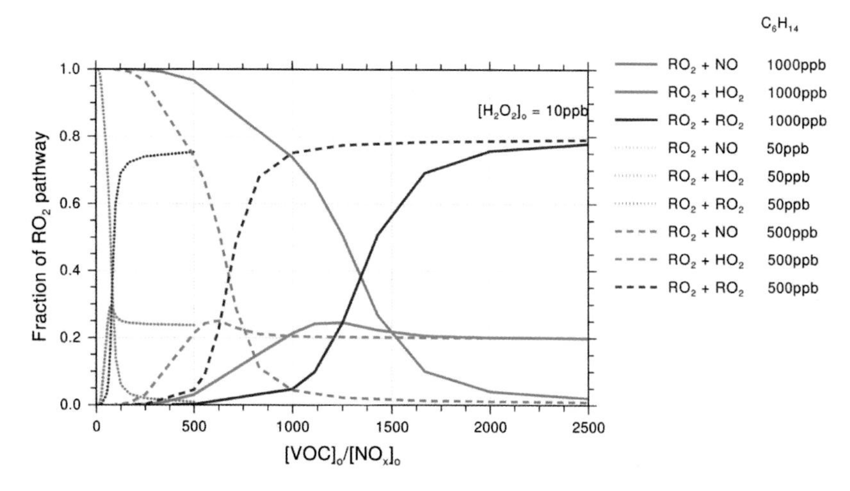

Fig. 1.10 Fate of the RO_2 radical as a function of $[VOC]_0/[NO_x]_0$ (after 4 h of reaction) in experiments starting with variable *n*-hexane and $[NO_x]_0$ concentrations each with 10 ppb of H_2O_2. The $RO_2 + HO_2$ pathway is shown in green, the $RO_2 + RO_2$ pathway in black, and the $RO_2 + NO$ pathway in blue. Different line styles are used for different initial *n*-hexane concentrations: solid lines for 1000 ppb, dots for 50 ppb, and dashes for 500 ppb.

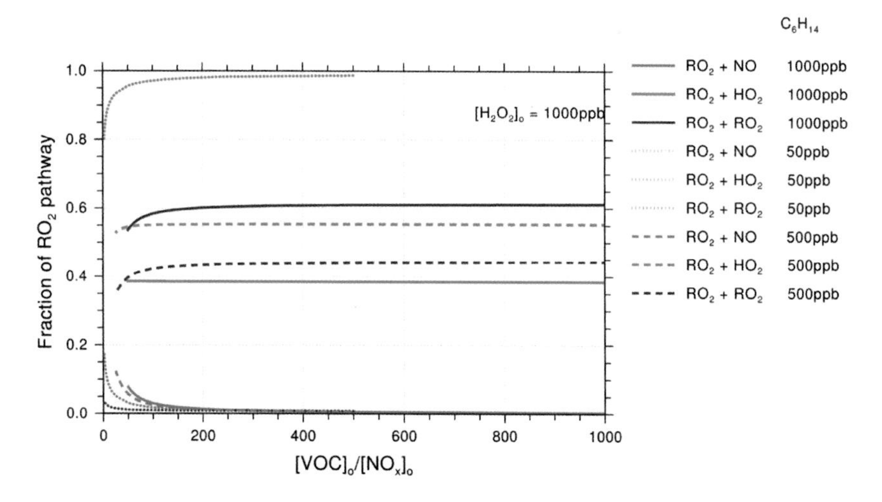

Fig. 1.11 Fate of the RO_2 radical as a function of $[VOC]_0/[NO_x]_0$ (after 4 h of reaction) in experiments starting with variable *n*-hexane and $[NO_x]_0$ concentrations each with 1000 ppb of H_2O_2. The $RO_2 + HO_2$ pathway is shown in green, the $RO_2 + RO_2$ pathway in black, and the $RO_2 + NO$ pathway in blue. Different line styles are used for different initial *n*-hexane concentrations: solid lines for 1000 ppb, dots for 50 ppb, and dashes for 500 ppb.

each initially with 10 ppb of H_2O_2. The transition from a RO_2 + NO regime to an RO_2 + RO_2 regime for these systems is not well predicted by the ratio $[VOC]_0/[NO_x]_0$. Depending on the initial VOC mixing ratio, the $[VOC]_0/[NO_x]_0$ ratio at which the transition occurs ranges from 75 for a low initial hexane concentration to 1500 for a high initial hexane concentration. Thus, $[VOC]_0/[NO_x]_0$ alone is not a reliable predictor of the fate of the RO_2 radical for these systems. For an initial H_2O_2 mixing ratio of 1000 ppb (Fig. 1.11), the RO_2 + NO pathway is negligible for almost all values of $[VOC]_0/[NO_x]_0$. The amount of initial VOC, not the $[VOC]_0/[NO_x]_0$ ratio, dictates those systems for which the peroxy radical reacts with RO_2 or with HO_2. For a low initial hexane mixing ratio of 50 ppb, the RO_2 radical reacts predominately with HO_2 at all $[VOC]_0/[NO_x]_0$ ratios. For higher initial hexane values, 500 ppb and 1000 ppb, roughly equal amounts of RO_2 react with HO_2 and RO_2.

For the systems in both Figs. 1.10 and 1.11, the different behavior at a particular $[VOC]_0/[NO_x]_0$ ratio does not occur simply as a result of the different initial VOC concentrations. Rather, the behavior at a particular $[VOC]_0/[NO_x]_0$ ratio changes both with initial VOC concentration and with initial H_2O_2 concentration. A more precise classification of the fate of the peroxy radical in an experiment in which H_2O_2 is the radical source is based on the ratio $[VOC]_0/[NO_x]_0$ *at a specific* $[H_2O_2]_0/[VOC]_0$. That is, for a specified $[H_2O_2]_0/[VOC]_0$ ratio, the fate of the RO_2 radical can then be distinguished on the basis of the ratio $[VOC]_0/[NO_x]_0$. Figure 1.12 shows the results of simulated photooxidation at two initial hexane levels, each at $[NO_x]_0$ levels ranging from 0.1 to 20 ppb. For each of the simulations, $[H_2O_2]_0/[VOC]_0 = 20$. The same trend is observed for each simulation with this ratio. Thus, specifying $[H_2O_2]_0/[VOC]_0$ is necessary in order to classify the behavior of a system based on the ratio, $[VOC]_0/[NO_x]_0$. For $[H_2O_2]_0/[VOC]_0 = 20$, RO_2 + NO is the dominant reaction below a $[VOC]_0/[NO_x]_0$ ratio of ~0.5. Above a $[VOC]_0/[NO_x]_0$ ratio of ~20, the $RO_2 + HO_2$ reaction dominates. Between $[VOC]_0/[NO_x]_0$ ratios of 0.5 and 20, a transition region exists in which both peroxy radical pathways are significant. Figure 1.13 shows the results of simulations of two

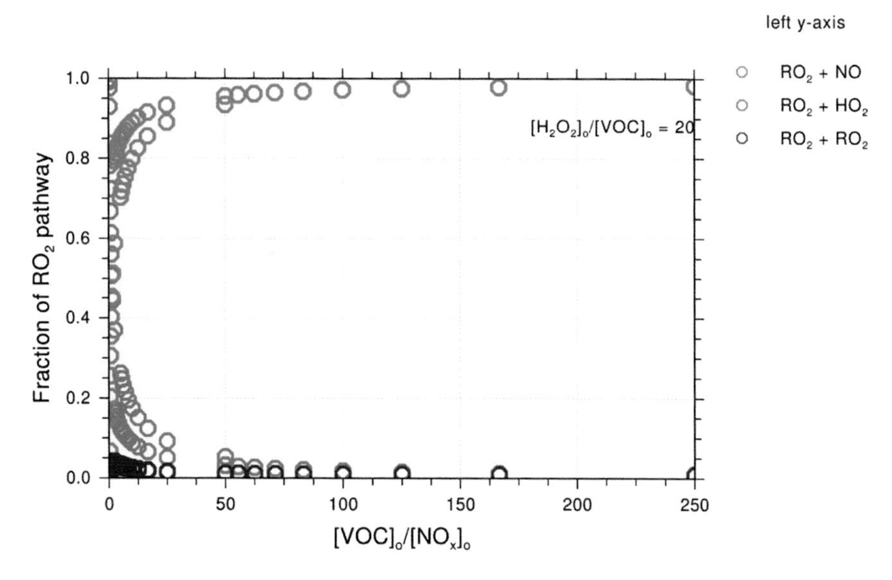

Fig. 1.12 Fate of the RO_2 radical as a function of $[n\text{-hexane}]_0/[NO_x]_0$ at $[H_2O_2]_0/[n\text{-hexane}]_0 = 20$ after 4 h of reaction. Two different combinations were simulated to produce this ratio, both under a large range of $[NO_x]_0$ conditions: 1000 ppb H_2O_2/50 ppb n-hexane and 100 ppb H_2O_2/5 ppb n-hexane.

initial hexane concentrations for a range of $[NO_x]_0$ values, both with an $[H_2O_2]_0/[VOC]_0$ ratio of 0.2. For this $[H_2O_2]_0/[VOC]_0$ ratio, $RO_2 + NO$ dominates below a $[VOC]_0/[NO_x]_0$ ratio of ~25. Above $[VOC]_0/[NO_x]_0$ ~150, the $RO_2 + RO_2$ reaction dominates, although the $RO_2 + HO_2$ reaction contributes about 20%.

For both of these $[H_2O_2]_0/[VOC]_0$ ratios, a $[VOC]_0/[NO_x]_0$ ratio is sufficient to describe the fate of the peroxy radical for a system. These conditions can be generalized into the following two rules of thumb: (1) the $RO_2 + HO_2$ reaction can be isolated by using an initial H_2O_2 concentration at least an order of magnitude higher than the initial VOC concentration at a moderately high $[VOC]_0/[NO_x]_0$ ratio (2) the $RO_2 + RO_2$ reaction can be isolated by using an initial H_2O_2 concentration at least an order of magnitude lower than the initial VOC concentration at a high $[VOC]_0/[NO_x]_0$ ratio. However, the latter result should be viewed with caution due to the uncertainty and variability in the rate constants for $RO_2 + RO_2$

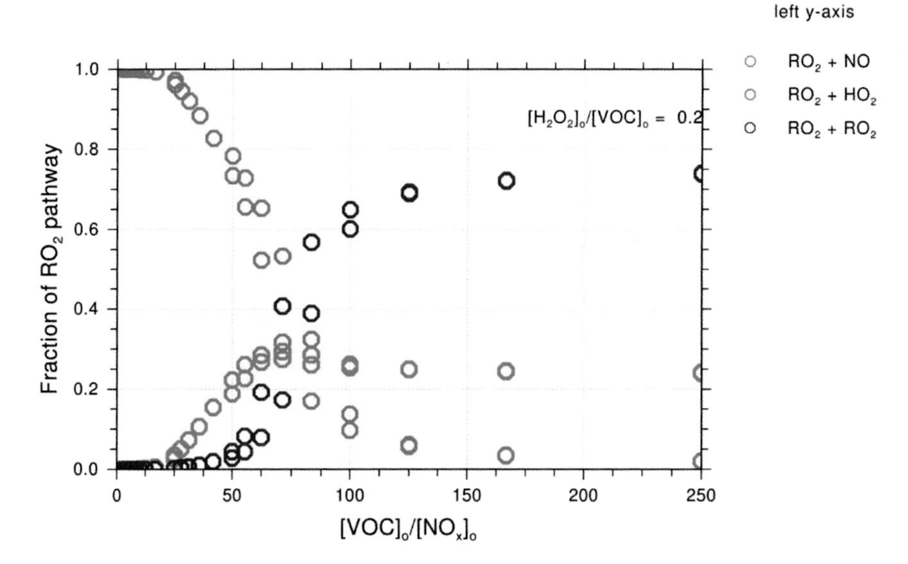

Fig. 1.13 Fate of the RO_2 radical as a function of $[VOC]_0/[NO_x]_0$ at $[H_2O_2]_0/[VOC]_0 = 0.2$ after 4 h of reaction. Two different combinations were simulated to produce this ratio, both under a range of $[NO_x]_0$ conditions: 10 ppb H_2O_2/50 ppb *n*-hexane and 100 ppb H_2O_2/500 ppb *n*-hexane.

reactions. Owing to the large number of permutation reactions of $RO_2 + RO_2$, MCM assumes that each peroxy radical reacts with a pool of all other peroxy radicals with a generic rate constant calculated from the structure of the organic group and general trends of peroxy reactivity.[10] Thus it is difficult to state with certainty when the $RO_2 + RO_2$ reaction will be dominant. The $RO_2 + NO$ reaction can be isolated for any combination of initial H_2O_2 and VOC concentrations but only at very low $[VOC]_0/[NO_x]_0$ ratios. With these cautionary considerations, these conditions enable the design of experiments to isolate the three main pathways of the peroxy radical.

For chamber conditions similar to those used in the simulations for H_2O_2 and HONO, when CH_3ONO is the OH source, sufficient NO is generated via photolysis that the $RO_2 + NO$ reaction is dominant throughout the experiment even with no additional NO_x. For most of the duration of the experiment, $RO_2 + NO$ is essentially the sole peroxy fate. After 8–10 h, as CH_3ONO is depleted and NO

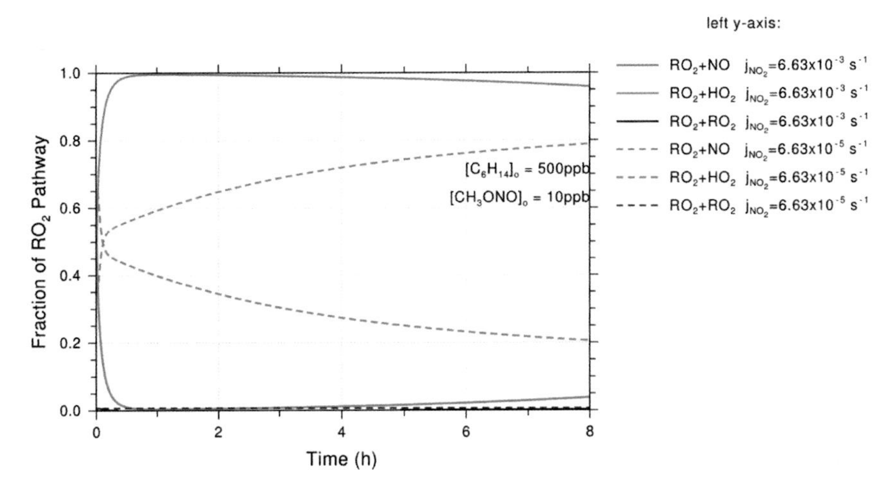

Fig. 1.14 Temporal profile of the fate of the RO_2 radical for a system with 500 ppb n-hexane and 10 ppb CH_3ONO.

diminishes, the $RO_2 + HO_2$ reaction increases to ~5% of the peroxy reactions (Fig. 1.14, $j_{NO_2} = 6.63 \times 10^{-3}\,s^{-1}$). If the photolysis rate of NO_2 is reduced, the formation of NO can be suppressed, allowing other reactions to contribute significantly. For several simulations using CH_3ONO as the radical source, the photolysis rates for all species were calculated by scaling the chamber irradiance data by 1% ($j_{NO_2} = 6.63 \times 10^{-5}\,s^{-1}$). The lower UV intensity produces HO_2 and NO mixing ratios of ~10 ppt and ~50 ppt, respectively, similar conditions to those generated in Ref. [7] by using only a single UV bulb ($j_{NO_2} = 2.8 \times 10^{-5}\,s^{-1}$). The effect of this lower UV intensity is shown in Fig. 1.14 by comparing a simulation with identical initial conditions to that discussed in Sec. 1.4.2 but where $j_{NO_2} = 6.63 \times 10^{-5}\,s^{-1}$. For this lower intensity, the $RO_2 + HO_2$ reaction constitutes close to half of the total peroxy reactions for the first few hours of experiment and decreases to 20% towards the end of the experiment. Similar behavior for both UV intensities is predicted for different initial hexane and CH_3ONO concentrations. The $RO_2 + NO$ and $RO_2 + HO_2$ reactions are expected to co-exist in pristine and tropical atmospheres, and the photolysis of CH_3ONO

at a lower UV intensity provides a convenient method to study these conditions.

It is useful to compare the chemical lifetime of the peroxy radical in each reaction, where the lifetime against reaction with NO is $(k_{\mathrm{RO2+NO}}[\mathrm{NO}])^{-1}$, etc. The overall lifetime for a peroxy radical is then

$$\tau_{\mathrm{overall}} = (k_{\mathrm{RO_2+NO}}[\mathrm{NO}] + k_{\mathrm{RO_2+HO_2}}[\mathrm{HO_2}] + k_{\mathrm{RO_2+RO_2}}[\mathrm{RO_2}])^{-1}$$

Lifetimes for representative initial conditions for each of the categories considered, calculated using concentrations at 4 h, are given in Table 1.3. The dominant reaction for each set of conditions is that with the shortest lifetime, and these data reinforce the conclusions drawn from the simulations.

It is also apparent that RO_2 lifetimes can vary dramatically within systems characterized by same peroxy pathway. For conditions in which $RO_2 + NO$ is the dominant reaction, the lifetime for reaction with NO can vary from 0.3 s at a high concentration of HONO to 18 s at low initial concentrations of both H_2O_2 and NO_x. Using CH_3ONO at a lower UV intensity generates lifetimes similar to those of remote chemistry (~30–60 s) and allows study of these reactions. Therefore, in addition to simply choosing initial conditions to isolate a particular peroxy pathway, it may also be necessary to design experiments to yield a particular RO_2 lifetime.

1.4.3. *Ozone–Alkene Oxidation and OH Scavenging*

Ozone–alkene reactions are important in gas-phase atmospheric chemistry and are a pathway to formation of SOA. The gas-phase reaction of ozone with alkenes proceeds by the addition of ozone across the C=C double bond to form an energy-rich primary ozonide, followed by decomposition of the primary ozonide to produce an energized carbonyl oxide species, known as the Criegee intermediate, and an aldehyde or ketone product.[11,12] Further unimolecular decay of the Criegee intermediate leads to formation of OH radicals. The OH yield from ozonolysis depends on the alkene structure, increasing from ~10% for ethene via the simplest Criegee intermediate CH_2OO

Table 1.3 Lifetimes of n-hexane peroxy radicals against reaction with NO, HO_2, and RO_2 for various conditions of n-hexane oxidation by OH, as calculated by MCM. Lifetimes are calculated at 4 h of photooxidation

OH source/I.C.s	C_6H_{14} (ppb)	NO_x (ppb)	τ_{RO_2+NO}(s)	$\tau_{RO_2+HO_2}$(s)	$\tau_{RO_2+RO_2}$(s)	$\tau_{Overall}$(s)
H_2O_2/VOC = 20						
High NO_x: VOC/ NO_x = 0.5	5	10	17.8	66	3930	14
Low NO_x: VOC/ NO_x = 10	50	5	162	10	1270	9.6
H_2O_2/VOC = 0.2						
High NO_x: VOC/NO_x = 25	50	2	13.8	497	4690	13.4
Low NO_x: VOC/NO_x = 100	500	5	2365	733	103	87.2
HONO						
HONO = 10 ppb	500	0	3.3	2360	22980	3.3
HONO = 100 ppb	500	0	0.3	15010	230240	0.3
CH_3ONO						
CH_3ONO = 10 ppb						
j_{NO_2} = 6.63×10^{-5} s^{-1}	50	0	83	168	3370	54.7
j_{NO_2} = 6.63×10^{-3} s^{-1}	50	0	5.1	650	9200	5.1
CH_3ONO = 100 ppb						
j_{NO_2} = 6.63×10^{-5} s^{-1}	50	0	28	44	2210	17
j_{NO_2} = 6.63×10^{-3} s^{-1}	50	0	1.9	563.6	14290	1.9

to >60% for trans-2-butene, which proceeds through the Criegee intermediate CH_3CHOO.

Because the OH radical itself is highly reactive toward alkenes, an experiment intended to isolate ozone–alkene chemistry will require removal of OH from the system, via a molecular OH scavenger. The necessary characteristics of an OH scavenger are twofold: (1) a molecule that is itself chemically inert toward alkenes, and (2) a molecule that upon reaction with OH does not generate products that are themselves reactive toward alkenes or that mimic products of the ozone–alkene chemistry. Hydroxyl radical scavengers that have been commonly used include cyclohexane, CO, alcohols, and aldehydes. Understanding the chemical role played by the OH scavenger is important in separating the effects of the scavenger itself from that of the intrinsic ozone–alkene reactions. Differences in observed products when different scavengers are used provide important clues to the gas-phase chemistry occurring in the system.

To illustrate the technicalities associated with choice of an OH scavenger in an ozone–alkene reaction, we consider the ozone-cyclohexene system.[13] The reaction of ozone with cyclohexene forms OH radicals at a yield of ~0.6. Ozonolysis of cyclohexene in the presence of the OH scavengers, cyclohexane, 2-butanol, and CO, and in the absence of an OH scavenger was carried out in the precursor to the current Caltech chambers. These experiments, carried out in the presence of ammonium sulfate seed aerosol, were directed at measuring the formation of SOA in this system. Each OH scavenger was injected at sufficient concentration that the reaction rate of OH radicals with the scavenger exceeded that of OH with the cyclohexene by a factor of ~100. The ratio of the mass concentration of SOA to the mass concentration of reacted cyclohexene, the so-called SOA yield (see Sec. 1.5), was used as the metric for the behavior of the system. The use of cyclohexane as the OH scavenger resulted in the smallest SOA yield. 2-Butanol scavenger led to a higher SOA yield than that of cyclohexane. When no OH scavenger was used, the SOA yield was similar to that when 2-butanol was used, and when CO was used as the scavenger, the largest SOA yield occurred, even larger than that in the absence of any scavenger.

Understanding the reasons for the observed effects of the different OH scavengers on the observed aerosol yields provides a clue to the chemistry occurring in the system. One possible explanation lies in reactions of the stabilized Criegee intermediate (SCI) with the scavenger, which could potentially form low volatility products. In the case of cyclohexene ozonolysis, however, such reactions are unlikely to occur to an appreciable extent, as little SCI forms. Criegee intermediates from endocyclic alkenes are formed with more energy than those from exocyclic alkenes and so are less likely to be stabilized; SCI yields from cyclohexene ozonolysis are measured at only ~3%. Also, it is unlikely that the reaction of the Criegee intermediate with CO would form products of lower volatility than those of the Criegee-2-butanol reaction. One concludes that reactions of the OH scavengers themselves with the Criegee intermediate probably do not affect the SOA yield appreciably.

A more likely explanation for the observed effects of the different scavengers on SOA yield lies in the radical products formed in the OH-scavenger reactions, especially the effect on production of HO_2 and RO_2. In the case of CO, the CO + OH reaction produces only HO_2. By contrast, when cyclohexane is used as the OH scavenger, an alkyl peroxy radical is formed:

OH + cyclohexane $\xrightarrow{O_2}$ cyclohexylperoxy radical (–O–O•) + H_2O

When 2-butanol is used as the OH scavenger, either HO_2 or RO_2 is formed, which in the case of OH abstraction of a H atom from two of the different carbon atoms in 2-butanol leads to:

OH + 2-butanol → radical (OH) $\xrightarrow{O_2}$ ketone (O) + HO_2

OH + 2-butanol → radical (OH) $\xrightarrow{O_2}$ peroxy radical (O–O•, OH)

For these reactions, formation of HO_2 is the major channel, with a branching ratio of ~70%.

The HO_2/RO_2 ratio from each of the scavengers (CO > 2-butanol > cyclohexane) matches that of the SOA yields, suggesting that increased formation of HO_2 and/or decreased formation of RO_2 favors SOA formation. In short, the RO_2 radical from the ozonolysis reaction and the HO_2 and/or RO_2 from the OH-scavenger reaction promote the ensuing radical chemistry. In the absence of NO_x, the chemistry consists largely of self- and cross-reactions of peroxy species, i.e., $HO_2 + HO_2$, $RO_2 + RO_2$, and $HO_2 + RO_2$.

The cyclohexene-O_3 results show that selection of an OH scavenger in an ozone–alkene system must be made with careful attention to the subsequent chemistry that occurs following the scavenger-OH reaction.

1.4.4. *NO_3 Oxidation*

An important class of chamber oxidation studies involves the nitrate radical NO_3 as the oxidant. Such studies are especially germane to nighttime chemistry. Because of its strong oxidizing capacity and its relatively high nighttime concentration, the NO_3 radical can play an important role in the nighttime removal of alkenes. Although the reaction of NO_3 with alkenes is 10–1000 times slower than that with OH, NO_3 can be present in sufficiently high concentrations at night so that the overall consumption of alkenes can be comparable for OH and NO_3. Alkenes react predominantly with NO_3 through addition to the double bond, yielding an alkyl radical that rapidly adds O_2 to produce a nitratoalkyl peroxy radical. The high alkyl nitrate yields lead to significant loss of atmospheric NO_x.

Two routes used to generate NO_3 *in situ* in chambers are: (1) decomposition of N_2O_5; and (2) reaction of O_3 and NO_2. After initial NO_3 reaction with the organic of interest, the fate of the nitratoalkyl peroxy radical depends on the composition of the mixture in the chamber. Either of these two routes of generation of NO_3 optimizes conditions for $RO_2 + NO_3$ or $RO_2 + RO_2$. Some HO_2 is expected to form from later-generation chemistry in these chamber experiments, but this is not sufficient to establish a dominant regime of $RO_2 + HO_2$ chemistry. This regime is expected to be important in nighttime chemistry, since HO_2 is not removed at night as effectively as OH. For example, in the BEARPEX field campaign, Mao *et al.*[14] estimated HO_2 mixing ratios at night to be ~4 ppt, while NO_3 mixing ratios were estimated as ~1 ppt.[15] These observations suggest that $RO_2 + HO_2$ chemistry is important in the nighttime.

In order to optimize for the $RO_2 + HO_2$ pathway, formaldehyde (CH_2O) can be injected into the chamber along with O_3 and NO_2. Formaldehyde will react with NO_3 to form HCO, and HCO will react immediately with O_2 to form HO_2:

$$O_3 + NO_2 \rightarrow NO_3 + O_2$$

$$NO_3 + CH_2O \rightarrow HNO_3 + HCO$$

$$HCO + O_2 \rightarrow HO_2 + CO$$

$$HO_2 + NO_2 + M \rightleftharpoons HO_2NO_2 + M$$

$$NO_2 + NO_3 + M \rightleftharpoons N_2O_5 + M$$

Production of NO_3 and HO_2 are coupled through the formaldehyde reaction in such a manner that the ratio of NO_3 to HO_2 can be controlled throughout the experiment. It is still necessary to employ kinetic simulation to estimate the relative pathways of reaction of the RO_2 radicals produced by the initial NO_3 reaction. The high aldehyde concentrations can also impact aqueous/aerosol phase chemistry.

1.5. SOA Formation

SOA forms when VOCs are oxidized to yield products of sufficiently low volatility to condense into the particulate phase. This transformation rarely occurs in a single reaction step; rather, oxidation products undergo progressive oxidation steps, leading to products of decreasing volatility. The atmospheric evolution that leads to SOA can be described in terms of the number of oxidation steps undergone, or the "generation number" of products formed. In an alternative route to SOA formation, low molecular weight, water-soluble VOCs can dissolve in droplets or particles wherein they undergo oxidation to products that remain in the aqueous phase as dissolved SOA. When a droplet containing dissolved SOA eventually evaporates, a residual aerosol particle rich in oxidized organics remains. While experiments addressed at understanding aqueous-phase pathways to SOA formation have tended to employ "beaker scale" systems, studies with moist aerosols and in chambers are starting to emerge.

In the absence of a seed aerosol, with a sufficiently high concentration of VOC, low volatility VOC oxidation products may accumulate in the chamber until a point is reached at which homogeneous nucleation of these products occurs. (The point at which nucleation occurs in a VOC system depends crucially on the VOC itself and the nature of its oxidation products.) Experiments carried out in the absence of a seed aerosol are useful for determining the density of the pure SOA, since the particles will consist exclusively of organic oxidation products. In the absence of a seed aerosol, the nucleated particles tend to concentrate in a range of relatively small particle sizes that may present measurement challenges. In addition, smaller sized particles are more prone to loss by deposition on the wall of the chamber than larger particles (see Sec. 1.6).

From the point of view of measurement of the aerosol size distribution, it is often useful to employ a seed aerosol (for example, ammonium sulfate, ammonium bisulfate, sulfuric acid, sodium chloride, etc.) to maintain particle sizes in the region above 100 nm diameter, which is where conventional aerosol particle-size instrumentation is most accurate. Seeded experiments also facilitate condensation of oxidized organics more readily than when vapors must accumulate until homogeneous nucleation occurs. The earlier particle growth stimulated by the presence of seed aerosol also promotes condensation of vapors onto growing particles as opposed to the walls of the chamber. (SOA formation in the atmosphere also occurs in the presence of pre-existing aerosol.) The use of aerosol mass spectrometry allows direct measurement of the organic-to-inorganic mass ratio, which allows one to quantify the extent of particle deposition onto the walls of the chamber through a mass balance on the suspended inorganic concentration. By control of relative humidity with an inorganic seed aerosol, if an aqueous phase is desired in the particles, experiments can be conducted at an RH above the deliquescence RH of the inorganic seed.

Ordinarily, an SOA formation chamber experiment is continued for several hours after the parent VOC is consumed, since the oxidation products themselves generally continue to react with the oxidant (usually OH) forming additional low volatility products and SOA. At the termination of the experiment, the *SOA yield* is customarily determined as the ratio of the mass concentration of SOA formed to the mass concentration of the parent VOC reacted. This single quantity is the manifestation of all the phenomena involved: gas-phase chemistry, particle-phase chemistry, compound volatility, etc. SOA yields exhibit a complex dependence on VOC-to-NO_x ratio, VOC concentration and volatility, and oxidant exposure. While the definition of the SOA yield is straightforward, its determination is complicated by two phenomena that are a consequence of the presence of chamber walls. First, during the course of an experiment, the particles, which contain both the inorganic seed and condensed VOC oxidation products, deposit irreversibly onto the walls of the chamber. Thus, the aerosol suspended in the chamber at the termination of an experiment does not reflect fully the extent of SOA formation, since a portion of the SOA that formed is on the walls of the chamber. In computing the actual SOA yield, it is necessary to account for the SOA that has deposited on the chamber walls. Section 1.6 addresses particle deposition onto chamber walls and the computation of the SOA mass concentration accounting for that deposition. Secondly, the low volatility gas-phase oxidation products that condense on the particles to constitute SOA are themselves subject to deposition on the walls of the chamber (Sec. 1.7). In the absence of walls, these compounds would contribute to the SOA mass formed, so it is necessary to estimate the effect of vapor-wall deposition to determine the actual SOA yield.

In chamber experiments it is desirable to mimic the degree of atmospheric oxidant exposure, the integral of the oxidant (e.g., OH) concentration and the reactor residence time. Environmental chambers operate at OH concentrations in the range of $10^6 - 10^7$ molecules cm^{-3}, which are roughly equal to ambient daytime concentrations. Chamber experiments are ultimately limited in duration by wall deposition of particles and vapors to the order of 24–36 h or so. As a result, SOA chamber formation experiments tend to simulate an oxidant exposure of about a day or two.

1.6. Particle Deposition onto Chamber Walls

1.6.1. *Experimental Determination of Particle Wall Deposition Rates*

Suspended particles in the chamber are transported to the walls by diffusion, gravitational settling and electrostatic forces.[16,17] The particle wall deposition rate can be measured directly in wall deposition experiments, which is a routine procedure in most environmental chambers. Calibration particles over a range of sizes are often generated by atomizing aqueous ammonium sulfate solution ($(NH_4)_2SO_4$) into the chamber. Figure 1.15 shows the hourly-averaged temporal profiles of the particle number distribution as a result of particle deposition to the chamber walls in the Caltech chamber. For a population of particles with diameters ranging from 10 to 1000 nm, those at an intermediate size exhibit the minimum deposition rate, which is a consequence of the combined effects of turbulent mixing in the chamber, particle diffusivity, and particle settling volocity.

The decay of particle number concentration in each size range is subsequently fitted to a first-order loss model in terms of a

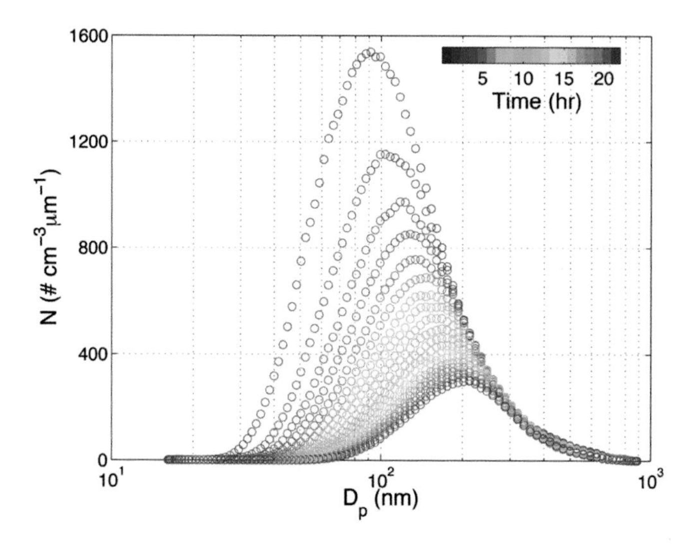

Fig. 1.15 Temporal profiles of particle number distribution as a result of particle wall deposition.

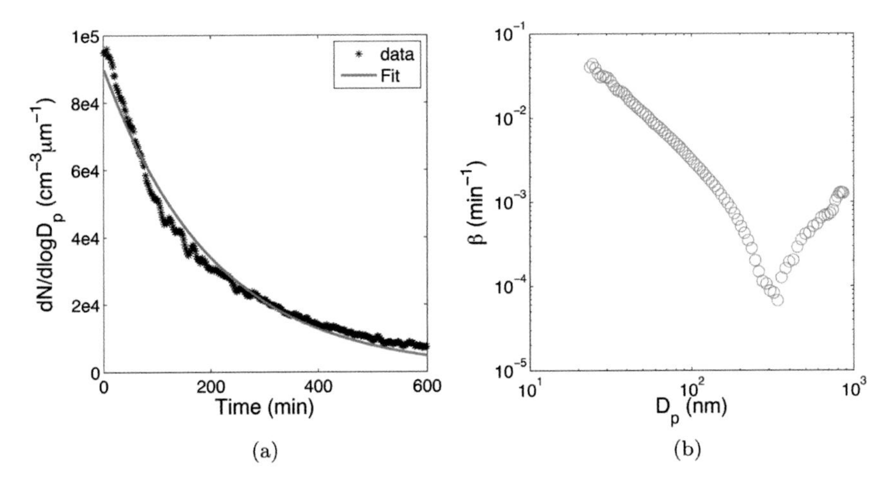

Fig. 1.16 Determination of the particle-wall deposition rate constant. Panel a: Observed and fitted particle number concentration decay at $D_p = 80\,\text{nm}$. Panel b: Best-fit β as a function of particle diameter.

size-dependent wall loss rate coefficient, β_i,

$$n_{s,i} = n_{0,i} \times \exp(-\beta_i t), \tag{1.1}$$

where $n_{s,i}$ is the suspended particle number distribution in size bin i at time t, and $n_{0,i}$ is the initial particle number distribution in size bin i. Figure 1.16(a) shows the measured as well as fitted deposition rate of particles at $D_p = 80\,\text{nm}$ as an illustration. Extending this analysis to particles over the complete range of relevant sizes will give the size-dependent particle wall deposition rate coefficients, as shown in Fig. 1.16(b). In the Caltech chamber, particle wall deposition rate coefficients are of the order of 10^{-4} to 10^{-2} min^{-1}, that is, the half-life with respect to particle wall deposition ranges from ~70 min to ~5 d, depending on particle size. It is advantageous when carrying out experiments to generate SOA to employ seed aerosol that causes the size distribution of the SOA-containing particles to focus in the vicinity of the minimum of $\beta_i(D_p)$.

Throughout an experiment, the volume of the chamber decreases due to sampling, but the surface area of the walls remains the same. The increasing surface area-to-volume ratio might lead to an increase in the particle wall loss rates. The duration of a typical wall loss experiment is 18–24 h, shorter than that of the longest SOA aging experiments. To confirm that wall loss rates do not vary significantly as chamber volume decreases, an additional wall loss calibration experiment can be performed. These calibration experiments were conducted following the same protocol as a typical wall loss calibration; however, before ammonium sulfate seed aerosol is injected, approximately 8 m^3 of air was removed from the chambers to simulate conditions at the end of an 18 h experiment. The wall loss rates determined from these lower-volume experiments were within the range of those observed in the fully inflated chambers.

1.6.2. *SOA Yields Accounting for Particle Wall Deposition*

As noted earlier, to determine the total SOA mass concentration at the termination of a chamber experiment, particle wall losses must be taken into account. An inherent uncertainty exists in this accounting: while the particle deposition is irreversible, the extent to which the deposited particles continue to interact with the gas phase in the chamber remains uncertain. This uncertainty has given rise to two limiting assumptions concerning the estimate of the extent to which particle wall deposition affects the SOA yield (see, for example, Refs [18–22]).

In one limit, particles deposited on the wall are assumed to cease interaction with suspended vapors after deposition. In this case, the amount of organic material in the deposited particles does not change after deposition, and these particles remain at the same size at which they deposited for the remainder of the experiment. In the other limit, particles on the wall are assumed to continue to interact with vapors in the chamber after deposition as if they had remained suspended. Thus, in this case, the amount of organic material in the particles after deposition changes at the same rate as the amount of organic material in the suspended particles, and the deposited particles are assumed to continue to grow throughout the remainder of the experiment. This limit is analogous in theory to that of a chamber without walls. In either limit, the material on the walls is added to that which remains suspended at the end of the experiment to obtain the total amount of SOA formed. The extent to which salts that have deposited to chamber walls may continue to interact with chamber contents is unknown.

During particle growth, if an inorganic seed is present, the organic mass fraction of the suspended particles increases. In the first limit, the organic mass fraction of deposited particles ceases changing after particle deposition; therefore, this case produces a lower limit for SOA mass. In the second limit, deposited particles are assumed to continue growing; therefore, this case represents an upper limit for SOA mass. These two limits of wall loss corrected SOA mass are referred to as the *lower bound* and *upper bound*, respectively.

The first limiting assumption has been widely used in analyzing data in chamber experiments. The size-dependent wall deposition rate coefficient (β_i) estimated earlier can be applied to chamber experiments for particle wall loss correction. (A key assumption here is that particles of the same size, regardless of the chemical nature, exhibit the same deposition rate. It should be noted that particles of different phase states might have different affinities for the wall. For examples, solid particles might bounce off the wall instead of sticking onto the walls with unity accommodation coefficient.) For each size bin i, the number distribution of particles deposited to the wall ($n_{w,i,j}$) over the time increment Δt from time step j to time step $j+1$, is

$$n_{w,i,j} = n_{s,i,j}[1 - \exp(-\beta_i \Delta t)], \tag{1.2}$$

where $n_{s,i,j}$ is the suspended particle number distribution in size bin i at time step j. The number distribution of deposited particles ($n_{w,i,j}$) is added to the suspended particle number distribution ($n_{s,i,j}$) to give the total particle number distribution ($n_{\text{tot},i,j}$),

$$n_{\text{tot},i,j} = n_{w,i,j} + n_{s,i,j}. \tag{1.3}$$

The total number concentration in size bin i at time step j ($N_{\text{tot},i,j}$) can be calculated by converting the number distribution based on $d(\ln D_p)$ to $d(D_p)$,

$$N_{\text{tot},i,j} = \frac{n_{\text{tot},i,j}}{D_{p,i}\ln 10}(D_{p,i+} - D_{p,i-}), \tag{1.4}$$

where $D_{p,i}$ is the median particle diameter for size bin i, $D_{p,i+}$ is the upper limit of particle diameter for size bin i, and $D_{p,i-}$ is the lower limit of particle diameter for size bin i. Assuming spherical particles, the total volume concentration at time step $j(V_{\text{tot},j})$ is:

$$V_{\text{tot},j} = \sum_i \frac{\pi}{6} D_{p,i}^3 \times N_{\text{tot},i,j}. \tag{1.5}$$

The total organic mass $(\Delta M_{0,j})$ at time step j can be obtained by,

$$\Delta M_{0,j} = \rho_p(V_{\text{tot},j} - V_{\text{seed}}), \tag{1.6}$$

where ρ_p is the average particle density, and V_{seed} is the volume concentration of seed particles injected at the beginning of the experiment.

The upper bound limit on SOA mass is calculated by combining data from on-line aerosol mass spectrometry (the Aerodyne AMS) and particle size distribution measurement (e.g., the DMA, see Sec. 1.11). In experiments that utilize seed particles containing sulfate, the only process that decreases sulfate concentration in the suspended phase is wall deposition. The initial sulfate concentration is determined from the initial seed volume concentration. (Collection efficiency in the AMS increases as organic content of the particles increases, and because the seed particles usually do not contain organic material, they are more susceptible to bounce in the instrument and exhibit a collection efficiency that is less than unity.[23]) To calculate the mass of sulfate in the seed, m_{SO4}, the following equation is used,

$$m_{\text{SO4}} = V_{\text{seed}}\rho_{\text{seed}}\frac{MW_{\text{SO4}}}{MW_{\text{seed}}}, \tag{1.7}$$

where ρ_{seed} is the density of the seed particles, MW_{SO4} is the molecular weight of sulfate, and MW_{seed} is the molecular weight of the seed particles. For dry AS seed, $\rho_{\text{seed}} = 1.77\,\text{g cm}^{-3}$. In the upper bound limit, both suspended and deposited particles are assumed to gain or lose organic material at the same rate; therefore, the organic-to-sulfate ratio of all particles of the same size is the same, and this ratio can be determined from unit mass resolution AMS data. (Differences in the organic-to-sulfate ratio, r_{OS}, between unit mass resolution and high-resolution data are less than 5%, except possibly during the early period of growth when organic loading is low.) To obtain the SOA mass, r_{OS} is multiplied by the initial mass of sulfate in the seed particles,

$$\Delta M_0 = m_{\mathrm{SO4}} r_{\mathrm{OS}} \tag{1.8}$$

This equation is valid if the organic-to-sulfate ratio does not vary with particle size or if particle wall loss rates are constant over the particle size range of interest. When particle wall deposition rates depend on particle size, as they usually do, the latter assumption is not valid. Depending on the condensation behavior of the SOA, r_{OS} may depend on particle size.[19,24]

The aerosol parameter estimation (APE) model[25] has been employed to compute the wall deposition corrected SOA yield under the upper limit assumption.[22] The suspended particle population evolves as a result of three processes: coagulation, condensation, and wall deposition,[26]

$$\frac{\partial n_s(D_p,t)}{\partial t} = \left(\frac{\partial n_s(D_p,t)}{\partial t}\right)_{\mathrm{coagulation}} + \left(\frac{\partial n_s(D_p,t)}{\partial t}\right)_{\mathrm{condensation}} + \left(\frac{\partial n_s(D_p,t)}{\partial t}\right)_{\mathrm{wallloss}}, \tag{1.9}$$

where $n_s(D_p,t)$ is the suspended particle number distribution. The change of suspended particle number distribution due to coagulation is,

$$\left(\frac{\partial n_s(D_p,t)}{\partial t}\right)_{\mathrm{coagulation}} = \frac{1}{2}\int_0^{D_p} K\left((D_p^3-q^3)^{\frac{1}{3}},q\right) \times n_s\left((D_p^3-q^3)^{\frac{1}{3}},t\right) n_s(q,t)\mathrm{d}q - n_s(D_p,t)\int_0^{\infty} K(q,D_p)n_s(q,t)\mathrm{d}q, \tag{1.10}$$

where $K(D_{p1}, D_{p2})$ is the coagulation coefficient for collision of particles of diameters D_{p1} and D_{p2}. The first term of Eq. (1.10) represents the formation of particles with diameter D_p from the coagulation of smaller particles, and the second term represents the loss of particles with diameter D_p due to collisions with all available particles. The change of the suspended particle number distribution owing to wall deposition is represented by the first-order loss model as described earlier. The continuous form of this model can be written as,

$$\left(\frac{\partial n_s(D_p,t)}{\partial t}\right)_{\text{wallloss}} = -\beta(D_p)n_s(D_p,t), \tag{1.11}$$

The change of suspended particle number distribution due to condensation is given by,

$$\left(\frac{\partial n_s(D_p,t)}{\partial t}\right)_{\text{condensation}} = -\frac{\partial}{\partial D_p}[I(D_p,t)n_s(D_p,t)], \tag{1.12}$$

where $I(D_p, t)$ is the change in particle diameter due to condensation,

$$I(D_p,t) = \frac{dD_p}{dt} = \sum_i \frac{4\mathcal{D}_i MW_i}{RTD_p\rho_p} f(\text{Kn},\alpha)(p_{\infty,i} - p_{s,i}) \tag{1.13}$$

where $\mathcal{D}_i$ is the diffusion coefficient of species i in air, MW_i is the molecular weight of species i, R is the gas constant, T is the temperature, $f(\text{Kn}, \alpha)$ is a correction factor for non-continuum transport and imperfect surface accommodation, $p_{\infty,i}$ is the partial pressure of species i in the bulk gas, $p_{s,i}$ is the vapor pressure of species i over the particle surface, Kn is the Knudsen number $(2\lambda/D_p)$, α is the accommodation coefficient for the species on the particle surface, and λ is the mean free path of the diffusing species. (See Table 1.4 for a listing of parameters and variables.) The mean free path of air molecules is ~70 nm at 298 K, and diameters of particles of interest in chamber studies range from 10 nm to 1 μm.

Table 1.4 Variables and parameters important in the environmental chamber

Symbol	Description
$\beta(D_p)$	Particle wall deposition rate coefficient, as a function of particle diameter D_p
β_i	Particle wall deposition rate coefficient, for diameter range D_p
$n_{w,i,j}$	Number distribution of particles in size bin i deposited to the wall over time step t_j to t_{j+1}
$n_{s,i,j}$	Number distribution of suspended particles in size bin i at time step t_j, or $n_s(D_p, t)$
$n_{\text{tot},i,j}$	Total number distribution of particles in size bin i at time step t_j
$N_{\text{tot},i,j}$	Total number concentration of particles in size bin i at time step t_j
$D_{p,i}$	Median particle diameter for size bin i
$\Delta M_{0,j}$	Total organic mass concentration at time t_j
$V_{\text{tot},j}$	Total particle volume concentration at time t_j
V_{seed}	Volume concentration of seed aerosol
ρ_p	Average particle density
m_{so4}	Mass concentration of sulfate in seed aerosol
ρ_{seed}	Density of seed aerosol
MW_{so4}	Molecular weight of SO_4
MW_{seed}	Molecular weight of seed aerosol
r_{os}	Organic to sulfate ratio
ΔM_0	SOA mass concentration
$K(D_p, D_p')$	Brownian coagulation coefficient
$I(D_p, t)$	Rate of growth of particle of diameter D_p by condensation of vapor
MW_i	Molecular weight of species i
R	Gas constant
T	Temperature
λ	Mean free path (for a formula see Table 1.5)
Kn	Knudsen number ($2\lambda/D_p$)
$f(\text{Kn}, \alpha_p)$	Correction factor for non-continuum transport (for formula see Table 1.5)
$\alpha_{p,i}$	Accommodation coefficient for vapor species i on a particle
$\alpha_{w,i}$	Vapor-wall accommodation coefficient
$p_{\infty,i}$	Partial pressure of species i in the bulk gas
$p_{s,i}$	Vapor pressure of species i over the particle surface
C_w	Equivalent absorbing organic mass on the chamber wall
$C_{v,i}$	Concentration of vapor species i in the well-mixed core of the chamber
$C_{w,i}$	Concentration of vapor species i in the thin layer adjacent to the wall
A/V	Surface area-to-volume ratio of chamber
$\mathcal{D}_i$	Molecular diffusivity of species i in air
K_e	Coefficient of eddy diffusion in chamber
$C_{\text{tot},i}$	Total concentration of vapor species i generated
$K_{W,i}$	Gas-aerosol partitioning coefficient of species i
$k_{w,i}$	First-order vapor-wall deposition rate constant for species i
$\bar{c}_i$	Molecular mean speed (for formula see Table 1.5)

Substituting Eq. (1.15) into Eq. (1.12) gives,

$$\left(\frac{\partial n_s(D_p,t)}{\partial t}\right)_{\text{condensation}} = -F_c\left[\frac{1}{D_p^2}n_s(D_p,t) + \frac{1}{D_p}\frac{\partial n_s(D_p,t)}{\partial D_p}\right], \tag{1.14}$$

where the parameter F_c can be considered as a characterization of the condensation rate, and the expression of F_c includes all the unknown parameters in Eq. (1.14),

$$F_c = \sum_i \frac{4\mathcal{D}_i MW_i}{RT\rho_p} f(\text{Kn}, \alpha)(p_{\infty,i} - p_{s,i}) \tag{1.15}$$

The value of F_c can be obtained by optimal fitting of the APE model predictions to the measured particle size distribution at each time step over the course of the experiment. Once the F_c values are estimated, they can be applied to parameterize the growth of particles on the walls due to condensation of gaseous vapor and deposition of suspended particles (note that coagulation of particles deposited on the walls is not considered). A factor ϕ can be introduced to describe the extent of interaction between deposited particles and suspended vapors and is applied when summing aerosol masses in the chamber core and on the walls. As noted above, the upper limit estimate accounts for the maximum vapor transport to particles deposited on chamber walls by assuming the interaction of deposited particles with vapors is the same as that for suspended particles. The magnitude of vapor-wall condensation on deposited particles is potentially overestimated under this assumption because the effect of a finite time-scale for vapor transport through the boundary layer adjacent to the wall is not accounted for (see Sec. 1.7).

Figure 1.17 gives an example of the SOA yield (expressed as the total organic aerosol mass formed, ΔM_0, vs. the total hydrocarbon mass reacted, ΔVOC) from the photooxidation of toluene under low-NO conditions with both upper- and lower-bound particle wall deposition corrections estimated by the APE model. Specifially, the DMA measured particle number distribution at each timestep is used to optimize the single adjustable parameter F_c in the APE model. Thus F_c is regarded as a characterization of (1) the diffusion of

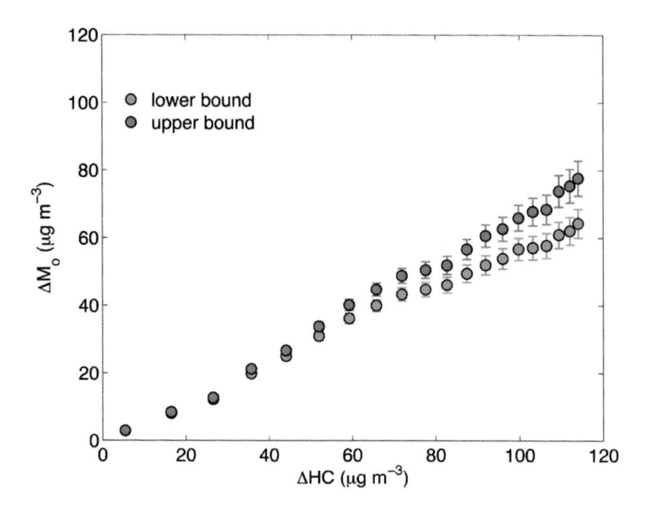

Fig. 1.17 SOA yield (expressed as the total organic aerosol mass formed, ΔM_0, vs. the total hydrocarbon mass reacted, ΔHC) from the photooxidation of toluene under low-NO conditions with both upper- and lower-bound particle wall deposition corrections. The error bars represent the 95% confidence interval from the particle wall deposition rate constant β fitting to experimental deposition data.

gas-phase organic vapors to the particle surface and (2) subsequent uptake by particles at different sizes. When applying the best-fit F_c value to represent interactions of organic vapors with particles deposited on the chamber wall, the scaling factor ϕ is employed. When $\phi = 0$, particles deposited on the walls cease to interact with gas-phase molecules and the particle size distribution remains the same as at the moment they deposited on the chamber wall. Summing up particle masses in the chamber air and on the chamber wall gives SOA yields with lower bound correction. When $\phi = 1$, the growth rate of deposited particles due to interaction with organic vapors is assumed to be idential with that for suspended particles. The overall SOA masses on the chamber wall result from two processes: deposition of suspended particles and condensation of organic vapors. As a result, SOA yield with upper bound correction can be derived as if no chamber wall exists. For the system shown in Fig. 1.17, the SOA yields with upper bound corrections are, on average, 1.13 times higher than those corrected under the lower limiting assumption.

1.7. Vapor Deposition on Chamber Walls

Vapor molecules generated in the course of VOC oxidation can be removed upon contact with chamber walls. Whereas the rate of particle wall deposition depends exclusively on particle size, deposition of vapor molecules on the wall varies in a complex, and generally unknown, manner depending on the chemical nature of the molecule and the nature of the wall itself. One hypothesis is that organic oxidation products can deposit to form a coating on the wall that acts as a medium for further absorption. In the case of a Teflon-walled chamber, it has also been hypothesized that the Teflon film itself can act as an absorbing medium, in a process akin to the sorption of small molecules by organic polymers. In order to formulate a model of vapor-wall deposition, the microscopic nature of vapor-wall interactions need not be understood in detail, as the physico-chemical parameters that arise in the theory of vapor-wall deposition must ultimately be determined from experimental data.

1.7.1. *Theoretical Description of Vapor-wall Deposition*

Vapor molecules are transported from the well-mixed core of a chamber through a thin layer adjacent to the walls by a combination of molecular and turbulent diffusion. The transport rate across this layer depends on both the molecular properties of individual compounds, as well as the extent of mixing in the chamber. As a vapor molecule (species i) encounters the chamber wall, the fraction of those encounters that lead to uptake is represented by the vapor-wall accommodation coefficient ($\alpha_{w,i}$), which depends, in principle, on the nature of the wall surface as well as the chemical nature of the vapor molecule. Species deposited on the walls may, in principle, re-evaporate, eventually leading to equilibrium between the gas phase and the wall.

The rate at which gas-wall equilibrium is approached can be evaluated with a simple kinetic model that assumes a concentration of the deposited vapor in the wall layer can be defined. We define for a species A:

A_g: concentration of species A in the gas phase

A_w: concentration of species A in the wall

A_{Tot}: total concentration of species A, in both the gas-phase and the wall

$\left(\frac{A_g}{A_{\text{Tot}}}\right)_{\text{eq}}$: ratio of species present in the gas phase to the total at equilibrium

k_F: forward rate constant for vapor-wall deposition

k_R: reverse rate constant for evaporation from wall

The equations describing the dynamics of A_g and A_w are:

$$\frac{dA_g}{dt} = -k_F A_g + k_R A_w$$

$$A_w = A_{\text{Tot}} - A_g$$

$$\frac{dA_g}{dt} = -(k_F + k_R)A_g + k_R A_{\text{Tot}}$$

The general solution is

$$\frac{A_g}{A_{\text{Tot}}} = \left(\frac{A_g}{A_{\text{Tot}}}\right)_{\text{eq}} + \left(1 - \left(\frac{A_g}{A_{\text{Tot}}}\right)_{\text{eq}}\right) e^{-(k_F + k_R)t}.$$

The time scale governing the approach to gas-wall equilibrium is

$$\tau_{gw} = \frac{1}{k_F + k_R}.$$

The chamber walls have been characterized by a parameter defined as the equivalent absorbing organic mass on the wall (C_w),[27] which is analogous to the concentration of absorbing aerosol mass, in gas-particle equilibrium partitioning theory.[28] Then the time scale for approach to gas-wall equilibrium is

$$\tau_{gw} = \frac{1}{k_F\left(1 + \frac{C^*}{C_w}\right)},$$

where C^* is the effective equilibrium mass concentration of the species at the temperature of interest. The time at which the wall concentration of deposited vapor reaches 90% of its equilibrium value is given by $-\log(0.1)/(k_F + k_R)$.

The above development assumes that a concentration of the vapor species in the wall layer can be defined. If C_w is sufficiently large, the wall presents essentially an infinite absorbing medium, and the concentration of the vapor species in the wall is effectively zero. In this case, vapor-wall deposition is ultimately an irreversible process, and the reversible analysis above does not apply. It is important to note that the concept of an 'equivalent absorbing organic mass' does not necessarily imply that an actual layer of organic material exists on the chamber wall. In the case of a Teflon-walled chamber, the quantity C_w can be regarded simply as a proxy for the equilibrium solubility of individual organic molecules in FEP Teflon polymer.

The simple kinetic model above can be replaced by one that defines the quantities involved in the wall deposition process more explicitly. A dynamic balance on the concentration of vapor species i in the chamber, $\overline{C}_{v,i}$ that is undergoing solely wall deposition is,[29,30]

$$\frac{d\overline{C}_{v,i}}{dt} = \left(\frac{A}{V}\right)\left(\frac{\alpha_{w,i}\bar{c}_i/4}{\pi\alpha_{w,i}\bar{c}_i/8(\mathcal{D}_i K_e)^{1/2}+1}\right)\left(\frac{\overline{C}_{\mathrm{tot},i}-\overline{C}_{v,i}}{K_{w,i}C_w}-\overline{C}_{v,i}\right), \tag{1.16}$$

where this equation expresses the rate of change of the concentration in the core of the chamber as a result of transport through the wall layer at a rate determined by the difference between the concentration of the species immediately above the wall and that in the core of the chamber. The fraction of encounters at the wall that lead to uptake is represented by the vapor-wall accommodation coefficient ($\alpha_{w,i}$). K_e is the coefficient of eddy diffusion in the well-mixed chamber, and $\mathcal{D}_i$ is the molecular diffusivity of species i in air. $K_{w,i}$ is the gas-aerosol partitioning coefficient of species i. $\bar{c}_i$ is the molecular mean speed of species i.

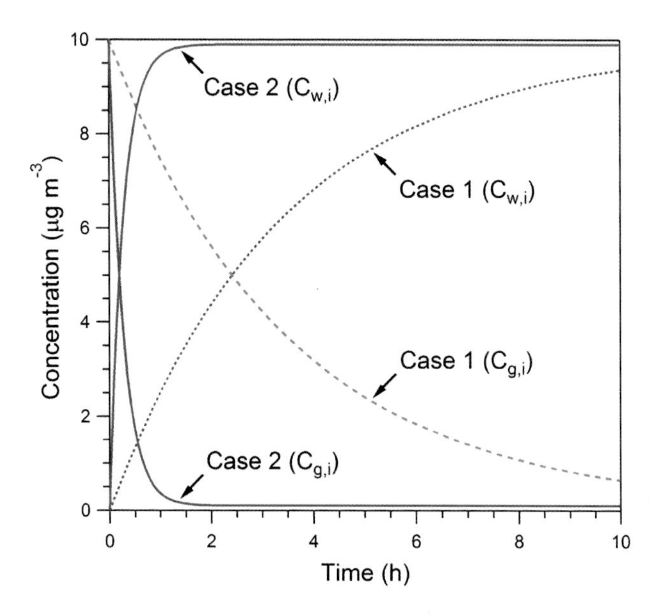

Fig. 1.18 Computed temporal profiles of semivolatile organic compound i in the gas phase ($C_{g,i}$) and on the chamber wall ($C_{w,i}$) during 10-h vapor-wall interaction. Conditions for Case 1: $\alpha_{w,i} = 10^{-6}$ and $C_w = 1\,\text{mg}$; and Case 2: $\alpha_{w,i} = 10^{-2}$ and $C_w = 1\,\text{mg}\,\text{m}^{-3}$.

Figure 1.18 shows two case studies on the simulated wall-induced decay profiles of a semivolatile organic compound i ($C^* = 10\,\mu\text{g}\,\text{m}^{-3}$) under conditions in which the vapor wall interaction is controlled either by $\alpha_{w,i}$ or C_w. For case 1, the wall accommodation of compound i is the limiting step that ultimately governs the overall deposition rate. The timescale for approaching gas-wall equilibrium is 18 h. This case has been confirmed by experimental data in the Caltech chambers. For case 2, the chamber wall accommodation of organic vapors is rapid and the vapor flux to the wall can be approximated as a constant that is related to the turbulent mixing state of the chamber. Yeh and Ziemann,[31] in experiments carried out in their chamber, observe that vapor-wall equilibrium is established rapidly and therefore that C_w is the sole governing parameter.

In the case in which $\alpha_{w,i}$ is the sole governing parameter, Eq. (1.16) becomes,

$$\frac{d\overline{C}_{v,i}}{dt} = -\left(\frac{A}{V}\right)\left(\frac{\alpha_{w,i}\bar{c}_i/4}{\pi\alpha_{w,i}\bar{c}_i/8(\mathcal{D}_i K_e)^{1/2}+1}\right)\overline{C}_{v,i}, \tag{1.17}$$

where the first-order vapor-wall deposition rate constant for species i is,

$$k_{w,i} = \left(\frac{A}{V}\right)\left(\frac{\alpha_{w,i}\bar{c}_i/4}{\pi\alpha_{w,i}\bar{c}_i/8(\mathcal{D}_i K_e)^{1/2}+1}\right). \tag{1.18}$$

Equation (1.18) reveals that two competing physical processes govern the vapor-wall deposition rate: (1) gas-phase transport by molecular diffusion and turbulent mixing and (2) transfer across the gas-wall interface. For small $\alpha_{w,i}$ which leads to the inequality $\pi\alpha_{w,i}\bar{c}_i/8(\mathcal{D}_i K_e)^{1/2} \ll 1$, the denominator of Eq. (1.18) approaches unity, and Eq. (1.18) simplifies to Eq. (1.19a) below. In this case, one expects that the flux of vapor molecules to the chamber wall is limited by the accommodation of organic vapors by the wall surface. For large $\alpha_{w,i}$ which leads to the inequality $\pi\alpha_{w,i}\bar{c}_i/8(\mathcal{D}_i K_e)^{1/2} \gg 1$, Eq. (1.18) becomes Eq. (19b) below. In this case, vapor molecules that encounter the wall are efficiently accommodated by the wall surface, and the supply of vapor molecules from the well-mixed core of the chamber becomes the limiting step,

$$k_{w,i} = \begin{cases} \left(\frac{A}{V}\right)\left(\frac{\alpha_{w,i}\bar{c}_i}{4}\right) & \text{(1.19a)} \\ \frac{\pi}{2}\left(\frac{A}{V}\right)(\mathcal{D}_i K_e)^{1/2} & \text{(1.19b)} \end{cases}$$

The maximum vapor-wall deposition rate is eventually approached for highly oxygenated, extremely low-volatility compounds (which are precisely those compounds that are most prone to form SOA). For values of $\alpha_{w,i} > 10^{-5}$, the rate of transfer to the wall is limited by the rate of diffusion to the wall rather than surface accommodation.[30,32]

1.7.2. *Experimental Characterization of Vapor-wall Deposition*

Chamber wall-induced decay of individual organic vapors can be measured experimentally. Organic vapors are directly injected by evaporating a known amount of chemical standards into the chamber or produced from photochemistry of the corresponding parent VOC.[27,30] Determining the initial vapor concentration is crucial in generating the entire temporal profile of organic vapors due to wall deposition. For the external vapor injection method, the mixing timescale of the chamber is required to be much shorter than that for establishing gas-wall partitioning equilibrium. If organic vapors are generated via photochemistry, a short reaction duration will minimize the interactions of vapors with the chamber wall prior to the onset of vapor-wall deposition. Given the measured vapor-wall deposition profile, values for the two unknown parameters in Eq. (1.16), $\alpha_{w,i}$ and C_w, can be obtained by optimal fitting of Eq. (1.16) to the observations.

1.7.3. *Correction of SOA Formation Data for Vapor-wall Deposition*

In order to accurately predict SOA formation in chemical transport models, experimental data must be corrected for the effects of vapor-wall deposition. Because the effects of vapor-wall deposition change with species volatility, it is not possible to simply apply a first-order deposition coefficient as in the case of particle wall loss. Rather, an SOA formation experiment must be simulated with a full dynamic model that accounts for time-dependent gas-phase chemistry, vapor-particle condensation, and vapor-wall loss. Examples of such models include the Statistical Oxidation Model (SOM)[33,34] and the Generator for Explicit Chemistry and Kinetics of Organics in the Atmosphere (GECKO-A).[35,36] In this section, the basic procedure of correcting SOA formation data will be explained, and then a brief example will be given using the SOM.

The model chosen to correct the data must include time-dependent chemistry. This chemistry can be in the form of a condensed chemical mechanism that has been fit to the experimental data (as in the SOM) or an explicit chemical mechanism (as in the GECKO-A). The model must track the evolving concentrations of individual species, as well as their volatility. The model must additionally include dynamic condensation to particles. Dynamic condensation rather than equilibrium partitioning must be employed because the relative timescales of gas-particle condensation and gas-wall deposition significantly impact the overall effects of this wall deposition (see Sec. 1.10). The rate of condensation to particles is controlled by the gas-phase diffusivity and the vapor-particle accommodation coefficient. This accommodation coefficient must be estimated or constrained by modeling, as described below. Condensation to particles is limited by gas-particle equilibrium, which depends on individual species volatilities and the concentration of absorbing aerosol mass. Finally, the model should include reversible vapor-wall deposition, characterized by the first-order wall loss rate, k_w, and the rate of evaporation from the wall, which depends on the species volatility and the absorbing organic mass of the wall, C_w. Both k_w and C_w must be determined experimentally, but can be constrained by modeling the data, as described below.

The model can be used to generate curves of organic aerosol concentration C_{OA} as a function of time. These curves can then be compared with the experimental results. Sensitivity tests should be conducted for the three unknown parameters: the vapor-particle accommodation coefficient, α_p, the first-order wall loss rate k_w, and the absorbing organic mass of the wall C_w to determine the optimal values that give the best fit for C_{OA}. When the predicted C_{OA} curve matches well with the experimental C_{OA} curve, the model can then be used to correct for the effects of vapor-wall deposition. By setting vapor-wall deposition within the model to zero but leaving all other parameters the same, the model can predict C_{OA} in the absence of vapor-wall deposition. Yields determined from this curve are then considered to be the true yields that would be observed in the absence of chamber walls and are thus the appropriate yields to incorporate into chemical transport models.

The approach described above was used in Ref. [32] to correct SOA yields using the SOM. The SOM represents the multi-generational oxidation of a volatile organic compound and includes dynamic gas-particle condensation and vapor-wall loss. Species are represented using a grid of number of carbon atoms and number of oxygen atoms, with an assigned volatility per grid box. The SOM has six tunable parameters to simulate the gas-phase chemistry: the probability that a reaction leads to fragmentation, the probabilities that a functionalization reaction leads to 1, 2, 3, or 4 oxygen atoms added, and the decrease in vapor pressure per each oxygen added.[33,34] Two additional tunable parameters are the vapor-particle accommodation coefficient and the first-order wall loss rate k_w. In Ref. [32], these parameters were varied to provide the optimal fit to experimental C_{OA} data. When the optimal values were determined, the model was re-run with vapor-wall loss set to zero to generate C_{OA} curves in the absence of vapor-wall deposition (see Ref. [32, Figs. S8, S9, or S10)].

1.8. The Continuously Mixed Flow Reactor (CMFR)

1.8.1. *The Nature of the CMFR*

The CMFR is a well-mixed environmental chamber with continuous feed of reactants and continuous withdrawal of reactor contents. After an initial start-up phase, the reactor is presumed to reach steady state conditions in which the reactor contents are no longer changing with time. For example, in the use of a CMFR to study SOA formation, a VOC/seed aerosol mixture is fed continuously to the reactor, and a comparable flow rate of unreacted VOC and reaction products, including seed aerosol coated with condensed oxidation products, is withdrawn from the reactor. An advantage of the CMFR is that the chamber contents can be sampled for as long a period of time as desired. The mean residence time of air in the chamber is the ratio of the volume of the chamber to the volumetric flow rate of air.

As in batch mode, particles and vapors in a CMFR are subject to deposition on the walls of the reactor. Wall deposition of particles is irreversible, so this process needs to be accounted for as in the batch chamber. If vapor-wall deposition is reversible, wall concentrations of deposited compounds increase until equilibrium between the gas phase and the wall is reached. The time scale associated with the approach of vapor-wall deposition to equilibrium is not necessarily the same as the residence time of fluid in the chamber. In order to assess the time scale for approach of the reactor contents to steady state in the presence of vapor-wall loss, it is necessary to account for the dynamic processes involving vapors occurring in the chamber. A key parameter in this assessment is C_w, which controls the wall concentrations of deposited species at equilibrium. Depending on the magnitude of C_w, vapor-wall equilibrium may or may not be achieved at steady state in the CMFR. For low values of C_w and semivolatile oxidation products, vapor-wall equilibrium is established within a reasonable timeframe. As the walls approach equilibrium, the net rate of transfer of species to the walls slows and the gas-phase concentrations increase. Therefore, vapor-wall equilibrium is a prerequisite for the CMFR to achieve steady state. If, instead, C_w is sufficiently large and the products have low volatilities, vapor-wall deposition is essentially irreversible. If the walls never saturate, vapor-wall loss depresses the SOA yield even at steady state since condensable species are continually being removed by the wall.

The variables, parameters, and governing equations of a CMFR model (in which coagulation is neglected) are given in Table 1.5. We will illustrate the application of this model using idealized gas-phase chemistry in which a VOC (denoted A) reacts to form products B and C of lower vapor pressures according to A $\rightarrow$ B $\rightarrow$ C. To demonstrate that steady state can be achieved even in the absence of vapor-wall equilibrium, we present the results of idealized simulations of the start-up period of a CMFR using the model in Table 1.5 and parameter values in Table 1.6. Wall uptake of vapors is modulated by the value of the equivalent absorbing organic mass on the wall, C_w. In batch chambers, C_w has been inferred to lie in the range of 2–24 $\mathrm{mg\,m^{-3}}$ based on individual measured vapor decay rates. Particle wall deposition is represented by a size-dependent first-order loss coefficient, as detailed in Sec. 1.6.1. For sufficiently low particle number concentrations, the characteristic time scale for coagulation

Table 1.5 Dynamic CMFR model

Variables	Definition	Units
A_i^j	Organic aerosol concentration of species i in size bin j $\frac{dA_i^j}{dt} = -\frac{A_i^j}{\tau} + J_i^j n^j$	$\mu g\, m^{-3}$
$\bar{c}_i$	Mean molecular speed $\bar{c}_i = (8RT/\pi M_i)^{1/2}$	$m\, s^{-1}$
D_p^j	Diameter of particles in size bin j $D_p^j = \left(\frac{6}{\pi\rho_p}\left(\frac{\pi}{6}\rho_p (D_{p0}^j)^3 + \frac{\sum_i A_i^j}{n^j}\right)\right)^{1/3}$	nm
$f(\mathrm{Kn}, \alpha_p)$	Correction factor for non-continuum transport $\frac{0.75\alpha_p(1+\mathrm{Kn})}{\mathrm{Kn}^2 + \mathrm{Kn} + 0.283\mathrm{Kn}\alpha_p + 0.75\alpha_p}$	
G_i	Gas-phase concentration of species i $\frac{dG_i}{dt} = \frac{G_{i0}}{\tau} - \frac{G_i}{\tau} - k[\mathrm{OH}]G_i + k[\mathrm{OH}]G_{j-1} - k_{wall,on,i}G_i + k_{wall,off}W_i - \sum_j J_i^j n^i$	$\mu g\, m^{-3}$
$G_{i,eq}^j$	Equilibrium gas-phase concentration of species i in size bin j $G_{i,eq}^j = \frac{A_i^j C_i^*}{\sum_k A_k^j + M_{\mathrm{init}}^{\mathrm{tot}} \frac{n_0^i}{n_0^{\mathrm{tot}}}}$	$\mu g\, m^{-3}$
J_i^j	Rate of condensation of species i onto a particle of diameter D_p $J_i^j = 2\pi D_i D_p^j (G_i - G_{i,eq}^j) f$	$\mu g\, m^{-3}\, s^{-1}$
Kn_i^j	$\mathrm{Kn}_i^j = 2\lambda_i / D_p^j$	
K_w	Gas-wall partitioning coefficient $K_w = RT/M_w \gamma_w p_i^0$	$m^3\, \mu g^{-1}$
$k_{wall,on,i}$	First-order wall deposition coefficient for species i $k_{wall,on,i} = \left(\frac{A}{V}\right) \frac{\frac{\alpha_{wall,i}\bar{c}_i}{4}}{1 + \frac{\pi}{2}\left[\frac{\alpha_{wall,i}\bar{c}_i}{4(K_e D_i)^{0.5}}\right]}$	s^{-1}
$k_{wall,off,i}$	First-order wall desorption coefficient for species i $k_{wall,off,i} = \frac{k_{wall,on,i}}{K_w C_w} = k_{wall,on,i}\left(\frac{C_i^* M_w \gamma_w}{C_w M_p \gamma_p}\right)$	s^{-1}
λ_i	Mean free path $\lambda_i = 3D_i/\bar{c}_i$	M
n^j	Number concentration of particles in size bin j $\frac{dn^j}{dt} = \frac{n_0^j}{\tau} - \frac{n^j}{\tau} - \beta^j n^j$	cm^{-3}
p_i^0	Vapor pressure of species i $p_i^0 = C_i^* RT/M_p \gamma_p$	Pa
W_i	Wall concentration of species i $\frac{dW_i}{dt} = k_{wall,on,i}G_i - k_{wall,off,i}W_i$	$\mu g\, m^{-3}$

Table 1.6 CMFR simulation parameters

Parameter	Definition	Base value
α_p	Accommodation coefficient of vapor species on particles	0.001
α_{wall}	Accommodation coefficient of vapor species on walls	10^{-5} [a]
C_w	Effective wall organic aerosol concentration	$10\,\mathrm{mg\,m^{-3}}$
v	Coefficient of eddy diffusion in chamber	$0.015\,\mathrm{s^{-1}}$ [b]
$k_{wall,on,i}$	First-order vapor-wall loss coefficient	$3.66 \times 10^{-4}\,\mathrm{s^{-1}}$ (not varied)
τ	Average residence time in chamber	1, 3, 5 h
A/V	Surface-area-to-volume ratio of the chamber	$3.0\,\mathrm{m^{-1}}$
C_i^*	Saturation concentration for species	$[10^0\ 10^{-1}]\ \mu\mathrm{g\,m^{-3}}$
D_i	Gas-phase diffusivity of species i	$3 \times 10^{-6}\,\mathrm{m^2\,s^{-1}}$
G_{A0}	Feed parent VOC concentration (mixing ratio)	$654\,\mu\mathrm{g\,m^{-3}}$ (80 ppb)
γ_n	Activity coefficient in the particle	Cancels with γ_w
γ_w	Activity coefficient in the wall layer	Cancels with γ_n
M_i	Species molecular weight	$200\,\mathrm{g\,mol^{-1}}$
$M_{\mathrm{init}}^{\mathrm{tot}}$	Absorbing organic material in seed aerosol	$0.01\,\mu\mathrm{g\,m^{-3}}$
M_n	Average molecular weight of the organic aerosol	Cancels with M_w
M_w	Effective molecular weight of the absorbing wall material	Cancels with M_p
$k_{wall,on,i}$	First-order vapor-wall loss coefficient	$3.66 \times 10^{-4}\,\mathrm{s^{-1}}$
R	Ideal gas constant	8.314 J $\mathrm{mol^{-1}\ K^{-1}}$
ρ_p	Particle density	1700 kg $\mathrm{m^{-3}}$
T	Temperature	298 K

[a]Estimated experimentally from semivolatile alkanes, alkenes, alcohols, and ketones.[27]
[b]For actively mixed chambers K_e has been estimated as 0.02–0.12 $\mathrm{s^{-1}}$; for chambers that are not intentionally mixed, K_e has been estimated as 0.015–0.075 $\mathrm{s^{-1}}$.[29,30]

(see Sec. 1.10.2) is much longer than typical reactor residence times, so coagulation can be neglected. Though the vapor-particle accommodation coefficient $\alpha_{p,i}$ is, in general, species-dependent, for convenience a single value α_p is used in the simulations. The increase in particle size from the seed particle diameter to that in the chamber is calculated numerically using a moving-bin size distribution based on the total amount of organic aerosol condensed in each size bin (see equation in Table 1.5).

The transient start-up period of the CMFR begins when parent VOC and seed particles are introduced in the flow into the chamber at $t = 0$. The seed aerosol distribution is assumed to be lognormal with a mean diameter of 50 nm and a geometric standard deviation of 1.5, with number concentration = 8000 cm^{-3}. The base values of the variable parameters, k_e, α_w, and α_p are taken as those obtained by fitting SOA data in Ref. [32] C_w is set to 10 mg m^{-3}. The inverse of the e-folding time for each step of progressive oxidation of the parent VOC and its oxidation products is taken as 10^{-3} s^{-1} (time scale ~20 min). A reactor residence time of 3 h is considered. Figure 1.19 shows the evolution of the parent VOC (species A) and two progressive oxidation products (species B and C) in the gas phase, aerosol phase, and wall phase. The gas-phase and aerosol-phase concentrations, which would be sampled in the outflow of the CMFR, each reach steady state after about 20 h. Even though the outflow of the CMFR is at steady state, the wall concentrations are still far from equilibrium and are continuing to increase. For these low volatility products and a large C_w, vapor-wall loss appears to be irreversible and still affects the SOA yield even at steady state. Therefore, steady state in a CMFR does not necessarily imply that vapor-wall loss is at equilibrium and no longer affects the SOA yield. These results suggest that experiments are useful to constrain the nature of wall deposition of vapors in a CMFR. These experiments can clarify the extent to which vapor-wall equilibrium is established during a typical experiment, or if vapor-wall deposition is essentially irreversible and must be taken into account when interpreting SOA yields.

1.8.2. *Comparison of the SOA Yield Obtained in Batch vs. CMFR Chambers*

For a given SOA system, it is informative to ask: How do the SOA yields compare between batch and CMFR systems? For instance, CMFR studies have tended to achieve higher SOA yields than those

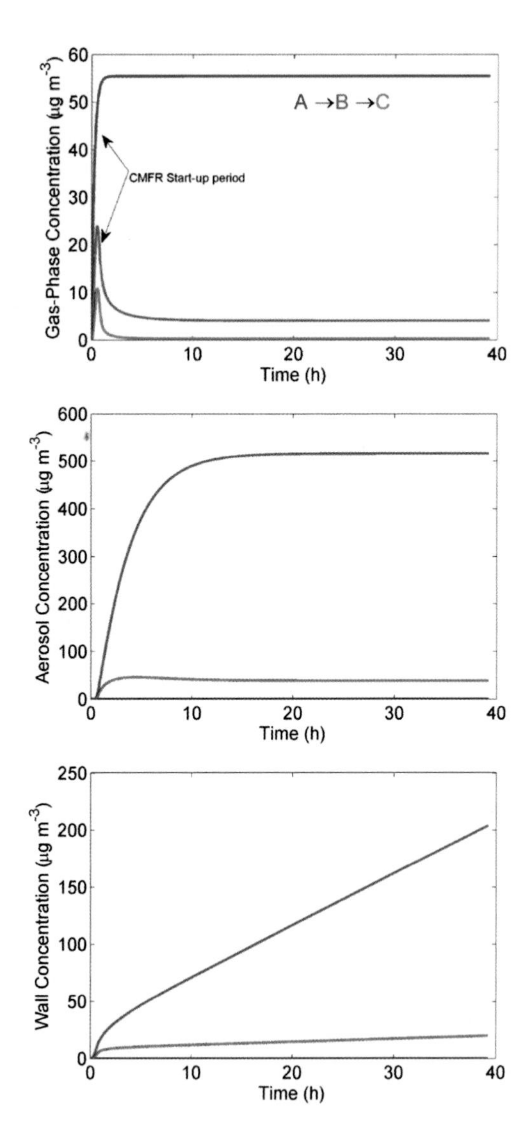

Fig. 1.19 Evolution of the parent VOC (species A) and two progressive oxidation products (species B and C) in the gas phase (top panel), aerosol phase (middle panel), and wall phase (bottom panel). Saturation mass concentrations for species B and C are $1\,\mu g\,m^{-3}$ and $0.1\,\mu g\,m^{-3}$, respectively.

observed in batch systems. This has been attributed to the walls of the CMFR reaching equilibrium.[37] However, as we have shown, this may or may not be the case. If the walls do not reach equilibrium, why then would a CMFR achieve a higher SOA yield than a batch reactor? To address this question, we present the results of simulations using the batch and CMFR versions of the model in Table 1.5. Parameter values from Table 1.6 are used with some exceptions. The simulations explore the effect of the parameters on the SOA yield. The base values of the parameters for this comparison are: $\alpha_p = 0.01$; $C_w = 10\,\mathrm{mg\,m^{-3}}$; $\alpha_w = 10^{-5}$; $\beta = 1.94 \times 10^{-5}\,\mathrm{s^{-1}}$ (assumed independent of particle size); $\tau = 5.2\,\mathrm{h}$. To simplify the comparison, the gas-phase oxidation chemistry is represented as $A \rightarrow B$, with a 50% yield, where B is a low volatility product, with mass saturation concentration $C^* = 0.01\,\mu\mathrm{g\,m^{-3}}$. The SOA yield in the batch reactor is taken as the maximum value achieved over the course of the experiment, whereas the CMFR SOA yield is taken as that when the CMFR is at steady state conditions. The results of the comparison are shown in eight panels in Fig. 1.20, in which the SOA yield is compared between the two reactor configurations. In each panel one parameter is varied, while the remaining parameters are held at their nominal base values. We discuss the results panel by panel.

Figure 20(a): Effect of $\alpha_p(0.01 - 1.0)$

SOA yield is larger in the CMFR for all values of α_p, with the two yields approaching each other as α_p approaches 1. As α_p approaches 1, the resistance to vapor-particle mass transfer is lowered, the impact of vapor-wall loss is lessened, and SOA yields approach 0.5, the highest possible yield since species B is produced with a 50% yield in the gas phase.

Figure 20(b): Effect of seed particle number concentration

SOA yield is larger in the CMFR for all values of the aerosol number concentration. At the beginning of a batch chamber experiment, the seed aerosol is free of condensed organics, so there is a delay in growth, during which time vapor-wall deposition is occurring. Once steady state is reached in a CMFR, there is no "start-up" effect, and the seed aerosol has the steady state level of organics.

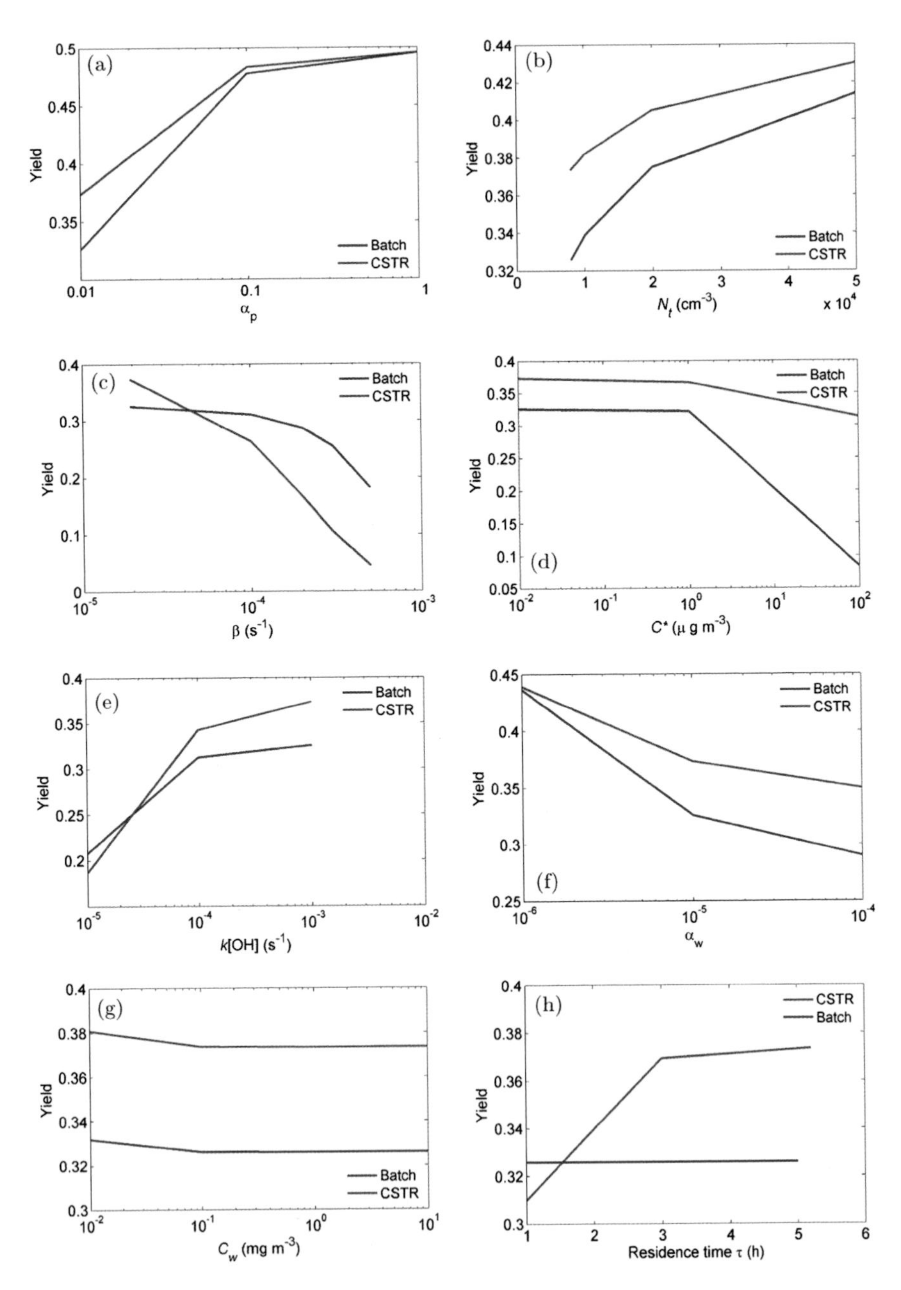

Fig. 1.20 Comparison of the performance between a batch chamber and a CMFR in SOA formation. See text for explanation of panels a–h.

Figure 20(c): Effect of particle wall deposition rate

For a relatively fast wall deposition rate of particles, the SOA yield in the batch reactor exceeds that in the CMFR. In the batch reactor, vapor condensation occurs as soon as oxidation takes place when wall deposition of particles has not yet had an appreciable effect on particle concentrations, whereas in the steady state CMFR particles are effectively removed over the relatively long residence time in the reactor. Only when the particle wall deposition rate is very slow does the yield in the CMFR exceed that in the batch chamber.

Figure 20(d): Effect of the volatility of the SOA oxidation product

Yields in both reactors decrease as C^* increases, as expected, but that in the batch system decreases much faster as C^* increases. For $C^* = 100\,\mu\text{g m}^{-3}$ and only one condensable product, the equilibrium gas-phase concentration is $100\,\mu\text{g m}^{-3}$. Because species B is continually being formed in a CMFR, the gas-phase concentration remains slightly above $100\,\mu\text{g m}^{-3}$ and the yield slowly increases over an exceedingly long time until steady state is achieved. In contrast, species B is rapidly depleted in a batch reactor since no new A is supplied to the chamber. Thus, the concentration of B decreases below $100\,\mu\text{g m}^{-3}$ and condensation to the particle phase no longer occurs. It is important to note that the simulated time for the CMFR to achieve steady state for $100\,\mu\text{g m}^{-3}$ is so long that this result would not be observed experimentally.

Figure 20(e): Effect of the rate of oxidation (k [OH])

For a relatively slower rate of oxidation, the yield in the batch chamber exceeds that in the CMFR due to the shorter residence time in the CMFR. For a slower reaction rate and a fixed residence time, the extent of reaction is lowered in the CMFR. Less species B is produced, lowering the driving force for condensation on the particles and reducing the yield. As the reaction rate increases, the CMFR eventually supplants the batch reactor.

Figure 20(f): Effect of the vapor-wall accommodation coefficient α_w

For a low vapor-wall accommodation coefficient, e.g., $\alpha_w \sim 10^{-6}$, there is very little vapor-wall loss and the SOA yields in both reactors are essentially equal. As α_w increases, proportionately more vapor product is lost early in the batch reactor. The larger impact of vapor-wall deposition on the batch chamber is due to this "startup" effect. When the reaction begins in the batch chamber, there is no organic aerosol present. As species B forms and condenses on seed particles, it is also being lost to the walls, leading to a steadily decreasing driving force for condensation. In contrast, in a CMFR, owing to immediate dilution of B, wall loss occurs but has less impact as the CMFR evolves toward steady state. At steady state the concentration of B is a constant, and the driving force for condensation is sustained.

Figure 20(g): *Effect of* C_w

The large C_w values considered here signify that at vapor-wall equilibrium, the concentration of species B in the wall will greatly exceed that in the gas phase, and the reverse flux of species B from the wall to the gas phase is essentially negligible. This remains true even when reducing C_w by two orders of magnitude. Therefore the SOA yield is unaffected, and the CMFR achieves higher yields than the batch chamber regardless of the value of C_w.

Figure 20(h): Effect of CMFR residence time

At relatively short CMFR residence times, the reaction does not have time to proceed sufficiently versus that in the batch chamber. As the CMFR residence time increases, the CMFR yield eventually exceeds that in the batch reactor.

Although comparison of the batch chamber with the CMFR depends on a number of specific design features, a few general conclusions can be drawn. The CMFR generally exhibits a higher SOA yield except when particle wall losses are high, VOC oxidation is slow, and the residence time in the CMFR is relatively short. The lower SOA yield exhibited in the batch chamber is the result of the wall deposition of condensable vapors at the beginning of the experiment that limits the ultimate amount of vapor available for SOA growth. Once steady state is reached in a CMFR, there is no corresponding "start-up" effect. In the absence of vapor-wall deposition, SOA yield is generally higher in a batch chamber; as the residence time in a CMFR is increased, the SOA yield in the CMFR will approach that in the batch chamber.

1.9. The Flow Tube Reactor

The flow tube reactor is an alternative to the batch environmental chamber. This reactor offers the ability for a wide range of oxidant exposures over relatively short residence times and with reduced wall effects. One principal motivation for the development of flow tube reactors for studying SOA formation is the limitation on OH levels that can be generated in batch chambers. In flow tube reactors, it is possible to generate OH concentrations of the order of 10^9 molecules cm^{-3}[37–45] and thereby to study SOA formation and evolution under conditions equivalent to multiple days of atmospheric OH exposure. The advent of flow reactors for the study of SOA formation can be considered to have begun with the introduction of the Potential Aerosol Mass (PAM) reactor.[38,39] Other laboratory flow reactor systems have been used for a variety of SOA formation studies.[37,41–43,46–50] Modeling studies have investigated the radical chemistry in the oxidation flow reactor.[44]

As noted, a key aspect of the flow tube reactor is the ability to generate OH radical concentrations substantially exceeding those in environmental chambers. In a number of flow tube reactors designed for large OH exposure, low-pressure Hg lamps producing wavelengths of 185 and 254 nm are used to generate OH radicals under continuous flow conditions.[44] OH radicals are produced by photolysis of H_2O and by photolysis of O_3 formed from O_2 photolysis:

$$H_2O + h\nu(185\,\text{nm}) \rightarrow OH + H$$

$$O_2 + h\nu(185\,\text{nm}) \rightarrow 2O \rightarrow 2O_3$$

$$O_3 + h\nu(254\,\text{nm}) \rightarrow O_2 + O(^1D)$$

$$O(^1D) + H_2O \rightarrow 2OH$$

Under typical operating conditions, about 10% of $O(^1D)$ reacts with H_2O to form 2 OH, the majority undergoing quenching to $O(^3P)$, reforming O_3. In H_2O photolysis, virtually all the H atoms formed react with O_2 to form HO_2. The H_2O vapor concentration determines the relative importance of the formation pathways. Overall HO_x loss tends to be dominated by $OH + HO_2$, with a minor contribution by $HO_2 + HO_2$.

The OH concentration can be quantified by measuring the decay of SO_2 by $SO_2 + OH$, for which the reaction rate constant is accurately known. OH concentrations, for example, can be generated in the range of $2\times10^8-2\times10^{10}$ molecules cm^{-3}. At a reactor residence time of 100 s, the corresponding OH exposures are 2×10^{10} to 2×10^{12} molecules cm^{-3} s, equivalent to about 0.2–17 days of atmospheric OH exposure. Li *et al.*[44] present an OH estimation equation as a function of experimental parameters. Lambe *et al.*[42] have shown that the composition of SOA produced in a flow reactor by OH oxidation of gas-phase VOCs and in chambers is the same within experimental accuracy. One must be cautious, however, that the intense oxidation conditions may produce sufficiently high RO_2 abundances that $RO_2 + RO_2$ chemistry may dominate in a way that is uncharacteristic of the atmosphere.

1.9.1. *Design Considerations: Introduction*

A flow tube reactor comprises three sections: an inlet/mixing section, a reaction section, and an exit section (Fig. 1.21). We illustrate these components with three designs currently employed to study atmospheric chemistry. For example, the PAM reactor employed by Kang *et al.*[38] and Lambe *et al.*[41] utilizes a 46 cm length × 22 cm diameter cylindrical design into which aerosol is introduced via standard 6.35 mm tubing and removed from the reactor via a large exhaust (Fig. 1.21a). A characteristic of this system is the generation of high OH exposures in order to simulate multi-day atmospheric processing. The PAM achieves this high-intensity oxidation using four mercury lamps with peak emission wavelength at $\lambda = 254$ nm. The system is operated at a flow rate of 8.5 L min^{-1}, yielding a plug-flow residence time of 106 s and an average velocity of 0.37 cm s^{-1}. The Reynolds number (see Sec. 1.9.2) for this system is 55.[41]

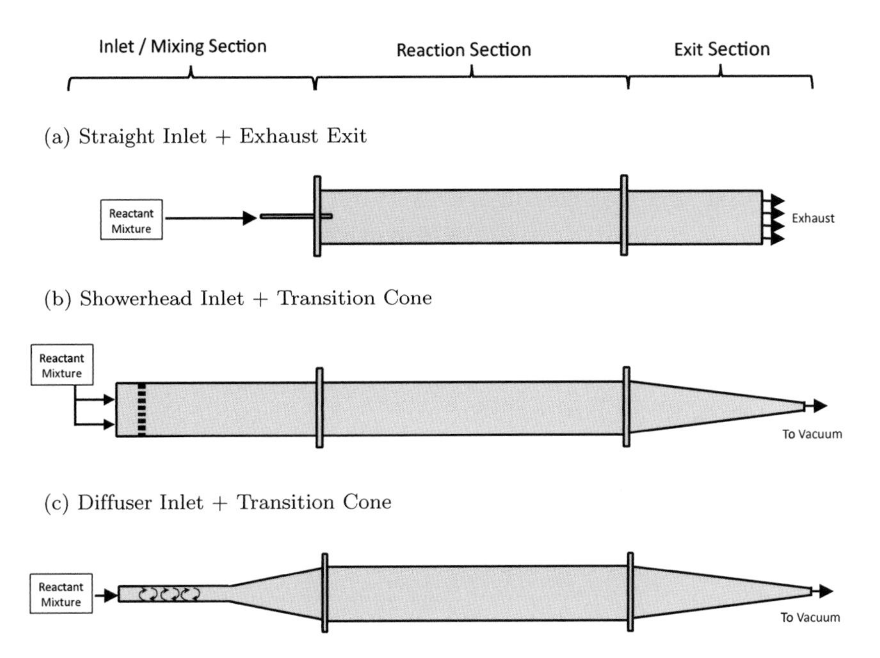

Fig. 1.21 Flow tube reactor designs. The reactors presented here comprise inlet, reaction, and exit sections.

The PAM oxidation flow reactor of Li *et al.*[44] is a 13 L cylindrical aluminum vessel with two or more low-pressure Hg lamps producing 185 nm and 254 nm light inside the reactor. Total OH exposure levels in the reactor can be varied from 10^{10} to 10^{12} molecules cm^{-3} s by changing the UV light intensity, absolute humidity, and residence time. The Hg lamps are enclosed in sleeves through which N_2 flows to remove the heat generated by the lamps. Two types of sleeves are used. Teflon sleeves transmit both 185 and 254 nm light, allowing the direct photolysis of O_2 and H_2O. Quartz sleeves allow only 254 nm light to be transmitted. An important parameter governing the behavior of the reactor is the ratio of 185 and 254 nm photon intensities. The photon flux at 185 nm is determined from the observed O_3 mixing ratio in the reactor, and the photon flux at 254 nm is estimated from the ratio of the fluxes at 185 nm and 254 nm. Reactor flow is ordinarily 3.1

standard liters per minute, giving a residence time in the reactor of 4.2 min.

The UC Irvine flow reactor described by Ezell *et al.*[47] utilizes a design employing a showerhead inlet and a transition cone exit section (Fig. 1.21b). This system introduces gas and particles evenly along the cross- section of the reaction tube, designed to minimize turbulence and particle interactions with the walls. This 8.5 m length $\times$46 cm diameter flow tube reactor when operated at a flow rate of 20 L min^{-1} yields a plug-flow residence time of 60 min and an average velocity of 0.2 cm s^{-1}. The Reynolds number (see below) for this system is 61.[47]

Figure 1.21c is a schematic of the Caltech Photooxidation Tube. This reactor is designed to gently introduce reactants with a diffuser cone while providing prolonged exposure to UV light. The flow tube is 244 cm length $\times$15 cm diameter and is typically operated at a flow rate of 1 L min^{-1}, yielding a plug-flow residence time of 44 min and an average velocity of 0.09 cm s^{-1}. The Reynolds number is $\sim$9.

1.9.2. *Design Considerations: The Nature of the Flow in the Reaction Section*

Flow tube reactors can be operated in laminar or turbulent flow regimes. The radiation source can be external to the tube or inside the tube itself. If the radiation source is inside the reactor itself, one must consider the effect of the imbedded heat source on the velocity profile in the reactor. Here, we focus on the design of the reaction section since this is the component of the flow tube apparatus that is most susceptible to thermal mixing induced by heat emitted from the UV lights.[47,51]

The essential dimensionless group that differentiates laminar vs. turbulent flow is the Reynolds number,

$$\text{Re} = \frac{\rho U D}{\mu}$$

where ρ is the fluid density, U is a characteristic velocity, μ is the fluid viscosity, and D is the tube diameter. For cylindrical tubes, $\mathrm{Re} < 2100$ is the condition for laminar flow.

An advantage of laminar flow is that deposition of gases and particles on the tube wall is minimized; a disadvantage is that the residence time of fluid elements along independent streamlines is different. Nonetheless, the average residence time in laminar flow is precisely known. In contrast, the velocity profile in turbulent flow is uniform across the tube (so-called plug flow), however the transport of material to the wall is greatly enhanced (see Sec. 1.9.3). For photochemical flow tube reactors in which the radiation source is inside the tube itself, the effect of natural convection induced by heating must be considered. For example, Khalizov *et al.*[51] modeled the effects of temperature differentials for a variety of flow conditions and system geometries. The authors found that even for flow tubes with small diameters, a radial temperature differential of only 1K can induce convective mixing within the tube.

The Richardson number, Ri, is a measure of the relative contributions of natural and forced convection,[47,51]

$$\mathrm{Ri} = \frac{g\beta_a \Delta T D}{U^2},$$

where β_a is the thermal expansion coefficient of air, g is the acceleration due to gravity, U is the velocity of the fluid, and ΔT is the radial temperature differential within the tube. For $\mathrm{Ri} > 1$, natural convection will influence the velocity profile in the tube. For the PAM and UC Irvine flow tubes operated under the conditions described in Sec. 1.9.1, a value of $\mathrm{Ri} < 1$ exists only if radial temperature differences are $< 1.5 \times 10^{-3}\,\mathrm{K}$ and $3 \times 10^{-4}\,\mathrm{K}$, respectively. In essence, when the source of radiation is in the tube itself, free convective mixing is unavoidable. To minimize convective mixing, Khalizov *et al.*[51] recommend arranging flow tube reactors in a vertical position. This is feasible for reactor tubes that are short or have relatively small diameters, however this is impractical for larger systems.

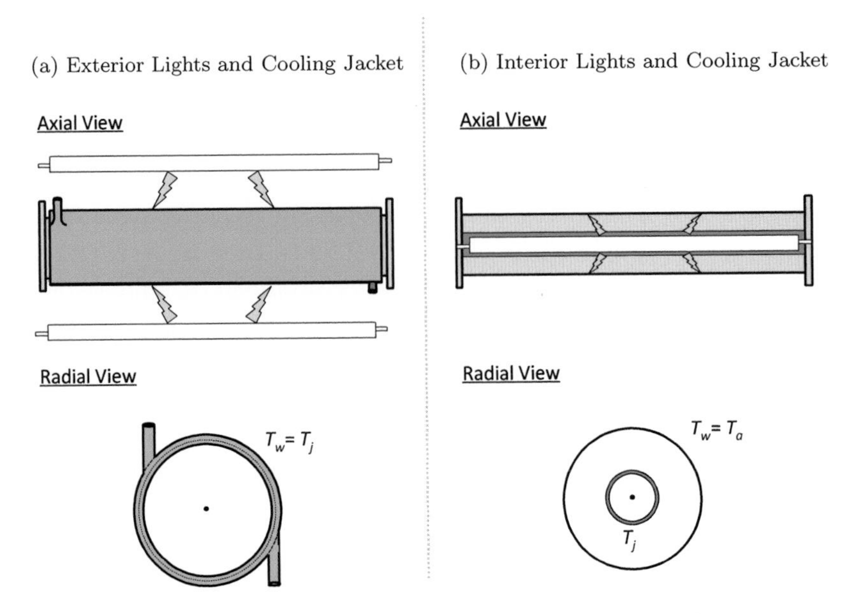

Fig. 1.22 Cooling jacket arrangements typically used to control the temperature within a flow tube reactor. Arrangement (a) employs an exterior cooling jacket with lights positioned on the outside of the tube. Arrangement (b) employs an interior cooling jacket with lights positioned on the inside of the tube. T_w, T_j, and T_a are the temperatures of the reactor wall, cooling jacket, and ambient, respectively.

Heat transfer control can be achieved using air or water-chilled jackets. Figure 1.22 demonstrates two radiation arrangements with jacketed components. Arrangement (a) is a representation of the Caltech flow tube jacket. With lights positioned on the outside of the tube, the exterior water jacket provides a heat transfer medium, while allowing UV radiation to drive photochemistry in the reactor. Since the wall temperature is constant, ΔT is solely dependent on the temperature of the reactant mixture. If the fluid recirculation in the jacket is sufficiently rapid, the axial temperature gradient in the cooling jacket is small. In arrangement (b), the lights are contained within the flow tube reactor. In this case, the reactor walls can be constructed from inexpensive, UV blocking materials (see Sec. 1.9.5). The jacket around the light source, however, must still allow UV penetration. Since the temperature at the reactor walls is affected by that of the surroundings, any jacket temperature maintained below or above ambient will lead to $\Delta T > 0$; thus, the temperature gradient in this arrangement is dependent on the temperature of the reactant mixture, reactor walls, and cooling jacket. As noted above, it may be difficult to avoid some degree of free-convective mixing with this radiation arrangement.

1.9.3. *Design Considerations: Particle Losses on Reactor Surfaces*

For flow tubes constructed to study aerosol chemistry, a consideration in the design of the reaction section is the extent of interaction between particles and the tube walls. As described above, the extent of particle deposition is a function of multiple processes including gravitational settling, particle diffusion, coagulation, and electrostatic interactions. Particle deposition may occur by impaction upon surfaces within the tube, such as sampling ports along the length of the reactor.

To evaluate the extent to which particle loss due to impaction on surfaces protruding into the flow may occur, one may evaluate the Stokes number, which is a measure of the tendency of a particle to follow streamlines of the flow or impact upon surfaces,[26]

$$\mathrm{St} = \frac{D_p^2 \rho_p C_c U}{18\mu L}, \quad \text{where } C_c = 1 + \frac{2\lambda}{D_p}\left[1.257 + 0.4\exp\left(-\frac{1.1D_p}{2\lambda}\right)\right],$$

where D_p is the particle diameter, ρ_p is the particle density, C_c is the Cunningham slip correction, and L is a characteristic length scale of the flow (e.g., sampling port diameter). The Cunningham slip correction factor accounts for non-continuum effects as the diameter of a particle approaches the mean free path of air (λ). When $\mathrm{St} \ll 1$, particles adapt to changes in fluid velocity quickly and impaction is unimportant. For a 500 nm ammonium sulfate particle (density $= 1770\,\mathrm{kg\,m^{-3}}$) travelling around a standard 6.35 mm sampling tube under typical operating conditions, for example, $\mathrm{St} \ll 1$.

Another potential factor contributing to particle loss is wall deposition. Lambe *et al.*[41] observed nearly identical particle transmission efficiencies in the PAM system (ratio of surface area to volume, $SA/V = 0.2\,cm^{-1}$) relative to the University of Toronto Photo-Oxidation Tube ($SA/V = 0.97\,cm^{-1}$) when operated at the same residence time. Both tubes exhibit significant losses for particles <100 nm (50% transmission) with improved transmission for particles >200 nm (80–90% transmission). The authors attribute this behavior to: (1) flow disturbances that enhance deposition to the walls (see Sec. 1.9.2); or (2) electrostatic deposition due to the non-conductive flow tube walls (see Sec. 1.9.5). If the reactor is constructed with stainless steel walls, electrostatic deposition is minimized. For example, in the UC Irvine flow tube reactor, Ezell *et al.*[47] report >98% transmission for particles <300 nm, 91% transmission for 800 nm particles, and 86% transmission for 1000 nm particles.

1.9.4. *Design Considerations: Entrance and End Effects*

When calculating key fluid mechanical parameters such as Re and Ri under laminar flow conditions, one assumes that the flow inside the tube is fully parabolic flow. This assumption implicitly neglects entrance effects. The flow tubes depicted in Fig. 1.21 introduce or remove reactants at the inlet and exit sections in a manner that minimizes turbulence or undesired mixing. This section addresses considerations in the design of the entrance region into a flow tube reactor.

For reactors that gently introduce reactants into the tube, there is an axial distance required for the flow to develop to the characteristic parabolic profile of laminar flow. This entrance length, L_e, is estimated to be $0.035\ D\ \mathrm{Re}$.[52] For the Caltech, PAM, and UC Irvine reactors, the entrance lengths at nominal operating conditions are approximately 0.11 m, 0.42 m, and 1 m, respectively.

A number of possible arrangements exist to introduce reactants into a flow tube reactor (Fig. 1.23). The benefit of a design that introduces gas and particles through a short injection tube, perhaps

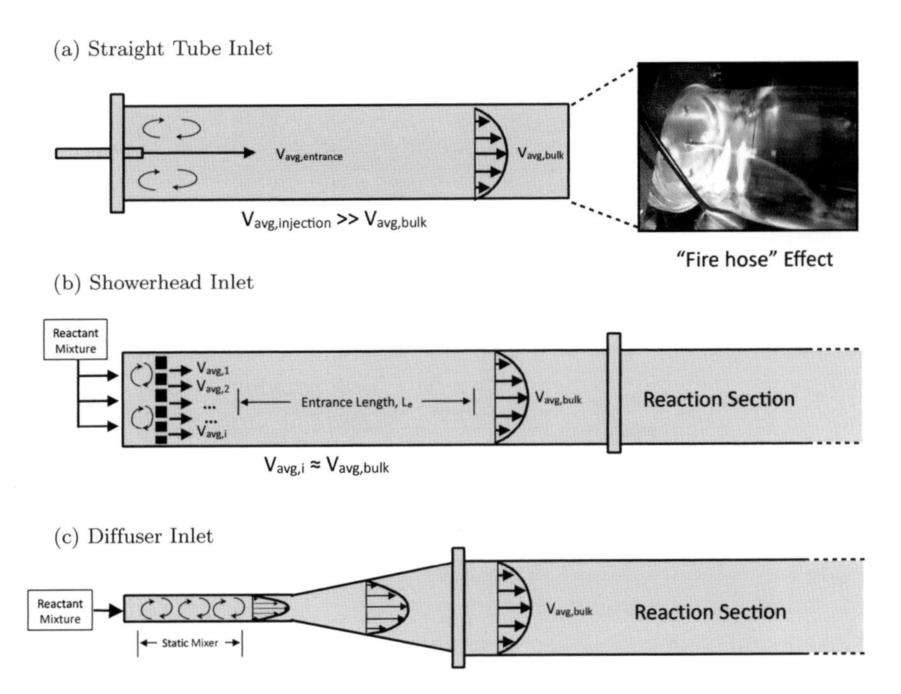

Fig. 1.23 Inlet designs for flow tube reactors.

surrounded by a sheath flow, is its simplicity (Fig. 1.23a); however, with this mode of injection it is challenging to distribute reactant mixtures evenly across the reactor cross-section. This inlet method was tested on the Caltech flow tube, and the flow pattern was visualized by the injection of smoke (Fig. 1.23a). With flow controlled by a vacuum line attached to the exit section (see Fig. 1.21), the gas-particle mixture is pulled into the reaction tube at a rate that is dictated by mass conservation. Smoke injection studies illustrate that the mixture concentrates in a plug focused at the center of the reactor. This "fire hose" effect can be attributed to the enhanced velocity at the exit of the injection tube ($V_{\text{avg, injection}}$). When operating the flow tube with an overall flow rate of $1\,\text{L min}^{-1}$, the average velocity exiting a standard 6.35 mm tube (ID = 3 mm) is $2.35\,\text{m s}^{-1}$. This is nearly 2500 times the average velocity of the flow within the reactor ($V_{\text{avg, bulk}} = 0.09\,\text{cm}\,\text{s}^{-1}$). As discussed by *Lambe et al.*,[41] this injection method has the potential to induce dead volume near the entrance of the reaction section and reactor-scale recirculation. Such behavior is typical for that occurring with a sudden expansion.[52] This can be alleviated by diverting the flow toward the outer edges of the chamber using a covered cap with holes drilled sideways.

The UC Irvine flow tube reactor utilizes a spoked-hub/showerhead disk inlet that distributes the reactants evenly about the reactor cross-section, provides sufficient mixing, and avoids the "fire hose" effect (Fig. 1.23b). Ezell *et al.*[47] designed the inlet with sufficient length to develop the laminar profile before gas and aerosol reach the reaction section. Here, we consider only the showerhead disk; we refer readers to Ezell *et al.*[47] for the complete inlet design. With a showerhead disk, the reactants can be mixed and introduced into the tube in a controlled, gentle manner. The disk itself is perforated with a number of holes such that the fluid velocity at each hole exit ($V_{\text{avg},i}$) approaches that of $V_{\text{avg, bulk}}$. To determine the number of holes in a showerhead design that would sufficiently slow the flow to $V_{\text{avg, bulk}}$, one can calculate $V_{\text{avg},i}$,

$$V_{\text{avg},i} = \frac{F_{\text{vol}}}{n_s \times A_c},$$

where F_{vol} is the total volumetric flow rate through the showerhead, n_s is the number of holes, and A_c is the cross-sectional area of each hole. To achieve a $V_{\text{avg},i}/V_{\text{avg, bulk}} = 25$ in the Caltech flow tube (a 100-fold reduction in velocity relative to a standard 3 mm pipe), a showerhead with 575 holes would be required. For perspective, the UC Irvine flow tube utilizes a showerhead with 940 equally spaced 3.2 mm diameter holes.

In the Caltech flow tube (Fig. 1.23c), reactants are injected via a diffuser cone design, after which a laminar flow profile develops. Like the showerhead disk, the advantage in the diffuser cone design is the reduction of the "fire hose" effect through the gradual decrease in the velocity profile. One consideration when designing a diffuser is flow separation from the wall of the cone, leading to concentration of reactants towards the center of the tube and recirculation within the reaction section. Fried and Idelchik[53] recommended that diffusers be designed with an angle of divergence $<7^\circ$ to avoid flow separation; alternatively, White[54] recommends an angle $<15^\circ$. Sparrow *et al.*[55] modeled the flow of fluid through diffuser cones at various Re. For further discussion about flow separation within diffusers, the authors recommend discussions by Tavoularis.[56] At the design stage, it is recommended that one simulate the diffuser cone numerically using a computational fluid dynamics code to determine if a particular design suits the needs of the reactor.

The exit section of a flow tube is primarily designed to minimize upstream disturbance. The UC Irvine and Caltech reactors utilize a transition cone in which the reactants are concentrated to a common sampling line that can be split among multiple instruments; thus, a representative measure of the entire tube cross-section is obtained. The PAM uses an exhaust exit, and sampling lines are directed in-line with the flow. Both a transition cone and exhaust configuration eliminate dead volume at the reactor exit; thus, flow recirculation due to exit effects is not expected.

1.9.5. *Design Considerations: Construction Materials*

Flow tube construction materials are strongly dependent on the purpose of the reactor. The most commonly used glass materials are Pyrex (borosilicate) and Fused Quartz, and the selection should be based on the desired wavelengths used to promote photolysis as well as the placement of the UV lights. Quartz is considerably more expensive and fragile than Pyrex, but provides properties essential for photochemical studies, assuming the lights are placed outside the reaction vessel (see Fig. 1.21a). Pyrex glass UV-wavelength cut-off is at 275 nm, whereas Quartz is transparent down to 170 nm (Fig. 1.24). If the lights are placed inside the tube (Fig. 1.21b), Pyrex and other more robust materials can be used to construct the flow reactor, but challenges related to mixing issues and structural considerations related to removing the heat produced by the light source need to be considered.

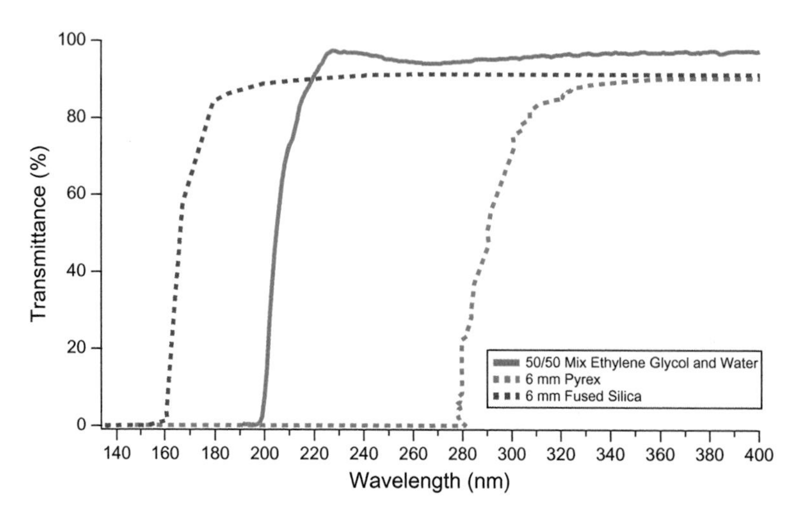

Fig. 1.24 UV-visible transmission spectra of glass flow tube materials (Quartz and Pyrex, dotted lines), and cooling liquid (50/50 mix of ethylene glycol/water, solid line). Pure water is transparent in the UV-visible range above 200 nm. Quartz and Pyrex transmission data are from the Martin Fleischmann Memorial Project (www.quantumheat.org).

As mentioned in Sec. 1.9.2, temperature-controlled studies require cooling of the flow reactor during irradiation; consequently, the cooling fluid should absorb as little UV radiation as possible while providing sufficient heat removal. Likewise, glass materials have a lower pressure threshold than other materials; thus, recirculation in the cooling jacket must be performed using a low-pressure pump. In the Caltech flow reactor (Fig. 1.21a), coolant recirculation in the jacket is performed with a magnetic-drive pump rated to operate under 10 psi, which is ~20 psi below the fracture point of 6 mm thick Quartz glass tubes. The coolant liquid is either a 50/50 mix of ethylene glycol/water, which does not significantly absorb in the UV range dictated by the 350 nm lights (transmission cut-off at around 225 nm) and provides sufficient cooling to maintain reactor temperatures between −10 and 35°C, or pure water which is transparent in the UV-visible range above 200 nm. Transmission spectra of the 50/50 mix of water/ethylene glycol are included in

Fig. 1.24. Alcohol coolants such as methanol can be used to reach very low temperatures, but the possible UV-absorption of the cooling liquid needs to be considered.

In the internal cooling arrangement in the UC Irvine flow reactor, the photolysis lamps are inside the Pyrex tube at the center of the reactor. Heat generation from the lights is removed with 460 L min^{-1} airflow, which enters the lamp housing in the middle of the reaction/photolysis section. With this cooling arrangement, the temperature inside the reactor is 4–6°C above the surrounding temperature. This system cools sufficiently to operate reactions at ambient temperatures.

In flow reactors designed for atmospherically relevant studies, commonly used light sources are broadband and narrowband black-lights and different arc lamps (xenon, argon, mercury), similar to those used in atmospheric chambers (see Sec. 1.3.1). The flow reactor construction should ideally allow changing of the lights to suit different photolysis needs, such as OH generation, photolysis at different wavelengths, etc. The PAM uses lights with variable intensity to produce a range of OH exposures within the reactor. The Caltech flow tube uses lights that have peak emission at 254 nm, 305 nm, or 350 nm, although they are interchangeable with other types of lights.

1.10. Time Scales and Regimes of Behavior for Chamber Processes

1.10.1. *Oxidation Time Scale*

In most chamber experiments involving oxidation of a volatile organic compound, the oxidation time scale is simply $(k\ [\text{Oxidant}])^{-1}$, where k is the reaction rate constant (cm^3 molecule^{-1} s^{-1}) and [Oxidant] is the oxidant (e.g., OH) concentration (molecules cm^{-3}). If the oxidant concentration changes over the course of an experiment, one can use an average concentration to estimate the reaction time scale. The reaction time scale can be tuned experimentally by adjusting the oxidant concentration in the reactor.

1.10.2. *Particle Coagulation Time Scale*

The coagulation time scale of a population of monodisperse particles of initial number concentration $N_0(\mathrm{cm}^{-3})$ undergoing Brownian coagulation can be estimated as $2(KN_0)^{-1}$, where K is the Brownian coagulation coefficient ($\mathrm{cm}^3\,\mathrm{s}^{-1}$) for particles of the given size.[26] Using this relation, the coagulation time scale for typical laboratory chamber conditions can be estimated. The coagulation coefficient for monodisperse seed particles of diameter 100 nm is $\sim 10^{-9}\,\mathrm{cm}^3\,\mathrm{s}^{-1}$. At an initial number concentration of $10^4\,\mathrm{cm}^{-3}$, the coagulation time scale is $\sim$55 h. This time scale exceeds the duration of typical chamber experiments, so the effect of particle coagulation can generally be neglected.

1.10.3. *Particle Wall Deposition Time Scale*

Wall deposition of particles is represented as a first-order process, with deposition rate coefficient of $\beta(D_p)$ (Sec. 1.6.1). The time scale is just the inverse of this first-order rate coefficient. A typical functionality of $\beta(D_p)$, as determined in the Caltech chambers, was shown in Fig. 1.16 (panel b). The longest particle wall deposition time scale is that for particles of the size corresponding to the minimum of $\beta(D_p)$. From Fig. 1.16, the minimum in $\beta(D_p)$ occurs at $\sim$400 nm diameter for which the time scale is $\sim$333 h.

1.10.4. *Vapor-Particle Equilibration Time Scale*

In cloud-free air, SOA forms via three possible mechanisms: (1) effectively irreversible condensation of very low volatility organic vapors produced by gas-phase VOC oxidation, (2) reversible absorption of semivolatile organic vapors into existing particles, and (3) absorption of semivolatile and volatile organic vapors into existing aerosol followed by particle-phase reactions to form effectively non-volatile products. A variety of studies have shown that SOA particles can exhibit the properties of semi-solids, in which case intra-particle diffusion of a species can be significantly retarded and does not occur instantaneously, as has been presumed for a liquid particle. This affects the time required for the dissolved species to equilibrate with the gas phase. Typical diffusivities D_b ($cm^2\,s^{-1}$) of organics are $10^{-10} - 10^{-5}$ for liquid, $10^{-20} - 10^{-10}$ for semi-solid, and $<10^{-20}$ for solid.

As semivolatile vapors condense into particles, equilibrium is eventually reached at which the partial pressure of the vapor is equal to the vapor pressure of the species over the particle. Estimation of the vapor-particle equilibration time scale must, in principle, account for three transport time scales (in the absence of particle-phase chemical reactions): (1) the characteristic time for the profile of the gas-phase concentration around the particle to relax to its new steady state following a perturbation of the bulk gas-phase concentration, (2) the characteristic time for interfacial equilibrium to be re-established following a perturbation, and (3) the characteristic time for particle-phase diffusion to establish a uniform concentration in the particle (in the absence of particle-phase chemical reaction). Depending on the specifics of a given situation, any of these three transport processes can govern the overall vapor-particle equilibration time scale. The rate of condensation is controlled by the rate of diffusion of vapor molecules to the surface of the particle and by the accommodation coefficient, α_p, of the vapor molecules at the particle surface. The accommodation coefficient α_p embodies empirically the net rate of uptake of molecules at the particle surface and has to be determined experimentally from vapor uptake (or particle growth) measurements.

With the recognition of the role of wall deposition of vapor molecules involved in SOA formation in chambers, the competition between suspended particles and the chamber walls for VOC oxidation products plays a crucial role in determining the measured SOA yield, and the relative time scales for equilibration of the vapors between the suspended particles and the walls of the chamber become important in determining the extent to which the SOA yield is affected by vapor-wall deposition. If the vapor-particle equilibration time scale is long compared to that for vapor-wall equilibration, then the effect of vapor-wall loss on SOA yield is exacerbated.

Shiraiwa and Seinfeld[57] performed a theoretical analysis of the equilibration timescale τ_{eq} of SOA partitioning in liquid, semi-solid, and solid particles using the numerical model KM-GAP, which resolves the mass transfer in both gas and particle phases. The model allows a systematic evaluation of the equilibration timescale τ_{eq} as a function of SOA volatility, particle-phase diffusivity, surface accommodation coefficient, and particle size. Species volatility can be represented by the effective saturation mass concentration, C^*, expressed in units of $\mu\text{g}\,\text{m}^{-3}$. Values of C^* define volatility in the following ranges: $10^3 - 10^5$ (intermediate volatility organic compounds, IVOC); $10^{-1} - 10^3$ (semivolatile organic compounds, SVOC); $10^{-4} - 10^{-1}$ (low volatility organic compounds,low volatility organic compounds, LVOC LVOC); $<10^{-3}$ (extremely low volatility organic compounds, ELVOC). The behavior of τ_{eq} can be summarized as follows:

(1) For liquid particles with diffusivities in the range of $10^{-8}\,\text{cm}^2\,\text{s}^{-1}$, τ_{eq} increases as C^* decreases. In this case, the partial pressure gradient between the gas phase and the particle surface is larger for smaller C^*, that is a less volatile species, and the equilibration time is correspondingly longer.
(2) For liquid particles as in (1) above, as the accommodation coefficient α_p decreases, τ_{eq} increases. At $\alpha_p = 1$, SOA growth is limited by gas-phase diffusion, but as α_p decreases, SOA growth eventually becomes limited by surface accommodation.
(3) For semi-solid particles, with particle-phase diffusivities in the range of 10^{-10} down to $10^{-20}\,\text{cm}^2\,\text{s}^{-1}$, the timescales for exchange between the surface and the particle bulk and diffusion in the bulk particle become longer than those for gas-phase diffusion and surface accommodation. In this case, τ_{eq} is insensitive to the value of α_p but sensitive to the value of the bulk diffusivity D_b. In this regime, decrease of D_b by an order of magnitude leads to roughly an order of magnitude increase in τ_{eq}. For a typical situation, τ_{eq} is the order of minutes for semi-solid particles with $D_b \sim 10^{-15}\,\text{cm}^2\,\text{s}^{-1}$, increasing to days for $D_b < 10^{-20}\,\text{cm}^2\,\text{s}^{-1}$.

For LVOC, the instantaneous gas-particle equilibrium model can overestimate the particle-phase concentration by an order of magnitude before equilibrium is established. The formation of oligomers and other multifunctional compounds in the particle phase with high molecular mass and low vapor pressure is one mechanism that can lead to high viscosity and low diffusivity. The occurrence of a semi-solid state and the associated effects may require a more detailed kinetic representation of SOA formation than has been the case for instantaneous equilibrium partitioning.[58]

1.11. Measurement of SOA Yield

The ratio of the mass of SOA produced to the mass of organic precursor VOC reacted, the SOA yield, is one of the key quantities that one seeks to determine in chamber experiments. Given the low precursor concentrations under simulated atmospheric conditions and the small volume of the environmental chamber, direct measurement of the aerosol mass concentration is impractical. Instead, the aerosol yield is deduced from measurements of the particle size distribution, $n(D_p)$, which is defined such that the number of particles with diameters between D_p and $D_p + dD_p$ is $dN = n(D_p)dD_p$. Assuming the particles to be spherical, with density ρ_p, the mass concentration can be determined by integrating over the entire size range.

$$M = \int_0^{D_{p,\max}} \rho_p \frac{\pi D_p^3}{6} n(D_p) dD_p, \tag{1.20}$$

where $D_{p,\max}$ is the maximum size measured and should be larger than the maximum size of the grown particles in the chamber. The density is calculated based upon an estimate of the aerosol composition.

In environmental chambers, seed particles are typically injected into the chamber before the onset of gas-to-particle conversion. These particles tend to be several tens of nanometers in diameter. Their presence promotes condensation of the low volatility oxidation products. Particles may grow to hundreds of nanometers in size, but few grow beyond 1 μm. The particle size distribution in this range is measured using a Differential Mobility Analyzer (DMA) to size select particles in a narrow interval of sizes, and then counting the classified particles with a CPC, see Fig. 1.25. The DMA separates charged particles in terms of their steady-state migration velocities,

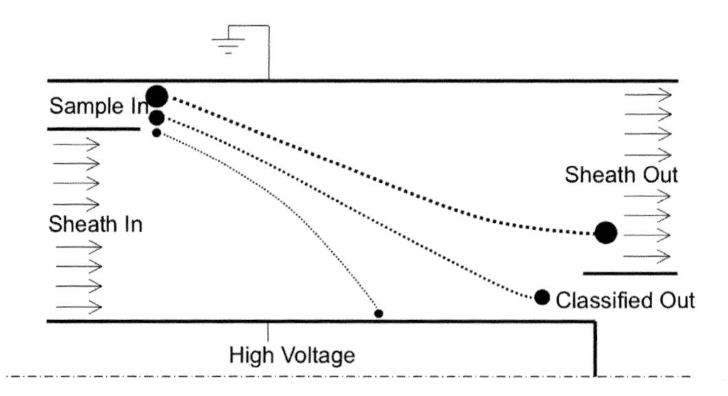

Fig. 25A Schematic of the differential mobility analysis system: Detailed schematic of the particle classification section.

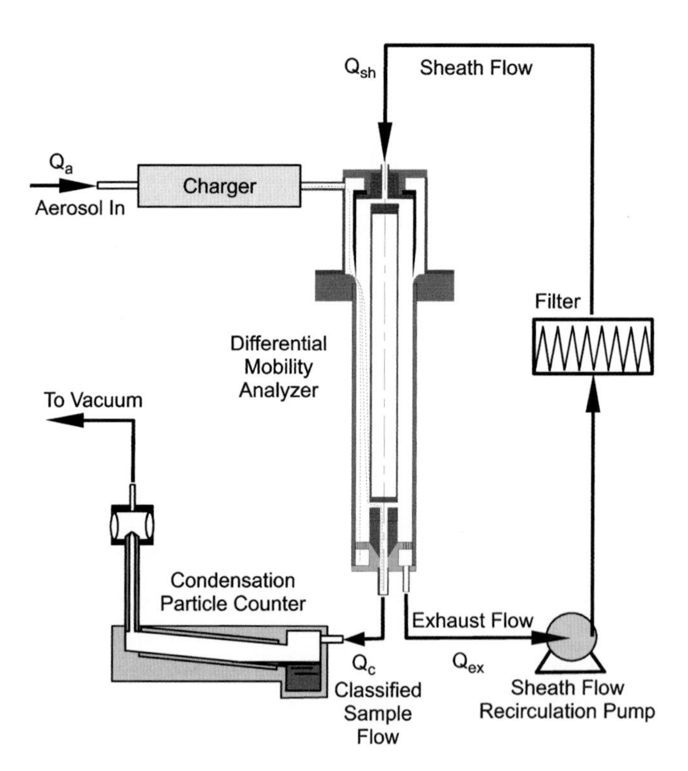

Fig. 25B Schematic of the differential mobility analysis system: Overall system.

v_{mig}, in an applied electric field, E. The property measured by the DMA,

$$Z_p = \frac{v_{\text{mig}}}{E} = \frac{n_e e C_c(\text{Kn})}{3\pi\mu D_p} \tag{1.21}$$

is called the electrical mobility. It is determined by the number, n_e, of elementary charges, e, on the particle, and its aerodynamic drag, which is calculated using Stokes' law with the slip correction, $C_c(\text{Kn})$.

Particles are customarily charged in two types of devices. The most common is the bipolar charger (also called an aerosol neutralizer). In this device, the aerosol is exposed to an electrically neutral ion cloud produced by a radioisotope source (^{85}Kr, ^{210}Po, or ^{241}Am), corona discharge, or soft X-ray generator. A steady state charge distribution on the aerosol results from this. At small sizes, only a small fraction of the particles carry any charge, and most of those that do are singly charged.

The measurement entails several steps: (i) bringing the aerosol to a known charge state, such that the probability that a particle of diameter D_p carries i charges is $p(D_p;\ i)$; (ii) mobility classification in the DMA, wherein the probability that a particle of mobility Z_p is transmitted from the sample entrance port of the DMA to its downstream classified aerosol exit port is $\Omega(Z_p; Z^*_{p;j})$; and (iii) counting the classified particles with efficiency η_{CPC} (D_p). The efficiency of transmission through plumbing within the sample lines and between the DMA and CPC, and within the DMA entrance and exit regions, $p(D_p)$, also affects the counts obtained. $Z^*_{p;j}$ is the mobility of the particle that is transmitted through the classification region of the DMA with the highest efficiency. Measurements are made at a series of target mobilities, $Z_{p;j}$, $j = 1, 2, \ldots, S$, that span the mobility/size range needed to determine the size distribution of the aerosol. The signal in the *jth* channel, usually CPC counts, is then

$$S_j = Q_\alpha \tau_c \int_0^\infty \sum_i n(D_p) p(D_p, i) \eta_p(D_p, i) \eta_{\text{CPC}}(D_p, i) \times \Omega(Z_p(D_p, i), Z^*_{p,j}) dD_p \tag{1.22}$$

where the possibility that η_{CPC} and/or η_p may depend upon the particle charge has been taken into account. The volumetric flow rate of the air stream containing aerosol entering the DMA system is Q_a. For each measurement channel, counts are accumulated over a counting time, τ_c, so the product $Q_a\tau_c$ is the volume of air from which particles are counted for each channel. We seek to determine $n(D_p)$ so that we can compute M, so the inverse of this Fredholm integral equation, Eq. (1.22) must be computed or estimated. This requires that we understand the performance of the DMA, CPC, and charging system, and the efficiencies. Each of these factors is discussed below.

1.11.1. *DMA*

The heart of the mobility analysis system is the DMA, see, for example, Ref. [59]. Figure 1.25 illustrates the most common type of DMA. This DMA, originally described by Liu and Pui[60] and Knutson and Whitby,[61] consists of coaxial cylindrical electrodes. Typically, the gap between the two electrodes is of order 10 mm. The aerosol flow, Q_a, is introduced through an annular slot entrance adjacent to the outer electrode of radius R_2, with a clean, particle-free sheath flow, Q_{sh}, filling the remaining space between the two electrodes. The flows carry the particles downstream while a voltage applied to the inner electrode, radius R_1, drives particles of appropriate polarity across the gap. Those that migrate across the gap in the time it takes to travel to a downstream sample extraction port in the inner electrode exit the channel in the classified-sample flow, Q_c. Generally, but not always, the aerosol inlet and classified sample flows are balanced, $Q_a = Q_c$. The remaining flow, Q_{ex}, carries low-mobility particles, and any that carry no charge out through an exhaust port.

When the DMA voltage is held constant at value, V_j,

$$Z_{p,j}^* = \frac{Q_{sh} + Q_{ex}}{4\pi LV} \ln \frac{R_2}{R_1}, \tag{1.23}$$

where L is the streamwise distance between the aerosol inlet and outlet ports. The DMA transmits particles with mobilities in a finite range around this value. The ability of the DMA to resolve the distribution of particles with respect to mobility and size is characterized in terms of the instrumental resolution, which is defined as

$$\mathscr{R} = \frac{Z_p^*}{\Delta Z_{\mathrm{fwhm}}} \tag{1.24}$$

where Z_p^* is the mobility at which particles are transmitted with highest efficiency, and Z_{fwhm} is the range of mobilities for which particles are transmitted with at least half the peak efficiency. In the high voltage limit, where Brownian diffusion is negligible, the resolution asymptotically approaches the ratio of the sheath flow rate to that of the aerosol, i.e., $\mathscr{R}_{\mathrm{nd}} = (Q_{sh} + Q_{ex})/(Q_a + Q_c)$. At very low voltage, where diffusion dominates,

$$\mathscr{R}_{el} \approx 0.425 \left[\frac{fV}{G_{mig}}\right]^{1/2}, \tag{1.25}$$

where f is a factor that accounts for nonuniformities in the electric field within the DMA, and G_{mig} accounts for the trajectory of the particles through the flows within the DMA.[59,62] For the cylindrical DMA, $f \approx (1 - R_1/R_2)/\ln(R_1/R_2)$. G_{mig} is calculated by evaluating the trajectory of the particles as they cross the nonuniform velocity field within the DMA. Its value is $G_{mig} = \mathcal{O}(2)$, for the TSI long column DMA, and a flow rate ratio $\beta = (Q_a + Q_c)/(Q_{\mathrm{sh}} + Q_{\mathrm{ex}}) = 5$, fully developed laminar flow leads to $G_{mig;fd} = 1.99$, while plug flow gives $G_{mig,plug} = 1.75$.[63] The transfer function for the instrument becomes[62]

$$\Omega(Z_p, Z_p^*, \beta, \delta, \tilde{\sigma}) = \frac{\tilde{\sigma}}{\sqrt{2}\beta(1-\delta)}\left[\varepsilon\left(\frac{\tilde{Z}-(1+\beta)}{\sqrt{2}\tilde{\sigma}}\right) + \varepsilon\left(\frac{\tilde{Z}-(1-\beta)}{\sqrt{2}\tilde{\sigma}}\right) - \varepsilon\left(\frac{\tilde{Z}-(1+\delta\beta)}{\sqrt{2}\tilde{\sigma}}\right) - \varepsilon\left(\frac{\tilde{Z}-(1-\delta\beta)}{\sqrt{2}\tilde{\sigma}}\right)\right], \tag{1.26}$$

where $\delta = (Q_c - Q_a)/(Q_c + Q_a)$, $\tilde{\sigma}$ the dimensionless broadening parameter,

$$\tilde{\sigma} = \frac{G_{mig}}{Pe^*_{mig}} \tilde{Z} \tag{1.27}$$

and

$$\varepsilon(y) = y\text{erf}(y) + \frac{1}{\sqrt{y}} e^{-y^2} \tag{1.28}$$

The migration Péclet number is the ratio of transport by migration to that by diffusion, which can be expressed as

$$\text{Pe}_{mig} = \frac{n_e eV}{kT} f \tag{1.29}$$

When the incoming aerosol flow equals the outgoing classified aerosol flow, the so-called balanced flow condition, the mean residence time of particles of mobility Z_p^* within the classification region of the DMA is

$$\tau_{r\text{DMA}} = \frac{\pi(R_2^2 - R_1^2)L}{Q_{sh}} \tag{1.30}$$

In experiments at the Caltech environmental chamber, a TSI long column DMA is used to make the size distribution measurements. Its dimensions are: $R_1 = 9.37\,\text{mm}$, $R_2 = 19.58\,\text{mm}$, and $L = 444.4\,\text{mm}$. The maximum voltage at which this DMA can be operated is about 10,000 V since higher voltages (field strength $E \sim 1\,\text{kV mm}^{-1}$) can lead to electrostatic breakdown, which has two deleterious effects: (1) the resulting arc creates large numbers of particles that confound the measurement, and (2) arcing creates carbon tracks on the Dacron screen in the instrument or creates pits on the electrodes, both of which increase the likelihood of arcing in the future and damage the instrument. Other DMAs may differ slightly in breakdown field strength and in damage mechanisms, but all instruments risk damage at excessive field strength.

In order to capture the entire growth of the aerosol and achieve accurate mass yield estimates, the size distribution in the chamber is measured over the 10 nm–1 μm size range. Given the DMA used, a sheath flow rate of $Q_{sh} = 2.5\,\mathrm{L\ min^{-1}}$ is employed to cover this range. To obtain adequate counting statistics, an aerosol flow rate of $0.5\,\mathrm{L\ min^{-1}}$ is used, i.e., $\beta = 0.2$, and $\mathscr{R}_{nd} = 5$. The mean time required for a particle to transit this DMA column under these flow rates is $\tau_{\mathrm{transit}} = 9\,\mathrm{s}$. One can operate the DMA by stepping through the 18 voltages needed to cover the mobility range, waiting at each voltage until the estimated number concentration reaches a steady-state value, and then counting for a finite time. Accounting for the transit time within the classification region means that a minimum of 75 s is spent waiting; additional delay is needed at each step to allow the CPC reading to reflect the concentration at its inlet. Ultimately, when measurements are made by stepping the DMA through voltages, more than 10 min was required for a size distribution measurement to be completed, making this mode of operation of the DMA too slow for many chamber experiments.

To accelerate the measurements, the DMA voltage can be scanned continuously in an exponential ramp, with counts continuously recorded in time bins, the basis of the scanning mobility particle sizer (SMPS).[64] The characteristic mobility for channel j is, at any instant of time,

$$Z_p^* = \frac{Q_{sh} + Q_{ex}}{4\pi R_1 L \overline{E}_j}, \tag{1.31}$$

where the steady electric field strength of Eq. (1.23), i.e., $E_1 = (V/R_1)\ln(R_2/R_1)$, is replaced by $\overline{E}_j$, the mean electric field strength that a particle experiences as it transits the classification region of the DMA and exits within measurement time interval j. In the high voltage, non-diffusive limit, the half-width of the transfer function is

$$\Delta Z_{\mathrm{fwhm},j} = \frac{Q_a + Q_c}{4\pi R_1 L \overline{E}_1}, \tag{1.32}$$

If the counting intervals are sufficiently short relative to the ramp time, these results are identical to those obtained for a DMA operated at constant voltage except that the electric field strength is the mean value that a particle experiences during its transit time. Figure 1.26

shows the time traces of particles that exit the classification column in the classified sample flow. The bottom trace is the applied voltage, for equal ramps of increasing (up-ramp) and decreasing (down-ramp) voltage. Above that are the time traces of the fraction of particles that are included in the classified aerosol flow, showing, from bottom to top: (i) the distribution of times at which the particles entered the classifier; (ii) the time at which they exited the classification region; (iii) the time at which they entered the CPC; and (iv) the times at which counts are recorded.

Fewer counts are recorded in the downscan due to increased losses owing to the long time that particles reside in the low velocity region near the classifier wall, which also distorts the transfer function. For this reason, measurements are usually made during the up-scan. The down-scan is then shortened to increase the frequency with which full size distribution measurements can be made.

Figure 1.26 also illustrates the effect of diffusion by showing three particle sizes: (i) 500 nm particles, for which diffusion is negligible (ii) 100 nm particles that diffuse slightly, and (iii) 20 nm particles that, for the conditions and instruments used in the Caltech chamber, are strongly affected by diffusion. Diffusion has two effects: broadening the range of particles that are transmitted through the DMA; and allowing loss to the walls of the classifier particles that would otherwise be transmitted, leading to reduced peak concentrations. The third plot shows the time at which those particles leave the classification region. The particles are delayed slightly by the plumbing time before they enter the mixing region of the CPC. The mixing region can be modeled as a continuously stirred tank reactor (CSTR), a volume that is perfectly mixed, such that particles exit the region randomly from the entire volume. This mixing "smears" the response of the CPC to transient aerosol inputs, such as occur when the CPC is used as a detector for the SEMS/SMPS. Different CPCs have different mixing times, ranging from a high value of a few seconds for some CPCs sold today, to as low as $\sim$100 ms. The top frame in Fig. 1.26 shows the effect of this mixing for the CPC used on the instrument in the Caltech chamber, with mix $\tau_{\mathrm{mix}} \approx 0.9$ s.

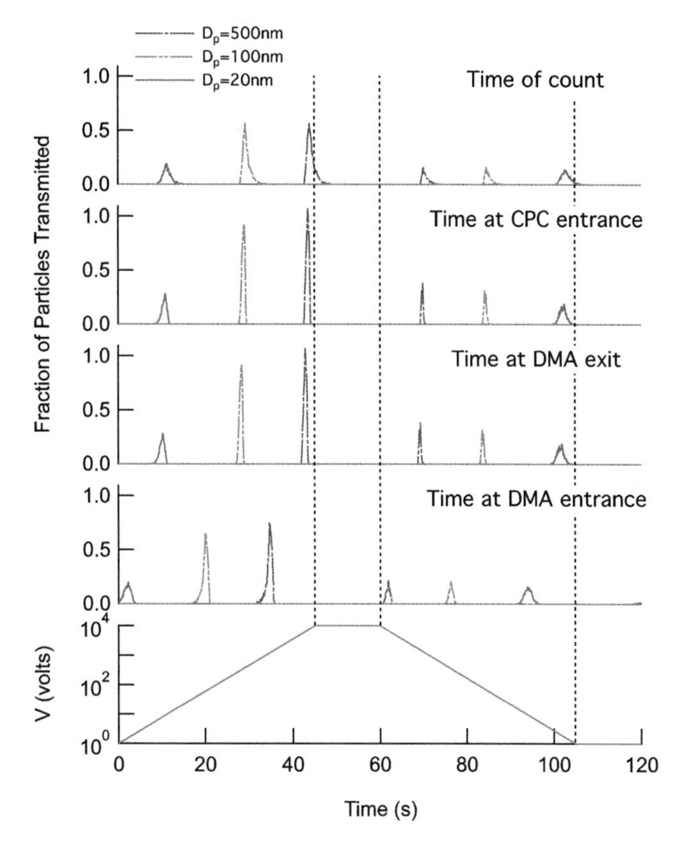

Fig. 1.26 Simulated transmission of particles through the SEMS/SMPS. The voltage profile through the scan, including up- and down-ramps and the holding periods when the voltage is constant are shown. The panels above show the instantaneous (integrating time of 0.1 s) fraction of particles that are ultimately included in the classified sample flow at the indicated times. Three particle sizes are shown: 500 nm particles (purple) for which diffusion is negligible; 100 nm particles (green) for which diffusion has a minor effect; and 20 nm particles (brown) for which diffusion significantly broadens and lowers the transmission probability.

Particle counts are recorded into time bins, but those are offset from the time interval of classification by the plumbing time, the time required to transit from the exit of the DMA to the CPC detection point. Thus, the plumbing time, τ_p, must be measured to determine which time bin corresponds to which particle size. Alternatively, one may calibrate with particles of known size, either polystyrene latex calibration particles, or with particles classified using a DMA operated at constant voltage. The SMPS eliminates the time required to clear the analysis column, thereby accelerating the measurement of size distributions dramatically.

1.11.2. *Particle Detection/Counting*

The classified particles must be counted to complete the measurement, usually using a CPC although a Faraday-cup electrometer can be used to measure the charge carried by the classified particles rather than counting them. The CPC operates by saturating a gas with vapor (the *working fluid*); that gas may, or may not contain the classified particles. In the laminar flow CPC illustrated in Fig. 1.26, the entire aerosol flows through a wet-walled saturator volume where the vapor is introduced. The vapor is then supersaturated, by cooling in the device shown in the figure, causing particles to activate and grow to sufficient size that they can be detected optically.

The most commonly used detector is the laminar flow CPC wherein the aerosol flows through a chamber whose walls are wetted with the working fluid to create the saturated state. The saturated vapor then passes through a cooled condenser to produce supersaturation in the presence of which particles that are larger than a critical size activate and grow. Most present-day CPCs use butanol as the working fluid, and activate particles larger than about 10 nm in diameter. The saturator in these instruments is relatively large, and contains recirculation zones that lead to a distribution of residence times between the entrance of the CPC and the point at which the particles are detected. If τ_r is the mean residence time in the mixed region, many display an exponential decay in residence times such that

$$dP = \exp\left[-\frac{t-\tau_p}{\tau_r}\right]\frac{dt}{\tau_r}$$

is the probability that a particle that exited the DMA at time $t-\tau_p$ will be detected in time interval t to $t+dt$.

The effect of this distribution of delay times on the times at which particles are counted is illustrated in Fig. 1.26 which simulates the classification of monodisperse (uniformly sized) particles, and shows the effects of the plumbing and detector response delays. The largest delay is that due to the flow time within the classifier, about 9 s. The plumbing between the DMA outlet and the inlet to the CPC detector is kept as short as possible, about 0.6 s for the simulated configuration. Large particles that do not diffuse significantly exit the classifier within a 2 s window. The mean residence time in the stirred region of the CPC is about 0.9 s for the commonly used TSI 3010 CPC. The residence time distribution within the CPC spreads the counts from particles of this one size over about 5 s. By delaying particle detection in up-scan measurements, small particles are recorded in the time intervals when larger ones are expected; down-scan measurements bias the counts toward smaller sizes. Empirically, it is found that, when the scan time is $\gtrsim$120 s, this "smearing" of the signals can be neglected, but this scan time is too long to capture transients seen in many experiments. Instead, one may accept the distortion in the measurements, and correct for during post-processing of the scan data using a deconvolution algorithm to reassign counts to the time interval corresponding to their exit from the DMA.[65]

Other CPCs have longer response times, though several respond more quickly. To extend the detection limit of the CPC below 10 nm, Stolzenburg and McMurry[66] introduced a small aerosol flow at the centerline of the condensation tube within a laminar flow CPC, with a larger vapor-laden flow surrounding the aerosol flow. Their design reduced diffusional losses, enabling activation of particles as small as 2.5 nm. It also reduced the response time of the CPC to $\sim$150 ms, though the very small aerosol flow leads to poor counting statistics when this ultrafine CPC is used as a detector for DMA measurements. Other commercially available CPCs attain similar response times while counting particles from larger aerosol flows, e.g., the water CPC,[67] and turbulent mixing designs.[68,69] In measurements of small particles ($\lesssim$100 nm) that do not require long migration times, size distributions can be acquired in tens of seconds with such detectors.

References

1. D. R. Cocker, R. C. Flagan and J. H. Seinfeld. State-of-the-art chamber facility for studying atmospheric aerosol chemistry. *Environ. Sci. Technol.* **35**, 2594–2601 (2001).
2. W. P. L. Carter, D. R. Cocker, D. R. Fitz, I. L. Malkina, K. Bumiller, C. G. Sauer, J. T. Pisano, C. Bufalino and C. Song. A new environmental chamber for evaluation of gas-phase chemical mechanisms and secondary aerosol formation. *Atmos. Environ.* **39**, 7768–7788 (2005).
3. W. P. L. Carter, G. Heo, D. R. Cocker and S. Nakao. SOA formation: Chamber study and model development. Final Report to the California Air Resources Board Contract No. 08-326. (CA, USA, 2012).
4. J. Wang, J. F. Doussin, S. Perrier, E. Perraudin, Y. Katrib, E. Pangui and B. Picquet-Varrault. Design of a new multi-phase experimental simulation chamber for atmospheric photosmog, aerosol and cloud chemistry research. *Atmos. Meas. Technol.* **4**, 2465–2494 (2011).
5. S. P. Sander, R. R. Friedl, J. R. Barker, D. M. Golden, M. J. Kurylo, P. H. Wine, J. P. D. Abbatt, J. B. Burkholder, C. E. Kolb, G. K. Moortgat, R. E. Huie and V. L. Orkin. *Chemical Kinetics and Photochemical Data for Use in Atmospheric Studies. Evaluation Number* 17. (Jet Propulsion Laboratory, Pasadena, CA, 2011).
6. R. Atkinson. Gas-phase tropospheric chemistry of organic compounds. *J. Phys. Chem. Ref. Data* **Monograph No. 2**, (1994).
7. J. D. Crounse, F. Paulot, H. G. Kjaergaard and P. O. Wennberg. Peroxy radical isomerization in the oxidation of isoprene. *Phys. Chem. Chem. Phys.* **13**, 13607–13613 (2011).
8. J. H. Kroll, N. L. Ng, S. M. Murphy, R. C. Flagan and J. H. Seinfeld. Secondary organic aerosol formation from isoprene photooxidation, *Environ. Sci. Technol.* **40**, 1869–1877 (2006).
9. S. M. Saunders, M. E. Jenkin, R. G. Derwent and M. J. Pilling. Protocol for the development of the Master Chemical Mechanism, MCM v3 (Part A): Tropospheric degradation of non-aromatic volatile organic compounds. *Atmos. Chem. Phys.* **3**, 161–180 (2003).
10. M. E. Jenkin, S. M. Saunders and M. J. Pilling. The tropospheric degradation of volatile organic compounds: A protocol for mechanism development, *Atmos. Environ.* **31**, 81–104 (1997).
11. J. G. Calvert, R. Atkinson, J. A. Kerr, S. Madronich, G. K. Moortgat, T. J. Wallington and G. Yarwood. *The Mechanisms of the Atmospheric Oxidation of the Alkenes.* (Oxford University Press, Oxford, UK, 2000).
12. D. Johnson and G. Marston. The gas-phase ozonolysis of unsaturated volatile organic compounds in the troposphere. *Chem. Soc. Rev.* **37**, 699–716 (2008).

13. M. D. Keywood, J. H. Kroll, V. Varutbangkul, R. Bahreini, R. C. Flagan and J. H. Seinfeld. Secondary organic aerosol formation from cyclohexene ozonolysis: Effect of OH scavenger and the role of radical chemistry. *Environ. Sci. Technol.* **38**, 3343– 3350 (2004).
14. J. Mao, X. Ren, L. Zhang, D. M. Van Duin, R. C. Cohen, J.-H. Park, A. H. Goldstein, F. Paulot, M. R. Beaver, J. D. Crounse, P. O. Wennberg, J. P. DiGangi, S. B. Henry, F. N. Keutsch, C. Park, G. W. Schade, G. M. Wolfe, J. A. Thornton and W. H. Brune. Insights into hydroxyl measurements and atmospheric oxidation in a California forest. *Atmos. Chem. Phys.* **12**, 8009–8020 (2012).
15. N. C. Bouvier-Brown, A. H. Goldstein, Gilman, J.B., W. C. Kuster and J. A. de Gouw. In-situ ambient quantification of monoterpenes, sesquiterpenes, and related oxygenated compounds during BEARPEX 2007: Implications for gas- and particle-phase chemistry. *Atmos. Chem. Phys.* **9**, 5505–5518 (2009).
16. J. G. Crump and J. H. Seinfeld. Turbulent deposition and gravitational sedimentation of an aerosol in a vessel of arbitrary shape. *J. Aerosol Sci.* **12**, 405–415 (1981).
17. P. H. McMurry and D. J. Rader. Aerosol wall losses in electrically charged chambers, *Aerosol Sci. Technol.* **4**, 249–268 (1985).
18. E. A. Weitkamp, A. M. Sage, J. R. Pierce, N. M. Donahue and A. L. Robinson. Organic aerosol formation from photochemical oxidation of diesel exhaust in a smog chamber. *Environ. Sci. Technol.* **41**, 6969–6975 (2007).
19. L. Hildebrandt, N. M. Donahue and S. N. Pandis. High formation of secondary organic aerosol from the photo-oxidation of toluene. *Atmos. Chem. Phys.* **9**, 2973–2986 (2009).
20. L. Hildebrandt, K. M. Henry, J. H. Kroll, D. R. Worsnop, S. N. Pandis and N. M. Donahue. Evaluating the mixing of organic aerosol componenets using high-resolution aerosol mass spectrometry. *Environ. Sci. Technol.* **45**, 6329–6335 (2011).
21. C. L. Loza, P. S. Chhabra, L. D. Yee, J. S. Craven, R. C. Flagan and J. H. Seinfeld. Chemical aging of m-xylene secondary organic aerosol: Laboratory chamber study. *Atmos. Chem. Phys.* **12**, 151–167 (2012).
22. C. L. Loza, J. S. Craven, L. D. Yee, M. M. Coggon, R. H. Schwantes, M. Shiraiwa, X. Zhang, K. A. Schilling, N. L. Ng, M. R. Canagaratna, P. J. Ziemann, R. C. Flagan and J. H. Seinfeld. Secondary organic aerosol yields of 12-carbon alkanes. *Atmos. Chem. Phys.* **14**, 1423–1439 (2014).
23. B. M. Matthew, A. M. Middlebrook and T. B. Onasch. Collection efficiencies in an Aerodyne aerosol mass spectrometer as a function of particle phase for laboratory generated aerosols. *Aerosol Sci. Technol.* **42**, 884–898 (2008).
24. I. Riipinen, J. R. Pierce, T. Yli-Juuti, T. Nieminen, S. Häkkinen, M. Ehn, H. Junninen, K. Lehtipalo, T. Petäjä, J. Slowik, R. Chang, N. C. Shantz, J. Abbatt, W. R. Leaitch, V.-M. Kerminen, D. R. Worsnop, S. N. Pandis, N. M. Donahue and M. Kulmala. Organic condensation: A vital link

connecting aerosol formation to cloud condensation nuclei (CCN) concentrations. *Atmos. Chem. Phys.* **11**, 3865–3878 (2011).

25. J. R. Pierce, G. J. Engelhart, L. Hildebrandt, E. A. Weitkamp, R. K. Pathak, N. M. Donahue, A. L. Robinson and S. N. Pandis. Constraining particle evolution from wall losses, coagulation, and condensation-evaporation in smog-chamber experiments: Optimal estimation based on size distribution measurements. *Aerosol Sci. Technol.* **42**, 1001–1015 (2008).
26. J. H. Seinfeld and S. N. Pandis. *Atmospheric Chemistry and Physics*, 3rd Ed. (John Wiley and Sons, New York, 2016).
27. A. Matsunaga and P. J. Ziemann. Gas-wall partitioning of organic compounds in a Teflon film chamber and potential effects on reaction product and aerosol yield measurements. *Aerosol Sci. Technol.* **44**, 881–892 (2010).
28. J. H. Seinfeld and J. F. Pankow. Organic atmospheric particulate material, *Ann. Rev. Phys. Chem.* **54**, 121–140 (2003).
29. P. H. McMurry and D. Grosjean. Gas and aerosol wall losses in Teflon film smog chambers. *Environ. Sci. Technol.* **19**, 1176–1182 (1985).
30. X. Zhang, R. H. Schwantes, R. C. McVay, H. Lignell, M. M. Coggon, R. C. Flagan and J. H. Seinfeld. Vapor wall deposition in Teflon chambers, *Atmos. Chem. Phys.* **15**, 4197–4214 (2015).
31. G. K. Yeh and P. J. Ziemann. Gas-wall partitioning of oxygenated organic compounds: Measurements, structure-activity relationships, and correlation with gas chromatographic retention factor. *Aerosol Sci. Technol.* **49**, 727–738 (2015).
32. X. Zhang, C. D. Cappa, S. Jathar, R. C. McVay, J. J. Ensberg, M. J. Kleeman and J. H. Seinfeld. Influence of vapor wall-loss in laboratory chambers on yields of secondary organic aerosol, *Proc. Natl. Acad. Sci. U.S.* **111**, 5802–5807 (2014).
33. C. D. Cappa and K.R. Wilson. Multi-generation gas-phase oxidation, equilibrium partitioning and the formation and evolution of secondary organic aerosol, *Atmos. Chem. Phys.* **12**, 9505–9528 (2012).
34. C. D. Cappa, X. Zhang, C. L. Loza, J. S. Craven, L. D. Yee and J. H. Seinfeld. Application of the Statistical Oxidation Model (SOM) to secondary organic aerosol formation from photooxidation of C12 Alkanes. *Atmos. Chem. Phys.* **13**, 1591–1606 (2013).
35. B. Aumont, S. Szopa and S. Madronich. Modeling the evolution of organic carbon during its gas-phase tropospheric oxidation: Development of an explicit model based on a self generating approach. *Atmos. Chem. Phys.* **5**, 2497–2517 (2005).
36. B. Aumont, M. Camredon, C. Mouchel-Vallon, S. La, F. Ouzebidour, R. Valorso, J. Lee-Taylor and S. Madronich. Modeling the influence of alkane molecular structure on secondary organic aerosol formation, *Faraday Discuss.* **165**, 105–122 (2013).
37. S. Chen, W. H. Brune, A. T. Lambe, P. Davidovits and T. B. Onash. Modeling organic aerosol from the oxidation of α-pinene in a Potential Aerosol Mass (PAM) chamber. *Atmos. Chem. Phys.* **13**, 5017–5031 (2013).

38. E. Kang, M. J. Root, D. W. Toohey and W. H. Brune. Introducing the concept of Potential Aerosol Mass (PAM). *Atmos. Chem. Phys.* **7**, 5727–5744 (2007).
39. E. Kang, D. W. Toohey and W. H. Brune. Dependence of SOA oxidation on organic aerosol mass concentration and OH exposure. *Atmos. Chem. Phys.* **11**, 1837–1852 (2011).
40. A. Keller and H. Burtscher. A continuous photo-oxidation flow reactor for a defined measurement of the SOA formation potential of wood burning emissions. *J. Aerosol Sci.* **49**, 9–20 (2012).
41. A. T. Lambe, A. T. Ahern, L. R. Williams, J. G. Slowik, J. P. S. Wong, J. P. D. Abbatt, W. H. Brune, N. L. Ng, J. P. Wright, D. R. Croasdale, D. R. Worsnop, P. Davidovits and T. B. Onasch. Characterization of aerosol photooxidation flow reactors: Heterogeneous oxidation, secondary organic aerosol formation and cloud condensation nucleus (CCN) activity measurements. *Atmos. Meas. Tech.* **4**, 445–461 (2011).
42. A. T. Lambe, P. S. Chhabra, T. B. Onasch, W. H. Brune, J. F. Hunter, J. H. Kroll, M. J. Cummings, J. F. Brogan, Y. Parmar, D. R. Worsnop, C. E. Kolb and P. Davidovits. Effect of oxidant concentration, exposure time, and seed particles on secondary organic aerosol chemical composition and yield *Atmos. Chem. Phys.* **15**, 3063–3075 (2015).
43. A. T. Lambe, T. B. Onasch, D. R. Croasdale, J. P. Wright, A. T. Martin, J. P. Franklin, P. Massoli, J. H. Kroll, M. R. Canagaratna, W. H. Brune, D. R. Worsnop and P. Davidovits. Transitions from functionalization to fragmentation reactions of laboratory Secondary Organic Aerosol (SOA) generated from the OH oxidation of alkane precursors. *Environ. Sci. Technol.* **46**, 5430–5437 (2012).
44. R. Li, B. B. Palm, A. M. Ortega, J. Hlywiak, W. Hu, Z. Peng, D. D. Day, C. Knote, W. H. Brune, J. A. de Gouw and J. L. Jimenez. Modeling the radical chemistry in an oxidation flow reactor: Radical formation and recycling, sensitivities, and OH exposure estimation equation. *J. Phys. Chem. A* **119**, 4418–4432 (2015).
45. J. G. Slowik, J. P. S. Wong and J. P. D. Abbatt. Real-time controlled OH-initiated oxidation of biogenic secondary organic aerosol. *Atmos. Chem. Phys.* **12**, 9775–9790 (2012).
46. E. A. Bruns, I. El Haddad, A. Keller, F. Klein, N. K. Kumar, S. M. Pieber, J. C. Corbin, J. G. Slowik, W. H. Brune, U. Baltensperger and A. S. H. Prevot. Inter-comparison of laboratory smog chamber and flow reactor systems on organic aerosol yield and composition. *Atmos. Meas. Tech.* **8**, 2315–2332 (2015).
47. M. J. Ezell, S. N. Johnson, Y. Yu, V. Perraud, E. A. Bruns, M. L. Alexander, A. Zelenyuk, D. Dabdub and B. Finlayson-Pitts. A new aerosol flow system for photochemical and thermal studies of tropospheric aerosols. *Aerosol Sci. Technol.* **44**, 329–338 (2010).
48. J. H. Kroll, J. D. Smith, D. L. Che, S. H. Kessler, D. R. Worsnop and K. R. Wilson. Measurement of fragmentation and functionalization pathways

in the heterogeneous oxidation of oxidized organic aerosol. *Phys. Chem. Chem. Phys.* **11**, 8005–8014 (2009).
49. A. M. Ortega, D. A. Day, M. J. Cubison, W. H. Brune, D. Bon, J. A. de Gouw and J. L. Jimenez. Secondary organic aerosol formation and primary organic aerosol oxidation from biomass-burning smoke in a flow reactor during FLAME-3. *Atmos. Chem. Phys.* **13**, 11551–11571 (2013).
50. J. D. Smith, J. H. Kroll, C. D. Cappa, D. L. Che, C. L. Liu, M. Ahmed, S. R. Leone, D. R. Worsnop and K. R. Wilson. The heterogeneous reaction of hydroxyl radicals with submicron squalane particles: A model system for understanding the oxidative aging of ambient aerosols. *Atmos. Chem. Phys.* **9**, 3209–3222 (2009).
51. A. F. Khalizov, M. E. Earle, W. J. W. Johnson, G. D. Stubley and J. J. Sloan. Modeling of flow dynamics in laminar aerosol flow tubes. *J. Atmos. Sci.* **37**, 1174–1187 (2006).
52. R. B. Bird, W. E. Stewart and E. N. Lightfoot. *Transport Phenomena*, 2nd Ed. (John Wiley, New York, 2002).
53. E. Fried and L. E. Idelchik. *Flow Resistance: A Design Guide for Engineers.* (Hemisphere Publishing Corporation, New York, 1989).
54. F. M. White. *Fluid Mechanics,* 6th Ed. (McGraw Hill, Berlin, 2008).
55. E. M. Sparrow, J. P. Abraham and W. J. Minkowycz. Flow separation in a diverging conical duct: Effect of Reynolds number and divergence angle. *Int. J. Heat Mass Transf.* **52**, 3079–3083 (2009).
56. S. Tavoularis. *Measurement in Fluid Mechanics.* (Cambridge University Press, Cambridge, UK, 2005).
57. M. Shiraiwa and J. H. Seinfeld. Equilibration timescale of atmospheric secondary organic aerosol partitioning. *Geophys. Res. Lett.* **39**, L24801 (2012).
58. R. A. Zaveri, R. C. Easter, J. E. Shilling and J. H. Seinfeld, Modeling kinetic partitioning of secondary organic aerosol and size distribution dynamics: Representing effects of volatility, phase state, and particle-phase reaction. *Atmos. Chem. Phys.* **14**, 5153–5181 (2014).
59. R. C. Flagan. Continuous-flow differential mobility analysis of nanoparticles and biomolecules. *Annu. Rev. Chem. Biomol. Eng.* **5**, 255–279 (2014).
60. B. Y. H. Liu and D. Y. H. Pui. A submicron aerosol standard and the primary, absolute calibration of the condensation nuclei counter, *J. Colloid Interface Sci.* **47**, 155–171 (1974).
61. E. O. Knutson and K. T. Whitby. Aerosol classification by electrical mobility: Apparatus, theory, and applications. *J. Aerosol Sci.* **6**, 443–451 (1975).
62. M. R. Stolzenburg. *An Ultrafine Aerosol Size Distribution Measuring System.* (University of Minnesota, 1988).
63. R. C. Flagan. On differential mobility analyzer resolution. *Aerosol Sci. Technol.* **30**, 556–570 (1999).
64. S. C. Wang and R. C. Flagan. Scanning electrical mobility spectrometer. *Aerosol Sci. Technol.* **13**, 230–240 (1990).
65. D. R. Collins, R. C. Flagan and J. H. Seinfeld. Improved inversion of scanning DMA data. *Aerosol Sci. Technol.* **36**, 1–9 (2002).

66. M. R. Stolzenburg and P. H. McMurry. An ultrafine condensation particle counter, *Aerosol Sci. Technol.* **14**, 48–65 (1991).
67. S. V. Hering, M. R. Stolzenburg, F. R. Quant, D. R. Oberreit and P. B. Keady. A laminar-flow water-based condensation particle counter (WCPC). *Aerosol Sci. Technol.* **39**, 659–672 (2005).
68. S. D. Shah and D. R. Cocker. A fast scanning mobility particle spectrometer for monitoring transient particle size distributions, *Aerosol Sci. Technol.* **39**, 519–526 (2005).
69. J. Wang, V. F. McNeill, D. R. Collins and R. C. Flagan. Fast mixing condensation nucleus counter: Application to rapid scanning differential mobility analyzer measurements. *Aerosol Sci. Technol.* **36**, 678–689 (2002).

Chapter 2

The Role of Water in Organic Aerosol Multiphase Chemistry: Focus on Partitioning and Reactivity

R. Zhao[*,‡,**], A. K. Y. Lee[*,§], C. Wang,[*,†] F. Wania,[*,†] J. P. S. Wong[*,¶], S. Zhou[*] and J. P. D. Abbatt[*,††]

[*]*Department of Chemistry, University of Toronto. 80 St. George Street, Toronto, ON, Canada, M5S 3H6*

[†]*Department of Physical and Environmental Sciences, University of Toronto Scarborough, 1095 Military Trail, Toronto, ON, Canada, MIC 1A4*

Current Affiliation

[‡]*Division of Chemistry and Chemical Engineering, California Institute of Technology, 1200 E California Blvd, Pasadena, CA, USA, 91125*

[§]*Department of Civil and Environmental Engineering, National University of Singapore, 1 Engineering Drive 2, Singapore 117576, Singapore*

[¶]*School of Earth and Atmospheric Sciences, Georgia Institute of Technology, 311 Ferst Dr. Atlanta, GA, USA, 30332*

[**]*rzhao@caltech.edu*

[††]*jabbat@chem.utoronto.ca*

2.1. Introduction

The chemical and physical processes associated with soluble inorganic aerosol particles are relatively well known. By contrast, the chemistry of the organic component of atmospheric aerosol is considerably more uncertain, in part because of the diversity of gas phase organic precursors to the particles but also because of the complexity of associated gas- and condensed-phase processes that determine their composition. Nevertheless, considerable progress in this field has been made in the past two decades or so.[1,2] It is now well recognized that organic material may be both primary and secondary in nature, with the latter thought to dominate sub-micron organic aerosol mass under many conditions. The primary component may dominate roadside locations, close to biomass burning sources, or perhaps in the marine boundary layer. The secondary component arises from oxidation of volatile organic compounds (VOCs) which are emitted both from biogenic and anthropogenic sources and are converted in the atmosphere into a range of molecules with varying degrees of oxidation, volatility, and size. In some cases, one or two oxidation steps of the VOCs are enough to make the species sufficiently involatile that they partition significantly to particulate matter.[3,4] In other cases, a number of oxidation steps are required, leading initially to increased functionalization of the precursor VOC and ultimately to fragmentation into smaller oxygenates. VOC oxidation takes place in the gas phase but may also occur within the condensed phase or at the gas–particle interface.

The term *multiphase chemistry* refers to these coupled processes, including reaction and diffusion in the gas phase, condensed phase and at the interface, and the dynamic partitioning of species between these three compartments.[5] As such, a quantitative description of the multiphase chemistry of organic aerosol requires knowledge of not only gas- and condensed-phase oxidation mechanisms but also estimates of mass transfer rates. These are easy to quantify in the gas phase but difficult in the condensed phase where the organic component may exhibit a wide range of viscosity,[6] and the rate at which molecules cross the gas–particle interface is highly uncertain.[7] Models of partitioning between these different phases are complicated in a thermodynamic sense by a lack of complete knowledge of the composition of the condensed organic phase.

The organic component of tropospheric aerosol is very commonly mixed to varying degrees with other constituents, such as various salts, mineral dusts, and elemental carbon. Indeed, while the term *organic aerosol* is frequently used to refer to organic particles, this term implicitly identifies the organic component of atmospheric particles, recognizing that other species are usually present. One ubiquitous species is condensed-phase water, which partitions to the particles from the gas phase to a degree dependent on the gas phase water activity or relative humidity (RH). In some circumstances, condensed-phase water can be the dominant contributor to particulate mass (see Sec. 2.2.1). Aerosol organics have some affinity for water, modulated to a large extent by their degree of oxidation and functionalization (see Sec. 2.2.2). As well, the inorganic salts, such as the ammoniated sulfates, have considerable hygroscopicity. As a result, the organic component of an atmospheric particle may be present in a homogeneous solution with aqueous salts, or phase separation may occur resulting in organic-rich and inorganic-rich components (see Sec. 2.3). Indeed, phase separation between hydrophobic organics and hydrophilic organics may also occur in the absence of inorganic salts, with its likelihood modulated by the RH.[8,9] In all cases, surfactant-like organics may also coat particles, whether they are insoluble, such as mineral dust and black carbon, or composed of soluble species.

The impact of condensed-phase water on the multiphase chemistry of the organic component of atmospheric aerosol particles is the focus of this review. This is an expansive, multi-faceted topic in atmospheric chemistry relevant to not only global processes and climate but also connected to the health effects associated with inhalation of organic aerosol. In that regard, the topics covered in this review are mostly associated with the sub-micron-sized fraction of atmospheric organic aerosol given that these particles are active in both direct and indirect forcing of climate, and the smallest particles are those believed to have the largest impacts on human health. The approach taken is to review the subject from a fundamental perspective, focussing on the processes at a molecular level. As discussed at the end of the review, one of the major challenges in this field is to connect this detailed fundamental understanding to true atmospheric behavior.

An overview of the role that water plays in the multiphase chemistry is provided in Fig. 2.1, illustrating aspects described in the main body of the review. Section 2.2 of the chapter addresses the degree to which water partitions to organic aerosol, noting the distinction between behavior for conditions below or above water saturation.

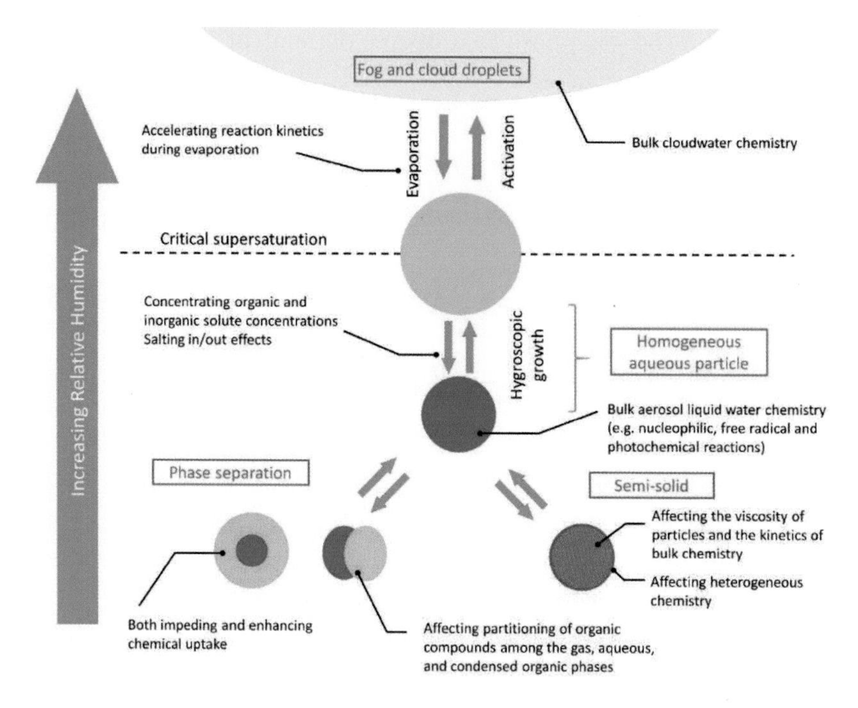

Fig. 2.1 Overview of the role of water in organic aerosol multiphase chemistry.

These topics have been reviewed extensively elsewhere[2] and so only a brief description is given. Section 2.3 describes the factors that control organic compounds partitioning to organic aerosol, focussing in detail on the behavior arising when a hygroscopic salt-rich phase is simultaneously present alongside an organic-rich phase. A newly formulated framework with associated initial results for assessing the propensity to partition to these two phases is presented. Section 2.4 describes in detail the reactivity of organic aerosol that contains water. Attention is given to the potential for OH radical-initiated, nucleophilic, evaporative, photochemical, and interfacial processes to occur. As well, we address how the water-mediated viscosity of organic aerosol can affect the rates of multiphase reactions. Processes occurring under cloudwater conditions are presented at times as a guide, given that analogous reactions may occur under aerosol water conditions. Section 2.5 concludes the review with a discussion of potential approaches to address the chemical complexity in this field, recognizing that the principal challenge is to find a pragmatic yet accurate collective representation of the chemistry of the hundreds to thousands of organic compounds that make up atmospheric organic aerosol.

Whereas other articles have comprehensively reviewed the field of organic aerosol formation, properties, and oxidation,[1,2,4] this review is specifically focussed on the role of water in the multiphase chemistry of organic aerosol. It is distinguished from the review of Ervens *et al.*,[10] which focussed largely on the formation of secondary organic aerosol (SOA) through aqueous-phase processes, by addressing instead the molecular processes that control the partitioning and reactivity of the constituents of organic aerosol particles. It acts as a complement to the recent overview article of McNeill on aqueous atmospheric organic chemistry by focussing in depth on the mechanisms for partitioning and reactivity that underpin the role of water in organic multiphase chemistry.[11] This review does not attempt to be exhaustive, attempting to provide an overview of the role of water in multiphase chemistry by combining recent results from laboratory experiments, field measurements and model simulations performed in the past decade or so. For novel progress in modeling multiphase-phase chemistry and up-to-date kinetic data, readers are referred to recent articles by Ervens[12] and Herrmann *et al.*,[13] respectively.

2.2. Effects of Organic Constituents on Aerosol Hygroscopicity

2.2.1. *Liquid Water Content (LWC)*

Liquid water can be a significant fraction of particulate matter. Using a general circulation model, Liao and Seinfeld[14] have shown that the mass of aerosol liquid water is typically 2–3 times the dry aerosol mass on a global scale. The amount of liquid water associated with an air mass is commonly described using LWC, the mass of liquid water residing in a unit volume of air. LWC is a pivotal parameter in determining the role that water plays in the multiphase chemistry, i.e., processes discussed in Secs. 2.3 and 2.4.

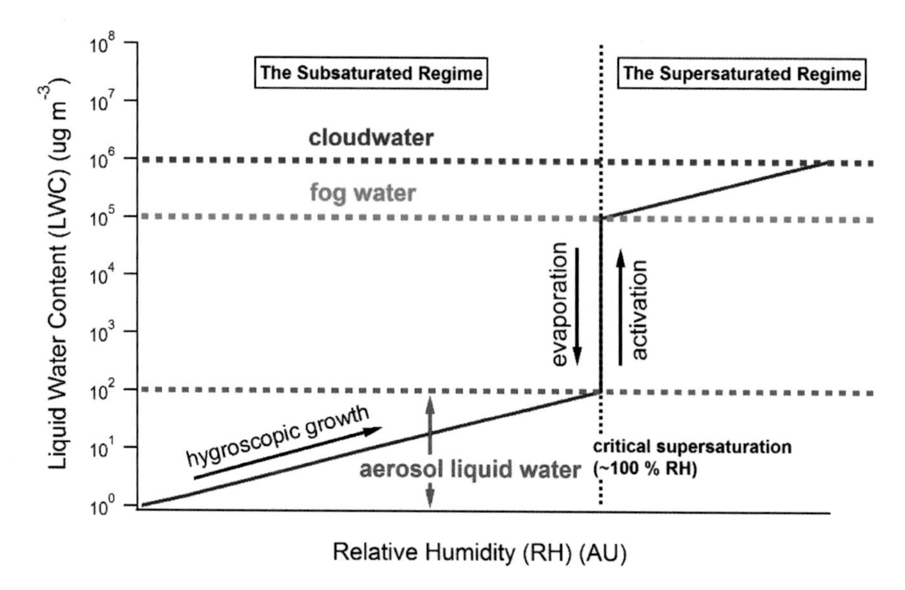

Fig. 2.2 Changes of LWC under the subsaturated and the supersaturated regimes. LWC typically observed for aerosol liquid water, fog water and cloudwater are displayed as the horizontal dotted lines. The curve of LWC should be viewed qualitatively only.

The LWC associated with aqueous aerosol varies with RH, and is typically at the μg m^{-3} level[15–17] but can be hundreds of μg m^{-3} in polluted regions.[18] When the air mass experiences supersaturated conditions, particles can activate to fog and cloud droplets. The activation process is associated with orders of magnitude increases in LWC, with typical fog and cloudwater LWC reaching 0.1–1 g m^{-3}.[17] A conceptual scheme of LWC changes associated with RH is shown in Fig. 2.2. Cloud/fog water and aerosol liquid water are usually considered as two distinct reaction media due to their differences in volume, LWC, surface area to volume ratio, and salinity.[10,19] Aerosol liquid water is particularly enriched with both organic and inorganic compounds compared to fog and cloudwater.[17,20]

Although less hygroscopic than inorganic compounds, organic compounds play an important role in determining the amount of water associated with a particle. The contribution of organic compounds under the subsaturated and supersaturated regimes is governed by different physical mechanisms. The detailed mechanisms and emerging research topics of the subsaturated and supersaturated regimes are discussed in Secs. 2.2.2 and 2.2.3, respectively.

2.2.2. *The Subsaturated Regime*

Organic constituents can have significant impact on the hygroscopicity of ambient aerosol under subsaturated RH conditions, i.e., RH $<$ 100%. Despite the fact that inorganic aerosol components play a key role in controlling the overall hygroscopic growth factor (GF, the ratio between the humidified and dry particle diameters at a reference RH) of ambient aerosol at high RH, it has been demonstrated that water-soluble organic compounds can contribute to water uptake most significantly at low RH where inorganic salts have crystallized (summarized by Kanakidou *et al*).[2] Many laboratory studies have confirmed that atmospherically relevant organics such as dicarboxylic acids can influence the deliquescence RH (DRH) and efflorescence RH (ERH) of inorganic particles.[21–23] In addition, a complex mixture of organics, such as that in ambient SOA, generally remains in the liquid form and retains water even at very low RH, i.e., there is a smooth water uptake with changing RH with no clear deliquescence and efflorescence.[24–26] Figure 2.3 shows the continuous water uptake of particles composed of SOA from monoterpene and oxygenated terpene photooxidation in the presence of ammonium sulfate ($(NH_4)_2SO_4$) within the subsaturated regime.[26]

In addition to thermodynamic considerations, organics can drive a variety of kinetic effects as well. Partitioning of organic compounds with low vapour pressure onto existing particles can result in the formation of organic coatings that partially or completely cover the particle surface. Of particular interest is the impact of hydrophobic organic coatings on particle hygroscopicity. These organic compounds can be immiscible with aqueous aerosol components, forming a physical barrier (i.e., solid shell or liquid film) to impede the water uptake kinetics. For example, Xiong *et al.*[27] showed that H_2SO_4 particles coated with long chain organic acids including lauric, stearic

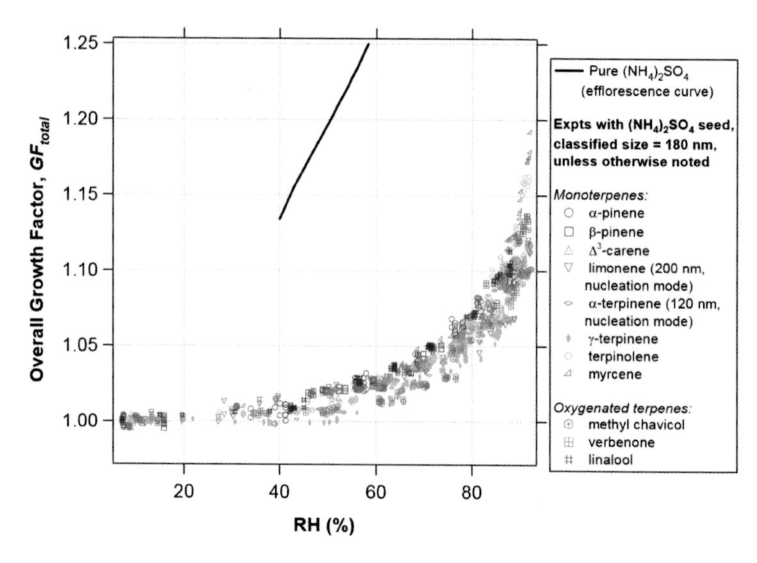

Fig. 2.3 Raw hygroscopic growth factors of SOA from monoterpene and oxygenated terpene photo-oxidation (with $(NH_4)_2SO_4$ seed). Reported data are from 180-nm diameter particles, unless otherwise noted. All the growth factors from various terpene precursors are contained within a relatively narrow envelope between 1.06–1.10 at 85%. Adapted from Varutbangkul *et al.*,[26] distributed under the Creative Commons Attribution 3.0 License.

and oleic acids required a longer equilibrium time compared to uncoated H_2SO_4 particles in hygroscopic tandem differential mobility analyzer (HTDMA) growth measurements. Similarly, using a single particle levitation technique, Chan and Chan[28] showed that octanoic acid coatings can reduce the mass transfer rate of water molecules depending on the coating thickness but only have minimal impacts on the phase transition behavior (i.e., DRH and ERH) of $(NH_4)_2SO_4$ particles. It is likely that the mass accommodation coefficient of water, a parameter describing the probability that a water molecule striking the particle surface will be taken up by the particle, to these organic coated sufaces is significantly smaller than unity, leading to the suppressed kinetics. Alternatively, the solubility of water in the coatings is sufficiently small that the overall mass transfer rates through to the hygroscopic core are slow.

Although not extensively explored, the presence of water-soluble organics in aerosol particles can also lead to mass transfer effects of water molecules during both water condensation and evaporation processes. Comparing the hygroscopic GF observed by HTDMA and single particle levitation techniques (i.e., with equilibrium times of seconds and hours, respectively) at high RH, Peng *et al.*[29] first reported the mass transfer limitation of the deliquescence process in pure glutaric acid particles. More recently, Bones *et al.*[30] investigated the influence of slow water diffusion in the condensed phase on the rates of water uptake and evaporation processes using NaCl–sucrose mixed droplets as a proxy for multicomponent aerosol, demonstrating that significant inhibition in mass transfer occurs for highly viscous aerosol particles, even with soluble components. Although current atmospheric models generally assume that aerosol particles are in equilibrium with the surrounding water vapour in the ambient air, there is growing evidence that SOA can exist as a highly viscous material (e.g., Refs. [31,32]). Such high viscosity of a particle can impede mass transfer processes (see Sec. 2.4.6), highlighting the need to better understand the kinetic limitations that may control water partitioning in ambient particles.[33] If the observation timescale is too short for equilibration, this mass transfer limitation can appear as a hysteresis in growth curves with slow uptake of water to viscous media occurring as RH is increased, and slow loss of water from the drying of equilibrated particles as the RH is rapidly decreased.

2.2.3. *The Supersaturated Regime*

Under supersaturated water vapour conditions (i.e., RH $>$ 100%), aerosol particles can activate as cloud condensation nuclei (CCN) and grow into cloud droplets. This process occurs via the condensation of water vapour and is described as an equilibrium process using Köhler theory.[34] Here, two competing factors affect the condensation of water, which occurs when ambient water vapour pressure exceeds the equilibrium water vapour pressure over the aerosol particle: the Raoult effect (i.e., lowering of vapour pressure of water by a solute) and the Kelvin effect (i.e., increased vapour pressure over a curved surface). A particle requires a minimum or critical supersaturation in order to act as a CCN. When exposed to the critical supersaturation, the particle must grow to a critical diameter to overcome the Kelvin effect. Following activation as a CCN, the droplet will grow spontaneously up to many μm in size. As such, the critical supersaturation and diameter are controlled by the particle size and its chemical affinity for water. Given the wide scope of this topic, the following discussion will only focus on how some properties of organic aerosols contribute to their hygroscopicity. We note that Sun and Ariya[35] provide a more comprehensive review of organic aerosol CCN. For a detailed description of Köhler theory, see Ref. [34].

Early organic aerosol CCN studies examined the hygroscopicity of selected organic compounds.[36–38] These studies demonstrated that highly water-soluble compounds do indeed give rise to good CCN by increasing the number of solutes. The CCN activity of these compounds can be predicted using the Köhler equation if their molecular properties are known. However, given that ambient organic aerosol particles contain a myriad of compounds, most of which are still unidentified and unquantified, their CCN activity cannot be predicted based on first principles. As a result, a modified Köhler theory was developed by Petters and Kreidenweis[39] where this extension of the Köhler theory uses one parameter (κ) to describe average molecular properties (e.g., molecular weight and solubility) that control the particle's water uptake characteristics. Numerous laboratory studies empirically determined κ values using organic aerosol composed of single compounds. For example, κ is 0 for the hydrophobic bis(2-ethylhexyl) sebacate[40] and 0.23 for the hydrophilic malonic acid.[38] In comparison, κ values in the atmosphere are 0.61 for $(NH_4)_2SO_4$ and 0.90 for H_2SO_4 particles.[41] Petters and Kreidenweis[39] provide a comprehensive list of κ values for both inorganic and organic compounds.

Field-based CCN closure studies suggest that the κ of organic aerosol can be as large as 0.3.[42,43] However, there is clearly a large variability in this value, with evidence that primary, urban organics studied having smaller values than secondary organics.[44] Organic aerosol particles also undergo chemical transformations during their atmospheric lifetimes (i.e., processes described in Sec. 2.4), which will change their properties and hygroscopicity.[3] Motivated by the need for a simple representation that captures this processing, Chang *et al.*[42] proposed that hygroscopicity is linearly related to the organic component's degree of oxygenation as expressed by the O:C ratio (O/C), i.e., more functionalized and/or smaller species will be more hygroscopic. Subsequent studies[45–47] have investigated the relationship between hygroscopicity and degree of oxygenation of organic aerosol, and results from both laboratory and ambient studies have been compiled by Rickards *et al.* (see Fig. 2.4). While there is indeed a general trend of increasing organic hygroscopicity with increasing O/C, considerable scatter in the correlation suggests that a single bulk property of the particles cannot completely capture the hygroscopic response of organic aerosol. For example, oxidation of SOA precursors with different molecular weight, gives rise to different number of oxidation products (i.e., different number of moles of solutes). The predictive power of the linear relationship between κ and O/C in Fig. 2.4 has yet to be fully tested using ambient organic aerosol, e.g., by comparison with measured CCN concentrations.

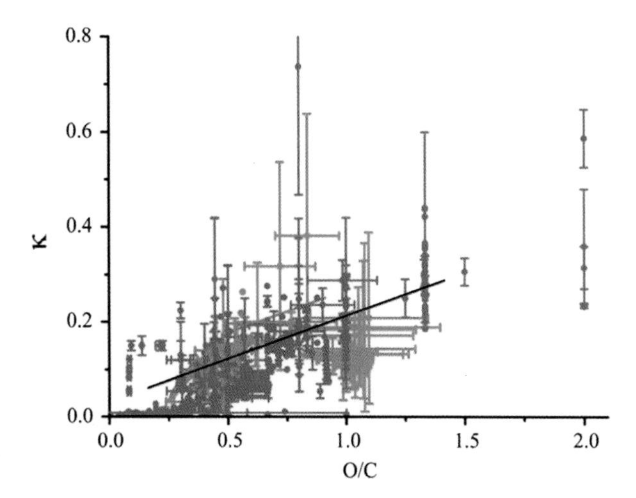

Fig. 2.4 Literature data from ambient measurements (red) and laboratory studies (grey), showing the relationship between κ and O/C ratio.[48] The linear relationship proposed by Chang *et al.*[42] is illustrated by the black line. Adapted with permission from Rickards *et al.*[48] Copyright 2013 American Chemical Society.

While the above discussion considers the contribution of organics to particle hygroscopicity through solubility, organic compounds can also affect CCN activity by altering surface tension. Surface active organic compounds, which may have both hydrophilic and hydrophobic moieties, will partition to the gas–particle interface, resulting in a reduction in the aerosol surface tension (i.e., reduction in the Kelvin effect) and a more CCN-active particle. Laboratory studies show that bulk solutions of atmospheric organics do indeed have lower surface tension compared to water.[49,50] These findings have also been corroborated by laboratory studies of organic aerosol composed of pure, slightly soluble compounds such as large organic acids.[51] As well, gas phase species with surfactant properties may enhance CCN activity.[52] Interestingly, the same organic acids that enhance CCN activity when present on their own actually decrease CCN activity when mixed with hygroscopic inorganic particles.[53,54] It is thought that the decrease in CCN activity in such mixtures arises from a decrease in soluble ions when the surfactant partitions to the surface.[55] This last point illustrates that uncertainties remain for the overall effect of surface active organics on CCN activity.

It is uncertain whether organics reduce the rate of water vapor transport during CCN activation. Changes to the droplet growth rates will change the amount of water in the droplet, thus affecting cloud droplet size distributions. For example, 1 to 2 seconds of delay in the growth of cloud droplets can reduce modelled cloud droplet number concentrations significantly.[56] The assumption that the activation of aerosol as CCN can be modelled purely as an equilibrium process has been shown to be invalid in some conditions.[57–59] While highly uncertain, it appears that organics can result in kinetic limitations due to slow solute dissolution associated with highly viscous organic particles[6,60,61] or that organics can form a film at the surface, affecting the mass transfer rate of water vapor.[62,63]

Slow droplet growth rates have been observed in the laboratory and during ambient measurements, yet the extent to which growth rate suppression occurs in the atmosphere remains unclear as only a limited number of studies have investigated this process, with contrasting results. Suppression in droplet growth kinetics is assumed to be responsible if the addition of an organic compound to an inorganic particle gives rise to a smaller droplet size compared to that of the bare inorganic particles, which have known rapid growth kinetics. Carboxylic acid particles,[56] and SOA from β-caryophylene ozonolysis[60] exhibit slow droplet growth kinetics in the laboratory. The formation of high molecular weight compounds via acid-catalyzed condensed-phase reactions (e.g., aldol condensation described in Sec. 2.4.2.5) can also result in the suppression in droplet growth kinetics.[64] In addition, kinetic limitations have been observed from particles of various sources/origins, including bovine emissions[65] and continental particles containing anthropogenic organic components.[66] However, using computational fluid dynamics models of the conditions inside a CCN counter, Raatikainen *et al.*[67] analyzed 10 ambient data sets and represented growth kinetics via an effective water uptake mass accommodation coefficient. This study constrained water mass accommodation coefficients between 0.1 and 1, indicating that rapid growth kinetics are globally prevalent. These contrasting results from laboratory and field data pose a challenge to determining whether the kinetic limitation of droplet growth is an important process, and to what degree it actually occurs in the atmosphere in specific situations.

2.3. Effects of Water on Organic Partitioning

The partitioning of organic compounds between the gas phase and the condensed phases present in aerosol is central to the formation of SOA and has effects on both particle phase and reactivity. For a long time, formation of SOA was described as a process whereby lower volatility products of the gas phase oxidation of precursor VOCs partitioned into a condensed phase made up of organic matter, without considering the presence of water or other inorganic species.[68–70] As discussed in Sec. 2.2.1, water is a major component of the atmosphere, and its presence has been shown to significantly influence the partitioning properties, physical phase state, overall mass and chemical composition of aerosol.[7,9]

It is now well established that small water soluble VOCs (≤C5), such as dicarbonyls (e.g., glyoxal and methylglyoxal) and small organic acids (e.g., oxalic acid), partition significantly to condensed water in the atmosphere and that aqueous-phase reaction products of low volatility contribute to the formation of SOA.[10] The first step in this process is gas-aqueous phase equilibrium partitioning of the small VOCs, described by the Henry's law constant (K_H). However, observations indicate that these compounds partition to aerosols in amounts larger than expected from K_H. This is usually attributed to further reactions in the condensed phase, such as hydration of aldehydes, dissociation of carboxylic acids, oligomerization or reaction with inorganic components of the aqueous phase;[17,71–73] see Sec. 2.4.2 for detailed descriptions of these condensed-phase reactions. An effective Henry's law constant (K_H^*), which is a Henry's law constant that accounts for such additional processing in the aqueous phase, is often used to describe the uptake of these organic compounds in aqueous aerosol.[10,15,74] Glyoxal, due to its high water solubility and reactivity, is an example of a small organic compound partitioning to water.[75] Glyoxal can undergo rapid hydration and oligomerization even in pure water, enhancing the K_H by a few orders of magnitude.[74] In $(NH_4)_2SO_4$ solutions, the K_H^* is even higher because of the promotion of oligomerization and hydration as well as the reaction with inorganic ions.[73,76]

With increasing awareness of the contribution that aqueous phase SOA can make to the total organic aerosol loading, more studies have focused on the potential role of water in the partitioning of organic compounds to aerosol.[9,77–81] The aqueous phase may be relevant even for organic substances that are considerably larger than those small water soluble compounds with less than six carbons, especially at higher LWC.[82] Larger, highly oxygenated multifunctional organic compounds are to some extent also water soluble and hydrophilic (due to their high degree of functionalization and polarity), and thus could partition significantly to the aerosol aqueous phase depending on the structure of the organics, the LWC, and the composition of the aqueous phase. The predicted organic aerosol mass and composition differ when including water, either phase separated from or homogeneously mixed with the organic phase.[9]

In the atmosphere, inorganic ions such as ammonium, sulfate, sodium, chloride, and nitrate are always associated with aerosol water. The inorganic salts not only contribute to the mass of aerosol particles but also influence the gas–particle partitioning of organic compounds.[83,84] Most organic compounds' partitioning to water is diminished in the presence of salts, i.e., they tend to be salted out of the aqueous phase to either the gas or another condensed phase. This is called the salting out effect. This effect increases with an organic molecule's size and decreases with its polarity, i.e., it is most pronounced for large hydrophobic chemicals.[83,84] The effect also differs for different salts, being approximately twice as large for $(NH_4)_2SO_4$ as for NaCl.[84] Salting out can eventually lead to liquid-liquid phase separation of the condensed phase, forming a predominantly organic phase and an aqueous electrolyte-rich phase.[9,84] This is the consequence of the non-ideality of the mixtures of both hydrophilic and hydrophobic organic compounds, water, and dissolved salts.[7,9] Some substances, in particular small water-soluble organic compounds such as glyoxal, have been found to partition more to the aqueous phase in the presence of salts ("salting in" effect). This is likely due to the influence of salts on the aldehyde's hydration equilibrium and due to reactions of salts with those substances.[72,74]

In order to illustrate the role of water on the atmospheric phase distribution of organic compounds we adopt here the two dimensional space defined by two equilibrium partition coefficients introduced recently.[82] The assumptions are: (1) the condensed aerosol phase separates into an aqueous phase (W) and a phase largely made up of organic matter (OM), (2) a quasi-instantaneous equilibrium is reached between the gas phase and these two condensed phases, (3) no reactions in the condensed phases are considered. Partitioning space plots illustrate the relative preference of organic compounds to partition to the organic or the aqueous phase of an aerosol. The two

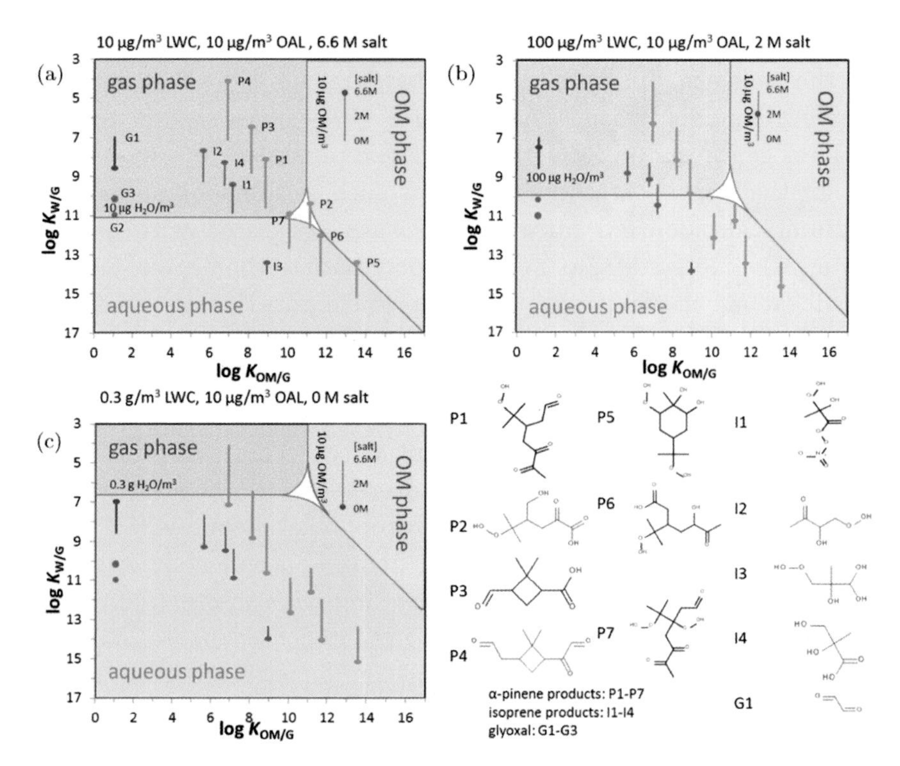

Fig. 2.5 Chemical partitioning space, showing in pink, blue and yellow the combinations of partitioning properties that lead to dominant equilibrium partitioning to the gas, aqueous, and organic matter (OM) phases, respectively. The lines indicate the 50% threshold for various liquid water contents, salt concentrations and one organic aerosol loading (OAL) ($10\,ug\,m^3$): (a) LWC $10\,\mathrm{ug\,m^{-3}}$ and $(NH_4)_2SO_4$ concentration of 6.6 M, (b) LWC $100\,ug\,m^{-3}$ and $(NH_4)_2SO_4$ concentration of 2 M, and (c) LWC $0.3\,\mathrm{g\,m^{-3}}$ and $(NH_4)_2SO_4$ concentration of 0 M. Selected oxidation products of α-pinene (blue), isoprene (red) and glyoxal (purple) were placed in the partitioning space to show their preference for a particular phase at different LWCs and salt concentrations.

axes of the plot (Fig. 2.5) are the equilibrium partitioning coefficients between OM and the gas phase ($K_{\mathrm{OM/G}}$), and between the aqueous and gas phase ($K_{\mathrm{W/G}}$).[82] The equilibrium partitioning coefficients are defined by the following equations:

$$K_{\mathrm{OM/G}} = C_{\mathrm{OM}}/C_{\mathrm{G}}, \tag{2.1}$$

$$K_{\mathrm{W/G}} = C_{\mathrm{W}}/C_{\mathrm{G}}, \tag{2.2}$$

where C_{OM}, C_{W} and C_{G} are the equilibrium concentrations of the organic compound in OM, water and the gas phase in units of $\mathrm{mol\,m^{-3}}$, respectively. $K_{\mathrm{W/G}}$ is related to K_{H} in units of $\mathrm{M\,atm^{-1}}$ through:

$$K_{\mathrm{W/G}} = K_{\mathrm{H}}RT, \tag{2.3}$$

where R is the ideal gas constant ($8.206 \times 10^{-2}\ \mathrm{L\,atm\,K^{-1}\,mol^{-1}}$) and T is absolute temperature in K.

The fraction Φ_X of a chemical in one of the three phases can be calculated using[80,85]:

$$\Phi_{\mathrm{G}} = 1/(1 + K_{\mathrm{W/G}} \cdot \mathrm{LWC} \cdot \rho_{\mathrm{W}} + K_{\mathrm{OM/G}} \cdot \mathrm{OAL} \cdot \rho_{\mathrm{OM}}), \tag{2.4}$$

$$\Phi_{\mathrm{W}} = 1\Big/\left(1 + \frac{1}{K_{\mathrm{W/G}} \cdot \mathrm{LWC} \cdot \rho_{\mathrm{W}}} + \frac{K_{\mathrm{OM/G}} \cdot \mathrm{OAL} \cdot \rho_{\mathrm{OM}}}{K_{\mathrm{W/G}} \cdot \mathrm{LWC} \cdot \rho_{\mathrm{W}}}\right), \tag{2.5}$$

$$\Phi_{\mathrm{OM}} = 1\Big/\left(1 + \frac{1}{K_{\mathrm{OM/G}} \cdot \mathrm{OAL} \cdot \rho_{\mathrm{OM}}} + \frac{K_{\mathrm{W/G}} \cdot \mathrm{LWC} \cdot \rho_{\mathrm{W}}}{K_{\mathrm{OM/G}} \cdot \mathrm{OAL} \cdot \rho_{\mathrm{OM}}}\right). \tag{2.6}$$

LWC and OAL are the aqueous phase volume and the organic aerosol loading (OM phase volume) in units of $\mathrm{g\,m^{-3}}$. The density of both the aqueous (ρ_{W}) and OM (ρ_{OM}) phases is assumed to be $10^6\ \mathrm{g\,m^{-3}}$.

The three colored fields in the plots of Fig. 2.5 indicate partitioning property combinations (of $K_{\mathrm{OM/G}}$ and $K_{\mathrm{W/G}}$) that lead to dominant ($\Phi_{\mathrm{X}} > 50\%$) partitioning into the gas (pink), OM (yellow) and aqueous phase (blue). The lines between the fields thus indicate partitioning property combinations that lead to equal distribution between two phases ($\Phi_{\mathrm{X}} = 50\%$). When drawing those fields and lines in the three plots of Fig. 2.5 OAL was assumed to be $10^{-5}\ \mathrm{g\,m^{-3}}$, whereas LWC was assumed to be $10^{-5}\ \mathrm{g\,m^{-3}}$ (panel a), $10^{-4}\ \mathrm{g\,m^{-3}}$ (panel b) or $0.3\ \mathrm{g\,m^{-3}}$ (panel c). The plots thus represent different atmospheric conditions corresponding to aerosol at low (panel a) or high (panel b) RH or a cloud (panel c), as described previously in Sec. 2.2.1. For example, at LWC and OAL of $10^{-5}\ \mathrm{g\,m^{-3}}$ (Fig. 2.5a), a compound with log $K_{\mathrm{W/G}}$ of 11 and log $K_{\mathrm{OM/G}}$ smaller than 10 is equally distributed between the gas and aqueous phases, while a compound with log $K_{W/G}$ higher than 11 and log $K_{OM/G}$ smaller than 10 is expected to be predominantly in the aqueous phase.

Glyoxal (G) and selected products formed during the oxidation of α-pinene (P1–P7) and isoprene (I1–I4), are located in the partitioning space according to their partitioning coefficients. Oxidation products were taken from Zuend and Seinfeld[9] (P1–P4), Valorso *et al.*[86] (P5–P7), Mouchel-Vallon *et al.*[87] (I1–I3) and Paulot *et al.*[88] (I4). $K_{OM/G}$ and $K_{W/G}$ at 15°C for α-pinene products were estimated using SPARC, and those for isoprene products with a polyparameter linear free energy relationship (ppLFER) by assuming the solvation properties of the OM phase can be approximated by those of structure B (Fig. 2.6) as described in Wania *et al.*[82] $K_{OM/G}$ for glyoxal was also predicted with the same ppLFER. $K_{W/G}$ for glyoxal was either the measured value at 25°C for a bulk solution (G1)[74] or the K_H^* values estimated at ambient temperature for aerosol particles (G2 and G3).[17,89] Note at 15°C $K_{OM/G}$ and $K_{W/G}$ for glyoxal would be slightly larger than those at 25°C (less than 0.5 log units different though).

By representing each chemical as a vertical line in Fig. 2.5, it is indicated how the presence of $(NH_4)_2SO_4$ will influence a chemical's partitioning into the aqueous phase. The salting-out effect of $(NH_4)_2SO_4$ on the oxidation products of α-pinene and isoprene was predicted with the ppLFER by Wang *et al.*[84] while the "salting-in effect" for glyoxal is displayed based on measurements by Kampf *et al.*[72] The influence of salt on log $K_{W/G}$ increases with the salt concentration as described by the Setschenow relationship.[84] The uppermost value (for α-pinene and isoprene products) or bottommost

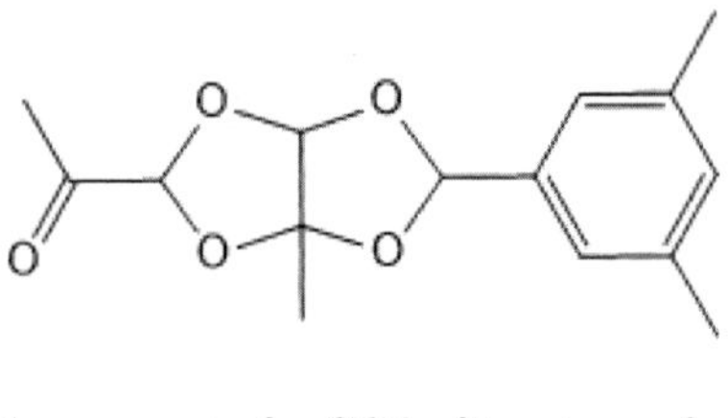

Fig. 2.6 Possible surrogate for SOA: Structure of molecule B.[82,315]

value (for glyoxal) on the vertical line indicates the $K_{W/G}$ in a 6.6 M supersaturated solution of $(NH_4)_2SO_4$, while the other end of the line indicates $K_{W/G}$ in pure water. The dot on the line shows the $K_{W/G}$ at the salt concentrations assumed to be prevalent in aerosol at low RH (6.6 M, panel a), at high RH (2 M, panel b) and in a cloud (0 M, panel c). The length of the line depends on a compound's Setschenow coefficient, increasing with molecular size and decreasing with polarity. Considering the presence of other salt species in addition to $(NH_4)_2SO_4$ would likely increase the salt effect.

Interestingly, the log $K_{W/G}$ of many compounds shown in Fig. 2.5 is larger than the K_H^* of glyoxal (i.e., are located further down in the partitioning space), and thus have a higher relative affinity for the aqueous phase than a compound long known to be associated with aqueous-phase SOA. This suggests that such larger, highly oxygenated organic products can potentially contribute to aqueous phase SOA. As LWC increases from $10\,\mu g\,m^{-3}$ (a) to $100\,\mu g\,m^{-3}$ (b) for aerosols, and to $0.3\,g\,m^{-3}$ (c) for clouds, larger fractions of many compounds (e.g., I3, P1, P2, and P7) are partitioning into the aqueous phase, which agrees with other studies.[10,17] For example, in cloudwater or aerosol with high LWC, even species with lower $K_{W/G}$ (e.g., P3, P4, I2, and I4) might partition to a significant extent to the aqueous phase. The aqueous phase not only provides a volume for organic compound uptake but also a medium for potential further reactions. Less volatile, multifunctional organic reaction products will generally have a higher K_H^* and partition more efficiently into aqueous phase.

The partitioning equilibrium between the aqueous and gas phases is shifted upwards (i.e., to lower $K_{W/G}$) for all α-pinene and isoprene oxidation products in the presence of $(NH_4)_2SO_4$, whereas glyoxal is salted in as expected. However, $K_{W/G}$ for glyoxal in a supersaturated solution of $(NH_4)_2SO_4$(6.6 M) is still lower than its K_H^* in ambient or lab-generated aerosol particles.[17,89] Obviously, the salting in effect alone cannot explain the high K_H^* for glyoxal in real aerosol samples. On the other hand, the salting out effect will shift the partitioning equilibrium to either the gas or OM phase for most organic compounds. This salt effect is much higher and thus crucial for aerosol water because of the high salt concentrations (Fig. 2.5(a)), but negligible for the much more diluted cloudwater (Fig. 2.5(c)). In highly saline aqueous aerosol (Fig. 2.5 panel a), P2, P5, and P7 are salted out into either the OM phase, gas phase or the transition area, while those compounds can partition predominantly into the aqueous phase if there were no salts. Even when the salting out effect is considered, some highly oxygenated organic compounds, such as I1, P2, P6, and P7, are still predicted to be in the aqueous phase or to fall in the transition areas of the three phases.

Overall, this again highlights the importance of including an aqueous phase and the salt effect of inorganic salts for an accurate prediction of SOA composition and phase distribution. It also emphasizes the need to better understand condensed phase reactions in both the aqueous and organic phases. Aqueous-phase reactivity and multiphase partitioning together will provide a better understanding of SOA formation.

Finally, we should note that equilibrium partitioning as discussed in this section applies to condensed particle phases with low viscosity. The assumption of quasi-instantaneous gas–particle partitioning equilibrium,[61,90] which is often made in large-scale modeling studies,[1] implies that diffusion within the particles is rapid compared to the timescales of other processes implicated in SOA formation.[91] However, recent laboratory and modeling studies suggest that some SOA particles can be highly viscous under some conditions, such as low RH.[7,31,32] As described in other sections of this review, mass transfer of semi-volatile compounds by diffusion in such semi-solid or solid particles can be slow, leading to reduced rates of heterogeneous reactions and hygroscopic growth. In these cases, a kinetic description of the uptake of gas phase compounds may be required when modeling SOA formation.

2.4. Organic Aerosol Reactivity in the Presence of Water

The sections above have illustrated that an organic aerosol particle is composed of a complex mixture of water and both inorganic and organic constituents. Reactions occurring within this complex mixture affect the partitioning of both water and organic compounds, through both thermodynamic and kinetic processes. As well, condensed-phase reactions affect the lifetimes of trace particulate components, the physico-chemical properties of the particles, and can lead to the formation of volatile products. This section addresses this complex chemistry in six sections. For convenience, different classes of reactions are identified — namely, OH radical, nucleophilic, evaporative, and photochemical reactions — recognizing that these processes can be closely interrelated. The radical and nucleophilic reactions are addressed in detail, given the large amount of work that has been completed in this area under cloudwater conditions and is useful to understanding processes occurring in aerosol water. The final two sections address the roles that absorbed and adsorbed water plays in bulk and interfacial reactions, respectively.

2.4.1. *OH Radical Reactions*

Radicals induce rapid oxidation of organic compounds in the aqueous phase. Given the difficulty in directly monitoring reactions occurring in aqueous aerosol, our current understanding of chemistry in aqueous aerosol is built on the extension of that in dilute aqueous solutions. Organic chemistry in the atmospheric aqueous phases has been recently reviewed. In particular, the kinetics of radical chemistry in the aqueous phase has been reviewed by Herrmann *et al.*[13,92] Lim *et al.*[93] have compiled aqueous-phase reaction mechanisms, focusing on glyoxal. Ervens *et al.*[10] have summarized the current state of understanding of aqueous-phase chemistry, integrating results from laboratory, field measurements and models. More recently, McNeill[11] has summarized chemistry taking place in cloudwater, aqueous aerosol, as well as gaps in our current understanding, while Ervens[12] has summarized recent progress in modeling multiphase chemistry. In this section, we focus on the reaction mechanisms of OH radical chemistry, given that the OH radical is the most important oxidant in atmospheric aqueous phases. Emerging topics in this area and the unique character of aqueous-phase mechanisms are discussed, following the general reaction mechanisms.

2.4.1.1. *Sources, Sinks and Concentrations of the OH Radical in Aqueous Aerosol*

2.4.1.1.1. Relative Importance of Gas–particle Partitioning and *In-situ* Formation of the OH Radical

Gas–particle partitioning and *in situ* formation can both be important sources of the OH radical in aqueous aerosol. The uptake of gas phase OH to an aqueous particle can be described by the 1st-order loss rate of gas phase OH radical (Eq. (2.7))[94]:

$$\frac{d[\mathrm{OH}]_{\mathrm{gas}}}{dt} = \left(\frac{r^2}{3D_{\mathrm{gas}}} + \frac{4r}{3\omega\gamma}\right)^{-1} [\mathrm{OH}]_{\mathrm{gas}}, \tag{2.7}$$

where $[\mathrm{OH}]_{\mathrm{gas}}$, D_{gas}, and ω are the gas phase concentration (mole cm^{-3}), the gas phase diffusion coefficient ($\mathrm{cm}^2\,\mathrm{s}^{-1}$), and the mean gas phase velocity of an OH radical (cm s^{-1}), respectively. The term r is the radius of the particle or droplet (cm) and γ is a dimensionless uptake coefficient. The uptake rate of OH radical ($\mathrm{M\,s}^{-1}$) as a function of particle diameter is shown in Fig. 2.7, assuming $[\mathrm{OH}]_{\mathrm{gas}} = 4 \times 10^6$ mole cm^{-3}, $D_{\mathrm{gas}} = 0.1\,\mathrm{cm}^{-2}\,\mathrm{s}^{-1}$, temperature at 298 K and three different γ values at 1, 0.1, and 0.01 to represent the highly variable condensed-phase composition.

Also included in Fig. 2.7 are *in situ* OH production rates measured in a variety of atmospheric aqueous phases. The production rate of OH radical in cloud, fog and rain water is commonly determined by adding an OH probe compound into the sample and monitoring the formation of a product compound with simulated or ambient irradiation.[95–99] For aerosol liquid water, there is currently no method to directly probe OH production, and the measurements have been done in the water extract of collected particles.[100–104] The *in situ* production rate of OH in atmospheric aqueous phases represents one of the largest uncertainties in modeling radical chemistry in these reaction media.[19]

As the particle size increases, the uptake rate decreases due to increasing particle volume and limitations incurred by gas phase diffusion. The *in situ* formation rate of OH radical also tends to be smaller in larger droplets due to dilution of OH precursors. The

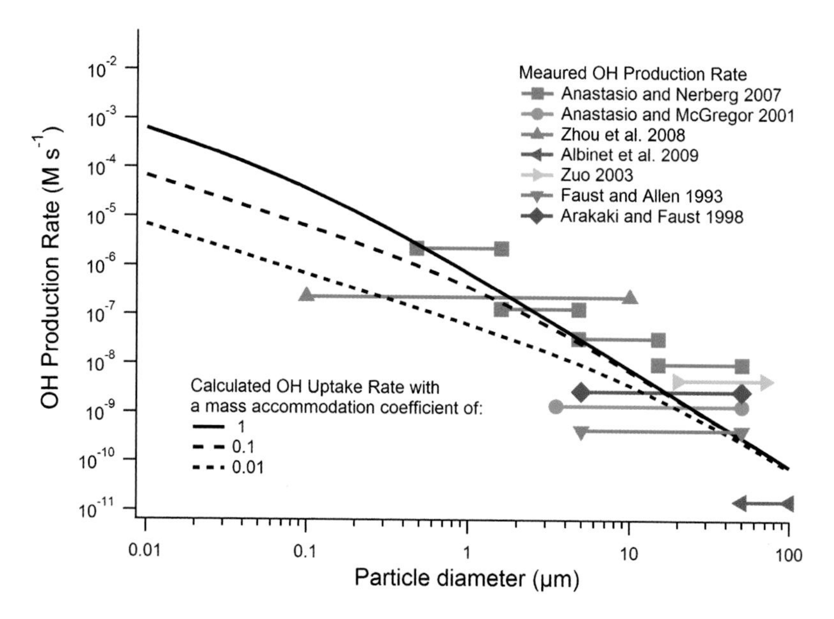

Fig. 2.7 Calculated gas–particle partitioning rates and measured *in situ* formation rates of the OH radical as a function of particle diameter. The calculated rates are based on three (1, 0.1, and 0.01) values of the uptake coefficient (γ), and the measured *in situ* formation rates are compiled based on a number of studies, the conditions of which are summarized in Table 2.1.

conditions employed in the studies shown in Fig. 2.7 are highly variable, as summarized in Table 2.1. Experimental photon flux in these studies has been normalized to ambient flux at different locations and times, therefore inter-comparison of the reported OH production rates is not straightforward. However, these results all point towards a conclusion that *in situ* formation of OH radical can be a significant, and at times the dominant, source of OH radical in atmospheric aqueous phases.

2.4.1.1.2. Mechanisms of *In situ* OH Radical Formation

Different mechanisms can lead to *in situ* formation of OH radical in the aqueous phase. O_3 can partition from the gas phase and react with HO_2 or its dissociated form (O_2^-, with a pKa value 4.9) to form OH in cloudwater.[105,106] However, this reaction is highly pH dependent, as O_3 reacts mostly with O_2^-.[107] Given the small LWC

Table 2.1 Conditions used in the measurements of *in situ* production rate of OH radical

Literature	Sample type	Particle size[a]	Light source	OH probe[b]	Normalization to ambient photon flux
98	Cloud/Fog	B	313 nm	B	Equinox, midday
97	Cloud	B	313 nm	B	Equinox, Zenith angle = 36°
96	Fog	M	313 nm	BA	Winter solstice, midday at Davis CA
99	Fog	A	313 nm	HMSA	Autumn solar noon
100	Sea salt	M	Xe lamp	BA	Summer solstice
104	Laboratory Sea Salt	A	sunlight	BA	Summer solstice, tropical midday
95	Rain	B	365 nm	B	Summer midday at 45°N

[a]M: Measured and specified in the original literature; A: Based on assumptions made in the original literature; B: Assumed from typical size ranges of corresponding atmospheric aqueous phases (5–50 μm for cloud and fog, >50 μm for rain).
[b]B: benzene; BA: benzoic acid; HMSA: hydroxymethanesulfonate.

of aerosol liquid water and its enhanced acidity, this chemistry may not be important in aqueous aerosol. Direct photolysis of inorganic and organic chromophores (i.e., NO_3^-, NO_2^-, H_2O_2, and organic hydroperoxides) can be an important OH source in atmospheric aqueous phases. OH radical production from these chromophores has been investigated in detail.[92,108,109] Iron-catalyzed OH generation via Fenton and photo–Fenton reactions is particularly important. Arakaki and Faust[97] have observed that Fenton chemistry is the most important *in situ* formation mechanism based on measurement using authentic cloudwater samples. Ervens *et al.*[19] have also shown in their model that Fenton chemistry is likely the most important mechanism in both cloudwater and aerosol liquid water.

While OH formation from dissolved organic compounds is more poorly understood, evidence for its importance as an OH source in atmospheric aqueous phases is emerging. Dissolved organic matter has been previously shown to be the major OH source in seawater.[110] Anastasio and Newberg[100] have shown that the contribution of NO_3^- as an OH source decreases in small sea salt aerosol particles, indicating an increasing contribution from organic chromophores. Arakaki *et al.*[102] have observed an unknown OH source exhibiting positive correlation with the concentration of dissolved organic carbon. More recently, Badali *et al.*[111] have observed OH production from the water extract of chamber-generated SOA. While the authors assumed organic hydroperoxides as likely candidates for the observed OH formation, the identity of the organic chromophores and the mechanisms of OH generation are currently unclear.

2.4.1.1.3. Aqueous-phase Diffusion and Reaction of OH Radical

Due to their significant reactivity, OH radicals introduced to a particle via gas phase partitioning may not diffuse through the entire particle. One approach to evaluate OH diffusion in the aqueous phase is to compare the typical time-scale of diffusion (τ_{daq}) to the typical time scale of chemical reaction (τ_{chem}), which can be calculated using Eq. (2.8) and Eq. (2.9), respectively.[75]

$$\tau_{\text{daq}} = \frac{r^2}{\pi^2 D_{\text{aq}}}, \tag{2.8}$$

$$\tau_{\text{chem}} = \frac{1}{k^{\text{I}}}, \tag{2.9}$$

where D_{aq} is the particle-phase diffusion coefficient of OH radical, and k^{I} is the 1st-order loss rate coefficient of OH radical in the particle phase. Using a modeling approach, Ervens *et al.*[19] have shown that τ_{daq} and τ_{chem} are comparable in cloud droplets with diameters larger than $10\,\mu$m. In smaller droplets and aqueous aerosol, however, τ_{chem} can be orders of magnitude smaller than τ_{daq}, indicating that OH radicals introduced via gas–particle partitioning may not travel far in the particle.

Alternatively, the diffusion of OH radicals can be characterized by the diffuso-reactive length (d_L) which describes the distance an OH radical typically travels before reacting; it is calculated using Eq. (2.10)[112]:

$$d_L = \sqrt{\frac{D_{\text{aq}}}{k^{\text{I}}}} = \sqrt{\frac{D_{\text{aq}}}{k^{\text{II}}_{\text{DOC}}[\text{DOC}]}}. \tag{2.10}$$

While the OH radical can react with inorganic compounds such as nitrate, SO_2, H_2O_2 and halogen species, the dominant sink of OH is dissolved organic compounds.[96,97,100] The 1st-order loss rate coefficient (k^{I}) of the OH radical can be approximated by the product of the concentration of dissolved organic carbon ([DOC]) and an averaged 2nd-order rate constant ($k^{\mathrm{II}}_{\mathrm{DOC}}$) of OH reacting with DOC. As the majority of DOC in atmospheric aqueous phases remains unspeciated,[113] Arakaki *et al.*[114] have proposed an averaged, carbon-based $k^{\mathrm{II}}_{\mathrm{DOC}}$ value of 3.8×10^8 mol-C $\mathrm{L}^{-1}\,\mathrm{s}^{-1}$. We used this value, along with a typical D_{aq}($1 \times 10^{-5}\,\mathrm{cm}^2\,\mathrm{s}^{-1}$, Seinfeld and Pandis[115]) to calculate d_L, assuming a wide range of DOC concentrations (Fig. 2.8). Also shown in Fig. 2.8 are the ranges of DOC concentration relevant to different types of atmospheric aqueous phases. The ranges

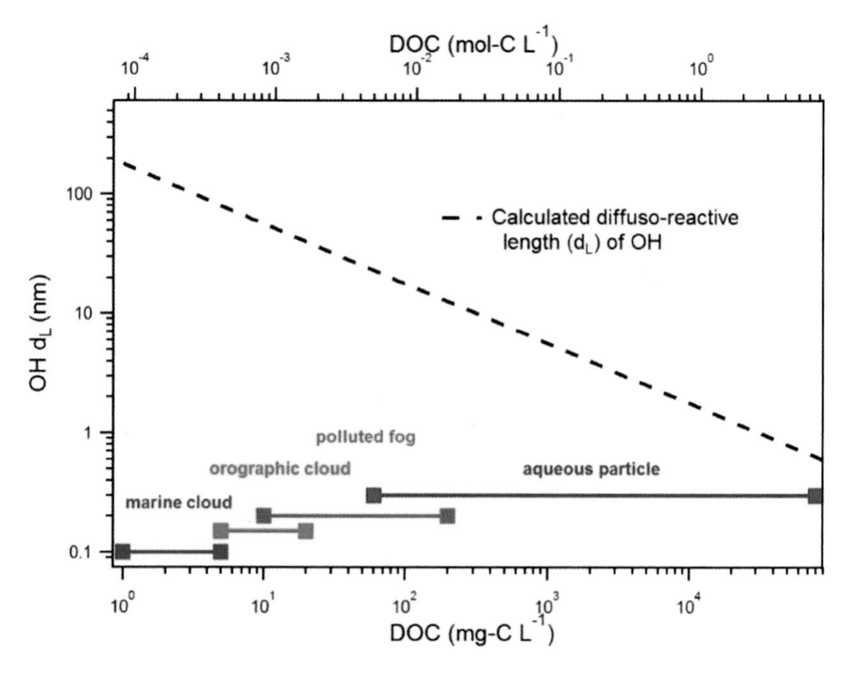

Fig. 2.8 The diffuso-reactive length (d_L) of OH as a function of dissolved organic carbon (DOC) concentrations. The loss rate of the OH radical is based on the averaged, carbon-based value reported in Arakaki *et al.*[114] The horizontal bars represent the approximate DOC range in different types of atmospheric waters.[17,19,113]

of marine clouds, orographic clouds and polluted fogs are adapted from Ref. [113] which have summarized measured [DOC] values across the globe. There is no direct measurement for [DOC] in aerosol liquid water, but it has been assumed to be at the molar level by several modeling studies.[15,17,19]

The values of d_L are orders of magnitude smaller than the radius of the atmospheric aqueous phases, implying that an OH radical introduced from the gas phase may reach only a small portion of a particle. Recent studies also revealed enhanced viscosity of organic aerosol[32,116,117] as well as liquid–liquid phase separation[8] at low RH conditions. When the diffusion of the OH radical is further hindered by these conditions, *in situ* formation of OH is expected to gain relative importance in aqueous-phase processing of organic compounds.

2.4.1.1.4. Steady State OH Concentration

The steady state concentration of OH radical in aqueous aerosol ($[OH]_{ss}$) can be estimated from the ratio of its production rate (P_{OH}) and its 1st-order loss coefficient (k^I) (Eq. (2.11)):

$$[OH]_{ss} = \frac{P_{OH}}{k^I}. \tag{2.11}$$

Arakaki *et al.*[114] recently summarized the $[OH]_{ss}$ values in different types of atmospheric aqueous phases (Fig. 2.9). Interestingly, although P_{OH} and k^I vary over orders of magnitude between cloudwater, rainwater, and aerosol liquid water, $[OH]_{ss}$ remains within a relatively narrow range: $0.5-7\times10^{-15}$ M. However, discrepancy exits between this measurement-based approach and other modeling-based approaches, where OH concentration varies by orders of magnitude among different atmospheric aqueous phases. The $[OH]_{ss}$ values obtained from a number of modeling studies are also summarized by Arakaki *et al.*[114] and shown in Fig. 2.9. The models generally predict much high $[OH]_{ss}$ values in cloudwater than in aerosol liquid water probably because of a missing OH sink. More data and improved techniques are required to better constrain this important parameter.

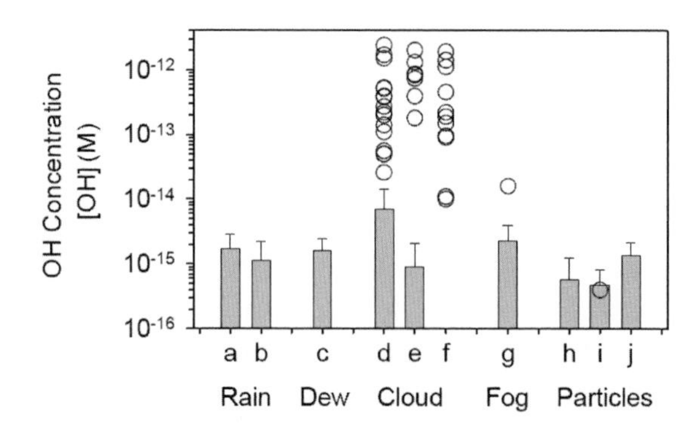

Fig. 2.9 Measured and modeled OH concentration ([OH]) in a variety of atmospheric aqueous phases. The bars illustrate [OH] calculated based on the measured sources and sinks of the OH radical. The circles represent modeled [OH] in modeling studies (Arakaki *et al.*[114] and references therein). Adapted with permission from Arakaki *et al.*[114] Copyright 2013 American Chemical Society.

2.4.1.2. *General OH Reactivity and Mechanisms*

2.4.1.2.1. Initiation of the Radical Chain Reaction

Although aerosol organic chemistry is still in its infancy, aqueous-phase OH oxidation of a wide variety of organic compounds has been investigated.[118] Figure 2.10 compiles the structures and OH reactivity (k^{II}_{OH}) of compounds relevant to atmospheric aqueous-phase processing. Similar to gas phase chemistry, the OH radical reacts with aromatic and unsaturated aliphatic compounds essentially at the diffusion limit through radical addition reactions (Fig. 2.11(a) and (b)). Reaction of OH radicals with saturated organic compounds proceeds via H-atom abstraction (Fig. 2.11(c)). Organic compounds relevant to the atmospheric aqueous phases usually contain highly oxygenated functional groups and react rapidly with the OH radical. As oxygenation proceeds further, the reactivity drops significantly due to lack of easily abstractable hydrogens (e.g., oxalic acid and pyruvic acid). Since dissolved molecular oxygen is abundant in atmospheric aqueous phases, reactions (a)–(c) also display the subsequent addition of O_2 after the initiation.

There is an ongoing effort to establish structure activity relationships (SAR) in the aqueous phase,[119–122] inspired by the

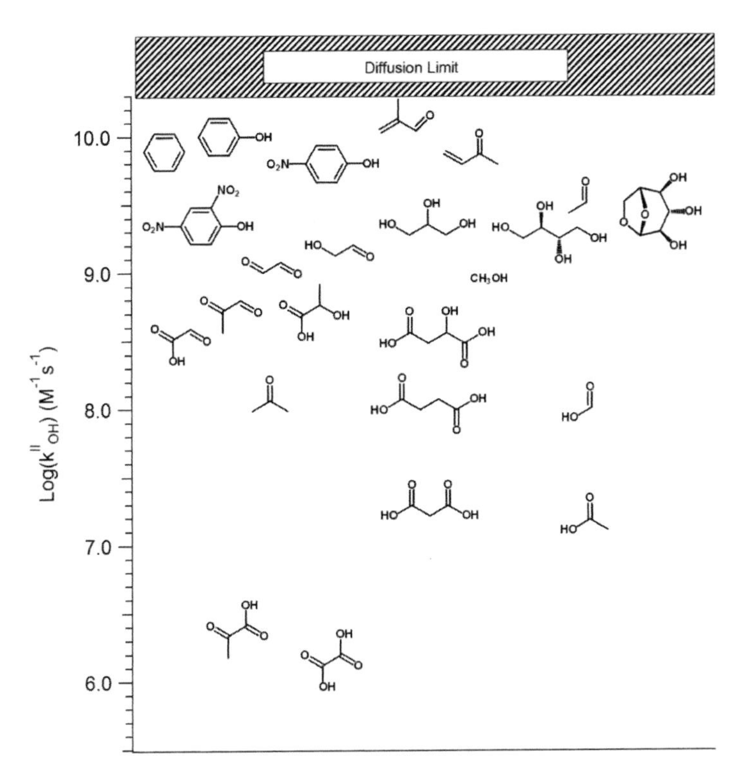

Fig. 2.10 Structures and OH reactivity (k^{II}_{OH}, 298 K) of organic compounds relevant to atmospheric waters.[92,118,130,133,140] The OH reactivity in the shaded area is diffusion limited.

gas phase SAR.[123] With the current expansion of the aqueous-phase database, such SARs are expected to gain accuracy and applicability.

2.4.1.2.2. Propagation and Termination of the Radical Chain

Due to low water solubility, nitrogen oxides (NO_x) do not play as pronounced a role in the aqueous phase as they do in the gas phase (Fig. 2.11(d)). Therefore, the radical propagation resembles the low-NO_x regime in the gas phase. In other words, the fate of peroxy radicals (RO_2) is dominated by reactions with another RO_2 or with a hydroperoxy radical (HO_2), rather than with NO. Peroxy radical chemistry in the aqueous phase is highly complex and has been reviewed in detail by von Sonntag *et al.*[124] The $RO_2 + RO_2$ reaction

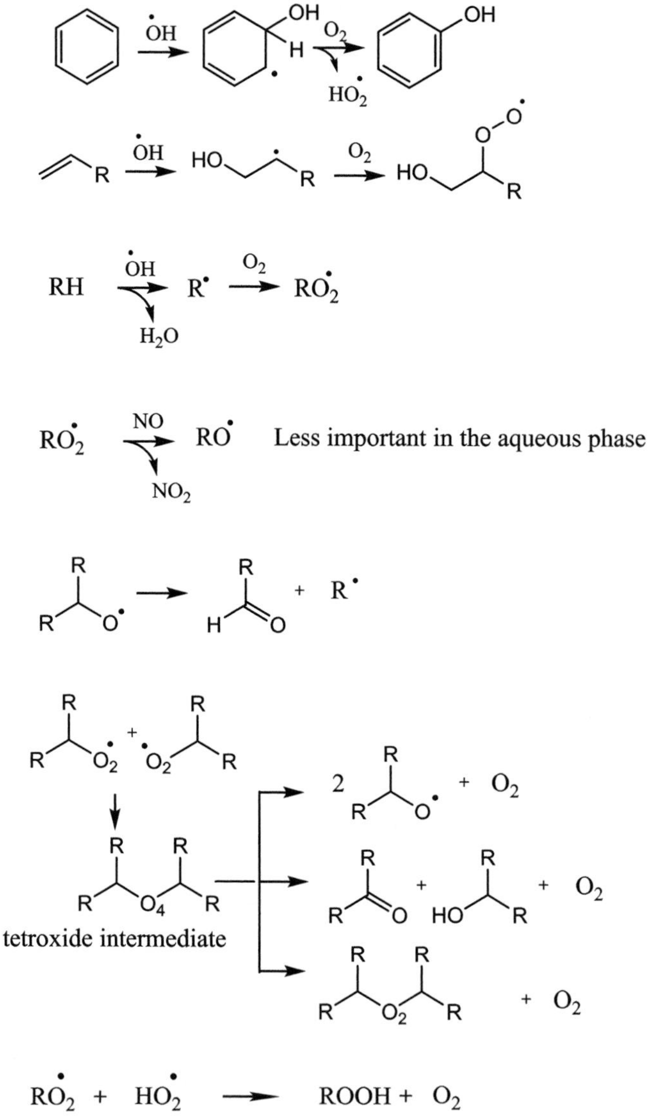

(a) $\bullet OH$; O_2; $HO_2^\bullet$

(b) $\bullet OH$; O_2

$$RH \xrightarrow[H_2O]{\bullet OH} R^\bullet \xrightarrow{O_2} RO_2^\bullet \quad \text{(c)}$$

$$RO_2^\bullet \xrightarrow[NO_2]{NO} RO^\bullet \quad \text{Less important in the aqueous phase} \quad \text{(d)}$$

(e) $\rightarrow \; + \; R^\bullet$

(f) $2 \; + \; O_2$

tetroxide intermediate

(g) $+ \; + \; O_2$

(h) $+ \; O_2$

$$RO_2^\bullet + HO_2^\bullet \longrightarrow ROOH + O_2 \quad \text{(i)}$$

Fig. 2.11 General radical chemistry mechanisms.

can lead to formation of alkoxy radicals which propagate the radical chain (Fig. 2.11(f)).

The $RO_2 + RO_2$ reactions can also terminate the radical chain via the Russell mechanism (Fig. 2.11(g)), giving rise to a carbonyl and an alcohol, as well as via formation of organic peroxides (ROOR) and organic hydroperoxides (ROOH) (Figs. 2.11(h) and 11(i)). In a modeling study, Ervens and Volkamer[15] demonstrated that aerosol liquid water favours formation of ROOR over ROOH. However, their results are significantly dependent on the self-decomposition rate of RO_2 radicals (e.g., Fig. 2.12, conversion of (f)–(g)), which is poorly constrained by experiments.

2.4.1.3. *Mechanisms Unique to the Aqueous Phase*

2.4.1.3.1. Efficient Carboxylic Acid Formation in the Aqueous Phase

Specific differences exist in reaction mechanisms in the aqueous and gas phases. An important case is the effective conversion of

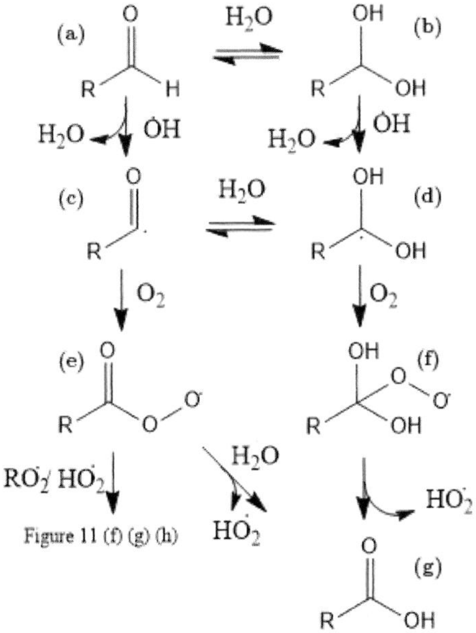

Fig. 2.12 Mechanisms of rapid carboxylic acid formation in the aqueous phase.

aldehydes to carboxylic acids, facilitated by hydration of the aldehyde functional group. Earlier studies have established that formic acid arises from OH oxidation of formaldehyde in the aqueous phase, contributing to cloudwater acidity.[115,125] Formation of a wider variety of carboxylic acids is of relevance due to their contribution to SOA formation.

A series of reactions leading to carboxylic acid formation is shown in Fig. 2.12. An aldehyde exists in equilibrium with its geminal diol in the aqueous phase (Figs. 2.12(a) and 12(b)). OH radicals can abstract a hydrogen atom from both the aldehydic and geminal diol forms, forming their respective alkyl radicals (Figs. 2.12(c) and 12(d)). These two types of alkyl radicals are also in a hydration equilibrium, as shown in the case of acetyl radical.[126] In the next step, oxygen molecules add to the alkyl radicals to form two types of peroxy radicals. The peroxy radical of the geminal diol (Fig. 2.12(f)) dissociates rapidly to form a carboxylic acid (Fig. 2.12(g)). Acylperoxy radicals (Fig. 2.12(e)) can participate in other types of chemistry (i.e., with RO_2 or HO_2) but can be also hydrated to form a carboxylic acid.[127]

Numerous laboratory studies have observed rapid and significant organic acid formation from glyoxal,[128–131] methylglyoxal,[131–133] and glycolaldehyde.[134,135] This formation pathway for carboxylic acid is absent in the gas phase where formation of geminal diols is much less likely. For this reason, organic acids, in particular oxalic acid, are considered tracers for cloudwater processing in field measurements.[136,137] In attempts to model aqueous-phase SOA formation, organic acids are commonly used to trace the yield and mass of aqueous SOA.[93,138,139] Laboratory experiments have also shown that formation of organic acids is likely more important under cloudwater-relevant conditions (i.e., low precursor concentrations), while oligomer formation gains importance in aqueous aerosols where precursor concentrations are high (also see Sec. 2.4.1.3.3).

2.4.1.3.2. Rapid OH Oxidation of Carboxylate

OH oxidation of carboxylate functional groups is more rapid compared to that of the non-dissociated carboxylic acid, resulting in

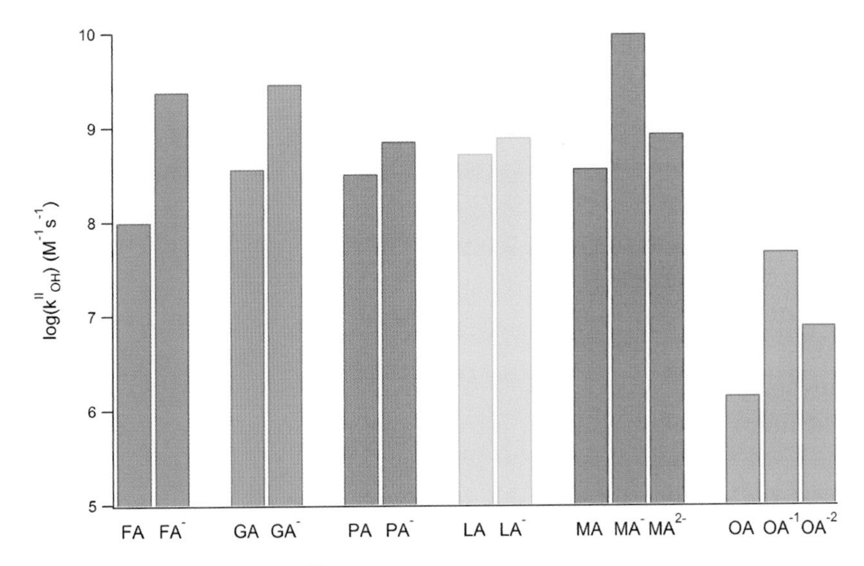

Fig. 2.13 OH reactivity (k_{OH}^{II}) of carboxylic acids and corresponding carboxylates. Acronym: formic acid (FA), glyoxylic acid (GA), pyruvic acid (PA), lactic acid (LA), malic acid (MA), oxalic acid (OA). References: FA, GA and OA,[130] PA,[316] LA and MA.[92]

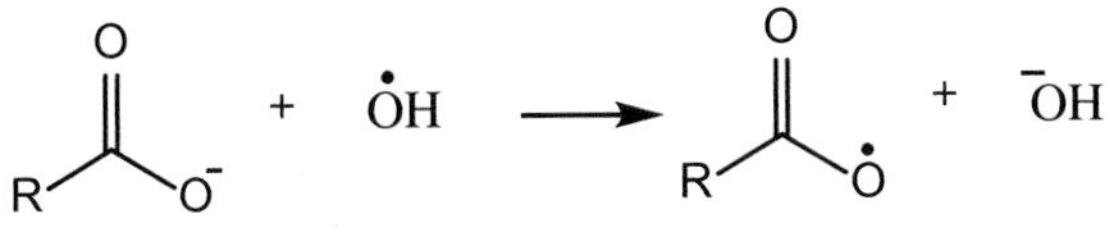

$$RC(O)O^- + {}^{\bullet}OH \longrightarrow RC(O)O^{\bullet} + {}^{-}OH$$

Fig. 2.14 Charge transfer reaction of carboxylate.

a pH dependence in the reactivity of organic acids. Figure 2.13 compiles the reactivity of a suite of carboxylic acids with that of their corresponding carboxylates, and this trend is observed for all the species shown in Fig. 2.13. More rapid OH oxidation of carboxylate is due to a charge transfer reaction (Fig. 2.14) in addition to H-abstraction.[140] The charge transfer reaction explains why the oxalate dianion, a compound without any hydrogen atoms, can also react with the OH radical. An interesting trend for diacids is that the monoanions tend to be the most reactive towards the OH radical.

Sorooshian *et al.*[136] have observed higher cloudwater oxalic acid concentrations in larger and less acidic cloud droplets and proposed that glyoxylic acid, the precursor of oxalic acid, reacts more rapidly in less acidic cloud droplets. This study implies that cloudwater acidity may alter the lifetime of organic acids.

2.4.1.3.3. Oligomer Formation via Radical Chemistry

Organic compounds with molecular weight up to and over 1000 Da are ubiquitous in cloud and fog waters.[113] The sources of such oligomeric compounds, also known as HUmic LIke Substances (HULIS), have remained elusive. There is growing evidence of aqueous-phase radical chemistry inducing oligomerization. We note that non-radical pathways can also give rise to oligomers, and these reactions are summarized in Sec. 2.4.2.

A radical–radical recombination mechanism has been postulated by Turpin and coworkers,[93,130,138] where two carbon-centered radicals form a new covalent bond and give rise to an oligomeric product (Fig. 2.15(a)). In OH oxidation of glyoxal and methylglyoxal, radical–radical recombination gives rise to products with larger carbon numbers than the reactants.[93,130,138] Requiring two carbon-centered radicals, the rate of radical–radical recombination is highly dependent to the reactant concentration. Using a modeling approach, Lim *et al.*[93,138] have shown that radical–radical recombination gains more importance in aqueous particles than in cloudwater. As shown in Fig. 2.16, the contribution of oligomers to the reaction products

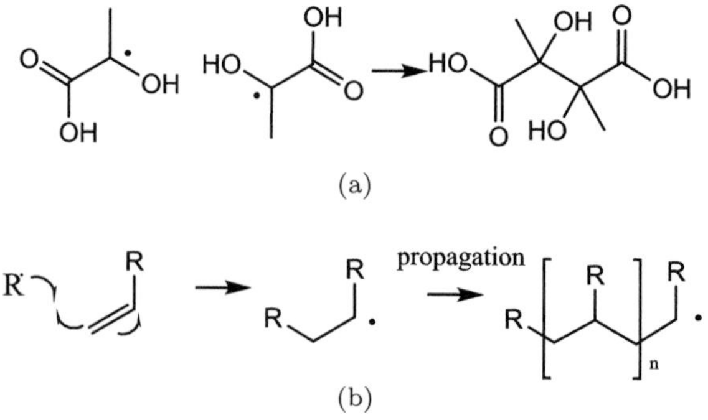

(a)

(b)

Fig. 2.15 Radical induced oligomerization mechanism, with one example each for radical–radical recombination (a) Refs. [93,225] and radical propagation on double bonds (b) Ref. [141].

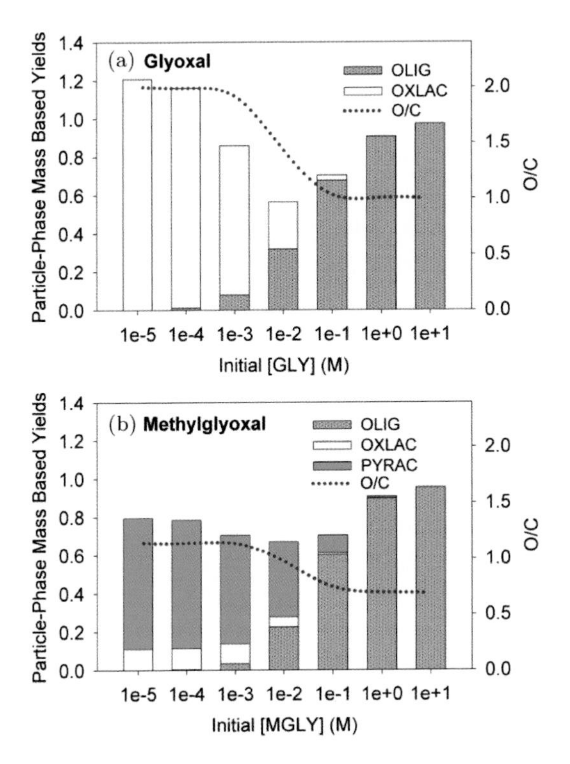

Fig. 2.16 Relative importance of organic acids and oligomers as reaction products of a): glyoxal (GLY) and b): methylglyoxal (MG) photooxidation. The relative contributions of monomeric and oligomeric products as a function of initial concentrations of GLY and MG are shown in the left axis. The average oxygen to carbon ratio (O/C) of the products is shown the right axis. Acronym: oligomers (OLIG), oxalic acid (OXLAC), pyruvic acid (PYRAC). Adapted from Lim *et al.*[138] distributed under the Creative Commons Attribution 3.0 License.

increases dramatically as the initial concentrations of glyoxal and methylglyoxal exceed millimolar levels. This conclusion was later experimentally confirmed by Renard *et al.*[141–143] where the authors found oligomers as the dominant products of methylvinylketone (MVK) OH oxidation when the MVK initial concentrations were 2 mM or higher. When the reactant contains an unsaturated aliphatic structure, radical addition to C=C double bond can proceed, propagating the radical chain and giving rise to oligomers (Fig. 2.15(b)). Renard *et al.*[141,143] proposed this mechanism for the OH oxidation of MVK. Kameel *et al.*[144] also proposed similar mechanisms in the OH oxidation of isoprene.

The radical-induced nature of these reactions has been confirmed by testing the role of dissolved molecular oxygen. Renard *et al.*[141] observed that the presence of oxygen suppresses oligomerization by forming RO_2 radicals with carbon-centered radicals, whereas an enhancement of oligomerization was seen as soon as dissolved oxygen was depleted in their reaction system, confirming the radical nature of the oligomerization.

Radical induced oligomerization has also been observed from aromatic compounds. Oligomerization mechanisms are initiated by both carbon-centered and phenolic radicals.[145–147] Oligomers from aromatic compounds contain extensive electron conjugation and are accompanied by significant light absorptivity,[147,148] which may be responsible for the light absorption observed for HULIS[149] and represent a class of atmospheric Brown Carbon.

2.4.1.3.4. Organosulfate Formation via Radical Chemistry

Radical chemistry can be responsible for organosulfate compounds observed in laboratory experiments and ambient particle samples. While non-radical pathways, such as esterification of alcohols[150,151] and nucleophilic addition of sulfate to epoxide,[152] can also be important, several laboratory experiments[153,154] observed organosulfate formation only under irradiated conditions, indicating the importance of radical chemistry.

Perri *et al.*[154] have proposed a radical–radical recombination mechanism, where an alkyl radical recombines with a sulfate radical to form organosulfates. The sulfate radical arises from H-abstraction of H_2SO_4 by the OH radical. This mechanism is analogous to the oligomerization via radical–radical recombination (Fig. 2.15(a)). More recently, Shindelka *et al.*[155] have proposed a sulfate radical addition mechanism, in analogy to Fig. 2.15(b). They have also shown that the amount of organosulfate formation scales with the number of laser pulses from which the sulfate radical is generated in their experiment. In a modeling study, McNeill *et al.*[139] found that organosulfate formation is especially important in aerosol liquid water, where reactant concentrations are high. As described in Sec. 2.2.2, organic compounds suppress phase transitions of inorganic components and lead to the retention of liquid water even at low RH, providing a high solute concentration environment where organic–inorganic interactions are facilitated. Therefore, organosulfates have been proposed as tracers for chemistry occurring in aqueous particles, as opposed to cloudwater.

2.4.2. *Nucleophilic Addition Reactions*

More than a decade ago, Jang *et al.*[156] proposed that acid-catalyzed nucleophilic addition reactions of carbonyl compounds (e.g., hydration, hemiacetal formation, and aldol condensation) represent a mechanism of particle-phase chemistry that is responsible for additional SOA growth. Nucleophilic addition also plays a central role in the aqueous phase, giving rise to processing of organic compounds via non-radical pathways. A classic example of aqueous-phase nucleophilic addition is the formation of hydroxyalkylsulfonate from S(IV) species and a variety of aldehydes.[157] This group of compounds acts as reservoirs of S(IV) compounds in cloudwater and eventually contributes to cloudwater acidity. In recent years, as the connections between aqueous-phase organic chemistry and the formation of SOA become clearer, a wide variety of nucleophilic addition reactions have been investigated, with important atmospheric implications. These reactions are expected to be particularly relevant in aerosol liquid water where reactant concentrations are orders of magnitude higher than in cloudwater.

While carbonyl compounds have been viewed as the most important electrophiles in atmospheric aqueous phases, epoxides are rapidly gaining attention as an important class of reactive intermediates in SOA formation via condensed-phase chemistry. In this section, nucleophilic addition reactions associated with carbonyl compounds are summarized first, followed by a brief discussion of epoxide chemistry.

2.4.2.1. *General Reaction Mechanism of Nucleophilic Addition to Carbonyls*

Due to the high electronegativity of oxygen, the electron density on a carbonyl group (C=O) is significantly shifted to the oxygen atom. The carbon is hence electrophilic and subject to attack by electron rich nucleophiles (Fig. 2.17(a)). The electrophilicity of the carbonyl group depends on the chemical nature of the adjacent functional groups (R) in terms of both steric and electronic effects.[158] The bulkier and more electron donating the adjacent functional groups, the more they stabilize the carbonyl compound from nucleophilic attack. For this reason, aldehydes (i.e., one of the R groups is a hydrogen) are much more reactive than ketones. Carboxylic acids are less reactive with nucleophiles, stabilized by carboxylate resonance structures, and are not discussed in this section.

2.4.2.2. *Importance of Acid-catalysis*

Since atmospheric aqueous phases are acidic, nucleophilic addition to carbonyl compounds is acid-catalyzed. As illustrated in Fig. 2.17(b), the oxygen on the carbonyl group can be protonated under acidic conditions. This protonated form of carbonyl is in turn in resonance with a carbocation form, as shown in Fig. 2.17(b). The positively charged carbon is even more electrophilic than a neutral carbonyl, hence the carbonyl is "activated" by protonation. Acid catalysis accelerates the reaction kinetics but does not, in principle, affect the equilibrium state. However, given the dynamic nature of organic chemistry in atmospheric aqueous phases, acid catalysis can significantly affect the rates of the products forming in reactions discussed in this section.

While it was generally believed that acid catalysis is initiated by strong acids such as H_2SO_4 and HNO_3,[156] Noziere and coworkers have proposed that amino acids,[76] the ammonium ion[76] and carbonate ion[159] can be important catalysts in atmospheric water. Given that inorganic salts can be at or above their saturation concentrations in aerosol liquid water,[20] catalysis by inorganic ions can be particularly important. Ammonium ion catalysis was

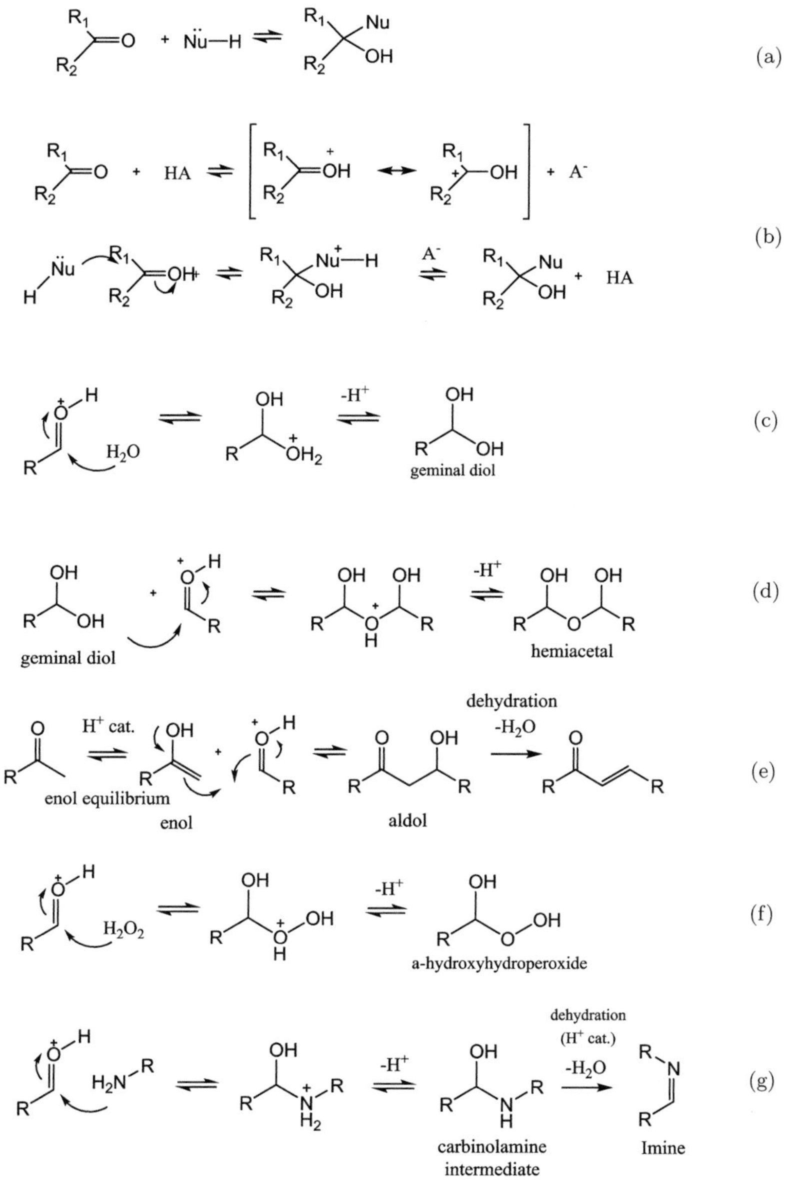

Fig. 2.17 Nucleophilic addition reactions associated with carbonyl compounds.

also observed in nucleophilic reactions of epoxides.[160] Assuming the universality of acid-catalysis in atmospheric aqueous phases, the carbonyl compounds demonstrated in Fig. 2.17 are in their protonated forms.

2.4.2.3. *The Water Nucleophile and Hydration*

Being the most abundant molecule in the aqueous phase, water is the most important nucleophile in atmospheric aqueous phases. Nucleophilic addition of water to carbonyl compounds is often referred to as the hydration reaction and gives rise to geminal diols (Fig. 2.17(c)). As hydration reactions are reversible, all the carbonyl compounds exist, to some extent, in equilibrium with their corresponding geminal diols. The degree of hydration directly affects the effective Henry's law constant (K_{H}^{*}) of an aldehyde, and hence its air-water partitioning.[161] The effects of highly concentrated salts on the hydration equilibria remain poorly constrained. Yu *et al.*[73] have shown that the hydration equilibrium of glyoxal shifts towards its dihydrated form in sodium sulfate and phosphate solutions. Also see Sec. 2.3 for a discussion of the distribution of glyoxal and its hydrates between the gas and the aqueous phases.

Hydration can affect the reactivity of carbonyl compounds in several ways. First, radical-induced oxidation of geminal diols can effectively convert aldehydes into carboxylic acids, as has already been discussed in Sec. 2.4.1.3. Second, geminal diols can act as nucleophiles themselves and add to other carbonyls to form oligomers. This type of reaction will be described in more detail later in this section. Lastly, geminal diol formation is accompanied by loss of the C=O double bond, hence leading to loss of light absorption related to the $n \rightarrow \pi^{*}$ band. This transition has implications for photolytic lifetimes of small aldehydes in the aqueous phase, and will also be discussed in Sec. 2.4.4.1.

2.4.2.4. *Alcohol Nucleophile and Hemiacetal Formation*

As shown in Fig. 2.17(d), hemiacetal formation proceeds via an alcohol functional group acting as the nucleophile. For small aldehydes with large hydration equilibrium constants, e.g., glyoxal, methylglyoxal and formaldehyde, the geminal diols of these compounds act as nucleophiles themselves, resulting in self-oligomerization.[76,93,162–167] Sugars and anhydrosugars contain multiple alcohol functional groups and can undergo hemiacetal formation. For example, the straight chain isomer of glucose, which contains an alcohol and an aldehyde on the two ends of the molecule, undergoes intramolecular hemiacetal formation to form cyclic isomers. In OH oxidation experiments of levoglucosan, a widely employed molecular marker for biomass burning,[168] Holmes and Petrucci[169,170] have shown that the reaction intermediates of levoglucosan oligomerize via hemiacetal formation.

2.4.2.5. *The Enol Nucleophile and Aldol Condensation*

Aldol condensation involves an enol nucleophile which is in equilibrium with its carbonyl form in the aqueous phase (Fig. 2.17(e)). Nucleophilic addition of an enol to an aldehyde first leads to an aldol intermediate, which dehydrates to form the final product. For ketones, or aldehydes that do not preferentially form geminal diols, aldol condensation becomes the major mechanism leading to oligomerization.[167,171,172] A large number of studies have observed a mixture of aldol condensation and hemiacetal formation products from α-dicarbonyl compounds such as glyoxal and methyglyoxal.[163–166,173,174] A noteworthy observation of glyoxal aldol condensation is that a carboxylic acid is produced as the product. This reaction represents an additional formation pathway of organic acids via non-radical chemistry.

2.4.2.6. *Similarities and Differences between Hemiacetal and Aldol Condensates*

Although both hemiacetal formation and aldol condensation have been proposed as important mechanisms leading to oligomer formation in the absence of radical chemistry, fundamental differences exist in the products arising from these two different mechanisms. The major difference is the dehydration step in aldol condensation, leaving the aldol condensates less oxygenated compared to the hemiacetals where all the oxygen atoms are retained. Instead, the dehydration step leaves a double bond on the aldol condensates. As a consequence, aldol condensates contain extensive π-conjugation that can absorb actinic radiation. Noziere and Esteve[172,175] have proposed the connection between aldol condensation products and atmospheric Brown Carbon. More recently, Nguyen *et al.*[176] have also proposed highly conjugated, strongly light absorbing oligomers arising from aldol condensation of limonene SOA extract. The important atmospheric implication arising from the difference between these two mechanisms is that hemiacetals are more oxygenated and less volatile, while aldol condensation may give rise to organic chromophores capable of absorbing actinic radiation.

2.4.2.7. *Hydroperoxide and Peroxyhemiacetal Formation*

Organic hydroperoxides undergo nucleophilic addition to aldehydes to form peroxyhemiacetals.[177] This type of product has been previously proposed to occur at particle surfaces[177–179] and is likely an important mechanism for SOA growth in pristine (i.e., low NO_x) environments.[180] Peroxyhemiacetal formation has been shown to proceed in the aqueous phase as well. Hydrogen peroxide (H_2O_2) can add to carbonyls to form the simplest form of peroxyhemiacetal: α-hydroxyhyderoperoxide (α-HHP) (Fig. 2.17(f)). Formation of α-HHP in the aqueous phase has been known since early studies.[181–183] More recently, α-HHPs from a wider variety of aldehydes, including glyoxal, methylglyoxal, and glycolaldehyde have been observed in laboratory studies.[129,131,184] Although the equilibrium constants of most of these reactions are not large enough to generate a significant amount of α-HHP under concentrations relevant to cloudwater, they may gain importance in aerosol liquid water or during cloud droplet evaporation.[184]

Interestingly, multiple studies have proposed formation of formic acid and acetic acid via α-HHPs. Pyruvic acid reacts with H_2O_2 to form an α-HHP species which decomposes to acetic acid.[185,186] Glyoxylic acid undergoes an analogous reaction to form formic acid.[129,130,184] Zhao *et al.*[131] proposed that α-HHPs arising from glyoxal and methylglyoxal form formic acid and acetic acid, respectively, via OH radical chemistry. Given that the source of formic acid and acetic acid are underestimated in a global chemical transport model,[187] α-HHPs may represent an unknown secondary source of these small organic acids.

Besides nucleophilic addition of H_2O_2 to aldehydes, α-HHPs alternatively form via hydration of Criegee intermediates arising from ozonolysis reactions.[188] In the bulk aqueous phase, α-HHPs arising from ozonolysis decompose to form aldehydes and H_2O_2 (the reverse reaction of Fig. 2.17(f)).[189,190] In chamber experiments, ozonolysis experiments operated under higher RH conditions exhibit higher yields of α-HHPs and H_2O_2 relative to dry conditions.[191,192]

2.4.2.8. *Nitrogen-Containing Nucleophiles and Atmospheric Brown Carbon*

Reduced nitrogen compounds with lone-pair electrons (e.g., ammonia, primary amines, and amino acids) undergo nucleophilic addition to form imines (Fig. 2.17(g)). The reaction proceeds via formation of a neutral carbinolamine intermediate and subsequent acid-catalyzed dehydration to form the final product. While this reaction requires acid-catalysis, excess acid protonates the nucleophiles, suppressing their nucleophilicity.[193,194] Therefore, this reaction is expected to proceed in a narrow pH window.

The crucial atmospheric importance of this reaction is the formation of atmospheric Brown Carbon. Imines can undergo subsequent reactions to form nitrogen-containing organic chromophores which represent a secondary source of Brown Carbon in the aqueous phase.[164,166,195] In particular, glyoxal can undergo serial addition reactions to form imidazole and its derivatives,[73,153,196] and the equivalent products have also been observed from methylglyoxal.[174,197]

Nizkorodov and coworkers have observed Brown Carbon formation from the water-soluble fraction of chamber-generated SOA. Particularly, they found that limonene ozonolysis SOA generated strongly colored products upon aging with nitrogen containing nucleophiles.[176,193,198] On the contrary, SOA generated via limonene OH oxidation or α-pinene ozonolysis did not result in significantly absorbing products.[176,199] Nizkorodov and coworkers proposed that ketolimononaldehyde, a second generation product in limonene ozonolysis is particularly efficient in generating colored products.[176,200]

2.4.2.9. *Competition between Different Nucleophiles*

Ambient atmospheric waters are highly complex, with multiple nucleophiles coexisting. As all the aforementioned reactions, or at least their initial addition steps, are reversible, competition among nucleophiles for addition sites can be expected. In fact, observation of nucleophilic addition products in the aqueous phase is already a sign that non-water nucleophiles are competing with water. Drozd and McNeill[197] have investigated the organic matrix effect in the kinetics of glyoxal and methylglyoxal with $(NH_4)_2SO_4$. They reported suppression of color formation rate with addition of polyols and sugars to the solution as surrogates for water-soluble organic carbon (WSOC). Although this type of competition may not be important in dilute aqueous phases, it may be important in highly concentrated aerosol liquid water or during droplet evaporation (see Sec. 2.4.3).

2.4.2.10. *Reactions of an Emerging Electrophile-Epoxide*

Epoxides have gained increasing attention in recent years as reactive intermediates in SOA formation. Isoprene epoxydiols (IEPOX), in particular, have been proposed as crucial reaction intermediates in SOA formation from isoprene,[88] and their condensed-phase chemistry is driven by a series of nucleophilic addition reactions.[152] Water plays a complicated role in these reactions. Mediated by the aqueous-phase, no reactive uptake of epoxides was observed when the seed aerosols are completely dry.[160] However, increasing LWC of seed aerosol can reduce reactive uptake due to dilution of nucleophile and proton activities.[160,201] Acid-catalyzed ring opening of epoxides plays a crucial role in initiating these reactions, with much more

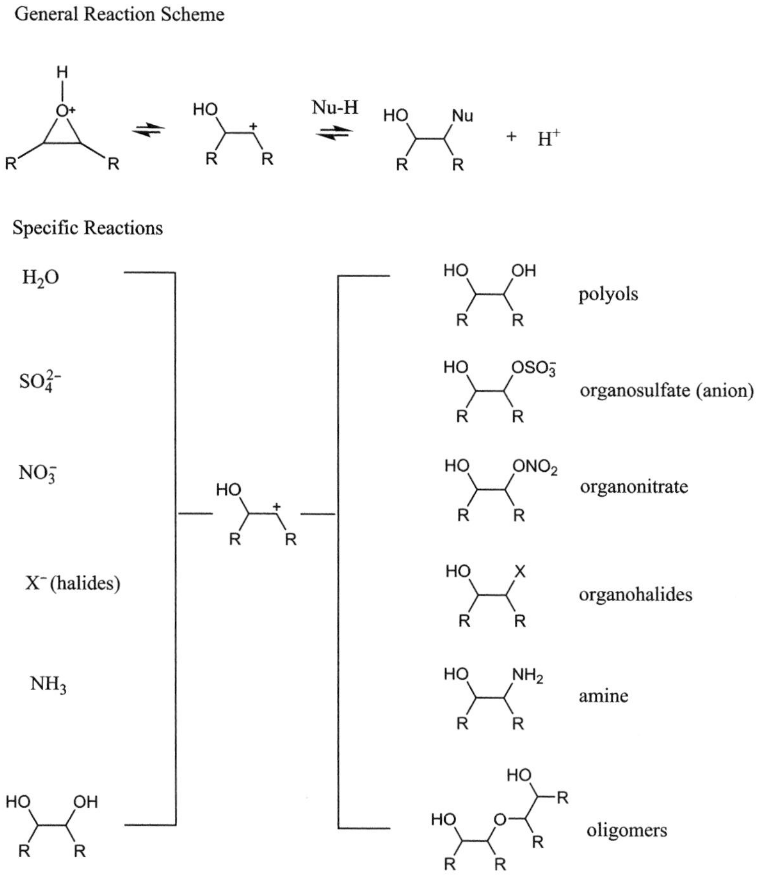

General Reaction Scheme

Nu-H

+ H^+

Specific Reactions

H_2O

SO_4^{2-}

NO_3^-

X^- (halides)

NH_3

polyols

organosulfate (anion)

organonitrate

organohalides

amine

oligomers

Fig. 2.18 Acid-catalyzed nucleophilic additions associated to epoxide.

efficient uptake of epoxides to acidified seed particles.[152,201,202] The general reaction mechanism of epoxide nucleophilic addition is shown in Fig. 2.18.[203] After the oxygen on the epoxide is protonated by an acid, the ring-opening reaction leads to formation of a carbocation to which the nucleophile adds to form the final product.

Specific nucleophiles react with the carbocation intermediate to form a variety of products as summarized in Fig. 2.18. Water can be an important nucleophile (this process is also referred to as hydrolysis of epoxides), forming polyols upon addition to the carbocation intermediate. In particular, IEPOX hydrolysis gives rise to tetrols which have been employed as tracers for isoprene SOA.[204] The kinetics of epoxide hydrolysis have been investigated in detail,[203,205] and a SAR has been developed by Cole-Filipiak *et al.*[205] Sulfate, nitrate and halides can add to epoxides to form organosulfates, organonitrates, and organohalides respectively.[152,202,203,206] Isoprene-derived organosulfates have been detected in ambient particles.[207] The possibility of ammonium addition to form amines has recently been proposed by Nguyen *et al.*[160] The alcohol group from another epoxide or its hydrated polyols can add to form oligomers.[152]

Intramolecular nucleophilic addition reactions, where an alcohol functional group from the same molecule adds to the carbocation intermediate have also been observed by Surratt and coworkers. Lin *et al.*[202] have observed 3-methyltetrahydrofuran-3,4-diols forming from IEPOX. Furthermore, Lin *et al.*[202] have proposed that hydrofuran formation from oligomers of IEPOX can give rise to light absorbing oligomers (i.e., atmospheric Brown Carbon).

Efforts to add epoxide chemistry into atmospheric models have begun. McNeill *et al.*[139] have incorporated IEPOX chemistry into the gas aerosol model for mechanism analysis (GAMMA) to simulate tetrol and organosulfate formation in aqueous aerosol. They observed that aqueous-SOA formation is dominant when the IEPOX chemistry is active (i.e., low NO_x, acidic particles and low RH). Pye *et al.*[208] have incorporated IEPOX chemistry into the community multiscale air quality (CMAQ) model and found an improvement in the predicted total organic carbon budget and the levels of the hydration product of IEPOX, 2-methyltetrol. As aqueous-phase SOA formation from epoxides requires biogenic precursors (e.g., isoprene) and anthropogenic input (i.e., particle acidity and sulfate), this mechanism represents a path by which formation of biogenic SOA is enhanced by anthropogenic emissions.[209]

2.4.3. *Reactions in Evaporating Droplets*

As we have seen in Sec. 2.3, atmospheric aqueous aerosol, fog and cloud droplets can be important sinks of water-soluble, volatile organic compounds. As well, as described in detail in Secs. 2.4.1 and

2.4.2, subsequent aqueous-phase reactions of the partitioned VOCs can lead to formation of low-volatility organic compounds and thus SOA.[10,11,209,210] In addition to photo-oxidative chemistry, molecules can interact strongly inside the shrinking droplets, which can provide a highly concentrated and supersaturated solute environment to speed up various chemical reactions. Atmospheric aerosol can undergo water uptake and evaporation cycles repeatedly in response to the surrounding RH during their lifetime (Figs. 2.1 and 2.2). Although the mechanisms of chemical reactions occurring in an evaporating droplet are essentially the same as those discussed Section 2.4.2, significant acceleration of the reaction rates between molecules inside evaporating droplets is an important mechanism for chemical processing of organic aerosols. In this section, we highlight the potential importance of this emerging research area.

As described above, glyoxal and methylglyoxal have been used as model compounds to study SOA production through aqueous-phase processing[15,93] due to their atmospheric abundance and high K_{H}^{*} values for dissolving in inorganic aerosol droplets (see Section 2.3).[74,211] In particular, a few studies have reported that glyoxal and methylglyoxal can undergo self- and/or cross-oligomerization to produce dimers and trimers upon droplet evaporation.[163,165,174] The active removal of water molecules during droplet evaporation drives the hydration equilibrium of the dicarbonyl monomers toward reactive unhydrated aldehyde structures that promote acetal formation and aldol condensation chemistry.[163,212–214]

De Haan *et al.*[212] performed a series of droplet evaporation experiments with solute concentrations relevant to atmospheric fog and cloud droplets, illustrating that about 33% of glyoxal and 19% of methylglyoxal remains in the aerosol phase mainly due to the self-oligomerization mechanisms. Galloway *et al.*[215] investigated the SOA formation potential of small aldehydes including glyoxal, methylglyoxal, hydroxyacetone upon droplet evaporation, and observed that water evaporation triggers aldehyde reactions that increased residual particle volumes regardless of the presence of $(NH_4)_2SO_4$, methylamine, or glycine. Furthermore, they reported that the amount of organic aerosol mass produced in the evaporating droplets was greater than the sum of all solutes and solvent impurities, indicating the additional presence of trapped water, likely caused by increasing aerosol-phase viscosity due to oligomer formation.

It is crucial to understand potential inorganic–organic interactions and reactions induced/accelerated by droplet evaporation. As described in Sec. 2.4.2.8, nucleophilic attack of ammonia (i.e., in equilibrium with ammonium ion (NH_4^+) in the aqueous phase) on the aldehyde group of glyoxal and methylglyoxal molecules can lead to formation of light absorbing organo-nitrogen species in bulk solutions if given sufficiently long reaction time, on the order of days.[73,76,153,166,174,194,196,216] By comparison, Lee *et al.*[194] show that drying of mixed glyoxal and $(NH_4)_2SO_4$ particles induces rapid formation of light absorbing organo-nitrogen compounds on the order of seconds, with the chemistry similar to that in the bulk solution (Fig. 2.19). They did not observe further light absorption

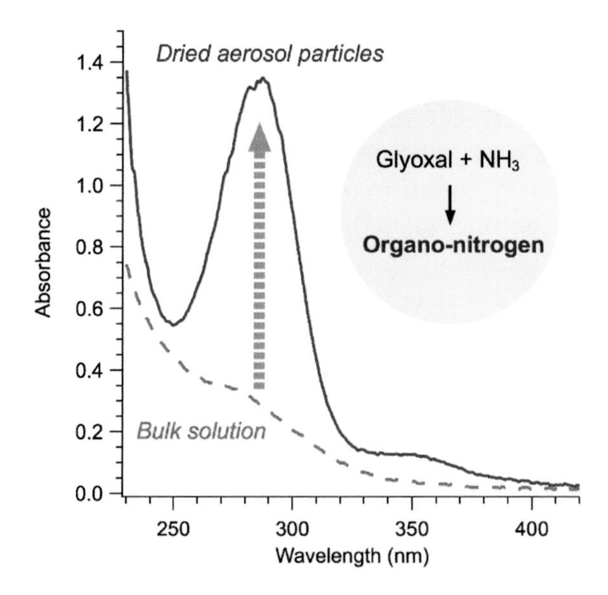

Fig. 2.19 The absorption spectra measured in glyoxal-$(NH_4)_2SO_4$ droplet evaporation experiments. The bulk solution under dilute condition was atomized to generate solution droplets, which were subsequently dried and collected on the filter. Blue solid line: aerosol filter extracts. Red dashed line: bulk solution for aerosol production. Adapted with permission from Lee *et al.*[194] Copyright 2013 American Chemical Society.

enhancement in the dried particles, highlighting the fact that water is an essential medium for the observed reactions inside the evaporating droplets. De Haan *et al.*[174] also reported that reactions of methylglyoxal with $(NH_4)_2SO_4$, amino acids, and methylamine can take place in simulated evaporating droplets to produce organo-nitrogen species and suggested the proposed reactions can be the source of light absorbing materials in the atmosphere.

To understand the impacts of water evaporation on chemical reactions in complex organic mixtures, Nguyen *et al.*[217] showed that water evaporation of SOA extracts generated via ozonolysis of *d*-limonene in the presence of $(NH_4)_2SO_4$ can enhance the formation rate of nitrogen-containing chromophores by at least three orders of magnitude compared to a similar reaction in bulk aqueous solution. Although the molar fraction of chromophores was less than 2%, they made a dramatic enhancement to the effective mass absorption coefficient of the residual organics. Furthermore, evaporation of SOA extracts without $(NH_4)_2SO_4$ resulted in the chromophore formation only when the SOA extract was acidified to pH at 2. The increased concentration of organic aerosol mass and acidity upon evaporation likely facilitates acid-catalyzed aldol condensation that generates water as a reaction product (Fig. 2.17(e)), followed by conversion into organosulfates. Overall, evaporation of atmospheric aqueous droplets containing dissolved complex organic matrices can serve as a potentially important source of light-absorbing compounds.

2.4.4. *Photochemical Reactions*

2.4.4.1. *Direct Photolysis of Carbonyl Compounds*

Direct photochemistry occurs when organic molecules absorb photons and are promoted to an electronically excited state from which the chemical transformation occurs. This section focuses on carbonyl compounds as they are water soluble and photo-reactive, hence comprising a class of important organic chromophores in the atmospheric aqueous phases. Direct photolysis of other important chromophores, such as iron-organic complexes and aromatic compounds, have been reviewed elsewhere.[218,219] Facilitated by an $n \rightarrow \pi^*$ transition, carbonyl compounds can be readily excited by actinic radiation. The subsequent photolysis of carbonyl compounds proceeds via the Norrish type I or type II mechanisms.[220] The Norrish type I mechanism, accompanied by an α-cleavage and formation of two radicals, is more relevant for small carbonyl compounds. The Norrish type II mechanism, which proceeds with a biradical intermediate, gives rise to non-radical products.[221]

In the aqueous phase, direct photolysis of carbonyl compounds can be significantly altered by the hydration reaction (Sec. 2.4.2.3), with the largest effects on small aldehydes whose hydration equilibrium constants are large. Early studies have investigated the case of formaldehyde in detail, as the gas phase photolysis of formaldehyde can be an important source of radicals in the troposphere.[222] The suppression of formaldehyde photolysis in the aqueous phase can affect the global budget of HO_x and O_3.[94] The effect of hydration is significant also for α-dicarbonyl compounds, such as glyoxal. While direct photolysis is the largest sink of glyoxal globally,[223] this sink is essentially removed in the aqueous phase because glyoxal exists almost entirely in its geminal diol forms.[73]

To provide fundamental insight into the contribution of carbonyl compounds to SOA formation and evolution, a number of recent studies have focused on the aqueous-phase photochemistry of individual carbonyl model compounds. In particular, direct photolysis of pyruvic acid has been investigated extensively.[224–226] These studies have observed formation of oligomeric products that do not form in the gas phase.[224,225] The magnitude of oligomer formation is dependent on the concentration of the precursor and dissolved oxygen, similar to the radical-induced oligomerization described in Sec. 2.4.1.3.3.

2.4.4.2. *Direct Photolysis of Complex Organic Compounds*

Direct photochemical aging of complex SOA material in the presence of UV radiation is starting to be investigated. The importance of photolysis as a process that can affect the reactivity of organic aerosol was first suggested in early studies of SOA yield, where exposure to direct UV radiation was observed to decrease α-pinene, *d*-limonene and isoprene SOA yield by as much as 20–40%.[227–229] In particular, studies by Seinfeld and co-workers observed that the fraction of peroxides in isoprene SOA decreased with UV irradiation time, demonstrating that certain species in organic aerosol are photolabile, and their photodegradation can be a significant loss process.[227,230] Studies by Nizkorodov and co-workers also confirmed that the photolysis of carbonyls contributes significantly to the photodegradation of *d*-limonene SOA.[231–233]

The quantum yield can be used to determine the efficiency of photolysis as an aging process for different types of organic aerosols/components. For the carbonyl component in limonene SOA, Bateman *et al.*[231] determined that the photolysis quantum yield was 0.03, where higher photolysis quantum yields were reported for cis-pinonic acid (0.5 ± 0.3; see Ref. [234]), a major compound in α-pinene SOA, and for the photolysis of α-pinene SOA itself (1.2 ± 0.2; (see Ref. [235]). These photolysis quantum yields indicate that photolysis can be an important loss process for specific types of organic aerosol constituents in the atmosphere. Historically, it was thought that photolysis of atmospheric organic aerosol is only significant in the gas phase but recent work by Lignell *et al.*[236] showed that the photolysis quantum yields of cis-pinonic acid are similar in both gas and condensed phases.

The contributions of secondary photochemical reactions to the measured yields described above are unclear. Such secondary photolysis reactions can be important to the reactivity of SOA. For example, the photolysis of peroxides, which are known constituents of SOA[178,231] can be a potential photochemical OH source, as discussed in Sec. 2.4.1.1.2. Using an aqueous radical trap, recent work has demonstrated that the photolysis of aqueous SOA material generated OH radicals. Specifically, relating the observed OH production rates to the composition of SOA, it was found that organohydroperoxides were the likely SOA component that was photolyzed into OH.[111] Formation of H_2O_2 has also been observed in direct photolysis of a variety of organic compounds,[237] arising from the recombination of secondary HO_2 radicals. H_2O_2 can generate the OH radical upon direct photolysis. In addition to the formation of OH, photosensitized reactions, another secondary photolysis process, can be important to SOA formation and aging.[218,219,238–240] The role of water in these processes is not well known.

While the studies discussed previously have demonstrated the importance of photolysis, only a narrow range of RH conditions have been explored experimentally. Knowing that the phase of organic aerosols may be affected by RH, it has been shown that the rate of mass loss from the photolysis of SOA is dependent on RH.[235] Recent work by Lignell *et al.*[234] has also shown that matrix effects, such as changes in viscosity, can have implications for the photolysis of an embedded substrate.

An emerging area of research is the effect of photolysis on the optical properties of organic aerosol, where photobleaching of Brown Carbon species has been observed.[198] While a class of Brown Carbon can be formed via nucleophilic addition in the aqueous phase (detailed discussion in Sec. 2.4.2.8), recent work suggests that these organic chromophores are susceptible to photodegradation.[241,242]

2.4.5. *Effects of Surficial Water on Interfacial Reactivity*

Given that experimental studies and molecular dynamic simulations have indicated that the detailed nature of water adsorbed on organic surfaces depends on the physico-chemical properties of the organic substrates,[243,244] it has long been questioned how surface water affects heterogeneous chemistry taking place at the interface. However, as illustrated below, there is no consensus as yet on the overall effects that arise. Specifically, water may slow down interfacial reactions by occupying adsorption sites on solids, may lead to enhanced partitioning of polar oxidants, and/or change the nature of the organic substrate. While interfacial processes have been investigated extensively in the past for organic layers at the aqueous interface,[245] this section focuses on the effects of water on the reactivity of organic aerosol constituents.

2.4.5.1. *Effects of Water on Heterogeneous Ozonolysis*

In early studies, Pitts *et al.* deposited five polycyclic aromatic hydrocarbons (pyrene, fluoranthene, benz(a)anthracene, benzo(a)pyrene (BaP) and benzo(e)pyrene) on glass fiber filters and ambient particles and exposed them to ozone at different RH.[246] They found no significant RH effects on the PAH degradation for ambient particles, whereas slower decays were observed for all the PAHs on glass fiber filters with RH increasing from 1% to 50%. Later work on the heterogeneous reaction of BaP coated on soot particles with ozone also reported a slower degradation of BaP at 25% RH compared to dry conditions ($<1\%$ RH).[247] Similar observations were also made for the ozonolysis of nicotine films[248,249] and unsaturated organics in meat-grease aerosol.[250] Because ozone heterogeneous reactions frequently proceed via the Langmuir–Hinshelwood mechanism,[75] with initial uptake of a gas phase species on the surface followed by a surface reaction between the adsorbed O_3 and condensed phase organics, the slow heterogeneous ozonation reactions at high RH could be due to competition between ozone and water molecules for surface adsorption sites.[246,247]

In contrast, studies of Kamens *et al.*[251] and McDow *et al.*[252] showed faster degradation of PAHs with ozone at high RH when diluted wood smoke and gasoline combustion emissions were exposed to sunlight in an outdoor chamber. It is possible that this is an effect arising from interfacial chemistry, but reactions in addition to those with O_3, such as those with OH radicals, or a modification of aerosol properties could be the cause of the enhancement (see Sec. 2.4.6). Moreover, it has been shown that the kinetics of the heterogeneous ozonolysis of PAHs deposited on different organic substances were either enhanced (e.g., azelaic acid aerosol particles)[253] or remained unchanged (e.g., anthrathene on Pyrex glass)[254] with RH, demonstrating the variable effect of the substrate on the heterogeneous losses of PAHs.[255] The elevated kinetics on azelaic acid particles at high RH[253] were attributed to the increased partitioning of ozone to water adsorbed on particles, with ozone-surface equilibrium constants increasing by a factor of more than 2 at 72% RH compared to 1% RH. Similar explanations have also been applied to rationalize the differences in the kinetics between heterogeneous ozonolysis of squalene and linoleic acid, with the former showing insignificant effects of RH[256,257] and the latter an enhancement in the reaction rate with RH.[258] Because linoleic acid is more hydrophilic than squalene, it favors water adsorption/absorption at high RH, and so more ozone partitioning results in elevated kinetics at higher RH.[258] Similarly, the higher reactive uptake of ozone on catechol thin films with RH was attributed to hydrogen bonding interactions between adsorbed water and catechol.[259] By contrast, another study of ozone reaction with linoleic acid found no discernible effect of RH on the kinetics.[260] The reason for the different RH dependent kinetics of linoleic acid reactions is not known. Segal-Rosenheimer and Dubowski[261] similarly found no dependence of the kinetics of the heterogeneous reaction of ozone with cypermethrin on RH. The different RH dependency in the loss kinetics between linoleic acid and erucic acid was attributed to the distinct hygroscopicity of these two carboxylic acids and their ozonolysis products.[262]

2.4.5.2. *Effects of Water on Hydrogen Halide Uptake*

Surface water has a variable effect on the uptake of gas phase hydrogen halides to organic substrates. Mass accommodation coefficients for HCl and HBr uptake on a hydrophilic ethylene glycol surface were reduced at high RH.[263] Whereas HBr and HI uptake on the more hydrophobic 1-octanol surface were also reduced at high RH, that for HCl increased with RH.[264] The uptake of organic substances, namely acetic acid, α-pinene, γ-terpinene, *p*-cymene, and 2-methyl-2-hexanol, on 1-octanol, was found to be independent of RH.[264,265] A model has been developed to propose a reasonable explanation for the observed results.[264]

2.4.5.3. *Effects of Water on the Uptake of Nitrogen Dioxide (NO_2)*

The kinetics of the heterogeneous reaction of NO_2 on soot[266] and pyrene surfaces[267] were reported to be RH independent. Contrary to these results, enhanced NO_2 uptake with RH has been observed on a variety of organic substrates, in some cases involving a photosensitized reaction: freshly prepared soot particles,[268] treated soot,[269] perchloric acid surfaces,[270] oxygenated aromatics,[271] humic acid,[272,273] and phenolic species.[274,275] The enhanced uptake of NO_2 could be due either to surface water participating in the reaction of NO_2 with organics through hydrolysis[273,274] or water involved in stabilizing the radical in the photochemical reaction.[275] Adsorbed water on the surface was also suggested to affect the heterogeneous reaction mechanism.[276,277]

2.4.5.4. *Effects of Water on OH Reactive Uptake*

In general, when OH radicals interact with an organic surface, reaction occurs readily with large uptake coefficients, i.e., the diffuso-reactive depth (d_L) is thought to be very short. While OH uptake on hydrophilic organic surfaces, such as glutaric acid, was found to increase with RH,[278,279] OH uptake on other organics, such as paraffin wax, pyrene, soot, hopanes, steranes, *n*-alkanes, and methyl-nitrocatechol particles was either independent of RH[278,279] or decreased with RH.[116,280] One reason discussed for the enhancement in the OH uptake by glutaric acid was that increased adsorbed water led to a larger degree of dissociation of glutaric acid. As pointed out in Sec. 2.4.1.3.2, OH reacts faster with organic acid anions, $RC(O)O^-$, than with their undissociated counterparts.[278,279] Most recently, the RH effect on the heterogeneous reaction of levoglucosan with OH radical was investigated by two research groups, with one reporting a linear decrease in the reaction rate with RH[281] and the other finding an elevated OH uptake with RH.[116] The reason for these contrary observations is not known.

Overall, these studies highlight the complexity of these interfacial processes, where adsorbed water has the potential to block reactive adsorption sites and to enhance partitioning of species to the surface. Moreover, as will be discussed in Sec. 2.4.6, water can also influence the heterogeneous reactions by affecting the phase and morphology of the organic substrate. Overall, there is no consensus on a single dominant role that adsorbed water plays in organic aerosol multiphase chemistry.

2.4.6. *Effects of Bulk Water on Chemical Reactivity*

Sections 2.4.1–2.4.4 focused largely on chemical processing occurring in the aqueous phase of organic aerosol, where water provides a reaction medium for the reactions. This section discusses other water-mediated effects on multiphase reactivity. In particular, water absorbed into the condensed organic phase can affect organic aerosol heterogeneous reactivity by acting as a reactant or by modification of the mobility of organics in the particulate phase.[61]

In the first case, water uptake to a crystalline organic aerosol can lead to deliquescence and formation of condensed water. For example, for the hydrolysis of N_2O_5, it was shown that the uptake coefficient on particles of malonic acid correlated with RH, consistent with water acting as a reactant in heterogeneous hydrolysis reactivity.[282] Absorbed water was also reported to react with Criegee intermediates formed from the heterogeneous reactions of ozone with unsaturated organic substances[283–285] and humic like substances.[286] Later work on N_2O_5 uptake to humic acid and mixed humic acid/$(NH_4)_2SO_4$ particles[287] demonstrated that the increase in N_2O_5 uptake coefficients with RH could not be explained by the water content of the aerosol particles alone. In addition, the mass accommodation of N_2O_5 at the surface was suggested to be the rate-limiting step in the uptake process.[287] Most recently, the study of Gaston *et al.*[288] on N_2O_5 uptake on a range of organics with a range of oxidation states suggested that the diffusivity, solubility and reactivity of N_2O_5 within organic substances depend on both RH and molecular composition of the organic materials.

Water can also affect the heterogeneous reactivity of aerosol particles by modifying in a complex manner the particle viscosity and therefore the diffusivity of substances within aerosol particles. For example, the reaction of ozone with maleic acid particles was found to be RH dependent with almost no reactivity under dry conditions (RH $<5\%$) and a dramatic increase when the RH increased to 50%, well below the deliquescence point for maleic acid.[289] By contrast, ozone uptake by linoleic acid films[260] and oleic acid aerosols[284] was not significantly affected by RH. This distinct RH dependent behavior for the unsaturated acids could be due to the different phases of maleic acid and oleic acid particles. Maleic acid is a crystalline solid at low RH; therefore, its ozonolysis will be confined to the particle surfaces and limited by the very slow diffusion of the reactants into the solid maleic acid particles. Exposure of maleic acid to higher RH resulted in more water in the particulate phase and a faster diffusion of the reactants in the particle bulk, hence an enhancement in the ozone reaction.[283] On the other hand, oleic acid and linoleic acid are liquids under all RH conditions, and so their ozone reactions are not limited by the diffusion of the reactants to the surface.[284]

Recent studies demonstrate that water absorption can modify the viscosity and diffusivity within organic aerosols to the point that the phase of the substrate is significantly transformed. As an example, Shiraiwa *et al.*[290] reported a pronounced increase in ozone uptake by amorphous protein with RH. This was explained by a transformation of the organic matrix from a glassy state to semi-solid state at high RH, leading to a decrease of viscosity and increase of diffusivity. Such moisture-induced phase transitions were also reported to influence the heterogeneous reactions on/within SOA. For example, Kuwata and Martin[291] studied gas phase ammonia uptake by α-pinene SOA and found a higher ammonium content in the particulate phase at high RH, but a greater LWC associated with high RH may also have enhanced the partitioning of ammonia. In another study, enhanced heterogeneous reactivity of BaP coated with SOA towards ozone was observed with increasing RH.[90] By simulating the kinetics with a kinetic multi-layer model for aerosol surface and bulk chemistry,[292] the diffusion coefficients (in $cm^2\,s^{-1}$) for BaP in α-pinene SOA were estimated to be 2×10^{-14}, 8×10^{-14} and $>1 \times 10^{-12}$ for $<5\%$ RH, 50% RH and 70% RH conditions, respectively.[90] For dry conditions, the estimate is a few orders of magnitude higher than other studies, e.g., for pyrene, the diffusion coefficient was estimated to be $2.5 \times 10^{-17}\,cm^2\,s^{-1}$[293] and $<10^{-17}\,cm^2\,s^{-1}$ for organics in

SOA.[31] For humid conditions, the values are in good agreement with the work of Renbaum-Wolff *et al.*[31]

As mentioned above, the photolytic loss of SOA from SOA particles mixed with $(NH_4)_2SO_4$ increased with increasing RH[235] (see Sec. 2.4.4.2). Given that the viscosity of α-pinene SOA decreases with an increasing RH,[31,294] it is possible that primary products from photolysis are recombining more readily in the more viscous environment at low RH, resulting in the observed slower decay of SOA components.

Reinmuth-Selzle *et al.*[295] also reported an effect of absorbed water on the phase and heterogeneous reactivity of organic aerosols; the reaction of allergenic protein with O_3/NO_2 was about an order of magnitude higher for aqueous solutions than for filter samples. Slade and Knopf[116] and Chan *et al.*[296] report an increase in OH uptake by levoglucosan and succinic acid aerosols with increasing RH, respectively, with OH uptake coefficients by levoglucosan increasing by a factor of 3 when the RH increased from 0% to 40%[116] and those by succinic acid being 41 times faster for concentrated aqueous droplets than dry aerosol particles.[296] In each case, it is likely that the oxidant can diffuse somewhat further into the particles with increasing substrate viscosity, enhancing the kinetics.

Overall, it can be concluded that increasing levels of absorbed water lead to increasing reactivity, either by acting as a reactant or by increasing mass transfer rates. In the examples where little effect is observed, the substrates are frequently liquid so that mass transfer is not rate determining.

2.5. Summary: Addressing Chemical Complexity

As illustrated in this review, significant progress has been made in determining the role that water plays in the multiphase chemistry of organic aerosol. In particular, as outlined in Sec. 2.2, hygroscopic growth studies have demonstrated that organic aerosol particles can take up appreciable amounts of water in subsaturated environments and may act as CCN, even in the absence of internally mixed inorganic components. Likewise, ambient measurements of organic aerosol, especially those made by on-line aerosol mass spectrometry, indicate that the degree of oxygenation is very high, consistent with the presence of highly soluble, highly functionalized molecules.[3] Indeed, measurements of WSOC with the particle-into-liquid sampler (PILS) technique indicate that a large degree of SOA is water soluble.[297]

A range of partitioning and reactive processes can occur within such a complex mixture of hydrophilic and hydrophobic organics, mixed with soluble inorganics. As outlined in Sec. 2.3, for a mixed inorganic–organic particle there is a competition between partitioning of gas phase organics into an inorganic-rich aqueous-phase that contains dissolved, highly functionalized organic compounds and an organic-rich phase that also contains small amounts of dissolved water. Under high RH or with highly oxygenated organic constituents, phase separation may not occur, leading to a more compositionally homogeneous particle.[8] Whereas the community has focussed largely on small dicarbonyls, such as glyoxal and methylglyoxal, we point out that much larger, highly functionalized oxidation products may also partition largely to the aqueous phase.

From a reactive perspective, different classes of reactions take place. In particular, radical oxidation occurs, arising from the input of gas phase radicals or from photolytic generation in the particles. Whereas Sec. 2.4.1 focussed exclusively on OH radical chemistry, there is also the possibility that chemistry driven by nitrate radical and organic triplet radical can occur.[92] A major issue though, for all oxidants, is lack of knowledge of their concentration as no on-line measurement methods are currently available.

Because of the very high density of reactive components in the condensed phase, reactions between closed-shell species also proceed that are generally too slow to be of importance in the gas phase. A set of nucleophilic reactions was described in Sec. 2.4.2 that illustrate the immense chemical complexity that can arise, including the formation of oligomers, organosulfates, and organo-nitrogen molecules. A framework for viewing the nucleophilic chemistry that occur in aerosol particles is provided by analogy to relevant cloudwater processes, although the latter proceed at much lower concentrations; products are frequently higher molecular weight species. An example of a subclass of nucleophilic reactions are those that occur when cloudwater conditions rapidly transition to aerosol water conditions, as cloud droplets evaporate when RH drops in the atmosphere. In this case, the rates of these nucleophilic reactions accelerate by orders of magnitude, as described in Sec. 2.4.3.

Lastly, photochemical processes, interfacial and bulk reactions proceed as well, although little is known about their nature in concentrated aerosol particle environments. One effect is that water can modify the nature of the substrate making it more accessible for reaction.

A number of uncertainties permeate the field, driven largely by the inherent chemical complexity. Unlike analogous multiphase chemistry that has been explored with much simpler inorganic aerosol particles, the complexity of organic systems is immense. As a graphic example, Fig. 2.20 is a chemical ionization mass spectrum of the water soluble component of SOA formed in the laboratory

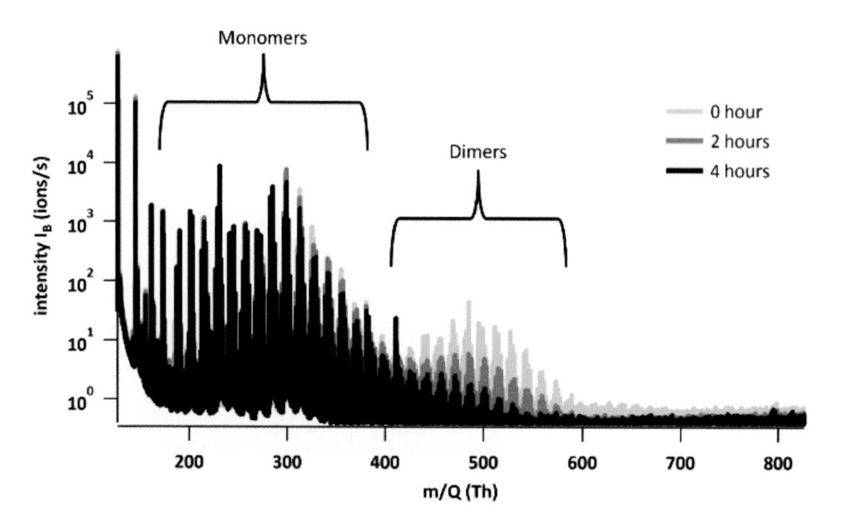

Fig. 2.20 Chemical ionization mass spectra of the water soluble α-pinene SOA that is photo-oxidized by OH radicals in solution, at different times. The technique of aerosol chemical ionization mass spectrometry is employed, using the iodide reagent ion. The spectra clearly illustrate loss of reactants during photo-oxidation. Monomer and dimer regions observed in the spectra are labelled. Adapted from Aljawhary *et al.*,[298] distributed under the Creative Commons Attribution 3.0 License.

by ozonolysis of only one precursor compound, α-pinene.[298] The ionization scheme is sufficiently gentle that each peak — of which there are more than 400 in this spectrum — is likely indicative of an individual compound. Whereas this example is a laboratory spectrum, ambient SOA, formed from many precursors, is much more complex. In addition, chemical processing occurs as the sun rises and sets each day, as other species are condensed onto particles, and as RH undergoes diurnal cycles.

How do we deal with such chemical complexity, in order to move forward towards a better understanding of multiphase chemistry? On the one hand, it is clear that studies with individual compounds can be fundamentally informative but potentially misleading. Consider, for example, the hygroscopic growth of malonic acid, a common dicarboxylic acid. This molecule is sufficiently soluble that it exhibits deliquescence-efflorescence phase transitions and CCN activity that mimic those of $(NH_4)_2SO_4$.[299] But, pure malonic acid particles do not exist in the atmosphere, and when present as only a trace component amongst many other species, malonic acid is unlikely to crystallize and deliquesce. Instead, the collective behavior of a representative mix of organics is more important to establish than the behavior of individual constituents. As described in Sec. 2, such complex mixtures do not exhibit such abrupt phase transitions as occur with individual species.

As well, performing experiments in bulk aqueous solutions may over-simplify the reaction system. While the bulk approach provides fundamental information about the chemical reactions, it does not fully represent the ambient conditions where aqueous phases are in a suspended form. Daumit *et al.*[300] have recently demonstrated that the gas–aqueous partitioning of organic compounds is significantly different between bulk aqueous solutions and suspended aqueous droplets, driven by differences in LWC and affecting SOA yield. In the future, it is important that experiments be conducted using suspended droplets, as has been done already by several recent studies.[235,301]

However, if we go to the other extreme and attempt to characterize the hygroscopicity of all organic aerosol using simple parameters, such as the O/C ratio, then there is interesting behavior that is not captured (see Fig. 2.4 and Sec. 2.2.3). This will be related to the chemical character of the different constituents, i.e., different precursors and their oxidation products will have varying hygroscopicities. Clearly, a balance needs to be established in our laboratory studies and theoretical calculations so that the systems under consideration are sufficiently chemically complex and diverse so as to appropriately describe atmospheric behavior, but not so complex that full characterization becomes intractable.

As far as reactivity is concerned, it is not realistic to map out every elementary reaction occurring within ambient particles, with known rate constants and product yields. The system is too complex for that. As mentioned, we do not accurately know the concentrations of oxidants in particles (see Sec. 2.4.1). One path taken is to follow the general character of changing constituents as chemical processing occurs using the power of advanced analytical instrumentation. A number of approaches have been developed to characterize complex natural organic matter using high resolution mass spectrometric techniques, as has been previously reviewed.[302] For example, van Krevelen diagrams have been used to analyze mass spectral data of organic aerosol, giving insight into processing and sources.[303] Other approaches that have gained applicability include the average carbon oxidation state (OS_c) vs. carbon number diagram,[304] and the Kendrick mass defect plot.[305,306]

To illustrate a more recent approach, consider a recently published example from our group on the aqueous OH oxidation of levoglucosan where we introduce a new mass spectrometric framework for mapping the nature of chemical change in a complex organic oxidation process.[307] In particular, levoglucosan is a polyol that undergoes reactions with OH that first convert alcoholic groups to carbonyl groups. Further functionalization converts carbonyls to carboxylic acids. Such transitions of functional groups have been reflected in a framework (Fig. 2.21), where OS_c is plotted against the average double bond equivalence to carbon number ratio (DBE/#C). The conversion of alcoholic groups to carbonyls is accompanied by an increase in both OS_c and DBE/#C. Therefore, the plot in Fig. 2.21

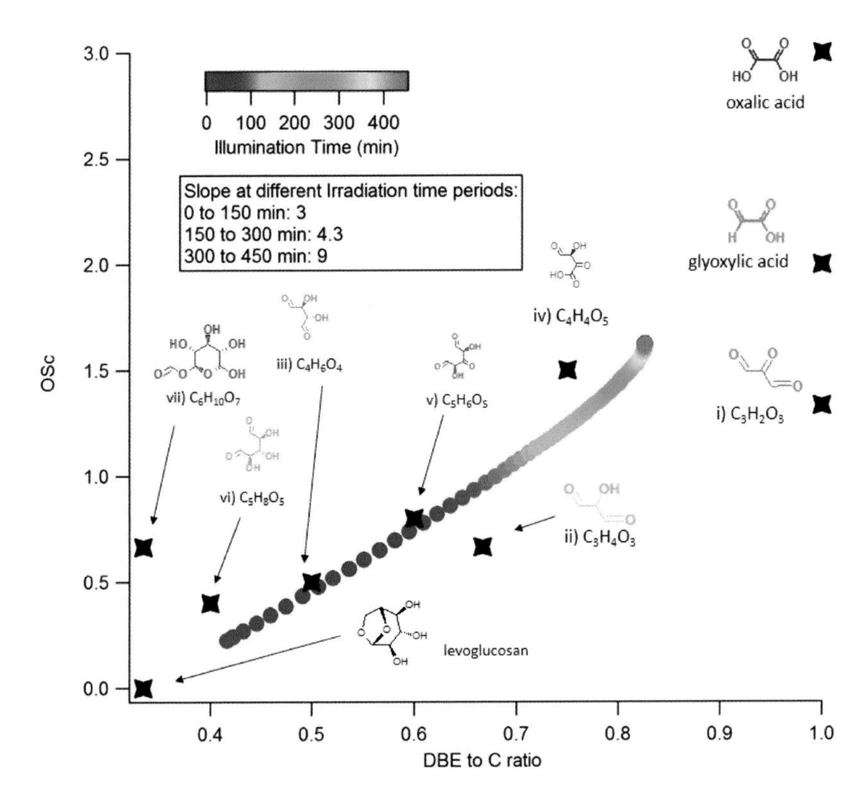

Fig. 2.21 Average carbon oxidation state (OS_c) vs double bond equivalence to carbon number ratio (DBE/#C) plot for a levoglucosan photooxidation experiment. The trajectory of the experiment is shown by the color coded line. The slopes of the trajectory during three experimental increments are shown in the inset text box. The coordinate of proposed major products on the plot is shown by the black markers. Adapted from Zhao *et al.*,[307] distributed under the Creative Commons Attribution 3.0 License.

exhibits rapid movement to the top right corner during the early stages of photooxidation. The later conversion of carbonyl groups to carboxylic groups is accompanied by an increase of OS_C only, with the DBE/#C ratio remaining constant. This transition is reflected in Fig. 2.21 by the gradually steepening slope in the latter stages of the photooxidation. This analysis approach takes advantage of high resolution mass spectrometry and represents a novel method to deduce transitions of functional groups in dynamic chemical systems. In some ways, this mimics the use of infrared absorption spectroscopy of organic aerosol samples, to monitor the changing nature of relevant functional groups.[308]

A second example that we are currently exploring is to establish a framework for characterizing the loss of oligomers in SOA material under oxidative conditions. As demonstrated in Secs. 2.4.1–2.4.3, oligomeric compounds preferentially form in aqueous aerosols and during droplet evaporation. For example, Fig. 2.20 illustrates that the dimers formed from α-pinene oxidation products are quite rapidly converted to monomeric oxidation products upon aqueous OH oxidation. More recently, Romonosky *et al.*[309] have also observed that photo-degradation of oligomeric compounds is a common trend among SOA extracts regardless of their precursors and generation conditions.

These examples illustrate the power of high resolution mass spectrometry to elucidate reaction pathways in such multiphase systems, so long as the character of these complex spectra can be converted into a simple framework. Similarly, computational approaches to characterize thermodynamic properties of organic aerosols can be tuned to handle this complexity. Whereas ab initio quantum mechanical approaches could in principle be used to calculate fundamental chemical properties related to partitioning and reactivity, this approach is unrealistic for all but the simplest chemical systems. Rather, Sec. 2.3 and related references[80,82] illustrate an approach where a hierarchy of modeling approaches has been used to establish the degree to which atmospheric oxygenates will partition to different organic or aqueous phases. Specifically, these methods predict phase partitioning for multi-functional compounds relevant to SOA formation with smaller uncertainties compared with traditionally used vapour pressure-based approaches. Using such methods, it is now feasible to model the behavior of hundreds, if not thousands, of molecules, and graphically illustrate multiphase partitioning, which can be applied to understand SOA formation and partitioning behavior.

Currently lacking from the community are strong connections between the fundamental studies described in this chapter and observations from the field. To make these connections, more emphasis should be placed on the identification of tracer compounds unique to reactive processes that only occur with water present. Ideally, the tracers would be detected using the same technique in both laboratory experiments studying the fundamental processes and relevant field measurements. The observation of specific organosulfates is a nice example of such compounds that form from specific SOA precursors.[310] A second approach is the use of ambient aerosol to better quantify the rates of key multiphase processes, in a laboratory style experiment. As an example, the reactive loss of N_2O_5 (see Secs. 2.4.5 and 2.4.6) has been measured on ambient aerosol under a set of different environmental conditions.[311] These experiments have illustrated that simple parameterizations of the reactivity of inorganic and organic particles that are based solely on laboratory experiments cannot fully explain the behavior of ambient particles of known organic content. There is considerable potential for studying other reactive and hygroscopic behavior of ambient particles under controlled laboratory conditions for a wide range of multiphase processes, whether done in flow tubes,[47,312] environmental chambers or using single particle techniques.

Finally, from a field measurement perspective, isolating key properties of both the gas and condensed phases under varying conditions of RH will provide pragmatic assessments of the potential importance of varying water levels in the particles. This approach was nicely applied to several field measurements conducted recently. A first example is measurement of the partitioning of WSOC between gas and particle phases as a function of RH in both Los Angeles and Atlanta.[313] Aerosol water was not observed to play as large a role in partitioning in Los Angeles as in Atlanta, perhaps reflecting the different VOC mixtures or else a more aged organic aerosol in Atlanta. A second example is a measurement of WSOC in the Sonoran Desert,[314] where the authors observed an enhancement of WSOC during the monsoon months compared to the proceeding dry months. With all the other parameters remaining the same, the enhancement of WSOC likely arises from higher RH and biogenic precursors carried to the Sonoran Desert during the monsoon months.

Acknowledgments

We thank Bob Christensen for helping to compile the references for this chapter.

Abbreviations and Acronyms

BaP	benzo[a]pyrene
CCN	cloud condensation nucleus
CIMS	chemical ionization mass spectrometry
CMAQ	Community Multiscale Air Quality model
CRH	crystallization relative humidity
C_X	equilibrium concentration in phase X
D	diffusion coefficient
DBE	double bond equivalence
d_L	diffuso-reactive length
DOC	dissolved organic carbon
DRH	deliquescence relative humidity
ERH	efflorescence relative humidity
GAMMA	Gas Aerosol Model for Mechanism Analysis
GF	growth factor
HTDMA	hygroscopic tandem differential mobility analyzer
HULIS	humic-like substance
IEPOX	isoprene epoxydiol
K_H	Henry's law constant
K_H^*	effective Henry's law constant
$K_{X/Y}$	partition coefficient between phases X and Y
LWC	liquid water content
MVK	methyl vinyl ketone
OAL	organic aerosol loading
O/C	oxygen to carbon ratio
OH	they hydroxyl radical
OM	organic matter
OS_c	carbon oxidation state
PAHs	poly-aromatic hydrocarbons
PILS	particle-into-liquid sampler
ppLFER	poly-parameter linear free energy relationship

P_X	production rate of species X
r	radius of droplet
RH	relative humidity
ROOH	organic hydroperoxides
ROOR	organic peroxide
SAR	structure-activity relationship
SOA	secondary organic aerosol
τ_X	lifetime with respect to process X
VOCs	volatile organic compounds
WSOC	water-soluble organic carbon
α-HHP	α-hydroxyhydroperoxide
γ	uptake coefficient
κ	kappa hygroscopicity parameter
φ_X	fraction in phase X
ω	mean gas phase velocity
#C	carbon number

References

1. M. Hallquist, J. C. Wenger, U. Baltensperger, Y. Rudich, D. Simpson, M. Claeys, *et al.* The formation, properties and impact of secondary organic aerosol: Current and emerging issues. *Atmos. Chem. Phys.* **9**(14), 5155–5236 (2009).
2. M. Kanakidou, J. H. Seinfeld, S. N. Pandis, I. Barnes, F. J. Dentener, M. C. Facchini, *et al.* Organic aerosol and global climate modelling: A review. *Atmos. Chem. Phys.* **5**, 1053–1123 (2005).
3. J. L. Jimenez, M. R. Canagaratna, N. M. Donahue, A. S. H. Prevot, Q. Zhang, J. H. Kroll, *et al.* Evolution of organic aerosols in the atmosphere. *Science.* **326**(5959), 1525–1529 (2009).
4. J. H. Kroll and J. H. Seinfeld. Chemistry of secondary organic aerosol: Formation and evolution of low-volatility organics in the atmosphere. *Atmos. Environ.* **42**(16), 3593–3624 (2008).
5. A. R. Ravishankara. Heterogeneous and multiphase chemistry in the troposphere. *Science.* **276**(5315), 1058–1065 (1997).
6. T. Koop, J. Bookhold, M. Shiraiwa and U. Pöschl. Glass transition and phase state of organic compounds: Dependency on molecular properties and implications for secondary organic aerosols in the atmosphere. *Phys. Chem. Chem. Phys.* **13**(43), 19238–19255 (2011).
7. M. Shiraiwa, A. Zuend, A. K. Bertram and H. J. Seinfeld. Gas–particle partitioning of atmospheric aerosols: Interplay of physical state, non-ideal mixing and morphology. *Phys. Chem. Chem. Phys.* **15**(27), 11441–11453 (2013).

8. Y. You, L. Renbaum-Wolff, M. Carreras-Sospedra, S. J. Hanna, N. Hiranuma, S. Kamal, *et al.* Images reveal that atmospheric particles can undergo liquid-liquid phase separations. *Proc. Natl. Acad. Sci.* **109**(33), 13188–13193 (2012).
9. A. Zuend and J. H. Seinfeld. Modeling the gas–particle partitioning of secondary organic aerosol: The importance of liquid-liquid phase separation. *Atmos. Chem. Phys.* **12**(9), 3857–3882 (2012).
10. B. Ervens, B. J. Turpin and R. J. Weber. Secondary organic aerosol formation in cloud droplets and aqueous particles (aqSOA): A review of laboratory, field and model studies. *Atmos. Chem. Phys.* **11**(21), 11069–11102 (2011).
11. V. F. McNeill. Aqueous organic chemistry in the atmosphere: Sources and chemical processing of organic aerosols. *Environ. Sci. Technol.* **49**(31), 1237–1244 (2015).
12. Ervens B. Modeling the processing of aerosol and trace gases in clouds and fogs. *Chem. Rev.* **115**(10), 4157–4198 (2015).
13. H. Herrmann, T. Schaefer, A. Tilgner, S. A. Styler, C. Weller, M. Teich, *et al.* Tropospheric aqueous-phase chemistry: Kinetics, mechanisms, and its coupling to a changing gas phase. *Chem. Rev.* **115**(10), 4259–4334 (2015).
14. H. Liao and J. H. Seinfeld. Global impacts of gas phase chemistry-aerosol interactions on direct radiative forcing by anthropogenic aerosols and ozone. *J. Geophys. Res. Atmos.* **110**(D18), (2005), doi:10.1029/2005JD005907.
15. B. Ervens and R. Volkamer. Glyoxal processing by aerosol multiphase chemistry: Towards a kinetic modeling framework of secondary organic aerosol formation in aqueous particles. *Atmos. Chem. Phys.* **10**(17), 8219–8244 (2010).
16. T. K. V. Nguyen, M. D. Petters, S. R. Suda, H. Guo, R. J. Weber and A. G. Carlton. Trends in particle-phase liquid water during the Southern Oxidant and Aerosol Study. *Atmos. Chem. Phys.* **14**(20), 10911–10930 (2014).
17. R. Volkamer, P. J. Ziemann and M. J. Molina. Secondary organic aerosol formation from acetylene (C_2H_2): Seed effect on SOA yields due to organic photochemistry in the aerosol aqueous-phase. *Atmos. Chem. Phys.* **9**(6), 1907–1928 (2009).
18. Y. X. Bian, C. S. Zhao, N. Ma, J. Chen and W. Y. Xu. A study of aerosol liquid water content based on hygroscopicity measurements at high relative humidity in the North China Plain, *Atmos. Chem. Phys.* **14**(12), 6417–6426 (2014).
19. B. Ervens, A. Sorooshian, Y. B. Lim and B. J. Turpin. Key parameters controlling OH-initiated formation of secondary organic aerosol in the aqueous-phase (aqSOA). *J. Geophys. Res. Atmos.* **119**(7), 3997–4016 (2014), doi: 10.1002/2013JD021021.
20. I. Tang, A. Tridico and K. Fung. Thermodynamic and optical properties of sea salt aerosols. *J. Geophys. Res. Atmos.* **102**(D19), 23269–23275 (1997).
21. M. Choi and C. Chan. The effects of organic species on the hygroscopic behaviors of inorganic aerosols. *Environ. Sci. Technol.* **36**(11), 2422–2428 (2002).

22. R. M. Garland, M. E. Wise, M. R. Beaver, H. L. DeWitt, A. C. Aiken, J. L. Jimenez, *et al.* Impact of palmitic acid coating on the water uptake and loss of ammonium sulfate particles. *Atmos. Chem. Phys.* **5**, 1951–1961 (2005).
23. A. A. Zardini, S. Sjogren, C. Marcolli, U. K. Krieger, M. Gysel, E. Weingartner, *et al.* A combined particle trap/HTDMA hygroscopicity study of mixed inorganic/organic aerosol particles. *Atmos. Chem. Phys.* **8**(18), 5589–5601 (2008).
24. C. Marcolli, B. P. Luo and T. Peter. Mixing of the organic aerosol fractions: Liquids as the thermodynamically stable phases. *J. Phys. Chem. A.* **108** (12), 2216–2224 (2004).
25. M. L. Smith, A. K. Bertram and S. T. Martin. Deliquescence, efflorescence, and phase miscibility of mixed particles of ammonium sulfate and isoprene-derived secondary organic material. *Atmos. Chem. Phys.* **12**(20), 9613–9628 (2012).
26. V. Varutbangkul, F. J. Brechtel, R. Bahreini, N. L. Ng, M. D. Keywood, J. H. Kroll, *et al.* Hygroscopicity of secondary organic aerosols formed by oxidation of cycloalkenes, monoterpenes, sesquiterpenes, and related compounds. *Atmos. Chem. Phys.* **6**, 2367–2388 (2006).
27. J. Q. Xiong, M. H. Zhong, C. P. Fang, L. C. Chen and M. Lippmann. Influence of organic films on the hygroscopicity of ultrafine sulfuric acid aerosol. *Environ. Sci. Technol.* **32**(22), 3536–3541 (1998).
28. M. N. Chan and C. K. Chan. Mass transfer effects on the hygroscopic growth of ammonium sulfate particles with a water-insoluble coating. *Atmos. Environ.* **41**(21), 4423–4433 (2007).
29. C. Peng, M. Chan and C. Chan. The hygroscopic properties of dicarboxylic and multifunctional acids: Measurements and UNIFAC predictions. *Environ. Sci. Technol.* **35**(22), 4495–4501 (2001).
30. D. L. Bones, J. P. Reid, D. M. Lienhard and U. K. Krieger. Comparing the mechanism of water condensation and evaporation in glassy aerosol. *Proc. Natl. Acad. Sci.* **109**(29), 11613–11618 (2012).
31. L. Renbaum-Wolff, J. W. Grayson, A. P. Bateman, M. Kuwata, M. Sellier, J. B. Murray, *et al.* Viscosity of α-pinene secondary organic material and implications for particle growth and reactivity. *Proc. Natl. Acad. Sci.* **110**(20), 8014–8019 (2013).
32. A. Virtanen, J. Joutsensaari, T. Koop, J. Kannosto, P. Yli-Pirila, J. Leskinen, *et al.* An amorphous solid state of biogenic secondary organic aerosol particles. *Nature.* **467**(7317), 824–827 (2010).
33. M. N. Chan and C. K. Chan. Mass transfer effects in hygroscopic measurements of aerosol particles. *Atmos. Chem. Phys.* **5**, 2703–2712 (2005).
34. H. R. Pruppacher and J. D. Klett. Microphysics of Clouds and Precipitation. 2nd Ed. (Kluwer Academic Publishers, Boston, MA, 1997).
35. J. Sun and P. A. Ariya. Atmospheric organic and bio-aerosols as cloud condensation nuclei (CCN): A review. *Atmos. Environ.* **40**(5), 795–820 (2006).

36. C. E. Corrigan and T. Novakov. Cloud condensation nucleus activity of organic compounds: A laboratory study. *Atmos. Environ.* **33**(17), 2661–2668 (1999).
37. C. N. Cruz and S. N. Pandis. A study of the ability of pure secondary organic aerosol to act as cloud condensation nuclei. *Atmos. Environ.* **31**(15), 2205–2214 (1997).
38. P. Pradeep Kumar, K. Broekhuizen and J. P. D. Abbatt. Organic acids as cloud condensation nuclei: Laboratory studies of highly soluble and insoluble species. *Atmos. Chem. Phys.* (3), 509–520 (2003).
39. M. D. Petters and S. M. Kreidenweis. A single parameter representation of hygroscopic growth and cloud condensation nucleus activity. *Atmos. Chem. Phys.* **7**(8), 1961–1971 (2007).
40. I. J. George, RY-W. Chang, V. Danov, A. Vlasenko and J. P. D. Abbatt. Modification of cloud condensation nucleus activity of organic aerosols by hydroxyl radical heterogeneous oxidation. *Atmos. Environ.* **43**(32), 5038–5045 (2009).
41. S. L. Clegg, S. Milioto and D. A. Palmer. Osmotic and activity coefficients of aqueous $(NH_4)_2SO_4$ as a function of temperature, and aqueous $(NH_4)_2SO_4$–H_2SO_4 mixtures at 298.15 K and 323.15 K. *J. Chem. Eng. Data.* **41**(3), 455–467 (1996).
42. RY-W. Chang, J. G. Slowik, N. C. Shantz, A. Vlasenko, J. Liggio, S. J. Sjostedt, *et al.* The hygroscopicity parameter (κ) of ambient organic aerosol at a field site subject to biogenic and anthropogenic influences: Relationship to degree of aerosol oxidation. *Atmos. Chem. Phys.* **10**(11), 5047–5064 (2010).
43. J. D. Yakobi-Hancock, L. A. Ladino, A. K. Bertram, J. A. Huffman, K. Jones, W. R. Leaitch, *et al.* CCN activity of size-selected aerosol at a Pacific coastal location. *Atmos. Chem. Phys.* **14**(22), 12307–12317 (2014).
44. K. Broekhuizen, R-W. Chang, W. R. Leaitch, S-M. Li and J. P. D. Abbatt. Closure between measured and modeled cloud condensation nuclei (CCN) using size-resolved aerosol compositions in downtown Toronto. *Atmos. Chem. Phys.* **6**(9), 2513–2524 (2006).
45. P. Massoli, A. T. Lambe, A. T. Ahern, L. R. Williams, M. Ehn, J. Mikkilä, *et al.* Relationship between aerosol oxidation level and hygroscopic properties of laboratory generated secondary organic aerosol (SOA) particles. *Geophys. Res. Lett.* **37**(24), (2010), doi:10.1029/2010GL045258/.
46. A. T. Lambe, T. B. Onasch, P. Massoli, D. R. Croasdale, J. P. Wright, A. T. Ahern, *et al.* Laboratory studies of the chemical composition and cloud condensation nuclei (CCN) activity of secondary organic aerosol (SOA) and oxidized primary organic aerosol (OPOA). *Atmos. Chem. Phys.* **11**(17), 8913–8928 (2011).
47. J. P. S. Wong, A. K. Y. Lee, J. G. Slowik, D. J. Cziczo, W. R. Leaitch, A. Macdonald, *et al.* Oxidation of ambient biogenic secondary organic aerosol by hydroxyl radicals: Effects on cloud condensation nuclei activity. *Geophys. Res. Lett.* **38**(22), (2011), doi:10.1029/2011GL049351.

48. A. M. J. Rickards, R. E. H. Miles, J. F. Davies, F. H. Marshall and J. P. Reid. Measurements of the sensitivity of aerosol hygroscopicity and the κ parameter to the O/C ratio. *J. Phys. Chem. A.* **117**(51), 14120–14131 (2013).
49. M. C. Facchini, M. Mircea, S. Fuzzi and R. J. Charlson. Cloud albedo enhancement by surface-active organic solutes in growing droplets. *Nature.* **401**(6750), 257–259 (1999).
50. M. L. Shulman, R. J. Charlson and E. James Davis. The effects of atmospheric organics on aqueous droplet evaporation. *J. Aerosol. Sci.* **28**(5), 737–752 (1997).
51. K. Broekhuizen, P. P. Kumar and J. P. D. Abbatt. Partially soluble organics as cloud condensation nuclei: Role of trace soluble and surface active species. *Geophys. Res. Lett.* **31**(1), (2004), doi:10.1029/2003GL018203.
52. N. Sareen, A. N. Schwier, T. L. Lathem, A. Nenes and V. F. McNeill. Surfactants from the gas phase may promote cloud droplet formation. *Proc. Natl. Acad. Sci.* **110**(8), 2723–2728 (2013).
53. M. Frosch, N. L. Prisle, M. Bilde, Z. Varga and Kiss G. Joint effect of organic acids and inorganic salts on cloud droplet activation. *Atmos. Chem. Phys.* **11**(8), 3895–3911 (2011).
54. N. L. Prisle, T. Raatikainen, A. Laaksonen and M. Bilde. Surfactants in cloud droplet activation: Mixed organic–inorganic particles. *Atmos. Chem. Phys.* **10**(12), 5663–5683 (2010).
55. R. Sorjamaa, B. Svenningsson, T. Raatikainen, S. Henning, M. Bilde and Laaksonen A. The role of surfactants in Köhler theory reconsidered. *Atmos. Chem. Phys.* **4**(8), 2107–2117 (2004).
56. N. C. Shantz, W. R. Leaitch and P. F. Caffrey. Effect of organics of low solubility on the growth rate of small droplets. *J. Geophys. Res. Atmos.* **108**(D5), (2005), doi: 10.1029/2002JD002540.
57. P. Y. Chuang, R. J. Charlson and J. H. Seinfeld. Kinetic limitations on droplet formation in clouds. *Nature.* **390**(6660), 594–596 (1997).
58. B. Ervens, G. Feingold and S. M. Kreidenweis. Influence of water-soluble organic carbon on cloud drop number concentration. *J. Geophys. Res. Atmos.* **110**(D18), (2005), doi:10.1029/2004JD005634.
59. A. Nenes, S. Ghan, H. Abdul-Razzak, P. Y. Chuang and J. H. Seinfeld. Kinetic limitations on cloud droplet formation and impact on cloud albedo. *Tellus. B.* **53**(2), 133–149 (2001).
60. A. Asa-Awuku and A. Nenes. Effect of solute dissolution kinetics on cloud droplet formation: Extended Köhler theory. *J. Geophys. Res. Atmos.* **112**(D22), (2007), doi:10.1029/2005JD006934.
61. E. Mikhailov, S. Vlasenko, S. T. Martin, T. Koop and U. Pöschl. Amorphous and crystalline aerosol particles interacting with water vapor: Conceptual framework and experimental evidence for restructuring, phase transitions and kinetic limitations. *Atmos. Chem. Phys.* **9**(24), 9491–9522 (2009).
62. G. Feingold and P. Y. Chuang. Analysis of the influence of film-forming compounds on droplet growth: Implications for cloud microphysical processes and climate. *J. Atmos. Sci.* **59**(12), 2006–2018 (2002).

63. J. Podzimek and A. N. Saad. Retardation of condensation nuclei growth by surfactant. *J. Geophys. Res.* **80**(24), 3386–3392 (1975), doi:10.1029/JC080i024p03386.
64. J. P. S. Wong, J. Liggio, S-M., Li, A. Nenes and J. P. D. Abbatt. Suppression in droplet growth kinetics by the addition of organics to sulfate particles. *J. Geophys. Res. Atmos.* **119**(21), 12222–12232 (2014), doi:10.1002/2014JD021689.
65. A. Sorooshian, S. M. Murphy, S. Hersey, H. Gates, L. T. Padro, A. Nenes, *et al.* Comprehensive airborne characterization of aerosol from a major bovine source. *Atmos. Chem. Phys.* **8**(17), 5489–5520 (2008).
66. N. C. Shantz, RY-W. Chang, J. G. Slowik, A. Vlasenko, J. P. D. Abbatt and W. R. Leaitch. Slower CCN growth kinetics of anthropogenic aerosol compared to biogenic aerosol observed at a rural site. *Atmos. Chem. Phys.* **10**(1), 299–312 (2010).
67. T. Raatikainen, A. Nenes, J. H. Seinfeld, R. Morales, R. H. Moore, T. L. Lathem, *et al.* Worldwide data sets constrain the water vapor uptake coefficient in cloud formation. *Proc. Natl. Acad. Sci.* **110**(10), 3760–3764 (2013).
68. N. M. Donahue, J. H. Kroll, S. N. Pandis and A. L. Robinson. A two-dimensional volatility basis set — Part 2: Diagnostics of organic-aerosol evolution. *Atmos. Chem. Phys.* **12**(2), 615–634 (2012).
69. J. R. Odum, T. Hoffmann, F. Bowman, D. Collins, R. C. Flagan and J. H. Seinfeld. Gas/particle partitioning and secondary organic aerosol yields. *Environ. Sci. Technol.* **30**(8), 2580–2585 (1996).
70. J. F. Pankow. An absorption model of the gas/aerosol partitioning involved in the formation of secondary organic aerosol. *Atmos. Environ.* **28**, 189–193 (1994).
71. S. A. K. Häkkinen, V. F. McNeill and I. Riipinen. Effect of inorganic salts on the volatility of organic acids. *Environ. Sci. Technol.* **48**(23), 13718–13726 (2014).
72. C. J. Kampf, E. M. Waxman, J. G. Slowik, J. Dommen, L. Pfaffenberger, A. P. Praplan, *et al.* Effective Henry's law partitioning and the salting constant of glyoxal in aerosols containing sulfate. *Environ. Sci. Technol.* **47**(9), 4236–4244 (2013).
73. G. Yu, A. Bayer, M. Galloway, K. Korshavn, C. Fry and F. Keutsch. Glyoxal in aqueous ammonium sulfate solutions: Products, kinetics and hydration effects. *Environ. Sci. Technol.* **45**(15), 6336–6342 (2011).
74. H. S. S. Ip, X. H. H. Huang and J. Z. Yu. Effective Henry's law constants of glyoxal, glyoxylic acid, and glycolic acid. *Geophys. Res. Lett.* **36**, (2009), doi:10.1029/2008gl036212.
75. J. P. D. Abbatt, A. K. Y. Lee and J. A. Thornton. Quantifying trace gas uptake to tropospheric aerosol: Recent advances and remaining challenges. *Chem. Soc. Rev.* **41**(19), 6555–6581 (2012).
76. B. Noziere, P. Dziedzic and A. Cordova. Products and kinetics of the liquid-phase reaction of glyoxal catalyzed by ammonium ions (NH_4^+). *J. Phys. Chem. A.* **113**(1), 231–237 (2009).

77. H. P. H. Arp, R. P. Schwarzenbach and K. U. Goss. Ambient gas/particle partitioning. 1. Sorption mechanisms of apolar, polar, and ionizable organic compounds. *Environ. Sci. Technol.* **42**(15), 5541–5547 (2008).
78. R. J. Griffin, K. Nguyen, D. Dabdub and J. H. Seinfeld. A coupled hydrophobic-hydrophilic model for predicting secondary organic aerosol formation. *J. Atmos. Chem.* **44**(2), 171–190 (2003).
79. J. F. Pankow, N. Niakan and W. E. Asher. Combinatorial variation of structure in considerations of compound lumping in one- and two-dimensional property representations of condensable atmospheric organic compounds. 1. Lumping by 1-D volatility with n(C) fixed. *Atmos. Environ.* **80**, 172–183 (2013).
80. F. Wania, Y. D. Lei, C. Wang, J. P. D. Abbatt and K. U. Goss. Novel methods for predicting gas–particle partitioning during the formation of secondary organic aerosol. *Atmos. Chem. Phys.* **14**(23), 13189–13204 (2014).
81. A. Zuend, C. Marcolli, T. Peter and J. H. Seinfeld. Computation of liquid-liquid equilibria and phase stabilities: implications for RH-dependent gas/particle partitioning of organic–inorganic aerosols. *Atmos. Chem. Phys.* **10**(16), 7795–7820 (2010).
82. F. Wania, Y. D. Lei, C. Wang, J. P. D. Abbatt and K-U. Goss. Using the chemical equilibrium partitioning space to explore factors influencing the phase distribution of compounds involved in secondary organic aerosol formation. *Atmos. Chem. Phys.* **15**(6), 3395–3412 (2015).
83. S. Endo, A. Pfennigsdorff and K. U. Goss. Salting-Out effect in aqueous NaCl solutions: Trends with size and polarity of solute molecules. *Environ. Sci. Technol.* **46**(3), 1496–1503 (2012).
84. C. Wang, Y. D. Lei, S. Endo and F. Wania. Measuring and modeling the salting-out effect in ammonium sulfate solutions. *Environ. Sci. Technol.* **48**(22), 13238–13245 (2014).
85. Y. D. Lei and F. Wania. Is rain or snow a more efficient scavenger of organic chemicals? *Atmos. Environ.* **38**(22), 3557–3571 (2014).
86. R. Valorso, B. Aumont, M. Camredon, T. Raventos-Duran, C. Mouchel-Vallon, N. L. Ng, *et al.* Explicit modelling of SOA formation from α-pinene photooxidation: sensitivity to vapour pressure estimation. *Atmos. Chem. Phys.* **11**(14), 6895–6910 (2011).
87. C. Mouchel-Vallon, P. Bräuer, M. Camredon, R. Valorso, S. Madronich, H. Herrmann, *et al.* Explicit modeling of volatile organic compounds partitioning in the atmospheric aqueous-phase. *Atmos. Chem. Phys.* **13**(2), 1023–1037 (2013).
88. F. Paulot, J. D. Crounse, H. G. Kjaergaard, A. Kurten, J. M. S. Clair, J. H. Seinfeld, *et al.* Unexpected epoxide formation in the gas phase photooxidation of isoprene. *Science.* **325**(5941), 730–733 (2009).
89. R. Volkamer, F. S. Martini, L. T. Molina, D. Salcedo, J. L. Jimenez and M. J. Molina. A missing sink for gas phase glyoxal in Mexico City: Formation of secondary organic aerosol. *Geophys. Res. Lett.* **34**(19), (2007), doi:10.1029/2007gl030752.

90. S. M. Zhou, M. Shiraiwa, R. D. McWhinney, U. Poschl and J. P. D. Abbatt. Kinetic limitations in gas–particle reactions arising from slow diffusion in secondary organic aerosol. *Faraday Discuss.* **165**, 391–406 (2013).
91. M. Shiraiwa and J. H. Seinfeld. Equilibration timescale of atmospheric secondary organic aerosol partitioning. *Geophys. Res. Lett.* **39**(24), (2012), doi:10.1029/2012GL054008.
92. H. Herrmann, D. Hoffmann, T. Schaefer, P. Braeuer and A. Tilgner. Tropospheric aqueous-phase free-radical chemistry: Radical sources, spectra, reaction kinetics and prediction tools. *Chem. Phys. Chem.* **11**(18), 3796–3822 (2010).
93. Y. B. Lim, Y. Tan, M. J. Perri, S. P. Seitzinger and B. J. Turpin. Aqueous chemistry and its role in secondary organic aerosol (SOA) formation. *Atmos. Chem. Phys.* **10**(21), 10521–10539 (2010).
94. J. Lelieveld and P. Crutzen. The role of clouds in Tropospheric Photochemistry. *J. Atmos. Chem.* **12**(3), 229–267 (1991).
95. A. Albinet, C. Minero and D. Vione. Photochemical generation of reactive species upon irradiation of rainwater: Negligible photoactivity of dissolved organic matter. *Sci. Total Environ.* **408**(16), 3367–3373 (2010).
96. C. Anastasio and K. G. McGregor. Chemistry of fog waters in California's Central Valley: 1. In situ photoformation of hydroxyl radical and singlet molecular oxygen. *Atmos. Environ.* **35**(6), 1079–1089 (2001).
97. T. Arakaki and B. C. Faust. Sources, sinks, and mechanisms of hydroxyl radical (OH) photoproduction and consumption in authentic acidic continental cloud waters from Whiteface Mountain, New York: The role of the Fe (r) (r = II, III) photochemical cycle. *J. Geophys. Res. Atmos.* **103**(D3), 3487–3504 (1998), doi: 10.1029/97JD02795.
98. B. C. Faust and J. M. Allen. Aqueous-phase photochemical formation of hydroxyl radical in authentic cloudwaters and fogwaters. *Environ. Sci. Technol.* **27**(6), 1221–1224 (1993).
99. Y. Zuo. Light-induced formation of hydroxyl radicals in fog waters determined by an authentic fog constituent, hydroxymethanesulfonate. *Chemosphere.* **51**(3), 175–179 (2003).
100. C. Anastasio and J. T. Newberg. Sources and sinks of hydroxyl radical in sea-salt particles. *J. Geophys. Res. Atmos.* **112**(D10), (2007), doi:10.1029/2006JD008061.
101. C. Anastasio and A. L. Jordan. Photoformation of hydroxyl radical and hydrogen peroxide in aerosol particles from Alert, Nunavut: Implications for aerosol and snowpack chemistry in the Arctic. *Atmos. Environ.* **38**(8), 1153–1166 (2004).
102. T. Arakaki, Y. Kuroki, K. Okada, Y. Nakama, H. Ikota, M. Kinjo, *et al.* Chemical composition and photochemical formation of hydroxyl radicals in aqueous extracts of aerosol particles collected in Okinawa, Japan. *Atmos. Environ.* **40**(25), 4764–4774 (2006).
103. S. N. Nomi, H. Kondo and H. Sakugawa. Photoformation of OH radical in water-extract of atmospheric aerosols and aqueous solution of water-soluble

gases collected in Higashi-Hiroshima, *Japan. Geochem. J.* **46**(1), 21–29 (2012).
104. X. Zhou, A. J. Davis, D. J. Kieber, W. C. Keene, J. R. Maben, H. Maring, *et al.* Photochemical production of hydroxyl radical and hydroperoxides in water extracts of nascent marine aerosols produced by bursting bubbles from Sargasso seawater. *Geophys. Res. Lett.* **35**(20), (2008), doi: 10.1029/2008GL035418.
105. D. Jacob. Chemistry of OH in Remove Clouds and Its Role in the Production of Formic-Acid Peroxymonosulfate. *J. Geophys. Res. Atmos.* **91**(D9), 9807–9826 (1986), doi: 10.1029/JD091iD09p09807.
106. W. McElroy. Sources of hydrogen peroxide in cloudwater. *Atmos. Environ.* **20**(3), 427–438 (1986).
107. K. Sehested, J. Holcman, E. Bjergbakke and E. J. Hart. A pulse radiolytic study of the reaction hydroxyl+ ozone in aqueous medium. *J. Phys. Chem.* **88**(18), 4144–4147 (1984).
108. S. Goldstein, D. Aschengrau, Y. Diamant and J. Rabani. Photolysis of aqueous H_2O_2: quantum yield and applications for polychromatic UV actinometry in photoreactors. *Environ. Sci. Technol.* **41**(21), 7486–7490 (2007).
109. R. Zellner, M. Exner and H. Herrmann. Absolute OH quantum yields in the laser photolysis of nitrate, nitrite and dissolved H_2O_2 at 308 and 351 nm in the temperature-range 278–353 K. *J. Atmos. Chem.* 411–425 (1990).
110. K. Mopper and X. Zhou. Hydroxyl radical photoproduction in the sea and its potential impact on marine processes. *Science.* **250**(4981), 661–664 (1990).
111. K. M. Badali, S. Zhou, D. Aljawhary, M. Antiñolo, W. J. Chen, A. Lok, *et al.* Formation of hydroxyl radicals from photolysis of secondary organic aerosol material. *Atmos. Chem. Phys.* **15**(14), 7831–7840 (2015).
112. D. R. Hanson, A. Ravishankara and S. Solomon. Heterogeneous reactions in sulfuric acid aerosols: A framework for model calculations. *J. Geophys. Res. Atmos.* **99**(D2), 3615–3629 (1994), doi: 10.1029/93JD02932.
113. P. Herckes, K. T. Valsaraj and J. L. Collett Jr. A review of observations of organic matter in fogs and clouds: Origin, processing and fate. *Atmospheric. Res.* **132**, 434–449 (2013).
114. T. Arakaki, C. Anastasio, Y. Kuroki, H. Nakajima, K. Okada, Y. Kotani, *et al.* A general scavenging rate constant for reaction of hydroxyl radical with organic carbon in atmospheric waters. *Environ. Sci. Technol.* **47**(15), 8196–8203 (2013).
115. J. H. Seinfeld and S. N. Pandis. Atmospheric chemistry and physics: From air pollution to climate change. 2nd Ed., pp. 1203. (Wiley, Hoboken, N.J. 2006).
116. J. H. Slade and D. A. Knopf. Multiphase OH oxidation kinetics of organic aerosol: The role of particle phase state and relative humidity. *Geophys. Res. Lett.* **41**(14), 5297–5306 (2014), doi: 10.1002/2014GL060582.
117. S. Zhou, A. Lee, R. McWhinney and J. Abbatt. Burial effects of organic coatings on the heterogeneous reactivity of particle-borne benzo [a] pyrene (BaP) toward ozone. *J. Phys. Chem. A.* **116**(26), 7050–7056 (2012).

118. G. V. Buxton, C. L. Greenstock, W. P. Helman and A. B. Ross. Critical review of rate constants for reactions of hydrated electrons, hydrogen atoms and hydroxyl radicals ($\cdot$OH/ $\cdot$O$^-$ in aqueous solution. *J. Phys. Chem. Ref. Data.* **17**(2), 513–886 (1988).
119. J-F. Doussin and A. Monod. Structure-activity relationship for the estimation of OH-oxidation rate constants of carbonyl compounds in the aqueous-phase. *Atmos. Chem. Phys.* **13**(23), 11625–11641 (2013).
120. A. Monod and J. Doussin. Structure-activity relationship for the estimation of OH-oxidation rate constants of aliphatic organic compounds in the aqueous-phase: Alkanes, alcohols, organic acids and bases. *Atmos. Environ.* **42**(33), 7611–7622 (2008).
121. A. Monod, L. Poulain, S. Grubert, D. Voisin and H. Wortham. Kinetics of OH-initiated oxidation of oxygenated organic compounds in the aqueous-phase: New rate constants, structure-activity relationships and atmospheric implications. *Atmos. Environ.* **39**(40), 7667–7688 (2005).
122. Y. Wang, J. Chen, X. Li, S. Zhang and X. Qiao. Estimation of aqueous-phase reaction rate constants of hydroxyl radical with phenols, alkanes and alcohols. *QSAR Comb. Sci.* **28**(11–12), 1309–1316 (2009).
123. E. S. C. Kwok and R. Atkinson. Estimation of hydroxyl radical reaction-rate constants for gas phase organic-compounds using a structure-reactivity relationship — an update. *Atmos. Environ.* **29**(14), 1685–1695 (1995).
124. C. Von Sonntag, P. Dowideit, X. Fang, R. Mertens, X. Pan, M. N. Schuchmann, *et al.* The fate of peroxyl radicals in aqueous solution. *Water Sci. Technol.* **35**(4), 9–15 (1997).
125. W. L. Chameides. The photochemistry of a remote marine stratiform cloud. *J. Geophys. Res. Atmos.* **89**(D3), 4739–4755 (1984), doi: 10.1029/JD089iD03p04739.
126. M. N. Schuchmann and C. von Sonntag. The rapid hydration of the acetyl radical — A pulse radiolosys study of acetaldehyde in aqueous-solution. *J. Am. Chem. Soc.* **110**(17), 5698–5701 (1988).
127. P. W. Villalta, E. R. Lovejoy and D. R. Hanson. Reaction probability of peroxyacetyl radical on aqueous surfaces. *Geophys. Res. Lett.* **23**(14), 1765–1768 (1996), doi: 10.1029/96GL01286.
128. A. G. Carlton, B. J. Turpin, K. E. Altieri, S. Seitzinger, A. Reff, H. J. Lim, *et al.* Atmospheric oxalic acid and SOA production from glyoxal: Results of aqueous photooxidation experiments. *Atmos. Environ.* **41**(35), 7588–7602 (2007).
129. A. K. Lee, R. Zhao, S. S. Gao and J. P. D. Abbatt. Aqueous-phase OH oxidation of glyoxal: Application of a novel analytical approach employing aerosol mass spectrometry and complementary off-line techniques. *J. Phys. Chem. A.* **115**(38), 10517–10526 (2011).
130. Y. Tan, M. J. Perri, S. P. Seitzinger and B. J. Turpin. Effects of precursor concentration and acidic sulfate in aqueous glyoxal-oh radical oxidation and implications for secondary organic aerosol. *Environ. Sci. Technol.* **43**(21), 8105–8112 (2009).

131. R. Zhao, A. K. Y. Lee and J. P. D. Abbatt. Investigation of aqueous-phase photooxidation of glyoxal and methylglyoxal by Aerosol Chemical Ionization Mass Spectrometry: Observation of hydroxyhydroperoxide formation. *J. Phys. Chem. A.* **116**(24), 6253–6263 (2012).
132. K. E. Altieri, S. P. Seitzinger, A. G. Carlton, B. J. Turpin, G. C. Klein and A. G. Marshall. Oligomers formed through in-cloud methylglyoxal reactions: Chemical composition, properties, and mechanisms investigated by ultra-high resolution FT-ICR mass spectrometry. *Atmos. Environ.* **42**(7), 1476–1490 (2008).
133. Y. Tan, Y. Lim, K. Altieri, S. Seitzinger and B. Turpin. Mechanisms leading to oligomers and SOA through aqueous photooxidation: Insights from OH radical oxidation of acetic acid and methylglyoxal. *Atmos. Chem. Phys.* **12**(2), 801–813 (2012).
134. D. L. Ortiz-Montalvo, Y. B. Lim, M. J. Perri, S. P. Seitzinger and B. J. Turpin. Volatility and yield of glycolaldehyde SOA formed through aqueous photochemistry and droplet evaporation. *Aerosol. Sci. Technol.* **46**(9), 1002–1014 (2012).
135. M. J. Perri, S. Seitzinger and B. J. Turpin. Secondary organic aerosol production from aqueous photooxidation of glycolaldehyde: Laboratory experiments. *Atmos. Environ.* **43**(8), 1487–1497 (2009).
136. A. Sorooshian, M-L. Lu, F. J. Brechtel, H. Jonsson, G. Feingold, R. C. Flagan, *et al.* On the source of organic acid aerosol layers above clouds. *Environ. Sci. Technol.* **41**(13), 4647–4654 (2007).
137. A. Sorooshian, N. L. Ng, A. W. Chan, G. Feingold, R. C. Flagan and J. H. Seinfeld. Particulate organic acids and overall water-soluble aerosol composition measurements from the 2006 Gulf of Mexico Atmospheric Composition and Climate Study (GoMACCS). *J. Geophys. Res. Atmos.* **112**(D13), (2007), doi: 10.1029/2007jd008537.
138. Y. B. Lim, Y. Tan and B. J. Turpin. Chemical insights, explicit chemistry, and yields of secondary organic aerosol from OH radical oxidation of methylglyoxal and glyoxal in the aqueous-phase. *Atmos. Chem. Phys.* **13**(17), 8651–8667 (2013).
139. V. F. McNeill, J. L. Woo, D. D. Kim, A. N. Schwier, N. J. Wannell, A. J. Sumner, *et al.* Aqueous-phase secondary organic aerosol and organosulfate formation in atmospheric aerosols: A modeling study. *Environ. Sci. Technol.* **46**(15), 8075–8081 (2012).
140. B. Ervens, S. Gligorovski and H. Herrmann. Temperature-dependent rate constants for hydroxyl radical reactions with organic compounds in aqueous solutions. *Phys. Chem. Chem. Phys.* **5**(9), 1811–1824 (2003).
141. P. Renard, F. Siekmann, A. Gandolfo, J. Socorro, G. Salque, S. Ravier, *et al.* Radical mechanisms of methyl vinyl ketone oligomerization through aqueous-phase OH-oxidation: On the paradoxical role of dissolved molecular oxygen. *Atmos. Chem. Phys.* **13**(13), 6473–6491 (2013).
142. P. Renard, F. Siekmann, G. Salque, C. Demelas, B. Coulomb, L. Vassalo, *et al.* Aqueous-phase oligomerization of methyl vinyl ketone through

photooxidation–Part 1: Aging processes of oligomers. *Atmos. Chem. Phys.* **15**(1), 21–35 (2015).
143. P. Renard, A. E. Reed Harris, R. J. Rapf, S. Ravier, C. Demelas, B. Coulomb, *et al.* Aqueous-phase oligomerization of methyl vinyl ketone by atmospheric radical reactions. *J. Phys. Chem. C.* **118**(50), 29421–29430 (2014).
144. F. R. Kameel, M. R. Hoffmann and A. J. Colussi. OH radical-initiated chemistry of isoprene in aqueous media. Atmospheric implications. *J. Phys. Chem. A.* **117**(24), 5117–5123 (2013).
145. J. D. Smith, V. Sio, L. Yu, Q. Zhang and C. Anastasio. Secondary organic aerosol production from aqueous reactions of atmospheric phenols with an organic triplet excited state. *Environ. Sci. Technol.* **48**(2), 1049–1057 (2014).
146. Y. Sun, Q. Zhang, C. Anastasio and J. Sun. Insights into secondary organic aerosol formed via aqueous-phase reactions of phenolic compounds based on high resolution mass spectrometry. *Atmos. Chem. Phys.* **10**(10), 4809–4822 (2010).
147. L. Yu, J. Smith, A. Laskin, C. Anastasio, J. Laskin and Q. Zhang. Chemical characterization of SOA formed from aqueous-phase reactions of phenols with the triplet excited state of carbonyl and hydroxyl radical. *Atmos. Chem. Phys.* **14**(24), 13801–13816 (2014).
148. J. L. Chang and J. E. Thompson. Characterization of colored products formed during irradiation of aqueous solutions containing H_2O_2 and phenolic compounds. *Atmos. Environ.* **44**(4), 541–515 (2010).
149. E. Graber and Y. Rudich. Atmospheric HULIS: How humic-like are they? A comprehensive and critical review. *Atmos. Chem. Phys.* 729–753 (2006).
150. J. D. Surratt, J. H. Kroll, T. E. Kleindienst, E. O. Edney, M. Claeys, A. Sorooshian, *et al.* Evidence for organosulfates in secondary organic aerosol. *Environ. Sci. Technol.* **41**(2), 517–527 (2007).
151. J. D. Surratt, Y. Gomez-Gonzalez, A. W. H. Chan, R. Vermeylen, M. Shahgholi, E. T. Kleindienst, *et al.* Organosulfate formation in biogenic secondary organic aerosol. *J. Phys. Chem. A.* **112**(36), 8345–8378 (2008).
152. J. D. Surratt, A. W. Chan, N. C. Eddingsaas, M. Chan, C. L. Loza, A. J. Kwan, *et al.* Reactive intermediates revealed in secondary organic aerosol formation from isoprene. *Proc. Natl. Acad. Sci.* **107**(15), 6640–6645 (2010).
153. M. Galloway, P. Chhabra, A. Chan, J. Surratt, R. Flagan, J. Seinfeld, *et al.* Glyoxal uptake on ammonium sulphate seed aerosol: Reaction products and reversibility of uptake under dark and irradiated conditions. *Atmos. Chem. Phys.* **9**(10), 3331–3345 (2009).
154. M. J. Perri, Y. B. Lim, S. P. Seitzinger and B. J. Turpin. Organosulfates from glycolaldehyde in aqueous aerosols and clouds: Laboratory studies. *Atmos. Environ.* **44**(21), 2658–2664 (2010).
155. J. Schindelka, Y. Iinuma, D. Hoffmann and H. Herrmann. Sulfate radical-initiated formation of isoprene-derived organosulfates in atmospheric aerosols. *Faraday Discuss.* **165**, 237–259 (2013).

156. M. Jang, N. M. Czoschke, S. Lee and R. M. Kamens. Heterogeneous atmospheric aerosol production by acid-catalyzed particle-phase reactions. *Science.* **298**(5594), 814–817 (2002).
157. T. M. Olson and M. R. Hoffmann. Hydroxyalkylsulfonate formation: Its role as a S (IV) reservoir in atmospheric water droplets. *Atmos. Environ.* **23**(5), 985–997 (1989).
158. J. McMurry. Organic chemistry. 6th Ed. (Thomson-Brooks/Cole, Belmont, CA, 2004).
159. B. Noziere, P. Dziedzic and A. Cordova. Inorganic ammonium salts and carbonate salts are efficient catalysts for aldol condensation in atmospheric aerosols. *Phys. Chem. Chem. Phys.* **12**(15), 3864–3872 (2010).
160. T. Nguyen, M. Coggon, K. Bates, X. Zhang, R. Schwantes, K. Schilling, *et al.* Organic aerosol formation from the reactive uptake of isoprene epoxydiols (IEPOX) onto non-acidified inorganic seeds. *Atmos. Chem. Phys.* **14**(7), 3497–3510 (2014).
161. E. A. Betterton and M. R. Hoffmann. Henry law constants of some environmentally important aldehydes. *Environ. Sci. Technol.* **22**(12), 1415–1418 (1988).
162. Z. Li, A. N. Schwier, N. Sareen and V. F. McNeill. Reactive processing of formaldehyde and acetaldehyde in aqueous aerosol mimics: Surface tension depression and secondary organic products. *Atmos. Chem. Phys.* **11**(22), 11617–11629 (2011).
163. K. W. Loeffler, C. A. Koehler, N. M. Paul and D. O. D. Haan. Oligomer formation in evaporating aqueous glyoxal and methyl glyoxal solutions. *Environ. Sci. Technol.* **40**(20), 6318–6323 (2006).
164. N. Sareen, A. N. Schwier, E. L. Shapiro, D. Mitroo and V. F. McNeill. Secondary organic material formed by methylglyoxal in aqueous aerosol mimics. *Atmos. Chem. Phys.* **10**(3), 997–1016 (2010).
165. A. N. Schwier, N. Sareen, D. Mitroo, E. L. Shapiro and V. F. McNeill. Glyoxal-methylglyoxal cross-reactions in secondary organic aerosol formation. *Environ. Sci. Technol.* **44**(16), 6174–6182 (2010).
166. E. L. Shapiro, J. Szprengiel, N. Sareen, C. N. Jen, M. R. Giordano and V. F. McNeill. Light-absorbing secondary organic material formed by glyoxal in aqueous aerosol mimics. *Atmos. Chem. Phys.* **9**(7), 2289–2300 (2009).
167. J. Zhao, N. P. Levitt, R. Y. Zhang and J. M. Chen. Heterogeneous reactions of methylglyoxal in acidic media: Implications for secondary organic aerosol formation. *Environ. Sci. Technol.* **40**(24), 7682–7687 (2006).
168. B. R. T. Simoneit. Biomass burning — A review of organic tracers for smoke from incomplete combustion. *Appl. Geochem.* **17**(3), 129–162 (2002).
169. B. J. Holmes and G. A. Petrucci. Oligomerization of levoglucosan by Fenton chemistry in proxies of biomass burning aerosols. *J. Atmos. Chem.* **58**(2), 151–66 (2007).
170. B. J. Holmes and G. A. Petrucci. Water-soluble oligomer formation from acid-catalyzed reactions of levoglucosan in proxies of atmospheric aqueous aerosols. *Environ. Sci. Technol.* **40**(16), 4983–4989 (2006).

171. M. T. Casale, A. R. Richman, M. J. Elrod, R. M. Garland, M. R. Beaver and M. A. Tolbert. Kinetics of acid-catalyzed aldol condensation reactions of aliphatic aldehydes. *Atmos. Environ.* **41**(29), 6212–6224 (2007).
172. B. Noziere and W. Esteve. Light-absorbing aldol condensation products in acidic aerosols: Spectra, kinetics, and contribution to the absorption index. *Atmos. Environ.* **41**(6), 1150–1163 (2007).
173. D. O. De Haan, A. L. Corrigan, K. W. Smith, D. R. Stroik, J. J. Turley, F. E. Lee, *et al.* Secondary organic aerosol-forming reactions of glyoxal with amino acids. *Environ. Sci. Technol.* **43**(8), 2818–2824 (2009).
174. D. O. De Haan, L. N. Hawkins, J. A. Kononenko, J. J. Turley, A. L. Corrigan, M. A. Tolbert, *et al.* Formation of nitrogen-containing oligomers by methylglyoxal and amines in simulated evaporating cloud droplets. *Environ. Sci. Technol.* **45**(3), 984–991 (2011).
175. B. Noziere and W. Esteve. Organic reactions increasing the absorption index of atmospheric sulfuric acid aerosols. *Geophys. Res. Lett.* **32**(3), (2005), doi: 10.1029/2011JD016944.
176. T. B. Nguyen, A. Laskin, J. Laskin and S. A. Nizkorodov. Brown Carbon formation from ketoaldehydes of biogenic monoterpenes. *Faraday Discuss.* **165**, 473–494 (2013).
177. H. Tobias and P. Ziemann. Thermal desorption mass spectrometric analysis of organic aerosol formed from reactions of 1-tetradecene and O_3 in the presence of alcohols and carboxylic acids. *Environ. Sci. Technol.* **34**(11), 2105–2115 (2000).
178. K. S. Docherty, W. Wu, Y. B. Lim and P. J. Ziemann. Contributions of Organic Peroxides to Secondary Aerosol Formed from Reactions of Monoterpenes with O_3. *Environ. Sci. Technol.* **39**(11), 4049–4059 (2005).
179. L. D. Yee, J. S. Craven, C. L. Loza, K. A. Schilling, N. L. Ng, M. R. Canagaratna, *et al.* Secondary organic aerosol formation from low-NO_x photooxidation of dodecane: Evolution of multigeneration gas phase chemistry and aerosol composition. *J. Phys. Chem. A.* **116**(24), 6211–6230 (2012).
180. P. J. Ziemann and R. Atkinson. Kinetics, products, and mechanisms of secondary organic aerosol formation. *Chem. Soc. Rev.* **41**(19), 6582–6605 (2012).
181. E. Hellpointner and S. Gab. Detection of methyl, hydroxymethyl and hydroxyethyl hydroperoxides in air and precipitation. *Nature.* **337**(6208), 631–634 (1989).
182. C. Satterfield and L. Case. Reaction of aldehyde and hydrogen peroxide in aqueous solution — kinetics of the initial reaction. *Ind. Eng. Chem.* **46**(5), 998–1001 (1954).
183. X. Zhou and Y. Lee. Aqueous solubility and reaction-kinetics of hydroxymethyl hydroperoxide. *J. Phys. Chem.* **96**(1), 265–272 (1992).
184. R. Zhao, A. K. Y. Lee, R. Soong, A. J. Simpson and J. P. D. Abbatt. Formation of aqueous-phase α-hydroxyhydroperoxides (α-HHP): Potential atmospheric impacts. *Atmos. Chem. Phys.* **13**(12), 5857–5872 (2013).

185. L. Schoene and H. Herrmann. Kinetic measurements of the reactivity of hydrogen peroxide and ozone towards small atmospherically relevant aldehydes, ketones and organic acids in aqueous solutions. *Atmos. Chem. Phys.* **14**(9), 4503–4514 (2014).
186. M. I. Stefan and J. R. Bolton. Reinvestigation of the acetone degradation mechanism in dilute aqueous solution by the UV/H_2O_2 process. *Environ. Sci. Technol.* **33**(6), 870–873 (1999).
187. F. Paulot, D. Wunch, J. D. Crounse, G. C. Toon, D. B. Millet, P. F. DeCarlo, *et al.* Importance of secondary sources in the atmospheric budgets of formic and acetic acids. *Atmos. Chem. Phys.* **11**(5), 1989–2013 (2011).
188. S. Gab, W. Turner, S. Wolff, K. Becker, L. Ruppert and K. Brockmann. Formation of alkyl and hydroxyalkyl hydroperoxides on ozonolysis in water and in air. *Atmos. Environ.* **29**(18), 2401–2407 (1995).
189. Z. M. Chen, H. L. Wang, L. H. Zhu, C. X. Wang, C. Y. Jie and W. Hua. Aqueous-phase ozonolysis of methacrolein and methyl vinyl ketone: A potentially important source of atmospheric aqueous oxidants. *Atmos. Chem. Phys.* **8**(8), 2255–2265 (2008).
190. H. L. Wang, D. Huang, X. Zhang, Y. Zhao and Z. M. Chen. Understanding the aqueous-phase ozonolysis of isoprene: Distinct product distribution and mechanism from the gas phase reaction. *Atmos. Chem. Phys.* **12**(15), 7187–7198 (2012).
191. P. Neeb, F. Sauer, O. Horie and G. Moortgat. Formation of hydroxymethyl hydroperoxide and formic acid in alkene ozonolysis in the presence of water vapour. *Atmos. Environ.* **31**(10), 1417–1423 (1997).
192. F. Sauer, C. Schafer, P. Neeb, O. Horie and G. Moortgat. Formation of hydrogen peroxide in the ozonolysis of isoprene and simple alkenes under humid conditions. *Atmos. Environ.* **33**(2), 229–241 (1999).
193. D. L. Bones, D. K. Henricksen, S. A. Mang, M. Gonsior, A. P. Bateman, T. B. Nguyen, *et al.* Appearance of strong absorbers and fluorophores in limonene-O_3 secondary organic aerosol due to NH_4^+-mediated chemical aging over long time scales. *J. Geophys. Res. Atmos.* **115**, D05203–D05203 (2010), doi:10.1029/2009JD012864.
194. A. K. Lee, R. Zhao, R. Li, J. Liggio, S-M. Li and J. P. Abbatt. Formation of light absorbing organo-nitrogen species from evaporation of droplets containing glyoxal and ammonium sulfate. *Environ. Sci. Technol.* **47**(22), 12819–12826 (2013).
195. M. H. Powelson, B. M. Espelien, L. N. Hawkins, M. M. Galloway and D. O. De Haan. Brown Carbon formation by aqueous-phase carbonyl compound reactions with amines and ammonium sulfate. *Environ. Sci. Technol.* **48**(2), 985–993 (2013).
196. C. J. Kampf, R. Jakob and T. Hoffmann. Identification and characterization of aging products in the glyoxal/ammonium sulfate system — implications for light-absorbing material in atmospheric aerosols. *Atmos. Chem. Phys.* **12**(14), 6323–6333 (2012).

197. G. T. Drozd and V. F. McNeill. Organic matrix effects on the formation of light-absorbing compounds from α-dicarbonyls in aqueous salt solution. *Environ. Sci-Process Impacts.* **16**(4), 741–747 (2014).
198. H. J. J. Lee, P. K. Aiona, A. Laskin, J. Laskin and S. A. Nizkorodov. Effect of solar radiation on the optical properties and molecular composition of laboratory proxies of atmospheric Brown Carbon. *Environ. Sci. Technol.* **48**(17), 10217–10226 (2014).
199. K. M. Updyke, T. B. Nguyen and S. A. Nizkorodov. Formation of Brown Carbon via reactions of ammonia with secondary organic aerosols from biogenic and anthropogenic precursors. *Atmos. Environ.* **63**, 22–31 (2012).
200. J. Laskin, A. Laskin, S. A. Nizkorodov, P. Roach, P. Eckert, M. K. Gilles, *et al.* Molecular selectivity of Brown Carbon chromophores. *Environ. Sci. Technol.* **48**(20), 12047–12055 (2014).
201. C. J. Gaston, T. P. Riedel, Z. Zhang, A. Gold, J. D. Surratt and J. A. Thornton. Reactive uptake of an isoprene-derived epoxydiol to submicron aerosol particles. *Environ. Sci. Technol.* **48**(19), 11178–11186 (2014).
202. Y-H. Lin, Z. Zhang, K. S. Docherty, H. Zhang, S. H. Budisulistiorini, C. L. Rubitschun, *et al.* Isoprene epoxydiols as precursors to secondary organic aerosol formation: Acid-catalyzed reactive uptake studies with authentic compounds. *Environ. Sci. Technol.* **46**(1), 250–258 (2012).
203. E. C. Minerath, M. P. Schultz and M. J. Elrod. Kinetics of the reactions of isoprene-derived epoxides in model tropospheric aerosol solutions. *Environ. Sci. Technol.* **43**(21), 8133–8139 (2009).
204. M. Claeys, B. Graham, G. Vas, W. Wang, R. V. P. Vermeylen, *et al.* Formation of secondary organic aerosols through photooxidation of isoprene. *Science.* **303**(5661), 1173–1176 (2004).
205. N. C. Cole-Filipiak, A. E. O'Connor and M. J. Elrod. Kinetics of the hydrolysis of atmospherically relevant isoprene-derived hydroxy epoxides. *Environ. Sci. Technol.* **44**(17), 6718–6723 (2010).
206. A. I. Darer, N. C. Cole-Filipiak, A. E. O'Connor and M. J. Elrod. Formation and stability of atmospherically relevant isoprene-derived organosulfates and organonitrates. *Environ. Sci. Technol.* **45**(5), 1895–1902 (2011).
207. M. N. Chan, J. D. Surratt, M. Claeys, E. S. Edgerton, R. L. Tanner, S. L. Shaw, *et al.* Characterization and quantification of isoprene-derived epoxydiols in ambient aerosol in the Southeastern United States. *Environ. Sci. Technol.* **44**(12), 4590–4596 (2010).
208. H. O. T. Pye, R. W. Pinder, I. R. Piletic, Y. Xie, S. L. Capps, Y-H. Lin, *et al.* Epoxide pathways improve model predictions of isoprene markers and reveal key role of acidity in aerosol formation. *Environ. Sci. Technol.* **47**(19), 11056–11064 (2013).
209. A. G. Carlton and B. J. Turpin. Particle partitioning potential of organic compounds is highest in the Eastern US and driven by anthropogenic water. *Atmos. Chem. Phys.* **13**(20), 10203–10214 (2013).
210. J. D. Blando and B. J. Turpin. Secondary organic aerosol formation in cloud and fog droplets: A literature evaluation of plausibility. *Atmos. Environ.* **34**(10), 1623–1632 (2000).

211. J. H. Kroll, N. L. Ng, S. M. Murphy, V. Varutbangkul, R. C. Flagan and J. H. Seinfeld. Chamber studies of secondary organic aerosol growth by reactive uptake of simple carbonyl compounds. *J. Geophys. Res. Atmos.* **110**(D23), (2005), doi:10.1029/2005JD006004.
212. D. O. De Haan, A. L. Corrigan, M. A. Tolbert, J. L. Jimenez, S. E. Wood and J. J. Turley. Secondary organic aerosol formation by self-reactions of methylglyoxal and glyoxal in evaporating droplets. *Environ. Sci. Technol.* **43**(21), 8184–8190 (2009).
213. H. E. Krizner, D. O. De Haan and J. Kua. Thermodynamics and kinetics of methylglyoxal dimer formation: A computational study. *J. Phys. Chem. A.* **113**(25), 6994–7001 (2009).
214. J. Kua, S. W. Hanley and D. O. De Haan. Thermodynamics and kinetics of glyoxal dimer formation: A computational study. *J. Phys. Chem. A.* **112**(1), 66–72 (2008).
215. M. M. Galloway, M. H. Powelson, N. Sedehi, S. E. Wood, K. D. Millage, J. A. Kononenko, *et al.* Secondary organic aerosol formation during evaporation of droplets containing atmospheric aldehydes, amines, and ammonium sulfate. *Environ. Sci. Technol.* **48**(24), 14417–14425 (2014).
216. M. Trainic, A. A. Riziq, A. Lavi and Y. Rudich. Role of interfacial water in the heterogeneous uptake of glyoxal by mixed glycine and ammonium sulfate aerosols. *J. Phys. Chem. A.* **116**(24), 5948–5957 (2012).
217. T. B. Nguyen, P. B. Lee, K. M. Updyke, D. L. Bones, J. Laskin, A. Laskin, *et al.* Formation of nitrogen- and sulfur-containing light-absorbing compounds accelerated by evaporation of water from secondary organic aerosols. *J. Geophys. Res. Atmos.* **117**(D1), D01207 (2012), doi:10.1029/2011JD016944.
218. C. George, B. D'Anna, H. Herrmann, C. Weller, V. Vaida, D. J. Donaldson, *et al.* Emerging areas in atmospheric photochemistry. In: *Atmospheric and Aerosol Chemistry.* (Springer, 2014), pp. 1–53.
219. D. Vione, V. Maurino, C. Minero, E. Pelizzetti, M. A. Harrison, R-I. Olariu, *et al.* Photochemical reactions in the tropospheric aqueous-phase and on particulate matter. *Chem. Soc. Rev.* **35**(5), 441–453 (2006).
220. R. G. W. Norrish and C. H. Bamford. Photodecomposition of aldehydes and ketones. *Nature.* **138**, 1016 (1936).
221. S. E. Paulson, D-L. Liu, G. E. Orzechowska, L. M. Campos and K. N. Houk. Photolysis of heptanal. *J. Org. Chem.* **71**(17), 6403–6408 (2006).
222. B. J. Finlayson-Pitts and J. N. Pitts Jr. Chemistry of the upper and lower atmosphere: Theory, experiments, and applications. Academic press; 1999.
223. T-M. Fu, D. J. Jacob, F. Wittrock, J. P. Burrows, M. Vrekoussis and D. K. Henze. Global budgets of atmospheric glyoxal and methylglyoxal, and implications for formation of secondary organic aerosols. *J. Geophys. Res. Atmos.* **113**(D15), (2008), doi:10.1029/2007JD009505.
224. E. C. Griffith, B. K. Carpenter, R. K. Shoemaker and V. Vaida. Photochemistry of aqueous pyruvic acid. *Proc. Natl. Acad. Sci.* **110**(29), 11714–11719 (2013).

225. M. Guzman, A. Colussi and M. Hoffmann. Photoinduced oligomerization of aqueous pyruvic acid. *J. Phys. Chem. A.* **110**(10), 3619–3626 (2006).
226. A. E. Reed Harris, B. Ervens, R. K. Shoemaker, J. A. Kroll, R. J. Rapf, E. C. Griffith, *et al.* Photochemical kinetics of pyruvic acid in aqueous solution. *J. Phys. Chem. A.* **118**(37), 8505–8516 (2014).
227. J. H. Kroll, N. L. Ng, S. M. Murphy, R. C. Flagan and J. H. Seinfeld. Secondary organic aerosol formation from isoprene photooxidation. *Environ. Sci. Technol.* **40**(6), 1869–1877 (2006).
228. A. A. Presto, K. E. Huff Hartz and N. M. Donahue. Secondary organic aerosol production from terpene ozonolysis. 1. effect of uv radiation. *Environ. Sci. Technol.* **39**(18), 7036–7045 (2005).
229. J. Zhang, K. E. Huff Hartz, S. N. Pandis and N. M. Donahue. Secondary organic aerosol formation from limonene ozonolysis: Homogeneous and heterogeneous influences as a function of NO_x. *J. Phys. Chem. A.* **110**(38), 11053–11063 (2006).
230. J. D. Surratt, S. M. Murphy, J. H. Kroll, N. L. Ng, L. Hildebrandt, A. Sorooshian, *et al.* Chemical composition of secondary organic aerosol formed from the photooxidation of isoprene. *J. Phys. Chem. A.* **110**(31), 9665–9690 (2006).
231. A. P. Bateman, S. A. Nizkorodov, J. Laskin and A. Laskin. Photolytic processing of secondary organic aerosols dissolved in cloud droplets. *Phys. Chem. Chem. Phys.* **13**(26), 12199–12212 (2011).
232. S. A. Mang, D. K. Henricksen, A. P. Bateman, M. P. S. Andersen, D. R. Blake, SA. Nizkorodov. Contribution of carbonyl photochemistry to aging of atmospheric secondary organic aerosol. *J. Phys. Chem. A.* **112**(36), 8337–8344 (2008).
233. M. L. Walser, J. Park, A. L. Gomez, A. R. Russell and S. A. Nizkorodov. Photochemical aging of secondary organic aerosol particles generated from the oxidation of *d*-limonene. *J. Phys. Chem. A.* **111**(10), 1907–1913 (2007).
234. H. Lignell, M. L. Hinks and S. A. Nizkorodov. Exploring matrix effects on photochemistry of organic aerosols. *Proc. Natl. Acad. Sci.* **111**(38), 13780–13785 (2014).
235. P. J. Wong, S. Zhou and J. P. D. Abbatt. Changes in secondary organic aerosol composition and mass due to photolysis: Relative humidity dependence. *J. Phys. Chem. A.* **119**(19), 4309–4316 (2014).
236. H. Lignell, S. A. Epstein, M. R. Marvin, D. Shemesh, B. Gerber and S. Nizkorodov. Experimental and theoretical study of aqueous cis-pinonic acid photolysis. *J. Phys. Chem. A.* **117**(48), 12930–12945 (2013).
237. T. Hullar and C. Anastasio. Yields of hydrogen peroxide from the reaction of hydroxyl radical with organic compounds in solution and ice. *Atmos. Chem. Phys.* **11**(14), 7209–7222 (2011).
238. M. E. Monge, B. D'Anna, L. Mazri, A. Giroir-Fendler, M. Ammann, D. J. Donaldson, *et al.* Light changes the atmospheric reactivity of soot. *Proc. Natl. Acad. Sci.* **107**(15), 6605–6609 (2010).

239. M. E. Monge, T. Rosenørn, O. Favez, M. Müller, G. Adler, A. A. Riziq, *et al.* Alternative pathway for atmospheric particles growth. *Proc. Natl. Acad. Sci.* **109**(18), 6840–6844 (2012).
240. A. J. Sumner, J. L. Woo and V. F. McNeill. Model analysis of secondary organic aerosol formation by glyoxal in laboratory studies: The case for photoenhanced chemistry. *Environ. Sci. Technol.* **48**(20), 11919–11925 (2014).
241. N. Sareen, S. G. Moussa and V. F. McNeill. Photochemical aging of light-absorbing secondary organic aerosol material. *J. Phys. Chem. A.* **117**(14), 2987–2996 (2013).
242. R. Zhao, A. K. Y. Lee, L. Huang, X. Li, F. Yang and J. P. D. Abbatt. Photochemical processing of aqueous atmospheric Brown Carbon. *Atmos. Chem. Phys.* **15**, 6087–6100 (2015).
243. M. Darvas, S. Picaud and P. Jedlovszky. Molecular dynamics simulations of the water adsorption around malonic acid aerosol models. *Phys. Chem. Chem. Phys.* **15**, 10942–10951 (2013).
244. S. G. Moussa, T. M. McIntire, M. Szori, M. Roeselova, D. J. Tobias, R. L. Grimm, *et al.* Experimental and theoretical characterization of adsorbed water on self-assembled monolayers: Understanding the interaction of water with atmospherically relevant surfaces. *J. Phys. Chem.* **113**, 2060–2069 (2009).
245. D. J. Donaldson and V. Vaida. The influence of organic films at the air-aqueous boundary on atmospheric processes. *Chem. Rev.* **106**(4), 1445–1461 (2006).
246. J. J. N. Pitts, H. R. Paur, B. Zielinska, J. Arey, A. M. Winer, T. Ramdahl, *et al.* Factors influence the reactivity of polycyclic aromatic hydrocarbons adsorbed on filters and ambient POM with ozone. *Chemosphere.* **15**, 675–685 (1986).
247. U. Poeschl, T. Letzel, C. Schauer and R. Niessner. Interaction of ozone and water vapor with spark discharge soot aerosol particles coated with benzo[a]pyrene: O_3 and H_2O adsorption, benzo[a]pyrene degradation, and atmospheric implications. *J. Phys. Chem. A.* **105**, 4029–4041 (2001).
248. L. Petrick, H. Destaillats, I. Zouev, S. Sabach and Y. Dubowski. Sorption, desorption, and surface oxidative fate of nicotine. *Phys. Chem. Chem. Phys.* **12**, 10356–10364 (2010).
249. L. Petrick, A. Svidovsky and Y. Dubowski. Thirdhand smoke: Heterogeneous oxidation of nicotine and secondary aerosol formation in the indoor environment. *Env. Sci. Technol.* **45**, 328–333 (2011).
250. E. A. Weitkamp, K. E. H. Hartz, A. M. Sage, N. M. Donahue and A. L. Robinson. Laboratory measurements of the heterogeneous oxidation of condensed-phase organic molecular markers for meat cooking emissions. *Env. Sci. Technol.* **42**, 5177–5182 (2008a).
251. R. M. Kamens, Z. Guo, J. N. Fulcher and D. A. Bell. Influence of humidity, sunlight, and temperature on the daytime decay of polycyclic hydrocarbons on atmospheric soot particles. *Env. Sci. Technol.* **22**, 103–108 (1988).

252. S. R. McDow, M. Vartiainen, Q. Sun, Y. Hong, Y. Yao and R. M. Kamens. Combustion aerosol water content and its effect on polycyclic aromatic hydrocarbon reactivity. *Atmos. Env.* **29**, 791–797 (1995).
253. N-O. A. Kwamena, J. A. Thornton and J. P. D. Abbatt. Kinetics of surface-bound benzo[a]pyrene and ozone on solid organic and salt aerosols. *J. Phys. Chem. A.* **108**, 11626–11634 (2004).
254. N.-O. A. Kwamena, M. E. Earp, C. J. Young and J. P. D. Abbatt. Kinetic and product yield study of the heterogeneous gas-surface reaction of anthracene and ozone. *J. Phys. Chem. A.* **110**, 3638–3646 (2006).
255. N.-O. A. Kwamena, M. G. Staikova, D. J. Donaldson, I. J. George and J. P. D. Abbatt. Role of aerosol substrate in the heterogeneous ozonation reactions of surface-bound PAHs. *J. Phys. Chem. A.* **111**, 11050–11058 (2007).
256. D. Fu, C. Leng, J. Kelly, G. Zeng, Y. Zhang and Y. Liu. ATR-IR study of zone initiated heterogeneous oxidation of squalene in an indoor environment. *Env. Sci. Technol.* **47**, 10611–10618 (2013).
257. L. Petrick and Y. Dubowski. Heterogeneous oxidation of squalene film by ozone under various indoor conditions. *Indoor Air.* **19**, 381–391 (2009).
258. G. Zeng, S. Holladay, D. Langlois, Y. H. Zhang and Y. Liu. Kinetics of heterogeneous reaction of ozone with linoleic acid and its dependence on temperature, physical state, RH, and ozone concentration. *J. Phys. Chem. A.* **117**, 1963–1974 (2013).
259. L. A. Woodill, E. M. O'Neill and R. Z. Hinrichs. Impact of surface adsorbed catechol on tropospheric aerosol surrogates: Heterogeneous ozonolysis and its effects on water uptake. *J. Phys. Chem. A.* **117**, 5620–5631 (2013).
260. T. Thornberry and J. P. D. Abbatt. Heterogeneous reaction of ozone with liquid unsaturated fatty acids: Detailed kinetics and gas phase product studies. *Phys. Chem. Chem. Phys.* **6**, 84–93 (2004).
261. M. Segal-Rosenheimer and Y. Dubowski. Heterogeneous ozonolysis of cypermethrin using real-time monitoring FTIR techniques. *J. Phys. Chem. C.* **111**, 11682–11691 (2007).
262. C. Leng, J. Hiltner, H. Pham, J. Kelley, M. Mach, Y. Zhang, *et al.* Kinetics study of heterogeneous reactions of ozone with erucic acid using an ATR-IR flow reactor. *Phys. Chem. Chem. Phys.* **16**, 4350–4360 (2014).
263. Y. Q. Li, H. Z. Zhang, P. Davidovits, J. T. Jayne, C. E. Kolb and D. R. Worsnop. Uptake of HCl(g) and HBr(g) on ethylene glycol surfaces as a function of relative humidity and temperature. *J. Phys. Chem. A.* **106**, 1220–1227 (2002).
264. H. Z. Zhang, Y. Q. Li, J-R. Xia, P. Davidovits, L. R. Williams, J. T. Jayne, *et al.* Uptake of gas phase species by 1-octanol. 2. uptake of hydrogen halides and acetic acid as a function of relative humidity and temperature. *J. Phys. Chem. A.* **107**, 6398–6407 (2003a).
265. H. Z. Zhang, Y. Q. Li, J-R. Xia, P. Davidovits, L. R. Williams, J. T. Jayne, *et al.* Uptake of gas phase species by 1-octanol. 1. uptake of α-pinene, γ-terpinene, p-cymene, and 2-methyl-2-hexanol as a function

of relative humidity and temperature. *J. Phys. Chem. A.* **107** 6388–6397 (2003b).
266. A. Gerecke, A. Thielmann, L. Gutzwiller and M. J. Rossi. The chemical kinetics of HONO formation resulting from heterogeneous interaction of NO_2 with flame soot. *Geophys. Res. Lett.* **25**, 2453–2456 (1998), doi: 10.1029/98GL01796.
267. M. Brigante, D. Cazoir, B. D'Anna, C. George and D. J. Donaldson. Photoenhanced uptake of NO_2 by pyrene solid films. *J. Phys. Chem. A.* **112**, 9503–9508 (2008).
268. C. A. Longfellow, A. R. Ravishankara and D. R. Hanson. Reactive uptake on hydrocarbon soot: Focus on NO_2. *J. Geophys. Res.* **104**, 13833–13840 (1999), doi: 10.1029/1999JD900145.
269. J. Kleffmann, K. H. Becker, M. Lackhoff and P. Wiesen. Heterogeneous conversion of NO_2 on carbonaceous surfaces. *Phys. Chem. Chem. Phys.* **1**, 5443–5450 (1999).
270. J. Kleffmann, H. K. Becker and P. Wiesen. Investigation of the heterogeneous NO_2 conversion on perchloric acid surface. *J. Chem. Soc. Faraday Trans.* **94**, 3289–3292 (1998).
271. C. George, R. S. Strekowski, J. Kleffmann, K. Stemmler and M. Ammann. Photoenhanced uptake of gaseous NO_2 on solid organic compounds: A photochemical source of HONO? *Faraday Discuss.* **130**, 195–210 (2005).
272. F. Arens, L. Gutzwiller, H. W. Gäggeler and M. Ammann. The reaction of NO_2 with solid anthrarobin (1,2,10-trihydroxy-anthracene). *Phys. Chem. Chem. Phys.* **4**, 3684–3690 (2002).
273. K. Stemmler, M. Ndour, Y. Elshorbany, J. Kleffmann, B. Danna, C. George, *et al.* Light induced conversion of nitrogen dioxide into nitrous acid on submicron humic acid aerosol. *Atmos. Chem. Phys.* **7**, 4237–4248 (2007).
274. Z. B. Alfassi, R. E. Huie and P. Neta. Substituent effects on rates of one-electron oxidation of phenols by the radicals chlorine dioxide, nitrogen dioxide, and trioxosulfate. *J. Phys. Chem.* **90**, 4156–4158 (1986).
275. Y. Sosedova, A. Rouvière, T. Bartels-Rausch and M. Ammann. UVA/Vis-induced nitrous acid formation on polyphenolic films exposed to gaseous NO_2. *Photochem. Photobiol. Sci.* **10**, 1680–1690 (2011).
276. J. Kleffmann and P. Wiesen. Heterogeneous conversion of NO_2 and NO on HNO_3 treated soot surfaces: Atmospheric implications. *Atmos. Chem. Phys.* **5**(1), 77–83 (2005).
277. M. S. Salgado Munoz and M. J. Rossi. Heterogeneous reactions of HNO_3 with flame soot generated under different combustion conditions. Reaction mechanism and kinetics. *Phys. Chem. Chem. Phys.* **4**, 5110–5118 (2002).
278. M. J. Molina, A. V. Ivanov, S. Trakhtenberg and L. T. Molina. Atmospheric evolution of organic aerosol. *Geophys. Res. Lett.* **31**, L22104 (2004), doi: 10.1029/2004GL020910.
279. J-H. Park, A. V. Ivanov and M. J. Molina. Effect of relative humidity on OH uptake by surfaces of atmospheric importance. *J. Phys. Chem. A.* **112**, 6968–6977 (2008).

280. E. A. Weitkamp, A. T. Lambe, N. M. Donahue and A. L. Robinson. Heterogeneous oxidation of condensed-phase organic molecular markers for motor vehicle exhaust. *Env. Sci. Technol.* **42**, 7950–7956 (2008b).
281. C. Lai, Y. Liu, J. Ma, Q. Ma and H. He. Degradation kinetics of levoglucosan initiated by hydroxyl radical under different environmental conditions. *Atmos. Env.* **91**, 32–39 (2014).
282. J. A. Thornton, C. F. Braban and J. P. D. Abbatt. N_2O_5 hydrolysis on sub-micron organic aerosols: The effect of relative humidity, particle phase, and particle size. *Phys. Chem. Chem. Phys.* **5**, 4593–3603 (2003).
283. P. J. Gillimore, P. Achakulwisut, F. D. Pope, J. F. Davies, D. R. Springa and M. Kalberer. Importance of relative humidity in the oxidative aging of organic aerosols: Case study of the ozonolysis of maleic acid aerosol. *Atmos. Chem. Phys.* **11**, 12181–12195 (2011).
284. J. W. L. Lee, V. Carrascon, P. J. Gallimore, S. J. Fuller, A. Bjoerkegren, R. D. Spring, *et al.* The effect of humidity on the ozonolysis of unsaturated compounds in aerosol particles. *Phys. Chem. Chem. Phys.* **14**, 8023–8031 (2012).
285. J. J. Najera, C. J. Percival and A. B. Horn. Kinetic studies of the heterogeneous oxidation of maleic and fumaric acid aerosols by ozone under conditions of high relative humidity. *Phys. Chem. Chem. Phys.* **12**, 11417–11427 (2010).
286. C. Baduel, M. E. Monge, D. Voisin, J-L. Jaffrezo, C. George, I. E. Haddad, *et al.* Oxidation of atmospheric humic like substances by ozone: A kinetic and structural analysis approach. *Env. Sci. Technol.* **45**, 5238–5244 (2011).
287. C. L. Badger, P. T. Griffiths, I. George, J. P. D. Abbatt and R. A. Cox. Reactive uptake of N_2O_5 by aerosol particles containing mixtures of humic acid and ammonium sulfate. *J. Phys. Chem. A.* **110**, 6986–6994 (2006).
288. C. J. Gaston, J. A. Thornton and N. L. Ng. Reactive uptake of N_2O_5 to internally mixed inorganic and organic particles: The role of organic oxidation state and inferred organic phase separations. *Atmos. Chem. Phys.* **14**, 5693–5707 (2014).
289. F. D. Pope, P. J. Gallimore, S. J. Fuller, R. A. Cox and M. Kalberer. Ozonolysis of maleic acid aerosols: Effect upon aerosol hygroscopicity, phase and mass. *Env. Sci. Technol.* **44**, 6656–6660 (2010).
290. M. Shiraiwa, M. Ammann, T. Koop and U. Poeschl. Gas-uptake and chemical aging of semisolid organic aerosol particles. *Proc. Natl. Acad. Sci.* **108**, 11003–11008 (2011).
291. M. Kuwata and S. T. Martin. Phase of atmospheric secondary organic aerosol affects its reactivity. *Proc. Natl. Acad. Sci.* **109**, 17354–17359 (2012).
292. M. Shiraiwa, C. Pfrang and U. Poeschl. Kinetic multi-layer model of aerosol surface and bulk chemistry (KM-SUB): The influence of interfacial transport and bulk diffusion on the oxidation of oleic acid by ozone. *Atmos. Chem. Phys.* **10**, 3673–3691 (2010).
293. E. Abramson, D. Imre, J. Beranek, J. Wilson and A. Zelenyuk. Experimental determination of chemical diffusion within secondary organic aerosols particles. *Physical Chem. Chem. Phys.* **15**, 2983–2991 (2013).

294. C. Kidd, V. Perraud, L. M. Wingen and B. J. Finlayson-Pitts. Integrating phase and composition of organic aerosol from the ozonolysis of α-pinene. *Proc. Natl. Acad. Sci.* **111**, 7552–7557 (2014).
295. K. Reinmuth-Selzle, C. Ackaert, C. J. Kampf, M. Samonig, M. Shiraiwa, S. Kofler, *et al.* Nitration of the birch pollen allergen Bet v 1.0101: Efficiency and site-selectivity of liquid and gaseous nitrating agents. *J. Proteome. Res.* **13**, 1570–1577 (2014).
296. M. N. Chan, H. F. Zhang, A. H. Goldstein and K. R. Wilson. The role of water and phase in the heterogeneous oxidation of solid and aqueous succinic acid aerosol by hydroxyl radicals. *J. Phys. Chem. C.* **118**, 28978–28992 (2014).
297. C. J. Hennigan, M. H. Bergin, A. G. Russell, A. Nenes and R. J. Weber. Gas/particle partitioning of water-soluble organic aerosol in Atlanta. *Atmos. Chem. Phys.* **9**(11), 3613–3628 (2009).
298. D. Aljawhary, A. K. Y. Lee and J. P. D. Abbatt. High-resolution chemical ionization mass spectrometry (ToF-CIMS): Application to study SOA composition and processing. *Atmos Meas Tech.* **6**(11), 3211–3224 (2013).
299. C. F. Braban, M. F. Carroll, S. A. Styler and J. P. D. Abbatt. Phase transitions of malonic and oxalic acid aerosols. *J. Phys. Chem. A.* **107**(34), 6594–6602 (2003).
300. K. E. Daumit, A. J. Carrasquillo, J. F. Huntera and J. H. Kroll. Laboratory studies of the aqueous-phase oxidation of polyols: Submicron particles vs. bulk aqueous solution. *Atmos. Chem. Phys.* **14**(19), 10773–10784 (2014).
301. T. B. Nguyen, M. M. Coggon, R. C. Flagan and J. H. Seinfeld. Reactive uptake and photo-Fenton oxidation of glycolaldehyde in aerosol liquid water. *Environ. Sci. Technol.* **47**(9), 4307–4316 (2013).
302. T. Reemtsma. Determination of molecular formulas of natural organic matter molecules by (ultra-) high-resolution mass spectrometry: Status and needs. *J. Chromatogr. A.* **1216**(18), 3687–3701 (2009).
303. C. L. Heald, J. H. Kroll, J. L. Jimenez, K. S. Docherty, P. F. DeCarlo, A. C. Aiken, *et al.* A simplified description of the evolution of organic aerosol composition in the atmosphere. *Geophys. Res. Lett.* **37**, L08803 (2010), doi:10.1029/2010gl042737.
304. J. H. Kroll, N. M. Donahue, J. L. Jimenez, S. H. Kessler, M. R. Canagaratna, R. K. Wilson, *et al.* Carbon oxidation state as a metric for describing the chemistry of atmospheric organic aerosol. *Nat. Chem.* **3**(2), 133–139 (2011).
305. C. A. Hughey, C. L. Hendrickson, R. P. Rodgers, A. G. Marshall and K. Qian. Kendrick mass defect spectrum: A compact visual analysis for ultrahigh-resolution broadband mass spectra. *Anal Chem.* **73**(19), 4676–4681 (2001).
306. P. J. Roach, J. Laskin and A. Laskin. Higher-order mass defect analysis for mass spectra of complex organic mixtures. *Anal. Chem.* **83**(12), 4924–4929 (2011).

307. R. Zhao, E. L. Mungall, A. K. Lee, D. Aljawhary and J. P. Abbatt. Aqueous-phase photooxidation of levoglucosan–a mechanistic study using aerosol time-of-flight chemical ionization mass spectrometry (Aerosol ToF-CIMS). *Atmos. Chem. Phys.* **14**(18), 9695–9706 (2014).
308. L. M. Russell, R. Bahadur, L. N. Hawkins, J. Allan, D. Baumgardner, P. K. Quinn, *et al.* Organic aerosol characterization by complementary measurements of chemical bonds and molecular fragments. *Atmos. Environ.* **43**(38), 6100–6105 (2009).
309. D. E. Romonosky, A. Laskin, J. Laskin and S. A. Nizkorodov. High-Resolution Mass Spectrometry and Molecular Characterization of Aqueous Photochemistry Products of Common Types of Secondary Organic Aerosols. *J. Phys. Chem. A.* **119**(11), 2594–2606 (2014).
310. H. F. Zhang, D. R. Worton, M. Lewandowski, J. Ortega, C. L. Rubitschun, J. H. Park, *et al.* Organosulfates as tracers for secondary organic aerosol (SOA) formation from 2-methyl-3-buten-2-ol (MBO) in the atmosphere. *Environ. Sci. Technol.* **46**(17), 9437–9446 (2012).
311. T. H. Bertram, J. A. Thornton, T. P. Riedel, A. M. Middlebrook, R. Bahreini, T. S. Bates, *et al.* Direct observations of N_2O_5 reactivity on ambient aerosol particles. *Geophys. Res. Lett.* **36**, (2009), doi:10.1029/2009gl040248.
312. J. G. Slowik, J. P. S. Wong and J. P. D. Abbatt. Real-time, controlled OH-initiated oxidation of biogenic secondary organic aerosol. *Atmos. Chem. Phys.* **12**(20), 9775–9790 (2012).
313. X. L. Zhang, J. M. Liu, E. T. Parker, P. L. Hayes, J. L. Jimenez, J. A. de Gouw, *et al.* On the gas–particle partitioning of soluble organic aerosol in two urban atmospheres with contrasting emissions: 1. Bulk water-soluble organic carbon. *J. Geophys. Res. Atmos.* **117** (2012), doi:10.1029/2012jd017908.
314. J-S. Youn, Z. Wang, A. Wonaschütz, A. Arellano, E. A. Betterton and A. Sorooshian. Evidence of aqueous secondary organic aerosol formation from biogenic emissions in the North American Sonoran Desert. *Geophys. Res. Lett.* **40**(13), 3468–3472 (2013), doi: 10.1002/grl.50644.
315. M. Kalberer, D. Paulsen, M. Sax, M. Steinbacher, J. Dommen, A. S. H. Prevot, *et al.* Identification of polymers as major components of atmospheric organic aerosols. *Science.* **303**(5664), 1659–1662 (2004).
316. T. Schaefer, J. Schindelka, D. Hoffmann and H. Herrmann. Laboratory kinetic and mechanistic studies on the OH-Initiated oxidation of acetone in aqueous solution. *J. Phys. Chem. A.* **116**(24), 6317–6326 (2012).

Chapter 3

Critical Review of Atmospheric Chemistry of Alkoxy Radicals

Theodore S. Dibble* and Jiajue Chai
Chemistry Department,
State University of New York-Environmental Science and Forestry,
1 Forestry Drive, Syracuse, NY USA 13210
**tsdibble@esf.edu*

Alkoxy radicals represent a critical branching point in the atmospheric oxidation of volatile organic compounds. The competition among reaction with O_2, isomerization reactions, and scission reactions influences ozone formation in polluted air. Reaction mechanisms are well understood, but major gaps exist in our knowledge of the branching ratios that governs the competition between various reaction pathways. Our knowledge is satisfactory for alkoxy radicals derived from alkanes, but mostly only at room temperature. Rate constants for isomerization and scission reaction will decrease dramatically between room temperature and the temperatures of the middle and upper troposphere, but the extents of these decreases are not generally known. Theoretical calculations indicate that functional groups dramatically change the rate constants for isomerization and scission; unfortunately, there is almost no experimental data available to quantify the influence of these groups on reaction rate constants.

3.1. Introduction

In the presence of actinic light and nitrogen oxides (NO_x), the degradation of volatile organic compounds (VOCs) leads to production of ozone, as illustrated in Fig. 3.1. This ozone production

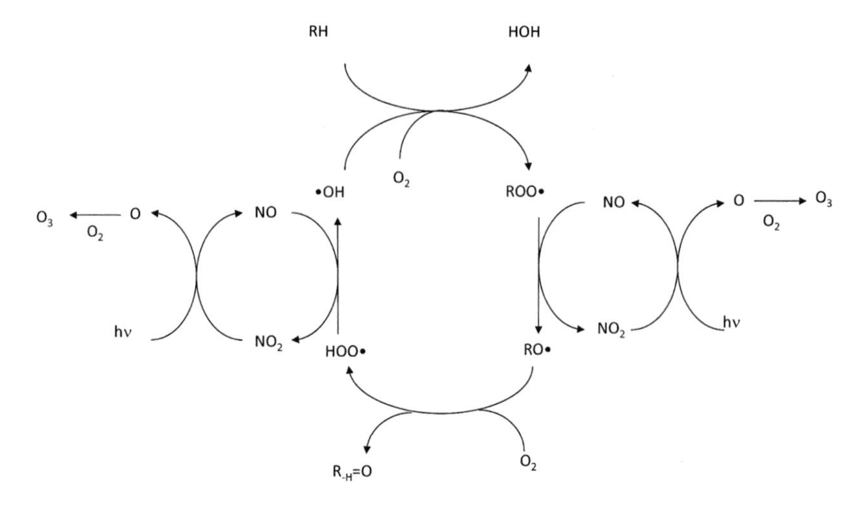

Fig. 3.1 Production of ozone during oxidation of an organic compound (RH) to a carbonyl compound (R_{-H}=O) while recycling HO_x and NO_x.

occurs via cycles that are catalytic in NO_x and hydrogen oxides (OH and HOO). Unfavorable meteorological conditions combined with large emissions of VOCs and NO_x can lead to production of concentrations of ground-level ozone that are harmful to the health of humans (as well as other animals and plant life). Notable reductions in ozone concentrations have been achieved in many metropolitan areas that once routinely experienced very harmful levels of ozone. Despite this fact, most of the population of the United States lives in areas where ozone concentrations are expected to cause health effects in vulnerable populations.[1] In the developing world many metropolitan areas routinely experience levels of ozone reminiscent of the experiences of decades ago in cites in the developed world.[2]

The role of alkoxy radicals (RO•, also called alkoxyl) in catalytic cycles producing ozone may first have been identified by Altshuller.[3] Alkoxy radicals are formed in polluted air by the reaction of peroxy radicals (ROO•) with NO, and alkoxy radical reactions lead to the formation of a carbonyl compound (represented by R_{-H}=O in Fig. 3.1). As is typical of illustrations of ozone-producing catalytic cycles, Fig. 3.1 does not show the fact that alkoxy radicals constitute an important three-way branching point in the oxidative degradation of organic compounds in the atmosphere.[4] Figure 3.2 illustrates the

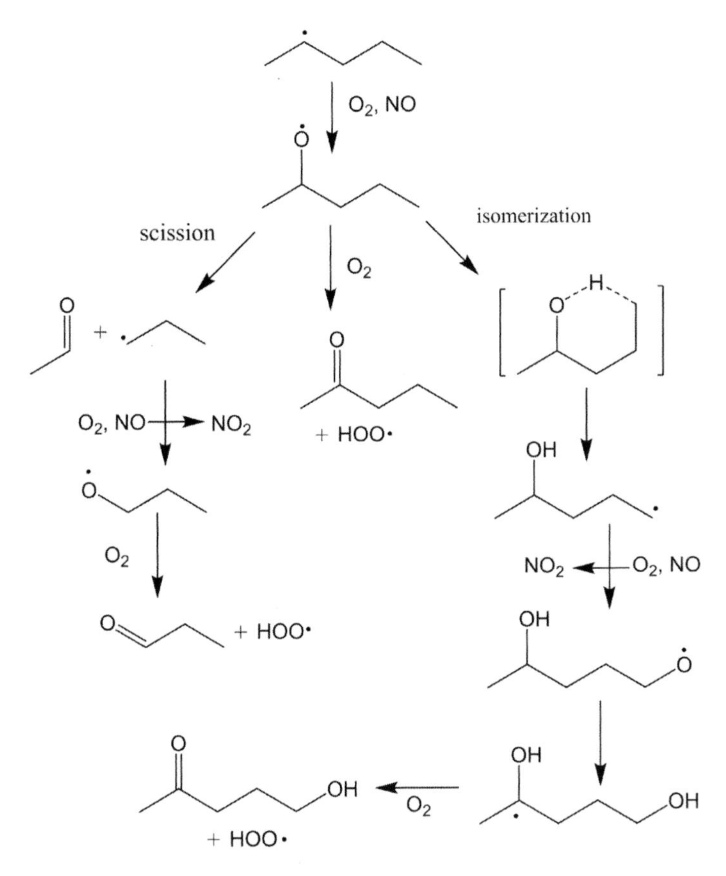

Fig. 3.2 Formation and potential degradation pathways of 2-pentoxy radical and subsequent organic radical chemistry under high-NO_x conditions. The use of "O_2, NO" as a reagent indicates the two-step process of O_2 addition to R• to form ROO• and the subsequent reaction of ROO• with NO to form RO• + NO_2.

three most common reactions of alkoxy radicals: isomerization, β-C–C scission, and reaction with molecular oxygen. As illustrated in Fig. 3.2, the isomerization and β-C–C scission reactions of RO• can propagate the organic radical chemistry while the RO• + O_2 reaction terminates the organic radical chemistry. Propagation of the organic radical chemistry leads to more ozone production than does termination. Thus, uncertainty in the competition between reactions pathways of RO• could lead to large uncertainties in the rate of production of ozone.

As ozone is not directly emitted to the atmosphere, efforts to reduce ozone concentrations in the lower atmosphere rely on the reduction of emissions of VOC and NO_x precursors to ozone production. To predict what emissions reductions are needed to achieve a given reduction of ozone, it is necessary to model emissions, transport, and chemical reactions in an episode of elevated ozone. Therefore, quantifying the effects of uncertainty in alkoxy radical kinetics on ozone formation is not trivial. Two modeling studies explicitly commented on this issue, quantifying the effect in terms of the photochemical ozone creation potential (POCP).[5,6] POCP is defined as the incremental ozone created by addition of emissions of a particular VOC vs. the incremental ozone created by additional emissions of an identical mass of ethylene, multiplied by 100. POCP is usually determined over the course of a multi-day ozone episode. Derwent *et al.* indicated that alcohols fall into two groups whose POCP differ by a factor of 2–3, depending on the potential for β-C–C scission of the alkoxy radicals formed in their degradation.[5] In another study, changes in the assumed kinetics of alkane-derived alkoxy radicals between mechanisms reduced the POCP of alkanes by 30–40%.[6]

In fact, Eulerian (3D box) simulations of realistic atmospheres spanning geographic regions as small as metropolitan areas generally do not *explicitly* include most alkoxy radicals. The chemical models in these simulations lump VOCs and VOC reaction products in ways that conserve mass but obscure molecular identity.[7,8] This is necessary to conserve computational resources. The omission of explicit representations of alkoxy radical chemistry in models of ozone pollution episodes may make it difficult for modelers to appreciate the potential effects of uncertainties in alkoxy radical chemistry on ozone formation.

Of course, ozone is not the only pollutant produced in the degradation of VOCs. For example, peroxyacyl nitrates ($RC(=O)OONO_2$) are important oxidants,[9] and the extent of their formation will depend, in part, on the extent of production of aldehydes versus ketones from alkoxy radical chemistry. Organic particulate matter (secondary organic aerosol, SOA) can be formed by the gas-to-particle conversion of organic compounds. Gas-to-particle conversion is enhanced when VOCs are oxidized to semi-volatile or non-volatile compounds. A useful approximation is that the lower the vapor pressure of a compound, the more likely it is to partition to the organic aerosol phase. Consider the stable products of the competing reactions of 2-pentoxy radical, as shown in Fig. 3.2. The C–C scission reaction leads, eventually, to two stable molecules, both of which have higher vapor pressure than the 2-pentanone and 5-hydroxy-2-pentanone produced from the O_2 and isomerization reactions, respectively. As a result, the isomerization and O_2 reactions of alkoxy radicals tend to lead to more organic-phase aerosol than does the scission reaction.

There have been several reviews of alkoxy radical kinetics in the context of atmospheric chemistry. We start with Atkinson's 1997 review, in which he proposed a set of structure-activity relationships (SARs) for the isomerization, C–C scission, and O_2 reactions of alkane- and alkene-derived RO•.[10] In 2003, Orlando *et al.* provided an extensive review of these and other reactions of alkoxy radicals, including those derived from oxygenated VOC (OVOC) and halogenated VOC.[11] They included discussion of HCl-elimination from chlorinated VOCs:

$$\mathrm{RCHClO\bullet} \rightarrow \mathrm{RC\bullet(=O)} + \mathrm{HCl} \tag{3.1}$$

and the α-ester rearrangement:

$$\mathrm{RC(=O)OCR'HO\bullet} \rightarrow \mathrm{RC(=O)OH} + \mathrm{R'C\bullet{=}O.} \tag{3.2}$$

They also discussed the role of chemical activation in the competition between C–C scission and reaction with O_2. A more limited review was published by Devolder in the same year.[12] Mereau *et al.*[13] and Curan[14] both suggested modifications to the SAR of Atkinson for C–C scission of alkoxy radicals derived from alkanes, while Johnson *et al.*[15] also included β-substituted alkoxy radicals. In 2007, Atkinson updated his previous SARs to include new results, effects of chemical activation, and OVOC-derived alkoxy radicals.[16] Vereecken and Peeters subsequently used computational chemistry to devise SARs for isomerization and C–C scission reactions for alkoxy radicals containing a wide variety of functional groups in a variety of positions.[17,18]

The goal of many reviews and original research papers is to build SARs. This is because the sheer variety of alkoxy radicals that are produced in the atmosphere far exceeds our capacity to carry out experiments or reliable calculations to determine their fate. The modest amount of experimental data published since 2007 does not require revision of the existing SARs, but many new insights have been generated in the last several years from both experiments and computational chemistry. In addition, older insights from computational chemistry have not been fully incorporated into previous reviews. As a result, this review is not intended as a comprehensive reference to alkoxy radical kinetics, but rather as a critical review of the literature together with suggestions of future directions for research.

The rest of this chapter is organized as follows: we start by describing the experimental and computational methods commonly used to investigate the kinetics and mechanisms of alkoxy radicals. Next, we review the kinetics and mechanisms of the major gas-phase reactions of alkoxy radicals important in the atmosphere (C–C scission, isomerization, and reaction with O_2) and under common experimental conditions (reaction with NO and NO_2). To this we add discussion of less-common reaction pathways that occur in OVOC, including the α-C–H scission and the α-ester rearrangement. We end with a separate section devoted to alkoxy radicals formed from isoprene and related compounds. Although there have been many experiments and calculations on alkoxy radicals relevant to chlorofluorocarbons and their replacements, this review mostly omits those studies.

3.2. Methods for Studying Alkoxy Radical Kinetics and Mechanisms

3.2.1. *Experimental Methods*

3.2.1.1. *Sources of Alkoxy Radicals for Experiments*

Alkoxy radicals can be produced in the laboratory by the same sequence of reactions as shown in Fig. 3.1. In this case, one needs a source of OH radicals. Direct sources of OH radicals include photolysis of HOOH or $HONO_2$. This approach, even using laser flash photolysis of the OH precursor, does not readily yield detectable concentrations of alkoxy radicals, but is valuable for studying OH recycling or product yields. As HOOH and $HONO_2$ possess low absorption cross-sections in the 300–400 nm region,[19] photolysis must be carried out using light near 200–250 nm. Continuous illumination at these wavelengths in chamber experiments would cause photolysis of stable oxidation products, thereby hindering efforts to determine product yields. As a result chamber experiments often use an indirect source of OH radical or use a different radical to initiate production of organic radicals.

OH radicals can be produced indirectly via photolytic generation of a small alkoxy radical from the corresponding alkylnitrite:

$$\mathrm{CH_3ONO} + h\nu \rightarrow \mathrm{CH_3O}\bullet + \mathrm{NO}, \tag{3.3}$$

followed by the sequence of reactions:

$$\mathrm{CH_3O}\bullet + \mathrm{O_2} \rightarrow \mathrm{CH_2{=}O} + \mathrm{HOO}, \tag{3.4}$$

$$\mathrm{HOO} + \mathrm{NO} \rightarrow \mathrm{OH} + \mathrm{NO_2}. \tag{3.5}$$

It is necessary to introduce NO to the reactor in addition to that produced in reaction (3.3) to drive the efficient production of OH in reaction (3.5). The time required for reaction (3.5) to occur, even at the elevated [NO] commonly used in experiments, is sufficiently long that this approach is typically used for continuous photolysis experiments rather than pulsed photolysis experiments. This sequence of reactions produces formaldehyde, which confounds efforts to detect formaldehyde as a product of alkoxy radical chemistry. To avoid this problem it is possible to use other small alkylnitrites, such as ethyl nitrite.

It is also possible to use radicals other than OH, most commonly atomic fluorine and chlorine, to initiate the degradation of the VOC of interest. F_2 and Cl_2 can be photolyzed to yield the corresponding atomic halogens, both of which are very efficient at abstracting hydrogen atoms from C–H bonds. Cl_2CO (phosgene) has been used as a source of Cl, albeit using shorter wavelengths than for photolysis of Cl_2. Some use has been made of ClNO as a radical source. Radiolysis of SF_6 is a good source of F atoms. The atomic halogens exhibit different preferences for abstraction from the varied sites on a VOC than does OH radical, as well as varying tendencies to add to C=C double bonds. As a consequence, the distribution of alkoxy radicals and the yields of the subsequent stable products will vary with the identity of the radical used to initiate VOC photooxidation. These differences need not impede the effort to understand the competition between the various fates of an alkoxy radical in product yield experiments, and can be exploited for understanding the overall mechanism.

There are often advantages to producing a single isomer of an alkoxy radical rather than, for example, the set of three alkoxy radicals that would be produced in the OH-initiated oxidation of *n*-pentane. It then becomes necessary to produce the desired alkoxy radical by photolysis of an organic precursor to the desired RO•. If it is not necessary for the photolysis event to *directly* produce RO•, then photolysis of alkyl halides (RX, X = Cl, Br, I) in the presence of NO and O_2 can be used as sources of the alkyl radical isomer that is the precursor of the desired alkoxy radical, e.g.,

$$CH_3CHXCH_2CH_3 + h\nu \rightarrow CH_3C{\bullet}HCH_2CH_3 + X, \qquad (3.6)$$

$$CH_3C{\bullet}HCH_2CH_3 + O_2 \rightarrow CH_3CH(OO{\bullet})CH_2CH_3, \qquad (3.7)$$

$$CH_3C(OO{\bullet})HCH_2CH_3 + NO \rightarrow CH_3CH(O{\bullet})CH_2CH_3 + NO_2. \qquad (3.8)$$

This approach has been determined to lead to secondary chemistry in some cases.[20] Some of that may arise from the atomic halogen reacting with the peroxy radical, or from production of HX + alkene in RX photolysis.[21]

Photolysis of alkylnitrites, such as shown in reaction (3.3), is the dominant method of generating alkoxy radicals with isomeric specificity. Based on studies of the dissociation dynamics of such compounds, dissociation occurs in 20–200 fs with cleavage of the RO–NO bond.[22–24] Photolysis of RONO produces RO• with an internal energy distribution that is significantly more energetic than a Boltzmann distribution near room temperature. This creates the potential for the energized radicals to undergo reaction prior to being quenched to a Boltzmann distribution of energy. Such "prompt" decomposition or isomerization of the alkoxy radical is expected to occur for reactions with particularly low activation barriers (critical energies). To date there is little evidence that this is a major problem for alkylnitrite photolysis at ~350 nm as opposed to ~250 nm.[25]

Alkylnitrites can be easily synthesized and isolated in multi-gram quantities.[26] The standard synthesis involves dropwise addition of a mixture of the corresponding alcohol with concentrated H_2SO_4 into aqueous $NaNO_2$. This approach yields a phase-separated system with alkylnitrite as the dominant component of the organic phase. Sulfuric acid will react with many functional groups, making this method problematic if one wants to produce functionalized RONO. Our laboratory had very modest success in producing functionalized RONO by including the alcohol in the aqueous $NaNO_2$ solution and introducing the acid directly into the aqueous phase from below the surface. This approach minimizes contact of the acid with RONO formed earlier in the course of sulfuric acid addition.

The group of T. A. Miller produced $HOCH_2CH_2ONO$, but found that an elaborate purification method was needed to isolate this water-soluble compound from the aqueous phase and minimize contamination by $ONOCH_2CH_2ONO$.[27] They were not successful in observing fluorescence from the desired $HOCH_2CH_2O\bullet$ radical upon 351 nm photolysis of $HOCH_2CH_2ONO$. Instead, they detected formaldehyde, which might be expected to form from prompt decomposition of the radicals. They reported a fluorescence excitation spectrum of $FCH_2CH_2O\bullet$ obtained from photolysis of the corresponding RONO, but did not succeed in obtaining signals attributable to the analogous species $ClCH_2CH_2O\bullet$ or $BrCH_2CH_2O\bullet$.[27]

Photolysis of CH_3OH is an efficient and fairly clean source of methoxy radical, but typically requires using light of wavelengths $\leq$200 nm to obtain reasonable radical concentrations. At these short wavelengths, competing photolysis pathways become increasingly important for larger alcohols.[28] Photolysis of peroxides (ROOR) and hydroperoxides (ROOH) can be used to generate alkoxy radicals, but photolysis typically requires wavelengths much shorter than 300 nm.[29] The use of such short wavelengths creates the potential for prompt reaction of the alkoxy radical, leading to low yields and secondary chemistry. Thermal decomposition of peroxides has also been used as a source of alkoxy radicals.[30]

3.2.1.2. *Chamber Experiments for Studies of Mechanisms and Relative Rates*

Chamber experiments have been, and continue to be, a critical source of information on alkoxy radical chemistry and kinetics. Analysis of product yields has been used to infer mechanisms and branching ratios. Branching ratios can be used to infer rate constant ratios, and this remains almost the sole source of our experimental knowledge of the kinetics of isomerization reactions. While detection methods for VOCs (both reactants and products) had, historically, been dominated by chromatographic techniques, electron-impact mass-spectrometry, and infra-red spectroscopy, the addition of a greater variety of instrumentation, especially chemical ionization techniques, have provided the ability to measure yields of products that were not previously measurable.

Environmental chambers use lamps or natural sunlight as a photolysis source to generate radicals. While some of the new chambers have been instrumented to measure concentrations of such radicals as OH and HOO, these chambers are never going to be able to detect alkoxy radicals directly in 1 atm of air. That is because the short lifetimes of alkoxy radicals lower their concentrations far below even that of OH radical.

When generating a single alkoxy radical, as in Figs. 3.2 or 3.3, it may be the case that each reaction pathway produces a unique stable product (or set of stable products), so that products can readily be assigned to specific reaction pathways of the alkoxy radical. In this case, one could potentially determine rate constant ratios $k_{\mathrm{iso}}/k_{\mathrm{C-C}}/k_{\mathrm{O_2}}[\mathrm{O_2}]$, where the subscripts refer to the different classes of reactions (isomerization, C–C scission, and $\mathrm{RO\bullet} + \mathrm{O_2}$). In the case of reaction with $\mathrm{O_2}$, it is the pseudo-first order rate constant, $k'_{\mathrm{O_2}}$ ($=k_{\mathrm{O_2}}[\mathrm{O_2}]$) that is measured. For most alkoxy radicals, one reaction pathway is negligible under atmospherically relevant conditions. A recently recognized complication is that 1,4 hydroxycarbonyls produced in isomerization reactions (see Fig. 3.2) have a tendency to cyclize to dihydrofurans via acid-catalyzed heterogeneous reactions:[31]

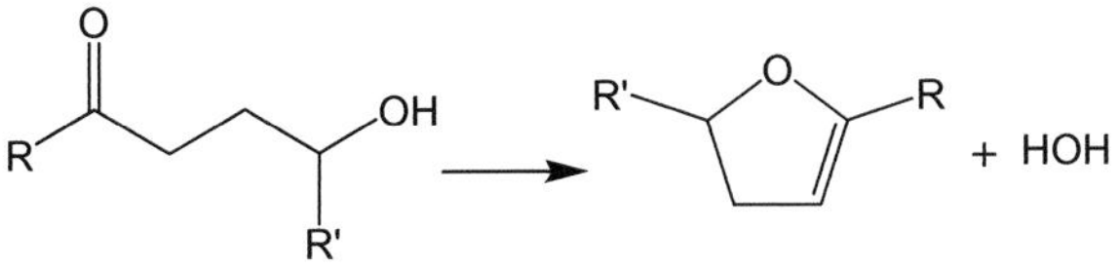

(3.9)

A major challenge facing chamber studies is identification and quantification of the numerous isomeric species produced in the oxidation of even modest-sized VOCs. This challenge is exacerbated by the lack of authentic standards for many of these species. Calls for the involvement of synthetic organic chemists in helping to meet this challenge date back at least 20 years.[32]

3.2.1.3. *Measuring Absolute and Direct Rate Constants*

Most absolute measurements of rate constants for alkoxy radical reactions have used laser flash photolysis (LFP) to generate the radical and pulsed laser induced fluorescence (LIF) to monitor radical concentration. These experiments are carried out in flow reactors. The principle of the experiment is that one photolyzes a direct precursor of the alkoxy radical and monitors the fluorescence emitted following absorption of a second laser pulse. This works well if the radical fluoresces, and if one knows which wavelengths of light lead to fluorescence. With the exception of a few halogenated species, fluorescence has been only been reported from alkane-derived alkoxy radicals. No fluorescence has been reported from alkoxy radicals that would be formed from such important classes of VOCs as alkenes, ethers, esters, or ketones. This reflects the major challenges facing absolute rate constant determinations, namely, the clean production and unambiguous identification of signal from individual isomers of particular alkoxy radicals.

Our group attempted to detect fluorescence of a variety of C_4 and C_5 alkoxy radicals following photolysis of the corresponding alkylnitrite at 355 nm. These experiments mostly did not produce signals that could readily be assigned to the corresponding alkoxy radical.[33] In the case of 1-butoxy and branched C_4 and C_5 1-alkoxy radicals, we observed clear evidence for formation of $CH_2{=}O$. The observation of clear fluorescence excitation spectra of C_4–C_{10} 1- and 2-alkoxy radicals by the Miller group in radicals cooled in a supersonic expansion[34] has not been followed up by any reports of fluorescence of these radicals in laboratory reactors. This severely limits our ability to determine absolute rate constants for these alkoxy radicals. These large 1- and 2-alkoxy radicals primarily undergo isomerization under atmospheric conditions,[16] so it would be natural to think that prompt isomerization was responsible for the failure to observe fluorescence from these radicals in flow reactors using LFP-LIF. By contrast, product yield studies by Cassanelli *et al.* indicated that prompt reaction was roughly 10% of the fate of 1-butoxy at 1 atm (when produced by photolysis of the corresponding nitrite).[35] Sprague *et al.* suggested that prompt isomerization was about 10% and 5%, respectively, of the fate of 1-butoxy and 2-pentoxy formed from 351 nm photolysis of the corresponding alkylnitrite at a total pressure of 670 Torr.[25] For prompt β-C–C scission, Sprague *et al.* were only able to obtain upper limits: <10% for 1-butoxy and <30% for 2-pentoxy.

As compared to our experiments, carried out at 6 Torr,[33] the much higher pressure employed by other authors. would tend to limit prompt unimolecular reaction. In the experiments of Miller and co-workers,[34] large 1-alkoxy radicals generated using photolysis of 1-alkylnitrites were observed to fluoresce. Miller and co-workers did not report observing formaldehyde, which would also have produced fluorescence signals. In these experiments fast cooling during the supersonic expansion would tend to minimize prompt decomposition or isomerization. Our observations of formaldehyde are not, therefore, inconsistent with the results of either Sprague *et al.*[25] Cassinelli *et al.*[35] or Miller and co-workers.[34]

From the spectra of 1-butoxy radicals obtained in the jet, it was clear that the observed conformers of 1-butoxy possessed extended conformations rather than bent conformations that would more readily lead to isomerization. This led the Miller group to suggest that isomerization was responsible for depleting the population of bent conformations.[36] By contrast, the result of Sprague *et al.* suggest that prompt isomerization is minimal (~5%) at 670 Torr.[25] We suggest, instead, the possibility that cyclic conformations are more likely than extended conformations to undergo (radiationless) internal conversion to the ground state. Radicals cooled in the jet would likely be locked into a particular conformation, while under the conditions of our flow tube experiments conformational changes would be quite rapid. These observations point to a possible resolution of the apparent contradictions between our results and these of Miller's group: if conformational change occurred with a time scale faster than several tens of nanoseconds in the electronically excited radicals produced in our experiments, fluorescence could have been quenched by internal conversion.

Hein *et al.* developed a method for inferring absolute rate constants that avoids the need for directly monitoring the concentrations of alkoxy radicals.[20,37,38] They accomplished this by photolyzing alkylbromides in the presence of NO and O_2 and measuring the time history of the absolute concentrations of NO_2 and OH. The reaction sequence is shown in Fig. 3.3 for 2-butoxy. The NO_2 is produced by reactions of both first- and second-generation peroxy radicals with NO; second-generation alkoxy radicals are formed following the isomerization or β-scission reactions of first-generation alkoxy radicals but not by their reaction with O_2. The logic is the same as that by which the unimolecular reactions of alkoxy radicals can lead to greater ozone production than the O_2 reaction (see Fig. 3.2). The time over which OH production occurs is extended by the unimolecular reactions, but the extent of production remains constant. Kinetic modeling of side reactions enabled Hein *et al.* to extract rate constants for the isomerization and O_2 reactions of 1-butoxy and 1-pentoxy and the scission and O_2 reactions of 2-butoxy and 3-pentoxy. Hein *et al.* identified their method as "direct", but while it provides absolute rate constants it relies heavily on kinetic modeling. We will discuss potential problems with their method in more detail in discussing isomerization reactions.

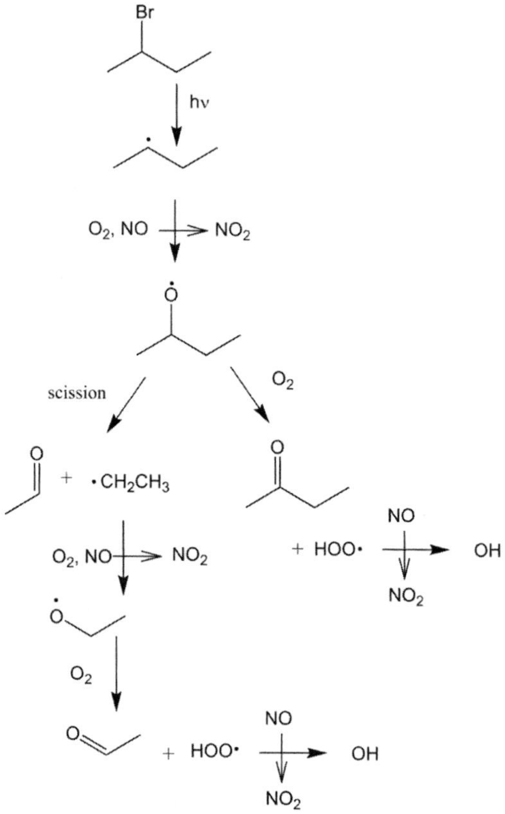

Br
hν
O_2, NO
NO_2
Ȯ
scission
O_2
O
+ $\cdot CH_2CH_3$
O
O_2, NO
NO_2
NO
+ HOO·
OH
NO_2
Ȯ
O_2
O
NO
+ HOO·
OH
NO_2

Fig. 3.3 Selective formation of 2-butoxy radical and subsequent chemistry in the experiments of Hein *et al.*[20,37,38] including formation of OH and NO_2 in the presence of large [NO].

Sprague *et al.* used cavity ringdown spectroscopy (CRDS) to measure time-resolved species abundance and determine the rate constant ratio k_{O_2}/k_{iso}.[25] CRDS is a multipass absorption technique in which one injects a pulse of light into a confocal optical cavity, and monitors the rate of decay of the intensity of light exiting the cavity.[39,40] Single-pass absorptions of 10^{-6} are readily measured by CRDS. One can use pulsed lasers as a light source despite their large shot-to-shot intensity variation, because one is measuring the e-fold decay time (ringdown time) rather than comparing absolute signal levels.

Sprague *et al.*[25] formed alkoxy radicals by photolysis of the corresponding alkylnitrite:

$$CH_3CH_2CH_2CH_2ONO + h\nu \rightarrow CH_3CH_2CH_2CH_2O\bullet + NO, \tag{3.10}$$

$$CH_3CH_2CH_2CH_2O\bullet \rightarrow \bullet CH_2CH_2CH_2CH_2OH, \tag{3.11}$$

$$\bullet CH_2CH_2CH_2CH_2OH + O_2 \rightarrow \bullet OOCH_2CH_2CH_2CH_2OH. \tag{3.12}$$

In their experiments, isomerization (reaction (3.11)) occurs on a time scale comparable to, or faster than, the ringdown time, so they could not extract rate information by monitoring the rate of appearance of the isomerization product. Instead they monitored the OH stretch of the δ-hydroxyperoxy radical formed in reaction (3.12) following isomerization. Reaction (3.12) goes rapidly to completion at even small concentrations of O_2, while the potentially interfering reaction with O_2:

$$CH_3CH_2CH_2CH_2O\bullet + O_2 \rightarrow CH_3CH_2CH_2CH{=}O + HOO\bullet, \tag{3.13}$$

occurs much more slowly. By varying $[O_2]$ and determining the relative yield of $\bullet OOCH_2CH_2CH_2CH_2OH$, they were able to determine k_{O_2}/k_{iso}. Only modest corrections for secondary chemistry were necessary in this approach.

3.2.2. *Computational Methods*

3.2.2.1. *Quantum Chemistry*

Quantum chemistry methods have become increasingly powerful and useful for predicting, confirming, or extending the results of experiments. Accessible introductions may be found in Refs. 41 and 42. Quantum chemical methods work by determining the total energy of all the electrons and nuclei in a species, whether it be a molecule, radical, or ion. Sophisticated algorithms enable optimization of a species' geometry to a potential energy minimum (stable structure) or to a first-order saddle point (transition state). From these geometries one can go on to compute the vibrational frequencies, which are needed determine the effects of temperature on thermodynamic properties. Having computed the total energy of a reactant and product (and their vibrational frequencies), one can compute enthalpies and Gibbs free energies of reaction versus temperature. Having computed the total energy of a reactant and a transition state, the critical energy and Gibbs free energy of activation can be determined.

For most quantum chemistry calculations, there are two dimensions which determine the quality: the treatment of electron correlation and the basis set. Electron correlation is tendency of electrons to minimize their instantaneous repulsion, not just their repulsion in the mean field of molecular orbitals. The basis set is the set of mathematical functions that define the flexibility of the atom-centered functions used to construct molecular orbitals. There are numerous theoretical methods for including electron correlation, each involving different approximations to treat electron correlation. Similarly, there are several different groups of basis sets that one might use, with each group spanning a range of flexibility. Both methods and basis sets are usually described by abbreviations or short-hand names, which tends to confuse the uninitiated. A major issue in most quantum chemical studies is to choose a method and basis set that yields a reliable result without excessive computational demands. This can be challenging because the computational time may vary as a large power (as high as 7, formally) of the number of electrons and size of the basis set.[43] As a result, these theoretical calculations are heavily influenced by very practical considerations.

A very popular set of methods is density functional theory (DFT) and the number of functionals (and abbreviations for functionals) has been growing rapidly.[44,45] These methods have gained popularity for their efficiency and robustness as compared to many other methods of similar accuracy. Another factor is that computing the energy of a species is much less computationally demanding than optimizing its geometry or computing its vibrational frequencies. As a result, it is common to use a modest level of theory, like DFT, to determine the structure and vibrational frequencies of species of interest, and then use a higher-level method to compute the energy at the geometry optimized using DFT. This relies on a degree of cancellation of error due to differences in the optimum geometries of species computed at the lower and higher levels of theory. The usual choice of the higher level is coupled cluster with single and double excitations with a perturbative estimate of the triple excitations (CCSD(T)).[43,46,47] Computing energies with this approach is computationally feasible and provides reasonably reliable energies in many cases. To achieve the accuracy of combining CCSD(T) with large basis sets without the computational expense, one can use composite methods[48–51] or new variants of CCSD(T).[52]

An underappreciated source of error in quantum chemistry papers is the failure to identify the lowest energy geometry of the reactants, products, and transition states, especially for species possessing multiple conformers. This can cause honest differences between pairs of results that use the identical implementation of the same method and basis set on the same problem.

3.2.2.2. *Computing Thermal Rate Constants*

An overview of methods for computing rate constants from quantum chemical data may be found in Ref. 42. Ordinary transition state theory (TST) describes the rate constants, $k(T)$, versus temperature, T, as:

$$k_{\mathrm{TST}}(T) = L\kappa(T)\frac{k_{\mathrm{B}}T}{h}\frac{Q^{\neq}(T)/V}{Q^{R}(T)/V}\exp(-E_0/k_BT), \qquad (3.14)$$

where L is the reaction path degeneracy, $\kappa(T)$ is the correction for quantum mechanical tunneling, k_B is Boltzmann's constant, h is Planck's constant, $Q^{\neq}$ and Q^{R} are the partition functions of the transition state and reactant(s), respectively, and E_0 is the critical energy of the reaction (the energy barrier per molecule at 0 K, including zero-point energy). Note that since Eq. (3.14) has more than one temperature dependent term, the critical energy, E_0, is different than the activation energy derived from Arrhenius fits to rate constants derived from experiments or computations.

For unimolecular reactions such as alkoxy radical isomerization or C–C scission, Eq. (3.14) is commonly cast into the simpler form of:

$$k_{\mathrm{TST}}(T) = L\kappa(T)\frac{k_{\mathrm{B}}T}{h}\exp(-\Delta G^{\neq}(T)/RT), \qquad (3.15)$$

where R is the gas constant and $\Delta G^{\neq}(T)$ is the molar Gibbs free energy of activation. For bimolecular reactions such as alkoxy radical reactions with O_2, Eq. (3.15) must be revised to:

$$k_{\mathrm{TST}}(T) = L\kappa(T)\frac{k_{\mathrm{B}}T}{c_0(T)h}\exp(-\Delta G^{\neq}(T)/RT), \qquad (3.16)$$

where $c_0(T)$ is the concentration corresponding to 1 atm total pressure. Gas-phase kinetics studies of atmospheric chemistry most commonly use concentration units of molecules cm^{-3}. Most quantum chemistry programs assume a standard state of 1 atmosphere, so that gas-phase atmospheric chemists would use c_0 corresponding to 2.46×10^{19} molecules cm^{-3} (at 298 K).

The tunneling correction term can be determined rather approximately but simply from variations of the Wigner formula[53]:

$$\kappa(T) = 1 - \frac{1}{24}\left(\frac{h\nu^*}{k_BT}\right)^2\left(1 + \frac{k_BT}{E_0}\right), \qquad (3.17)$$

where ν^* is the imaginary vibrational frequency corresponding to motion over the transition state (squaring ν^* yields a positive real number, so that κ is always greater than or equal to unity). A much better estimate of the tunneling correction can be obtained from numerical integration of the Eckart tunneling expression.[54,55] The Eckart tunneling correction can be found in programs like MESMER[56] and MULTIWELL[57–59] that are becoming widely used for atmospheric chemical kinetics. More sophisticated approaches to tunneling such as small-curvature tunneling (SCT) are available via the POLYRATE program.[60]

Most calculations of rate constants are done with the assumption that vibrations, including torsional motions, can be treated as harmonic oscillators. While this approximation is poor, errors in computed critical energies are commonly a somewhat larger source of error at atmospheric temperatures. As a result, errors arising from this assumption are not readily noticed. The quite common neglect of the existence of multiple conformers of reactants and transition states can introduce large errors in rate constants. For example, in the isomerization reaction of 1-butoxy radical, accounting for multiple conformers lowers the rate constant by a factor of five at 298 K.[61,62]

The errors in quantum chemical calculations mean that a calculation at high levels of theory, which also properly treats multiple conformers and anharmonicity, still has difficulty approaching the accuracy obtained by a good experiment.

3.2.2.3. *Kinetics Outside the High-pressure Limit*

There is a severe limitation of Eqs. (3.14) and (3.15) for unimolecular reactions: these equations only apply when the reaction is in the high-pressure limit. In the high pressure limit, collisions are sufficiently rapid to maintain the reactant in a Boltzmann distribution of energy despite that highly energetic reactants are rapidly reacting. This is well-known from the Lindemann–Hinshelwood model. At lower pressures (see Fig. 3.4, below) the steady-state population is severely depleted of highly energetic reactants. As a result, the reaction rate constant can be significantly reduced from the value at the high-pressure limit.

Prototypical isomerization and C–C scission reactions appear to be near the high-pressure limit at 1 atmosphere, but may be far from this limit at the lower pressures of some experiments or the upper troposphere. The lower the critical energy of a reaction, the more likely it is that the rate constant will deviate from the high-pressure limiting value. Note that alkoxy radicals often have very low critical energies for scission.

Experimental rate constants are often fitted to the Troe formula,[63] which allows a parameterization of experimental rate constants in the low pressure limit (third-order rate constant, k_0) and high pressure limit (second-order rate constant, k_∞), and the temperature dependences of each.[64] Empirically, the temperature dependence of the low-pressure limiting rate constant, $k_0(T)$, and

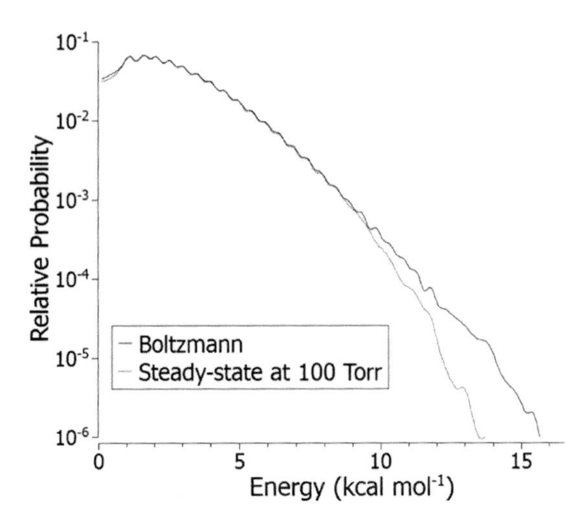

Fig. 3.4 Illustration of a Boltzmann and steady state distribution of energy for 1-butoxy radical simulated in 100 Torr N_2 at 298 K. The steady-state distribution is taken following 1 μs of simulation time, corresponding to ~1700 collisions of radicals with N_2. The critical energy of reaction was specified as 10.0 kcal mol^{-1}.

the high-pressure limiting rate constant, $k_\infty(T)$, are typically fit to power law expressions:

$$k_0(T) = k_0(300\,\mathrm{K})(T/300)^{-n}, \tag{3.18}$$

$$k_\infty(T) = k_\infty(300\,\mathrm{K})(T/300)^{-m}. \tag{3.19}$$

In the fall-off region between these two limits, the effective first-order rate constant, $k(T,[M])$, is given by:

$$k(T,[M]) = \left(\frac{k_0(T)[M]}{1 + k_0(T)[M]/k_\infty(T)}\right) F_{\mathrm{cent}}, \tag{3.20}$$

where F_{cent} is the broadening factor. The JPL recommendation[19] for F_{cent} is:

$$F_{\mathrm{cent}} = 0.6^{\{1+[\log_{10}(k_0(T)[M]/k_\infty(T))]^2\}^{-1}} \tag{3.21}$$

although IUPAC recommends a more complicated expression.[65] These equations are also applied to barrierless association reactions, such as RO• + •NO and RO• + •NO_2. The dissociation and association rate constants are coupled by the equilibrium constant.

To compute $k(T,[M])$ for unimolecular reactions one must determine the steady state energy distribution arising from the competition between reaction and collisional energy transfer The MULTWELL and MESMER programs, mentioned above, can be used to calculate $k(T,[M])$. Most of the needed inputs are computed in the course of the quantum chemistry calculations needed to use Eqs. (3.14)–(3.16). The major exceptions are the terms describing energy transfer between the reactant and the bath gas.[66] These are usually Lennard–Jones parameters needed to compute the collision rate constant and the average energy transferred per collision. Common approximations are available to estimate Lennard–Jones parameters,[67] but one must make incompletely justified assumptions about the efficiency of energy transfer.

Tunneling corrections to rate constants for isomerization will also depend on the steady state energy distribution, and are larger the further one is from the high-pressure limit.[68,69] Computing tunneling effects outside the high-pressure limits requires incorporating the energy-dependent transmission (tunneling) coefficient, $\Gamma(E)$, into a Master Equation calculation. For computing $k(T,[M])$, the values of $\Gamma(E)$ obtained from Eckart expressions are likely less accurate than $\kappa(T)$ applied in the high-pressure limit.

An additional challenge to theoretical predictions of branching ratios and interpretations of product yield data can occur when alkoxy radicals are produced in ROO• + NO reactions. These reactions are exothermic by 11–14 kcal mol^{-1}, and most of this excess energy is deposited in the alkoxy radical.[70,71] Where the energy, E, in the alkoxy radical exceeds the critical energy for one of its unimolecular reactions, the alkoxy radical is referred to as activated.[72] Depending on the rate of collisional quenching of this excess energy and the rate constant of reaction, $k(E)$, at a given energy, E, these radicals may be able to undergo reactions on sub-nanosecond timescales.[71,73] These "prompt or "chemically activated" reactions are far more rapid than the corresponding thermal reactions. The occurrence of prompt reactions has been inferred from experiment.[70–72] For example, consider the fate of $HOCH_2CH_2O$•, the prototype for β-hydroxyalkoxy radicals produced in the OH-initiated oxidation of alkenes:

$$HOCH_2CH_2OO\bullet + NO \rightarrow HOCH_2CH_2O\bullet^* + NO_2 \qquad (3.22)$$

$$HOCH_2CH_2O\bullet^* + M \rightarrow HOCH_2CH_2O\bullet + M \qquad (3.23)$$

$$HOCH_2CH_2O\bullet^* \rightarrow \bullet CH_2OH + CH_2{=}O \text{ (fast)} \qquad (3.24)$$

$$HOCH_2CH_2O\bullet \rightarrow \bullet CH_2OH + CH_2{=}O \text{ (slow)} \qquad (3.25)$$

$$HOCH_2CH_2O\bullet + O_2 \rightarrow HOCH_2CH{=}O + HOO \qquad (3.26)$$

$$\bullet CH_2OH + O_2 \rightarrow CH_2{=}O + HOO. \qquad (3.27)$$

where the * indicates the chemically activated radical. Since the extent of prompt decomposition depends, in part, on the rate of collisional quenching, the yield of scission products (from reactions (3.24) and (3.25)) versus products of O_2 reaction (reaction (3.26)) depends on pressure, and hence, on atmospheric conditions and altitude. At 293 K and 700 Torr, approximately 30% of $HOCH_2CH_2O\bullet$ radicals produced in reaction (3.22) undergo prompt scission (reaction (3.24)) in preference to quenching to a thermal distribution of energy (reaction (3.23)).[70]

By contrast to alkoxy radical production from ROO• + NO, the reactions of ROO• + ROO• and ROO• + HOO• to produce alkoxy radicals are usually nearly thermoneutral.[19,74,75] As a result, alkoxy radicals produced from ROO• + ROO• or ROO• + HOO• cannot undergo prompt reactions. Chemically activated reactions are well known to atmospheric chemists from production of activated Criegee intermediate in alkene ozonolysis.[76]

In the last several years, physical chemists have been excited by the idea of roaming mechanisms[77] for unimolecular reactions. Roaming occurs when an energized molecule very close to its dissociation energy begins to separate into two fragments, but does not completely dissociate. Instead, the two fragments orbit each other ("roam") and then one fragment can abstract an atom from the other. This may have relevance to HNO production in RO• + NO reactions,[78–81] and also to $RONO_2$ formation from ROO• + NO.[82,83] Transition states for roaming reactions have been found, but the application of TST has not yielded reliable rate constants for roaming reactions.[84]

3.3. Kinetics and Mechanisms of Gas-Phase Reactions of Alkoxy Radicals

3.3.1. *Reaction with O_2*

Absolute rate constants, k_{O_2}, for RO• + O_2 reactions have been measured as a function of temperature for a number of alkane-derived alkoxy radicals. Arrhenius parameters for k_{O_2} of some small alkoxy radicals are listed in Table 3.1. Absolute rate constants listed in Table 3.1 have been measured by LFP-LIF for all the non-chlorinated species, and also inferred from relative rate experiments for CH_3O• and CH_3CH_2O•. For the two chlorinated radicals listed in Table 3.1, Wu and Carr used pulsed photolysis followed by molecular-beam mass spectrometry for analysis.[85–87]

For alkoxy radicals derived from alkanes larger than methane, Atkinson recommended the following expression:[10]

$$k_{O_2} = 2.5\ 10^{-14} \times e^{-300/T}\ \text{cm}^3\ \text{molecule}^{-1}\ \text{s}^{-1}. \quad (3.28)$$

The corresponding pseudo-first order rate constant, k'_{O_2}, for reaction with O_2 in 1 atm of air at 298 K is ~4×10^4 s^{-1}. For methoxy radical

Table 3.1 Arrhenius pre-exponential factor (A), ratio of activation energy to the gas constant (E_a/R, in K), and rate constants at 298 K for reactions of selected alkoxy radicals with O_2

Radical	A^a	E_a/R	T (K)	$k(298)^a$	Ref.
CH_3O•	7.2×10^{-14}	1080	290–610	1.9×10^{-15}	65
	3.9×10^{-14}	900	295–608	1.9×10^{-15}	19
CH_3CH_2O•	2.4×10^{-14}	325	295–355	8.1×10^{-15}	65
$CH_3CH_2CH_2O$•	2.6×10^{-14}	253	220–380	1.0×10^{-14}	65
$CH_3CH(O$•$)CH_3$	1.5×10^{-14}	230	210–390	7.0×10^{-15}	65
$CH_3CH_2CH(O$•$)CH_3$	—	—	291–295	9×10^{-15}	88
CH_2ClO•	2.1×10^{-12}	940	265–306	9.0×10^{-14}	11[b]
$CFCl_2CH_2O$•	2.5×10^{-15}	960	251–341	1.0×10^{-16}	11[b]

[a]In units of (cm^3 molecule^{-1} s^{-1}).
[b]Reanalysis of results of Wu and Carr from Refs. 85–87.

the only unimolecular reaction pathway is loss of hydrogen atom, which is too slow to be atmospherically relevant.[89]

The results of Wu and Carr[85–87] for $CH_2ClO\bullet$ and $CFCl_2CH_2O\bullet$ yield activation energies and A-factors which are very different from those of alkane-derived alkoxy radicals, as shown in Table 3.1. The conditions of their experiments tend to contribute to secondary chemistry: relatively long reaction times (up to 80 ms) or with relatively high radical concentrations ($\sim 4 \times 10^{14}$ molecules cm^{-3}). As a result, additional measurements of these rate constants would be reassuring. Orlando *et al.*[11] called for computational chemistry studies to shed light on the cause of the large influence of the Cl- and CF_2Cl- groups on the rate constant, and we heartily second their suggestion.

To account for the effects of functional groups on k_{O_2}, Atkinson[10,16] suggested the following purely empirical dependence of k_{O_2} at 298 K on the enthalpy of reaction, ΔH_{O_2}, in kcal mol^{-1}:

$$k_{O_2}(298) = 4.0 \times 10^{-19}\ L\ \exp(-0.28\Delta H_{O_2})\ \mathrm{cm}^3\ \mathrm{molecule}^{-1}\,\mathrm{s}^{-1}, \tag{3.29}$$

where L is the reaction path degeneracy (the number of hydrogen atoms on the carbon bound to the radical center). For ether-derived alkoxy radicals of the structure RCH(O•)OR', ΔH_{O_2} is estimated to be -48 kcal mol^{-1}, much larger in magnitude than the values for alkane-derived alkoxy radicals (about -35 kcal mol^{-1}). As noted by Aschmann and Atkinson,[90] the application of Eq. (3.29) would lead to $k_{O_2}(298) = 3 \times 10^{-13}$ cm^3 molecule^{-1} s^{-1}. This value is twelve times higher than the A-factor Atkinson suggested in Eq. (3.28). As a result, Aschmann *et al.* suggested use of a value of $k_{O_2}(298)$ close to the value of the A-factor: 1.5×10^{-14} cm^3 molecule^{-1} s^{-1}. Atkinson still endorsed Eq. (3.29) in his 2007 review.[16]

It should be noted that Eq. (3.29) was based only on results for methoxy, ethoxy, and 2-propoxy radicals. As a result, extension of Eq. (3.29) to alkoxy radicals bearing functional groups may not be warranted. The value of $k_{O_2}(298)$ listed in Table 3.1 for $CH_2ClO\bullet$ is 50 times higher than that for $CH_3O\bullet$, while k_{O_2} for $CFCl_2CH_2O\bullet$ is 80 times higher than that for $CH_3CH_2O\bullet$. These chlorinated species, then, currently provide the only opportunity to test the validity of Eq. (3.29). To enable such a check, we computed ΔH_{O_2} for these two species at the CBS-QB3 level of theory.[91,92] This method is expected to obtain thermochemistry with typical uncertainties of $\sim$2 kcal mol^{-1}. We obtained ΔH_{O_2} of -36.9 kcal mol^{-1} for $CH_2ClO\bullet$ and -27.9 kcal mol^{-1} for $CFCl_2CH_2O\bullet$. Based on Eq. (3.29), this would lead to $k_{O_2}(298)$ of 1.2×10^{-14} cm^3 molecule^{-1} s^{-1} for $CH_2ClO\bullet$, which is about eight times lower than the value obtained by Wu and Carr but six times higher than that for $CH_3O\bullet$. For $CFCl_2CH_2O\bullet$, Eq. (3.29) predicts $k_{O_2}(298)$ of 9.9×10^{-16} cm^3 molecule^{-1} s^{-1}, which is ten times higher than the value observed by Wu and Carr but eight times lower than that for $CH_3CH_2O\bullet$. It appears that the sign of the effect of ΔH_{O_2} on k_{O_2} is accurately reflected by Eq. (3.29), even though Eq. (3.29) is not quantitatively reliable for $CH_2ClO\bullet$ and $CFCl_2CH_2O\bullet$. Given the expected accuracy of the computed thermochemistry, noted above, we conclude that Eq. (3.29) cannot be reliably applied to alkoxy radicals not derived from alkanes.

Table 3.1 does not include values of k_{O_2} obtained by Hein *et al.*[20,37,38] at room temperature for 1-butoxy, 2-butoxy, and 3-pentoxy radicals using the absolute rate method discussed in Sec. 3.2.1.3. Their values of k_{O_2} are close to those for ethoxy and 1- and 2-propoxy radicals. We discuss their approach in more detail in discussing isomerization reactions in Sec. 3.2.

Early work in our laboratory measuring $k_{O_2}(T)$ for 2-butoxy and 3-pentoxy had large uncertainties, probably due to secondary chemistry from high radical concentrations.[93,94] Later work in our laboratory on cyclohexoxy and related radicals used lower radical concentrations and obtained higher precision. The values of k_{O_2} obtained for 4-methylcyclohexoxy radical[95] for $228 \leq T \leq 292$ K are reasonably consistent with the measured absolute rate constants of other secondary alkoxy radicals. Curiously, cyclohexoxy radical, itself,[96] yielded very different Arrhenius parameters: the pre-exponential factor (A-factor) is higher by a factor of $\sim$200 and the activation energy is higher by 3.0 kcal mol^{-1} than values obtained for 2-propoxy. As we noted at the time, the combination of our $k_{O_2}(T)$

with k_{O_2}/k_{C-C} measured as a function of temperature by Orlando *et al.*[97] would imply an unusually large A-factor for β-C–C scission. This contradicted our own calculations (later confirmed by Welz *et al.*[98]) which indicated that C–C scission of cyclohexoxy radical has essentially the same A-factor as other C–C scission reactions.[96] At present, we have no explanation for this unusual result. Our results for perdeuterated cyclohexoxy radical (*cyc*-$C_6D_{11}O\bullet$) over the limited range of 228–267 K yield a lower E_a than normal cyclohexoxy radical, contrary to the expected effect of deuteration.[95]

The review by Orlando *et al.*[11] displays the complete set of available k_{O_2} data for all the above reactions in a series of Arrhenius plots. The interested reader is referred to their paper for more detail and analysis of experimental results. Figure 3.5 displays an Arrhenius plot for methoxy radical, including our recent relative rate results extending to temperatures below 296 K.[99] The Arrhenius

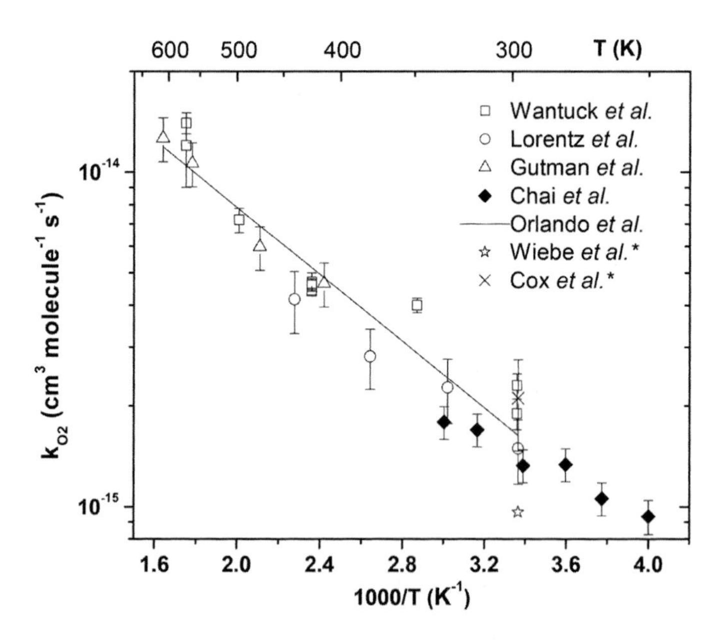

Fig. 3.5 Arrhenius plot for the $CH_3O + O_2$ reaction. The $*$ indicates relative rate data reanalyzed as described in Chai *et al.*[99] The Arrhenius fit is that of Orlando *et al.*[11] Other references are to Wantuck *et al.*,[100] Lorentz *et al.*,[101] Gutman *et al.*,[102] Wiebe *et al.*,[103] and Cox *et al.*[104]

fit shown in Fig. 3.5 is that previously suggested by Orlando *et al.*[11] The appearance of curvature in Fig. 3.5 is enhanced by our results, but it seems premature to suggest a three-parameter fit to this data.

From a kineticist's viewpoint, the A-factors of reactions of alkoxy radicals with O_2 (except, perhaps, cyclohexoxy and $CH_2ClO\bullet$) are curious for being much lower than is typical for direct hydrogen-abstraction reactions. Jungkamp and Seinfeld[105] carried out quantum chemistry calculations and proposed the following mechanism involving a trioxy radical intermediate:

$$CH_3O\bullet + O_2 \rightarrow CH_3OOO\bullet, \tag{3.30}$$

$$CH_3OOO\bullet \rightarrow [\bullet CH_2OOOH] \rightarrow CH_2{=}O + HOO. \tag{3.31}$$

Bofill *et al.*[106] found that Jungkamp and Seinfeld made an error which invalidated this mechanism. Also, we now know that trioxy radicals are tricky molecules for quantum calculations,[107,108] which would raise questions about the reliability of the energy differences computed by Jungkamp and Seinfeld. Bofill *et al.* showed that the low Arrhenius pre-exponential term for the $CH_3O\bullet + O_2 \rightarrow CH_2{=}O + HOO$ reaction could be explained by a cyclic transition state. This transition state, depicted in Fig. 3.6, involves a non-covalent interaction between the oxygen atom of methoxy and the non-reacting oxygen atom of O_2.

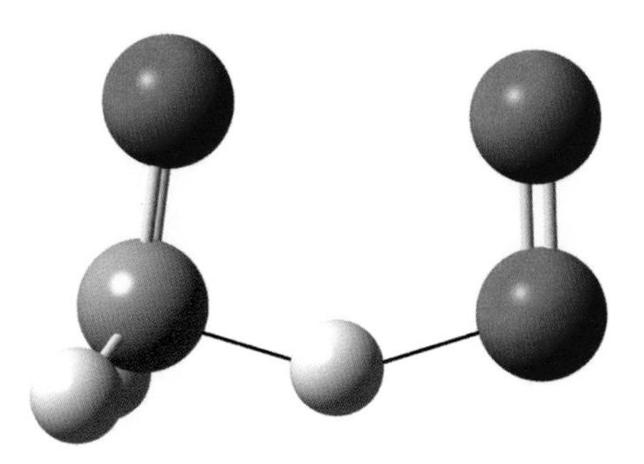

Fig. 3.6 Structure of the transition state for the $CH_3O\bullet + O_2$ reaction.

Setokuchi and Sato used the mechanism of Bofill *et al.* to compute k_{O_2} for methoxy, ethoxy and 2-propoxy radicals, and included a reliable treatment of tunneling.[109] They obtained very satisfying agreement with experiment for all three species. Setokuchi and Sato reported a tunneling correction, $\kappa(298)$ of about 2 for $CH_3O\bullet$, and lower values for the two larger radicals.

Since publication of the most recent review on alkoxy radicals, we attempted to validate a theoretical method for calculations of k_{O_2}.[108] The long term goal was to extend calculations to a wider variety of alkoxy radicals than those derived from alkanes, such as those derived from isoprene or OVOCs We first wanted to verify that the agreement of the theoretical k_{O_2} with experiment was not due to fortuitous cancellation of error. Our efforts included experimental studies of $k_{O_2}(T)$ for $CH_3O\bullet$ and $CD_3O\bullet$,[99] together with a determination of the temperature-dependent branching ratio, k_{32a}/k_{32b}, for the two reaction pathways of $CH_2DO\bullet$:[110]

$$CH_2DO\bullet + O_2 \rightarrow CHDO + HOO, \tag{3.32a}$$

$$CH_2DO\bullet + O_2 \rightarrow CH_2{=}O + DOO. \tag{3.32b}$$

At the same time, we carried out a series of calculations to determine what methods could accurately reproduce the experimental results.

Hu *et al.* measured the branching ratio k_{32a}/k_{32b} as a function of temperature ($250\,\text{K} \leq T \leq 333\,\text{K}$) by determining the yields of $CH_2{=}O$ and $CHD{=}O$ in a chamber by FTIR.[110] An Arrhenius plot of the experimentally determined branching ratio k_{32a}/k_{32b} exhibited little curvature. This suggested that tunneling plays only a small role in the rate constant, which was consistent with the results of Setokuchi and Sato.[109] This conclusion was confirmed and refined as a result of calculations by Hu and Dibble,[108] who found $\kappa(298) = 2.2$ for the production of $CHD{=}O + HOO$. Surprisingly, $\kappa(298)$ was only $\sim$10% lower for production of $CH_2{=}O + DOO$. Hu and Dibble also reproduced k_{32a}/k_{32b} to within one standard deviation of the experimental values over the range $250\,\text{K} \leq T \leq 333\,\text{K}$. The only computational method that produced a reasonable critical energy was that used by Setokuchi and Sato:[109] a composite method that uses spin-restricted open-shell CCSD(T) to remove the influence of the quartet state from the doublet wavefunction of the transition state.[108] Most calculations not using this approach are expected to overestimate the critical energy and grossly underestimate the rate constant for alkoxy + O_2 reactions.

The deuterium kinetic isotope effect (k_{33}/k_{34}) for:

$$CH_3O\bullet + O_2 \rightarrow CH_2{=}O + HOO \quad (3.33)$$

$$CD_3O\bullet + O_2 \rightarrow CD_2{=}O + DOO \quad (3.34)$$

was determined by Chai *et al.* over the range $277\,K \leq T \leq 333\,K$ from the results of two types of experiments.[99] The first experiment used product yield studies in a chamber to determine the ratio k_{O_2}/k_{NO_2} (separately for each isotopologue) at 700 Torr, where k_{NO_2} refers to:

$$CH_3O\bullet + NO_2 \rightarrow CH_3ONO_2, \quad (3.35)$$

$$CD_3O\bullet + NO_2 \rightarrow CD_3ONO_2. \quad (3.36)$$

A second experiment used LFP-LIF to determine k_{NO_2}. The inferred rate constant, k_{35}, at and above 298 K was in reasonable agreement with previous absolute measurements. Unfortunately, when combining $k_{35}(T)$ with $k_{33}(T)/k_{35}(T)$, Chai *et al.* obtain an A-factor and activation energy somewhat lower than previously reported for k_{33} (k_{O_2} for $CH_3O\bullet$). This difference, which is apparent in Fig. 3.5, cannot be explained by tunneling effects. The experimental KIE (k_{34}/k_{36}) does not show as much temperature dependence as might be expected (see Fig. 3.7), but the theoretical KIE was consistently 72–82% of the experimental value. The success of theory in reproducing and predicting experimental kinetic data suggests that it would be productive to use this approach for investigating the effects of functional groups on k_{O_2}.

It would be valuable to have direct, absolute measurements of k_{33} below 298 K, but its low value at 298 K already requires the use of large concentrations of O_2 (~50 Torr),[101] and O_2 is an efficient quencher of alkoxy radical fluorescence.[111] Experiments at lower

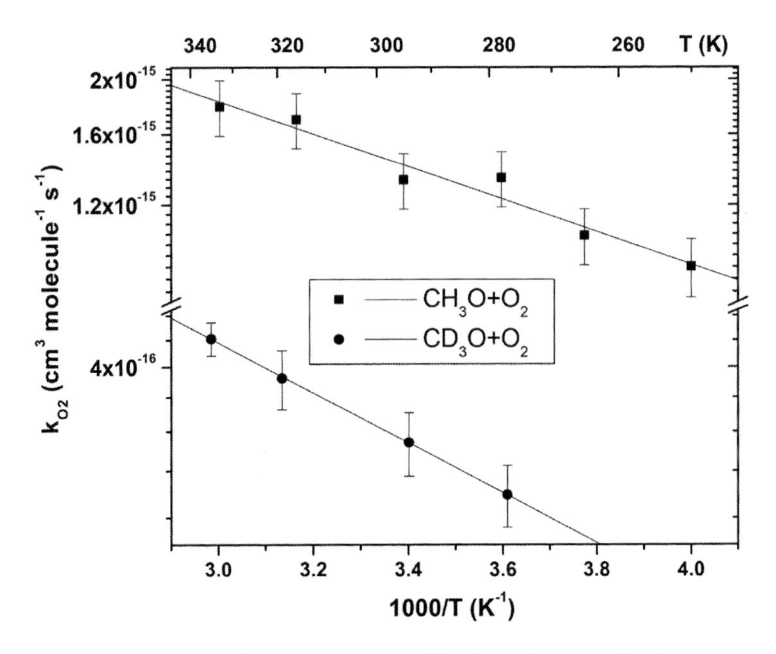

Fig. 3.7 Arrhenius plot for the reactions $CH_3O\bullet + O_2$ and $CD_3O\bullet + O_2$, from Ref. 99.

temperatures (or with deuterated methoxy radical) would require much higher $[O_2]$, thereby leading to significant degradation in signal to noise ratios.

3.3.2. *Isomerization*

Isomerization of alkoxy radicals in the context of VOC oxidation was first suggested by Carter *et al.*[112] The only absolute measurement of isomerization rate constants in the gas phase comes from Hein *et al.*[38] As was depicted in Fig. 3.3 in Sec. 3.2.1.3, they photolyzed alkylbromides to selectively generate a particular isomeric alkyl radical, and measured the time history of [OH] and $[NO_2]$ in the presence of O_2 and NO. Using kinetic modeling, they were able to determine $k_{\rm iso}$ and k_{O_2} for 1-butoxy radical and upper and lower limits to $k_{\rm iso}$ and k_{O_2} for 1-pentoxy radical. A key source of uncertainty is that while their modeling accounted for the loss of OH by reaction with the alkylbromide, e.g.,:

$$OH + C_4H_9Br \rightarrow \text{Products} \tag{3.37}$$

they did not try to account for secondary chemistry following reaction (3.37). Of course, such modeling would have required estimating a great number of rate constants, most critically, those for the O_2 reaction and unimolecular reactions of the various brominated alkoxy radicals produced in this secondary chemistry. We carried out this modeling[113] effort for the case of 1-butoxy radical, and found significant deviations from the time history of [OH] and [NO_2] observed experimentally. A partial experimental test of the effects of secondary chemistry in this type of experiment could be carried out by using photolysis of 1-bromopropane or 2-bromopropane to generate 1-propoxy or 2-propoxy radical, respectively. Since neither of these alkoxy radicals would undergo significant unimolecular reaction, deviations from time history of [OH] and [NO_2] predicted from the (known) rate constants for propoxy radicals + O_2 would be due to secondary chemistry.

Atkinson reviewed rate constants, k_{iso}, of 1,5 H-shift reactions in alkane-derived alkoxy radicals.[16] In addition to the absolute rate constants available from the experiments of Hein *et al.* at room temperature, relative rate constants, k_{iso}/k_{O_2}, have been measured as a function of temperature for four alkoxy radicals derived from alkanes: 1-butoxy,[35,114] 1-pentoxy,[115] 2-pentoxy,[115] and 5-methyl-2-hexoxy.[115] These rate constants are summarized in Table 3.2. No experimental data exists that would indicate the effect of functional groups on k_{iso} or k_{iso}/k_{O_2}. Just as in hydrogen abstraction by OH radicals, the nature of the site of abstraction strongly influences the rate constant for isomerization. In addition, β-substitution of the radical center appears to affect k_{iso}. In particular, it has been demonstrated that k_{iso} for 2-pentoxy is 2.5× larger than that for 1-butoxy, despite the fact that both isomerizations are 1,5 H-shifts in which a hydrogen atom is transferred from a methyl group.[115,116]

Table 3.3 summarizes recommendations of Atkinson[16] to be used in estimating rate constants depending on whether the hydrogen

Table 3.2 Experimentally derived Arrhenius pre-exponential factors (A), activation energies (as E_a/R, in K), and rate constants near 298 K and 1 atm (except where specified) for 1,5 H-shift reactions for specific alkoxy radicals. The site of abstraction is also listed.

Reactant	Site	A^a	E_a/R	$k(298)^a$	Ref.
1-Butoxy	$-CH_3$	3.1×10^9	2910	1.8×10^5	[16][b,c]
		5.0×10^9	3020	2.0×10^5	[35][b]
		—	—	1.5×10^5	[25][b]
		—	—	3.5×10^4	[38][c]
1-Pentoxy	$-CH_2-$	5.6×10^{10}	2900	3.3×10^6	[115][d]
2-Pentoxy	$-CH_3$	1.2×10^{11}	3830	3.1×10^5	[115][d]
		—	—	3.1×10^5	[25][b]
5-Methyl-2-hexoxy	>CH-[e]	4.1×10^{10}	2440	1.1×10^7	[115][d]

[a]In units of s^{-1}.
[b]Based on k_{O_2} recommended by Atkinson: $2.5 \times 10^{-14}\ e^{-300/T}\ cm^3\ molecule^{-1}\ s^{-1}$.
[c]At 50 mbar and 293 K.
[d]Recommendation of Atkinson based on multiple studies.
[e]Assumes no contribution from 1,6 H-shift from methyl groups.

atom is transferred from a primary, secondary or tertiary site. As a result of the difference between k_{iso} for 1-butoxy and 2-pentoxy, the recommended Arrhenius expression for abstraction from a methyl group listed in Table 3.3 is very different from that found in Table 3.2 for 1-butoxy radical.

Recall from Sec. 3.3.1 that k'_{O_2} is $\sim 4 \times 10^4\,s^{-1}$ in 1 atm of air at 298 K (for alkoxy radicals similar to alkanes). Therefore, the recommendations of Atkinson indicate that isomerization will outcompete the O_2 reaction by a large margin. As will be discussed later, the presence of functional groups near the abstraction site can greatly increase or decrease the rate constant for isomerization.

Aschmann *et al.* carried out product yield studies on some C_6 and larger cycloalkanes at 298 K.[117] They found k_{iso}/k'_{O_2} corresponding to roughly 10, 30, and 3 in 1 atm of air for cycloheptoxy, cyclooctoxy, and cyclodecoxy radicals, respectively. By comparison, k_{iso}/k'_{O_2} is $\sim$70 for 1-pentoxy radical. They suggested that the relatively low k_{iso}/k'_{O_2} for cycloheptoxy and cyclodecoxy could result from an energy penalty needed to deform the ring and reach the transition state for isomerization.

Table 3.3 Recommendations of Atkinson[16] for Arrhenius pre-exponential factor (A), ratio of activation energy to the gas constant (as E_a/R, in K), and rate constants at 298 K and 1 atm for 1,5 H-shift reactions from selected sites.

Abstraction site	A^a	E_a/R	$k(298)^a$
$-CH_3$	1.2×10^{11}	3825	3.2×10^5
$-CH_2-$	8×10^{10}	3010	3.3×10^6
$>CH-$	4×10^{10}	2440	1.1×10^7

[a]In units of s^{-1}.

A number of authors computed rate constants, k_{iso}, for 1,5 H-shift in 1-butoxy and 1- or 2-pentoxy radical. Several computational papers[116,118–121] reported using TST to compute A-factors for 1,5 H-shift which were almost an order of magnitude higher than the values suggested by Atkinson (in Table 3.3). By contrast, calculations by Vereecken and Peeters[61] and Xu *et al.*[62] that accounted for multiple conformers of 1-butoxy obtained A-factors in good agreement with those recommended by Atkinson.

No experimental studies have confirmed the occurrence of isomerizations via seven-member or larger rings in acyclic alkoxy radicals, although such isomerizations are often included in models of combustion chemistry of peroxy radicals. Transition states for 1,n H-shifts have ring-like structures analogous to the ring structures of cycloalkanes. There have been attempts to estimate the effects of ring-size on critical energies for 1,n H-shifts from the strain energies of cycloalkanes. By contrast, Davis and Francisco found that the ring strain of the transition state is likely not as high as that of a cycloalkane of the same size.[121] In fact, their quantum calculations find the barrier heights for 1,5 and 1,6 H-shifts to be nearly identical in 1-hexoxy and 1-heptoxy, whereas the strain energy of cycloheptane (the analogue of the transition state for a 1,6 H-shift) is thought to be ~6 kcal mol^{-1}, while that that of cyclohexane (the analogue ring for a 1,5 H-shift) is essentially zero.[122] In addition, Vereecken and Peeters found that, compared to reactions via 6- and 7-member ring transition states, reactions through 8- and 9-member rings (1,7 and 1,8 H-shifts) experience 2–3 kcal mol^{-1} penalties in critical energies.[18]

These values are much less than the 10–12 kcal mol^{-1} strain energy the corresponding cycloalkanes.[122] Note that the critical energy is not the only factor favoring a 1,5 H-shift over other 1,n H-shifts: the 298 K rate constant for 1,6 H-shift in 1-hexoxy or 1-heptoxy is predicted to be about one-half that for 1,5 H-shift.[121] The formation of a ring-like transition state decreases the entropy relative to the many configurations accessible to a chain.[123] A larger ring causes a larger entropy penalty for forming a ring-like structure. As a result, isomerization through ring-like transition states containing more than six atoms are disfavored entropically (in the A-factor) as compared to isomerization via a six-member ring.

Several computational reports indicate that that the rate constant for the 1,5 H-shift reaction of 1-butoxy radical is within ~15% of the high-pressure limiting value at 1 atm and 298 K.[13,116,120,124] From the graphical results provided by Somnitz and Zellner,[116] one can infer that this rate constant will be within a factor of ~2 of the high-pressure limiting value under upper tropospheric conditions (0.1 atm and 220 K).[125] All other things being equal, increasing molecular size leads to a more rapid approach to the high-pressure limit. Unfortunately, the H-shift from $-CH_2-$ and $>CH-$ groups have lower barriers than H-shift from $-CH_3$ groups, which will tend to lead to larger deviations from the high-pressure limit. For example, Mereau *et al.* calculated that the rate constant for 1,5 H-shift of 5-methyl-2-hexoxy radical was only one-third the high-pressure limiting value at 298 K and 1 atm.[119] To the extent functional groups lower or raise the barrier to isomerization, pressure effects on rate constants will be more or less important, respectively.

Table 3.4 collects reports of tunneling corrections, κ, from the literature. Values are listed in the high-pressure limit, except where noted. Somnitz included Eckart tunneling in computing pressure- and temperature-dependent rate constants, but did not provide detail about how the impact of tunneling varied with pressure.[120] Davis and Francisco calculated tunneling corrections for a variety of 1,n H-shifts for a range of 1-alkoxy radicals as a function of temperature,[121] and a selection of their results is included

Table 3.4 Selected reports of tunneling corrections, κ, for $1,n$ H-shift reactions of alkoxy radicals from various authors. Unless noted, these refer to values in the high-pressure limit.

Species	n	T(K)	Method	κ	Ref.
Various	5	298	Eckart	1.7–2.7	[119]
1-Butoxy[a]	5	300 (200)	Eckart	2.7 (13)	[120]
1-Butoxy	5	300 (200)	Wigner	5.9 (12.1)	[121]
1-Butoxy[b]	5	300 (200)	SCT	11.0 (500)	[62]
2-Pentoxy[a]	5	300 (200)	Eckart	2.2 (7)	[120]
1-Pentoxy	5	300 (200)	Wigner	5.4 (11)	[121]
1-Pentoxy	6	300 (200)	Wigner	6.2 (13)	[121]
1-Hexoxy	6	300 (200)	Wigner	5.5 (11)	[121]

[a]For total pressure of 1 atmosphere.
[b]SCT = small curvature tunneling.

in Table 3.4. Truhlar and co-workers published two papers[62,126] on 1-butoxy isomerization using a more elaborate and rigorous method for tunneling as opposed the approaches used by other authors. While their initial paper reported $\kappa = 40$ at 298 K,[126] their revised work reported a value of 11.[62] This lower value is still much higher than that computed by Somnitz[120] or Mereau *et al.*[13]

All the above discussion of isomerization reactions refers to alkane-derived alkoxy radicals, that is, those without any functional groups. The biggest challenge to experimental advances in determining how functional groups affect k_{iso} lies in cleanly preparing a single selected isomer of a functionalized radical. Luckily, theoretical calculations are not constrained by this issue.

Ferenac *et al.* computed barrier heights for 1,5 H-shift reactions abstracting from a methyl group in ketone-, ether-, and ester-derived analogs of 1-butoxy radical.[127] Subsequently, Vereecken and Peeters carried out extensive calculations to develop an SAR for a much wider range of $1,n$ H-shift reactions.[18] They included the effects of carbonyl and hydroxyl functional groups and considered different positions of the functional groups relative to the site of abstraction and a variety of ring-sizes for the transition state. Their SAR includes the effects of tunneling, strain energy, and the presence of multiple conformers. Pressure-dependencies of the rate constant were estimated from the work of Mereau *et al.*[119] The result is an SAR that predicts temperature dependence of rate constants at a pressure of 1 atm.

The work of Vereecken and Peeeters[18] indicates that the effect of the ring-size of the transition state on the critical energy in a wide range of alkoxy radicals is very similar to those observed in alkane-derived alkoxy radicals. The largest effect of functional groups was, as expected, connected with their influence on the site of H-abstraction. Illustrative results of effects of functional groups are shown in Table 3.5, largely from Ferenac *et al.*[127] and Vereecken and Peeters.[18] As different levels of theory predict different critical energies, Table 3.5 lists the barrier height relative to that in 1-butoxy:

$$\Delta E_0 = E_0(\text{radical})\text{-}E_0(\text{1-butoxy}) \tag{3.38}$$

as opposed to the absolute values of barrier heights. This approach both condenses the table and takes advantage of a tendency for cancellation of errors in individual quantum chemical methods.

The largest effect listed in Table 3.5 is an 11.5 kcal mol^{-1} increase in barrier height for an ester group, but only for one of the two positional isomers considered. A carbonyl group increases the barrier height except in the case that the hydrogen atom being transferred is aldehylic; in that case, the barrier is reduced, similar to what one sees in hydrogen abstraction by OH. An ether group can either raise or lower the barrier height, depending on its position relative to the site of abstraction.

β-hydroxyalkoxy radicals are formed in the OH-initiated oxidation of alkenes via the following set of reactions:

$$>\text{C=C}< + \text{OH} \rightarrow >\text{C(OH)C}\bullet<, \tag{3.39}$$

$$>\text{C(OH)C}\bullet< + \text{O}_2 \rightarrow >\text{C(OH)C(OO}\bullet\text{)}<, \tag{3.40}$$

$$>\text{C(OH)C(OO}\bullet\text{)}< + \text{NO} \rightarrow >\text{C(OH)C(O}\bullet\text{)}< + \text{NO}_2. \tag{3.41}$$

Thus, it is unfortunate that there is a discrepancy in reports of barriers to 1,5 H-shift in 2-hydroxy-1-butoxy radical. As indicated in Table 3.5, two calculations suggest a very small impact on the barrier to reaction (compared to 1-butoxy radical), while one indicates $\sim$2 kcal mol^{-1} increase in the barrier. As C–C scission is expected

Table 3.5 Suggested effects (ΔE_0, in kcal mol^{-1}) of functional groups on barrier heights for 1,5 H-shifts, relative to that for 1-butoxy radical. The level of theory and basis set is listed in the footnotes to the table

Reactant	ΔE_0	Ref.
$\bullet OC(=O)CH_2CH_2CH_3$	+1.4	[18][b]
$\bullet OCH_2C(=O)CH_2CH_3$	+2.1	[127][a]
	0.5	[18][b]
$\bullet OCH_2CH_2C(=O)CH_3$	+3.4	[127][a]
	+4.3	18[b]
$\bullet OCH_2CH_2CH_2CH(=O)$	−2.6	[18][b]
$\bullet OCH_2CH(OH)CH_2CH_3$	+2.0 (+2.3)	[128][c]
	+0.1	[18][b]
	+0.2	[129][d]
$\bullet OCH_2CH_2CH(OH)CH_3$	+1.6	[18][b]
$\bullet OCH_2CH_2CH_2CH_2(OH)$	−2.7	[18][b]
	−1.8	[16][e]
$\bullet OCH_2CH_2OCH_3$	−4.4	[127][a]
$\bullet OCH_2OCH_2CH_3$	+6.2	[127][a]
$\bullet OCH_2OC(=O)CH_3$	+11.5	[127][a]
$\bullet OCH_2C(=O)OCH_3$	+1.6	[127][a]

[a]B3LYP/6-311G(2df,2p).
[b]CBS-QB3.
[c]BAC-MP4 and (in parentheses) B3LYP/6-31G(d,p).
[d]Both G2 and G4 gave identical ΔE_0.
[e]SAR of Atkinson.

to be much more rapid than the 1,5 H-shift for 2-hydroxy-1-butoxy radical, the discrepancy is not a significant problem for models of atmospheric chemistry. One key species producing β-hydroxyalkoxy radicals is isoprene (see Sec. 3.3.6), but those β-hydroxyalkoxy are expected to mostly undergo scission. Isoprene also produces δ-hydroxyalkoxy radicals (**V** and **VI** in Fig. 3.8) for which C–C scission is slow. Barrier heights for these processes have been calculated at multiple levels of theory,[130] and agree that isomerization of the (Z) isomers has a much lower barrier than that of 1-butoxy. Isomerization of (E)-**V** appears to have essentially the same barrier as 1-butoxy radical.

In the course of developing their SAR for isomerization, Vereecken and Peeters examined tunneling for a range of 1,n H-shift

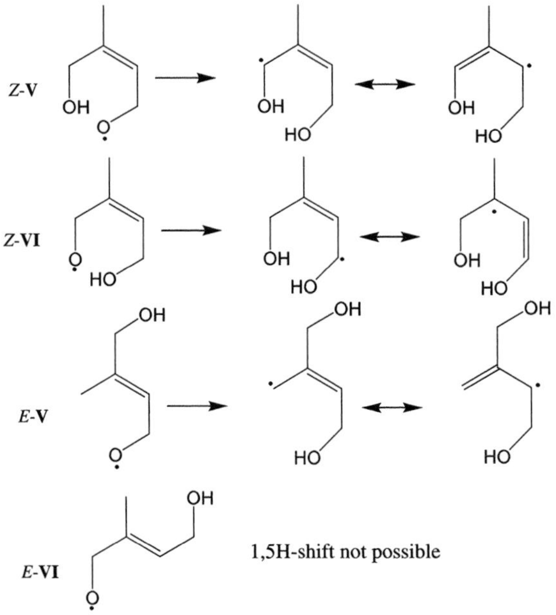

Fig. 3.8 Isomerization reactions of alkoxy radicals from the OH-initiated oxidation of isoprene. From Ref. 130.

reactions. They did not specifically report tunneling corrections, but did report how the presence of hydroxyl and carbonyl groups changed tunneling corrections from the value found for 1-butoxy. They conclude that the presence of these functional groups did not usually change κ by more than a factor of 2 from the value found for 1,5 H-shift in 1-butoxy, even for temperatures as low as 250 K. While the Eckart approach they used to compute κ is not reliably accurate, cancellation of errors likely makes their conclusion fairly robust.

Those seeking predictions of k_{iso} for ester- or ether-derived alkoxy radicals could combine the SAR of Vereecken and Peeters[18] with the results of Ferenac *et al.*[127] For considering the effects of C=C bonds on k_{iso}, Ref. 130 provides a starting point. Although calculations suggest little effect of pressure on the rate constant for 1,5 H-shift in 1-butoxy and 2-pentoxy radicals, there is little understanding of the effects of pressure on the rate constant for 1,n H-shifts possessing lower barriers than these two radicals. These pressure effects are potentially very important for understanding alkoxy radical chemistry in the upper troposphere.

3.3.3. *β-C–C Scission*

3.3.3.1. *Thermal β-Scission Reactions*

Unlike the case for isomerization, there have been direct measurements of absolute rate constants, k_{C-C}, for C–C scission. These have been carried out for ethoxy, 2-propoxy, 2-butoxy, *tert*-butoxy, and cyclohexoxy radicals using LFP-LIF Values of the high-pressure limiting rate constant, $k_{C-C,\infty}$ from these studies are listed in Table 3.6.

The attempt to define a general rate law for C–C scission, even in the high pressure limit, has been hindered by disagreement of a factor of 10 over the correct value of the *A*-factor. Most of these experiments used pressures as high as 60 bar, so experimental conditions were likely in the high-pressure limit. The disagreement over the *A*-factor has not been resolved.

All of the experiments cited in Table 3.6, except for the study of cyclohexoxy, examined the fall-off behavior of k_{C-C}. Most

Table 3.6 Arrhenius expression for the observed high-pressure limiting rate constants (except where noted) for C–C scission reactions of alkoxy radicals.

Reactant	$k_{C-C,\infty}$ (s^{-1})	Ref.
Ethoxy	$1.1 \times 10^{13} e^{-8460/T}$	[131]
	$1.0 \times 10^{14} e^{-9400/T}$	[132][a]
2-Propoxy	$1.2 \times 10^{14} e^{-7660/T}$	[133]
2-Butoxy	$1.1 \times 10^{14} e^{-6450/T}$	[88]
	$1.3 \times 10^{13} e^{-7000/T}$	[35][b,c]
Tert-butoxy	$1.4 \times 10^{13} e^{-6860/T}$	[134]
	$1.4 \times 10^{14} e^{-7280/T}$	[132]
Cyclohexoxy	$3.8 \times 10^{13} e^{-6030/T}$	[98]

[a]Reanalysis of results of Ref. 131.

[b]Based on k_{O_2} recommended by Atkinson: 2.5×10^{-14} $e^{-300/T}$ cm^3 $molecule^{-1}$ s^{-1}

[c]At 1 atm.

experiments agree that C–C scission is close to the high-pressure limit at 298 K and 1 atm of air. Since most of the experiments listed in Table 3.6 used He as a buffer gas, some authors carried out modeling using the Troe expression to predict rate constants in air.

Zellner and coworkers used the same method to measure k_{C-C}[20,37] as they used to measure k_{iso} (see Sec. 3.2). They obtained values of k_{C-C} for 2-butoxy radicals of $(3.5 \pm 2) \times 10^3\,s^{-1}$ and for 3-pentoxy radicals of $(5 \pm 2.5) \times 10^3\,s^{-1}$. These rate constants, obtained at 50 mbar N_2 and 293 K, are very close to, if perhaps slightly lower than, those predicted by Somnitz and Zellner for the identical conditions.[116] The rate constant they obtained for 2-butoxy is quite close to that obtained by Falgayrac *et al.* in He;[88] the rate constant should be higher in N_2 than in He. As noted in Sec. 3.2, secondary chemistry is probably distorting the time history of [OH] and [NO_2] in the experiments of Hein *et al.*

Product yield and relative rate studies have been carried out for a variety of radicals, and these were used by Atkinson to build SARs.[10,16] Rather than carry out yet another review of the same data, we will just note the temperature dependent studies. For alkane-derived alkoxy radicals, k_{C-C}/k'_{O_2} was determined for 2-methyl-2-butoxy and 2-methyl-2-pentoxy by Johnson *et al.*[135] and for cyclopentoxy and cyclohexoxy by Orlando *et al.*[97] For alkene-derived alkoxy radicals, the studies of 2-hydroxyethoxy and 2-hydroxypropoxy radicals deserve attention.[70,71] For acetonoxy radical ($CH_3C(=O)CH_2O\bullet$) C–C scission occurs so rapidly that no evidence of reaction with O_2 was observed.[136]

Atkinson extended previous Evans–Polanyi relationships for the high-pressure limiting rate constant, $k_{C-C,\infty}(T)$.[10,16] Recall that values of k_{C-C} are likely to be close to the high pressure limiting value at atmospheric pressure. Atkinson's goal was to enable estimates of rate constants for large alkoxy radicals with diverse functionality. His recommended equation was:

$$k_{C-C,\infty}(T) = L \times 5 \times 10^{13} e^{-E_d/RT} s^{-1}. \tag{3.42}$$

with $E_d = a + 0.40\Delta H^\circ_{C-C}$ and $a = 1.81 \times (IP/eV) - 4.92\,kcal\,mol^{-1}$, where IP is the ionization potential (in eV) of the alkyl leaving group and ΔH°_{C-C} is the standard state enthalpy change

Table 3.7 Evaluated Arrhenius activation energies (kcal mol^{-1}) of C–C scission reactions for alkane-derived alkoxy radicals. We present recommendations both at 1 atm (E_a, from Ref. 11) and in the high pressure limit (E_d from Atkinson).[16]

Alkoxy radical scission	E_a	E_d
CH_3–CH_2O•	17.3	18.1
RCH_2–CH_2O•	14.8 ± 0.3^a	12.2
CH_3–CHRO•	14.8 ± 0.3^a	15.1
RCH_2–CHR'O•	12.0 ± 0.3	11.9–13.1[b]
$(CH_3)_2CH$–CH_2O•	12.4	12.2
$(CH_3)_3C$–CH_2O•	9.6	10.5
CH_2CH_2–C(O•)$(CH_3)_2$	10.5	13.0

[a]Somnitz and Zellner treated these two cases as identical.
[b]Depends on specific structure of the alkyl radical (R•) and carbonyl (R′CH=O).

(at 298 K) for the C–C scission reaction. The expression reflects the fact that scission seems to be favored by low endothermicity of reaction and low IP (greater substitution) of the alkyl leaving group. Atkinson listed values of a and IP for several leaving groups. Table 3.7 lists selected values of E_d to convey some of the structure dependence incorporated into this SAR. It is interesting that Atkinson and co-workers recently suggested that scission of >C_6 cycloalkoxy radicals was much slower than predicted by their SAR.[117]

Mereau *et al.*[13] built an SAR for $k_{C-C,\infty}(T)$ that was analogous to that of Atkinson.[16] They computed A_∞ at two levels of theory, typically obtaining values of $L \times (1-2) \times 10^{14}\,s^{-1}$. They arrived at formulas for E_d that depended on ΔH°_{C-C} and the IP of the radical fragment. Their SAR differs from that of Atkinson by explicitly including terms in ΔH°_{C-C} that depend on whether the alkoxy radical is primary (RCH_2O•), secondary, RR′CHO•), or tertiary (RR′R″CO•). Mereau *et al.* also proposed a model for C–C scission of alkane-derived alkoxy radicals based on curve crossing in a valence-bond approach.[137] Their analysis has not been extended to functionalized alkoxy radicals.

Curran[14] suggested an Evans–Polanyi relationship for estimating k_{C-C} on the basis of formulae for the reverse reaction: association of an alkyl radical with a carbonyl compound. His equation for E_a is similar to that of Atkinson in depending on ΔH°_{C-C} and the IP. As this SAR was intended to cover temperatures relevant to both atmospheric chemistry and combustion, it uses a modified Arrhenius expression with three parameters. Extending this SAR to alkoxy radicals not explicitly listed in Curran's paper requires estimates of heat capacity versus temperature. It is important to note that Atkinson,[16] Curran,[14] and Mereau *et al.*[13] only used data for alkoxy radicals derived from alkanes.

Somnitz and Zellner[118] argued that the thermodynamics of C–C scission, which is a major part of most SARs, was not known with sufficient precision to build a reliable SAR. Somnitz and Zellner used computational chemistry to compute rate constants at 1 atm, $k_{C-C,1\,atm}(T)$, for scission of linear alkoxy radicals. Their model assigns E_a based on the structure of the fragments of scission.

Somnitz and Zellner suggested an A-factor at 1 atm of $1-2 \times 10^{13}\,s^{-1}$ (times the reaction path degeneracy) for C–C scission of a linear alkoxy radical when it is the sole or dominant unimolecular reaction of that alkoxy radical.[118] For C–C scissions that are not the sole or dominant unimolecular reaction, calculated A-factors at 1 atm vary considerably. For example, although C–C scission of 1-pentoxy and 1-propoxy both can be viewed as examples of $RCH_2CH_2O\bullet \rightarrow RCH_2\bullet + CH_2{=}O$ (see Table 3.7), the A-factor at 1 atm for 1-pentoxy is only $4 \times 10^{11}\,s^{-1}$: at least 25 times lower than the value for 1-propoxy.[118] The difference is that scission is the dominant channel for 1-propoxy, while isomerization (1,5 H-shift) is the dominant channel for 1-pentoxy. This decrease in A-factors in the presence of faster unimolecular reactions reflects the fact that, as shown in Fig. 3.4, when a reaction is not in high-pressure limit the population at energies above the critical energy will be depleted in comparison to a Boltzmann distribution. As a result, the rate constants for competing unimolecular reactions influence each other. The largest effect of this phenomenon is that the faster reaction depletes the population available to undergo the slower reaction.

Somnitz and Zellner framed this mostly as a decrease in A-factors, although they showed how values of E_a would also change on account of competing reactions.

Johnson *et al.* reviewed experimental and theoretical rate constants for scission,[15] and correlated these values with the arithmetic average of the IP of the radical (R•) and carbonyl products of scission:

$$\mathrm{IP_{ave}} = [\mathrm{IP(R\bullet)} + \mathrm{IP(carbonyl)}]/2. \tag{3.43}$$

Their correlation is shown in Fig. 3.9, below. They suggested an A-factor of $1 \times 10^{13}\,\mathrm{s}^{-1}$ (times the reaction path degeneracy, L) and the following formula for alkoxy radicals derived from alkanes:

$$\mathrm{E_a(kcal\ mol^{-1})} = 3.49(\mathrm{IP_{ave}}) - 19.7. \tag{3.44}$$

For alkoxy radicals with, chloro, hydroxyl, or carbonyl substituents in the β position, they suggested a rather different functional dependence of E_a on $\mathrm{IP_{ave}}$:

$$E_a(\mathrm{kcal\ mol^{-1}}) = 2.194(\mathrm{IP_{ave}})^2 - 31.74(\mathrm{IP_{ave}}) + 116.34 \tag{3.45}$$

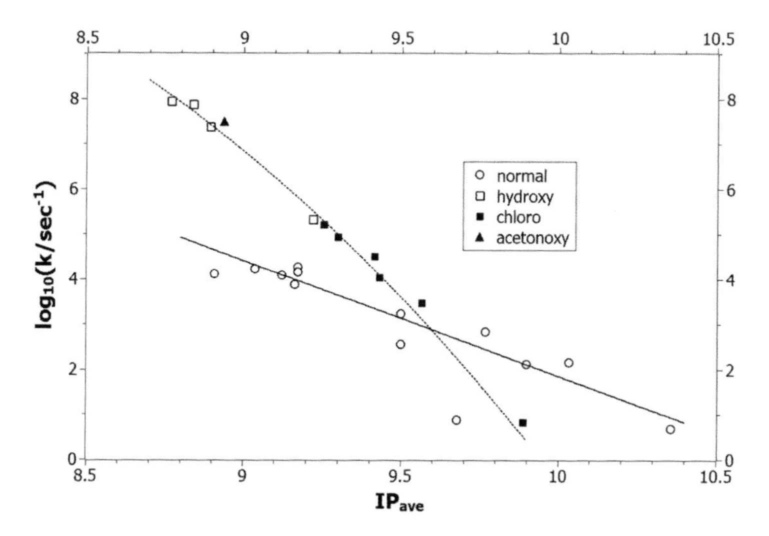

Fig. 3.9 Rate constants for C–C scission of alkane-derived and β-substituted alkoxy radicals and their reliationship to the average ionization potential of the reaction products. From Ref. 15.

This correlation is also shown in Fig. 3.9. The SAR of Johnson *et al.* was based on a mix of rate constants: some measured at one atmosphere and some measured in the high-pressure limit.[15] They did note the potential for errors in interpretation of their data due to chemically activated scission.

A broad overview of the expected fall-off behavior of several alkoxy radicals is provided in the theoretical study by Somnitz and Zellner.[116] The largest and most reactive alkoxy radicals they studied were pentoxy radicals. At both ground-level (298 K and 1 atm) and upper tropospheric conditions, k_{C-C} is predicted to be within a factor of 2 of the high-pressure limiting value. It would be good to determine the effect of pressure on alkoxy radicals with lower critical energies for scission than 2- and 3-pentoxy radical.

β-hydroxyalkoxy radicals are formed in the OH-initiated oxidation of alkenes as described in reactions (3.39–3.41) in Sec. 3.3.2. Given the ubiquity of alkenes, especially isoprene and terpenes, this class of radicals has received extensive study. While no direct measurements of absolute rate constants have been made, extensive chamber experiments and theory have provided significant insights. As noted by Atkinson,[16] critical energies for scission of β-hydroxyalkoxy radicals are about 2–3 kcal mol^{-1} lower than predicted by his SAR (Eq. (3.42)), despite the fact that this SAR qualitatively predicts E_d for β-hydroxyalkoxy radicals to be lower than that for the corresponding alkane-derived alkoxy radicals. The relatively low barriers for scission of β-hydroxyalkoxy radicals means that experiments that employ ROO• + NO to produce RO• are likely to lead to a considerable degree of prompt (chemically activated) reaction This has caused problems in interpreting experiments (see Sec. 3.3.3.2, below).

Vereecken and coworkers published an SAR for C–C scission of alkoxy radicals in 2004[138] and updated it in 2009.[17] Their SAR was based on computations, but was validated against experimental data. Given the values of E_0 predicted by their SAR (they used the notation E_b, for barrier height), one can generate an Arrhenius expression for $k_{\mathrm{C-C},\infty}(298)$ and $k_{\mathrm{C-C,1\,atm}}(298)$:

$$k_{\mathrm{C-C},\infty}(298) = L \times (1.8 \times 10^{13})\ \mathrm{e}^{-E_0/\mathrm{RT}}\ \mathrm{s}^{-1}, \tag{3.46}$$

$$k_{\mathrm{C-C,1\,atm}}(298) = L \times (1 \times 10^{14})\ \mathrm{e}^{-E_\mathrm{a}/\mathrm{RT}}\ \mathrm{s}^{-1}. \tag{3.47}$$

The difference between the critical energy, E_0, and the Arrhenius activation energy, E_d, is explicitly accounted for; they find $E_\mathrm{d} \approx E_0 + 1\,\mathrm{kcal\,mol^{-1}}$. Note that this approximation neglects the effects of competing reactions that was treated by Somnitz and Zellner and discussed above.[118] Vereecken and coworkers approximated the temperature dependence of the pre-exponential factor as $A(T) = A_{298}\ (T/298)^{1.7}$. The core of their SAR is the computation of E_0 from substituent effects, referenced to E_0 for ethoxy radical:

$$E_0 = E_0(\mathrm{CH_3CH_2O\bullet}) + \Sigma_\mathrm{s}(F_\mathrm{s} \times N_\mathrm{s}) \tag{3.48}$$

where the summation is over different substituents, s, of number N_s, each possessing an effect F_s on the barrier height. The presence of multiple substituents that lower E_0 leads to non-additive effects on the barrier height, and an approach to deal with these is described in their second SAR paper.[17] The influence, F_s, of a substituent on the barrier height depends on the position, and this effect is explicitly considered.

The second SAR paper included a large variety of functional groups, including R–, =O, –OH, –OR, $-\mathrm{ONO_2}$, and a few cyclic structures. Substitution of the radical product of scission reduces the barrier more than substitution of the carbonyl co-product; this is consistent with the Evans–Polanyi relationships of Atkinson, Mereau *et al.*, and Curran.[10–14,16] The decrease in E_0 caused by carbonyl substitution is position dependent. For example, when scission of a carbonyl-containing alkoxy radical produces an acyl radical (RC•(=O)), E_0 is lower than if the carbonyl group is retained in the closed-shell product of scission. While most substituents reduce E_0, a C=C bond in the α position, e.g.,:

$$\mathrm{CH_2{=}C(O\bullet)CH_3 \rightarrow CH_2{=}C{=}O + \bullet CH_3} \tag{3.49}$$

drastically increases the barrier height (by 21 kcal mol^{-1}) over that for ethoxy radical. The reader is encouraged to study the SAR for additional insights. It would be interesting to extend the correlation of Johnson *et al.*[15] between activation energies and IP_{ave} to the wide-ranging SAR of Vereecken and Peeters.[17]

Recently, the effect of β-NO_3 substitution was investigated by Yeh *et al.* in chamber experiments on 1-pentadecene oxidation initiated by NO_3.[139] Yeh *et al.* found that the ratio of β-scission to isomerization in the long chain alkoxy radicals was 2.5:1. Assuming that the isomerization rate constant was not affected by β-NO_3 substitution, they suggested that the β-NO_3 lowered the barrier to scission to ~9 kcal mol^{-1}, which is rather lower than the ~12 kcal mol^{-1} suggested by Vereecken and Peeters.[17] Given the importance of NO_3 for oxidation of alkenes and the tendency of nitrate-substituted compounds to partition to SOA, it would be good to resolve the contradiction between the SAR and this experiment.

Ferenac *et al.*[127] studied a more limited set of functional groups (ether, ester, carbonyl) of the structure $RCH_2O\bullet$ in order to contribute to building SARs. Most of the trends they found parallel those of Vereecken and Peeters.[17] Curiously, the two groups report opposite effects of ether substitution on the critical energy for $CH_3OCH_2CH_2O\bullet$ at the same level of theory. Ferenac *et al.* found a correlation between the reaction enthalpy (at 0 K), $\Delta_r H^\circ$, and the critical energy for scission. The IPs of the functionalized radicals, R•, produced in scission of the $RCH_2O\bullet$ radicals they studied are generally not known, but it would be interesting to compute these IPs and check how the results of Ferenac *et al.* fit with the correlation of Johnson *et al.*[15]

Davis and Francisco studied the decomposition of alkane-derived 1-alkoxy radicals and a number of hydroxyl-substituted 1-alkoxy radicals with the hydroxyl groups in various positions on the alkyl chain.[121,129] The results of Davis and Francisco are consistent with those of Vereecken and Peeters.[17]

While explicitly stated in reported SARs for C–C scission, we have tried to emphasize which SARs apply to the high-pressure limit and which estimate k_{C-C} at 1 atm. We noted above that the rate constant for scission of 3-pentoxy radical at 1 atm appears to be within a factor of two of the high-pressure limiting value. To the extent substitution lowers the barrier height for scission, the deviation from the high-pressure limit will increase. To the extent substitution increases the size of the alkoxy radical, reactions will tend to reach the high-pressure limit at lower pressures. The competition between these two factors needs to be investigated.

3.3.3.2. *Chemically Activated β C–C Scission Reactions*

We now turn to considering chemically activated C–C scission, an effect which can confound efforts to build SARs from experimental data. Both experimental and theoretical evidence indicates that alkoxy radicals produced from the exothermic ROO• + NO reaction, especially β-hydroxyalkoxy radicals, can undergo chemically activated scission.[70–73,140]

Experimentally, the occurrence of chemically activated unimolecular reactions can be determined from product yield data obtained by varying $[O_2]$ while holding the total pressure constant. For example, Orlando *et al.* studied the prompt decomposition of 2-hydroxyethoxy radical derived from the oxidation of ethene.[70] The mechanism shown here starts from the peroxy radical $HOCH_2CH_2OO\bullet$ and the symbol * indicates the chemically activated species:

$$HOCH_2CH_2OO\bullet + NO \rightarrow HOCH_2CH_2O\bullet^* + NO_2, \quad (3.22)$$

$$HOCH_2CH_2O\bullet^* + M \rightarrow HOCH_2CH_2O\bullet + M, \quad (3.23)$$

$$HOCH_2CH_2O\bullet^* \rightarrow \bullet CH_2OH + CH_2{=}O, \quad (3.24)$$

$$HOCH_2CH_2O\bullet \rightarrow \bullet CH_2OH + CH_2{=}O, \quad (3.25)$$

$$HOCH_2CH_2O\bullet + O_2 \rightarrow HOCH_2CH{=}O + HOO, \quad (3.26)$$

$$\bullet CH_2OH + O_2 \rightarrow CH_2{=}O + HOO. \quad (3.27)$$

Assuming no additional losses of $HOCH_2CH_2O\bullet$, the molar yield, Y $(0 \le Y \le 1)$, of glycolaldehyde ($HOCH_2CH{=}O$) from Reaction (3.26) depends only on the fraction, α, of the excited radicals that are collisionally quenched in reaction (3.23) (as opposed to undergoing prompt scission in reaction (3.24)) and the fraction of quenched radicals that react with O_2. Then Y can be expressed as:

$$Y = \alpha\{k_{26}[O_2]/(k_{26}[O_2] + k_{25})\} \tag{3.50}$$

Taking the reciprocal of Eq. (3.50) gives the following equation:

$$1/Y = (1/\alpha)(k_{25}/(k_{26}[O_2]) + 1). \tag{3.51}$$

Plotting $1/Y$ vs. $1/[O_2]$ should yield a straight line whose intercept equals $1/\alpha$ and whose slope equals $k_{25}/(k_{26}\alpha)$.

Figure 3.10 shows a sketch of the initial energy distribution of $HOCH_2CH_2O\bullet$ produced from $HOCH_2CH_2OO\bullet + NO$. Figure 3.10 also illustrates the strong energy dependence of the rate constant for scission and the competition between collisional quenching and prompt scission.

Figure 3.11, illustrates how chemically activated and thermal reactions occur on separate time scales. A significant fraction of $HOCH_2CH_2O\bullet$ (some of the $HOCH_2CH_2O\bullet^*$) undergoes scission in mere nanoseconds to produce $HOCH_2\bullet + H_2C{=}O$. The speed of this reaction is why the adjective "prompt" is used for such

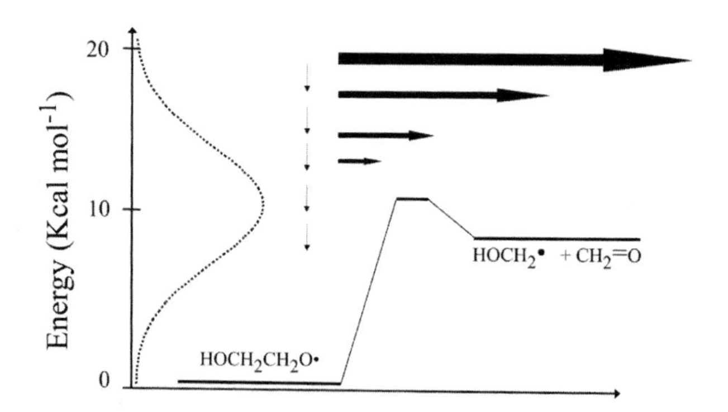

Fig. 3.10 Illustration of the potential energy profile, initial energy distribution, and competition between prompt reaction and quenching of $HOCH_2CH_2O\bullet$ produced from $HOCH_2CH_2OO\bullet + NO$. The dotted curve corresponds to the probability (on the x-axis) of $HOCH_2CH_2O\bullet$ being produced with the energy indicated on the y-axis. The areas of the bold horizontal arrows represent the rate constants for decomposition. The narrow vertical arrows illustrate energy loss via collisions.

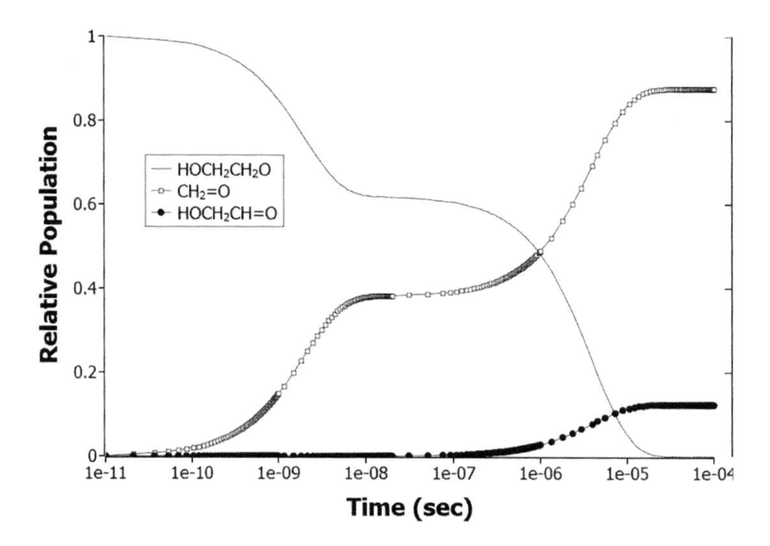

Fig. 3.11 Simulation of the time history of $HOCH_2CH_2O\bullet$ produced from $HOCH_2CH_2OO\bullet + NO$, along with products of prompt and thermal scission of $HOCH_2CH_2O\bullet$ and thermal reaction with O_2. Simulated from data in Refs. 70 and 73 at ~1 atm air and 298 K.

chemically activated reactions. Following this, one can see plateaus in the relative population of $HOCH_2CH_2O\bullet$ and of $H_2C{=}O$ from 10 to 100 ns. These plateaus indicate that almost no reaction occurs from 10 to 100 ns, reflecting that activated $HOCH_2CH_2O\bullet$ has been largely deactivated (quenched) by 10 ns. Thermal C–C scission does not start occurring to a significant extent until after 100 ns. Note that reaction of chemically activated alkoxy radicals with O_2 is feasible, but usually assumed unimportant. See, however, Ref. 141 regarding chemically activated alkyl radicals reacting with O_2.

The extent of prompt decomposition is lowered by increasing the size of the radical and by increasing the total pressure. As temperature has only a modest effect on the initial energy distribution illustrated in Fig. 3.10, prompt decomposition should be more favored in the upper troposphere than at ground level. Evidence has been presented for prompt C–C scission in CF_3CFHO,[72] $CH_3C({=}O)CH_2O\bullet$,[136] $HOCH_2CH_2O\bullet$,[70] $HOCH_2CH_2(O\bullet)CH_3$,[71] and alkoxy radicals from pinonaldedyde,[142] β-pinene,[143] butenes,[144] and isoprene.[144]

To date, there is only a little computational evidence for prompt isomerization reactions in alkoxy radicals produced from ROO• + NO reactions.[145] Because isomerization has a smaller *A*-factor than scission, isomerization will have a much smaller tendency than scission for prompt reaction (for the same critical energy barrier). Isomerization reactions with particularly low barriers might be subject to prompt reactions.

3.3.4. *Reactions with NO and NO_2*

The reactions of alkoxy radicals with NO and NO_2 are grouped together because they share a number of similarities in mechanism and kinetics. These reactions are unlikely to be important in the atmosphere, except perhaps for some tertiary alkoxy radicals (which cannot react with O_2) under the high $[NO_x]$ present in the near-source region of a power plant plume. These reactions are frequently used as reference reactions in relative rate studies in the laboratory, and can occur at the high levels of $[NO_x]$ used in some chamber experiments. These reactions have not been included in the review articles cited in the Introduction, but were discussed in a 1994 monograph.[146]

The reaction of RO• with NO_2 proceeds by addition to form an energized $RONO_2$ molecule which can either be deactivated by collisions or disproportionate (if it does not back-dissociate to reactants). The reaction sequence is listed below for methoxy radical:

$$CH_3O\bullet + NO_2 \rightarrow CH_3ONO_2^*, \tag{3.52}$$

$$CH_3ONO_2^* \rightarrow CH_3O\bullet + NO_2, \tag{-3.52}$$

$$CH_3ONO_2^* \rightarrow CH_2{=}O + HONO, \tag{3.53}$$

$$CH_3ONO_2^* + M \rightarrow CH_3ONO_2. \tag{3.54}$$

The net reactions can be written as:

$$CH_3O\bullet + NO_2 \rightarrow CH_3ONO_2, \tag{3.35}$$

$$CH_3O\bullet + NO_2 \rightarrow CH_2{=}O + HNO_2. \tag{3.55}$$

There is no evidence for formation of HNO_2 in the reactions of larger alkoxy radicals with NO_2.

Not included in the reactions (3.52)–(3.54), above, is the formation of an isomer of $RONO_2$, namely ROONO. Various lines of evidence suggest this isomer is thermally unstable, and when formed will largely back-react to RO• + NO_2.[147–150] The critical energies for CH_3OONO to form a carbonyl compound plus HONO appear to be far too large for these reactions to be important at atmospheric temperatures. There is no direct experimental evidence for formation of CH_3OONO or larger ROONO. ROONO may also isomerize to $RONO_2$.[82,83,151,152]

Direct kinetic investigations of the CH_3O• + NO_2 reaction (using LIF detection of CH_3O•) have been carried out at pressures ranging from 0.6 to 700 Torr over the temperature range 220–473 K, with Ar, CF_4, He or N_2 as bath gases.[153–160] McCaulley *et al.*[154] studied this reaction at low pressures (0.6–5 Torr) and inferred a temperature dependent rate constant expression for the disproportionation reaction as $k_{55} = 9.5^{+17.3}_{-2.7} \times 10^{-12} \exp[-(1150^{+550}_{-170})/T]$ cm^3 molecule^{-1}s^{-1}. This result was based on changes in the pressure-dependence of the overall rate constant at 390–473 K. It would be valuable to have a more direct measurement of this rate constant and its pressure dependence. Note that all these experiments monitored the loss of CH_3O•, so the observed rate constant was the sum of k_{35} and k_{55}. Subsequent studies computed the extent of disproportionation using McCaulley *et al.*'s rate expression for k_{55}, and concluded that disproportionation only competes with recombination at pressures less than 1 Torr. As a result, the rate constant measured for loss of CH_3O• in reaction with NO_2 is believed to essentially equal k_{35} over most atmospheric conditions.[153,158,159]

Biggs *et al.*[155] and Frost and Smith[156,157] obtained k_0 and k_∞ for the recombination reaction (reaction (3.35)) at room temperature. Wollenhaupt *et al.*,[158] Martínez *et al.*,[159] and Chai and Dibble[153] determined both the temperature- and pressure-dependence of the recombination reaction. Chai and Dibble found that the rate constant for methoxy + NO_2 clearly approaches the high-pressure limit at 700 Torr N_2. Table 3.8 lists Troe parameters obtained from absolute

Table 3.8 Troe parameters (Eqs. (3.18)–(3.19)) for the methoxy + NO_2 reaction. Cited errors are 2σ. Units for k_0 and k_∞ are $cm^6\ molecule^{-2}\ s^{-1}$ and $cm^3\ molecule^{-1}\ s^{-1}$, respectively. Note Refs. 19 and 65 are recommendations by JPL and IUPAC respectively. M represents the type of bath gas used as third body collider.

	Troe parameters				Experimental condition		
k_0 (298 K) ($\times 10^{-29}$)	n	k_∞ (298 K) ($\times 10^{-11}$)	m	M	P range (Torr)	T range (K)	Ref.
4.3 ± 0.4	1.7 ± 1.1	1.9 ± 0.03	1.1 ± 0.08	N_2	30–700	250–333	[153]
5.2 ± 0.3	4.4 ± 0.4	1.8 ± 0.05	1.9 ± 0.2	Ar	10–200	233–356	[158]
4.0 ± 0.05	1.7 ± 0.4	2.4 ± 0.02	0.9 ± 0.3	He	50–600	250–390	[159]
2.7 ± 0.8	4.5 ± 1.3			He	0.6–4.0	220–473	[154]
5.3 ± 0.2		1.4 ± 0.1		He	1.0–10	298	[155]
9.0 ± 1.9		1.9 ± 0.3		Ar	6–100	295	[156,157]
5.2 ± 1.9		2.1 ± 0.6		He	30–100	295	[156,157]
11.0 ± 3.0		2.0 ± 0.5		CF_4	30–125	295	[156,157]
5.3	4.0 ± 2.0	1.9	1.8	air			[19]
8.1	4.5	2.1	1.8	air			[65]
		1.4 ± 0.15		He	12–612	300	[160]

rate experiments. All of these studies reported similar values of $k_0(298)$ when the experimental data were fit using $F_{cent} = 0.6$, taking into account the effects of different bath gases. Unfortunately there is a significant disagreement in the value of the exponent, n, in the expression (Eq. (3.18)) for the temperature dependence of k_0. The values of n reported by us and by Martínez *et al.* were less than half of those obtained by McCaulley *et al.* and Wollenhaupt *et al.*, who employed lower pressures.

The high-pressure limiting rate constant at any one temperature, is, by definition, independent of the type of third body molecule. The value of $k_\infty(298\,K)$ of Biggs *et al.* is surprisingly lower than that reported in the other studies discussed here. This is probably because the pressure ($\leq$10 Torr) used by Biggs *et al.* is rather low for an accurate determination of k_∞. Martínez *et al.*'s value of $k_\infty(298\,K)$ is significantly larger than the rest, despite the fact that it was obtained using He rather than N_2 or Ar as the bath gas. This contradicts a large literature which indicates that He is less efficient at collisional quenching than N_2 or Ar. Thus, the results of Martínez *et al.* are an anomaly. There is reasonable agreement in the values of the exponent, m, in the expression (Eq. (3.19)) for the temperature dependence of k_∞ from the other three studies discussed here. In addition, our values of rate constants for the reaction $CD_3O\bullet + NO_2$ agreed very well with those for $CH_3O\bullet + NO_2$.

Absolute rate constant measurements of $RO\bullet + NO_2$ have also been measured for larger alkoxy radicals, including ethoxy,[156] 1-propoxy,[161] 2-propoxy,[161,162] 2-butoxy,[163] and *tert*-butoxy radicals.[164] The pressures investigated in these studies range from a few Torr to 100 Torr but no pressure dependences were observed. It is generally assumed this lack of pressure dependence is because the addition channels have reached the high-pressure limit at a few Torr. Table 3.9 lists k_∞ for these alkoxy radicals and its temperature dependence, where known. Slightly negative temperature dependences were found, and the results are given in the form of Arrhenius expressions in Table 3.9. Room temperature values of k_∞ for larger alkoxy reacting with NO_2 are in the range of $2.4{-}3.7 \times 10^{-11}$ cm^3 molecule^{-1} s^{-1}.

The reaction mechanism of $RO\bullet + NO$ is similar to that of $RO\bullet + NO_2$:

$$CH_3O\bullet + NO \rightarrow CH_3ONO*, \tag{3.56}$$

$$CH_3ONO^* \rightarrow CH_3O\bullet + NO, \tag{-3.56}$$

Table 3.9 High-pressure limiting rate constants at room temperature and Arrhenius parameters for alkoxy + NO_2 reactions. Units of k_∞ and A are 10^{-11} cm^3 molecule^{-1} s^{-1}.

	k_∞ (298 K)	M	P (Torr)	T(K)	A	E_a (kJ mol^{-1})	Ref.
C_2H_5O	2.8 ± 0.3	Ar	15 & 100	295			[156]
1-C_3H_7O	3.6 ± 0.4	He	5 & 40	296			[161]
2-C_3H_7O	3.3 ± 0.3	He	5 & 80	296			[161]
	3.7 ± 0.2	N_2	1 & 10	295–384	1.5 ± 0.8	-2.2 ± 1.3	[162]
2- C_4H_9O	3.5 ± 0.2	N_2	5–78	223–305	0.9 ± 0.3	-3.6 ± 0.3	[163]
tert-C_4H_9O	2.4 ± 0.1	N_2	5–80	223–305	0.4 ± 0.1	-4.6 ± 0.7	[164]

Table 3.10 Troe parameters for $CH_3O\bullet + NO \rightarrow CH_3ONO$. Cited errors are 2σ. Units for k_0 and k_∞ are cm^6 $molecule^{-2}$ s^{-1} and cm^3 $molecule^{-1}$ s^{-1}, respectively. M represents the type of bath gas as third body collider.

Troe parameters				Experimental condition			
k_0 (298 K) ($\times 10^{-29}$)	n	k_∞ (298 K) ($\times 10^{-11}$)	m	M	P (Torr)	T (K)	Ref.
1.35	3.8	3.6	0.6	Ar	3–100	296–573	[156]
1.8 ± 1.3	3.2			Ar	0.8–5	220–473	[166]
1.65	2.8	3.4 ± 0.4	0.8	He, Ar	0.5–500	248–273	[167]
		2.1 ± 0.1		SF_6	10–50	298	[165]
1.69 ± 0.69		2.5 ± 0.3		He	0.7–8.5	300	[168]
2.3	2.8	3.8	0.6	air			[19]

$$CH_3ONO* \rightarrow CH_2{=}O + HNO, \quad (3.57)$$

$$CH_3ONO * + M \rightarrow CH_3ONO. \quad (3.58)$$

A number of absolute rate constant measurements have been carried out on the $CH_3O\bullet + NO$ reaction and these are reviewed in Ref. 65. Troe parameters for the addition pathway are listed in Table 3.10. Sanders *et al.*[165] carried out the first absolute rate study of this reaction and detected HNO with LIF, proving the existence of the disproportionation channel. McCaulley *et al.*[166] studied the kinetics of this reaction using molecular beam mass spectrometry. They determined the temperature dependent rate constants for both the disproportionation channel ($(1.3 \pm 0.4) \times 10^{-12} \exp([(250 \pm 100)/T]$ cm^3 $molecule^{-1} s^{-1}$) and the addition channel in the temperature range 220–473 K at modest pressures. Frost and Smith[156] and Caralp *et al.*[167] measured the rate constant in a much larger pressure range, and they obtained values of $k_0(T)$ similar to those of McCaulley *et al.* In addition, these two studies[156,167] obtained similar values of $k_\infty(T)$, as shown in Table 3.10. Sanders *et al.* and Dobe *et al.* measured k_∞ at room temperature.[165,168] They obtained values of k_∞ that were 30–40% smaller than those obtained by Frost and Smith[156] and Caralp *et al.*[167] Since the latter two results were obtained at higher pressures, their values of k_∞ are preferred. The

Table 3.11 High-pressure limiting rate constants at room temperature and Arrhenius parameters for alkoxy + NO reactions. Units of k_∞(298 K) and A are 10^{-11} cm^3 molecule^{-1} s^{-1}.

	k_∞ (298 K)	M	P (Torr)	T(K)	A	E_a (kJ mol^{-1})	Ref.
C_2H_5O	3.1 ± 0.8	He	0.55–2	298			[169]
	4.4 ± 0.4	Ar	15 & 100	298			[156]
	3.7 ± 0.7	He	30–500	286–388	2.0 ± 0.7	-0.6 ± 0.4	[170]
1-C_3H_7O	3.8 ± 0.1	He	30–100	289–380	1.2 ± 0.2	-3.3 ± 0.5	[170]
i-C_3H_7O	3.3 ± 0.1	He	30–500	286–389	0.9 ± 0.07	-2.9 ± 0.4	[170]
	3.5 ± 0.3	N_2	1–50	295–378	1.2 ± 0.3	-2.6 ± 0.6	[162]
2-C_4H_9O	2.5 ± 0.6	N_2	50–175	266–311	0.8	-3.0 ± 0.5	[93]
	3.9 ± 0.3	N_2	5–78	223–305	0.9 ± 0.3	-3.5 ± 0.6	[163]
	3.2 ± 0.4	He	23–300	223–348	0.4 ± 0.06	-4.9 ± 0.3	[88]
tert-C_4H_9O	2.9 ± 0.1	N_2	5.0–80	223–305	0.8 ± 0.1	-3.2 ± 0.8	[164]
	4.2 ± 0.1	He	70–600	200–390	0.8 ± 0.2	-2.9 ± 0.3	[134]

JPL recommendations (included in Table 3.10) are based on Refs. 156 and 166.

The reactions of larger alkoxy radicals with NO, like those of larger alkoxy reacting with NO_2, are in the high-pressure limit at total pressures of only few Torr near room temperature.[88,93,134,156,157,162–164,169,170] These k_∞ values are listed in Table 3.11, and range from $2.5–4.4 \times 10^{-11}$ cm^3 molecule^{-1} s^{-1}. As is expected, slightly negative temperature dependencies were observed in all these studies. Pressure dependences of k_{NO} could only be investigated for ethoxy radical.[169]

There have been several theoretical investigations of the energetics and mechanisms of $CH_3O\bullet$ + NO_2.[147–151] A potential energy profile is shown in Fig. 3.12. Although there have been several theoretical studies of C_2–C_4 ROO• reacting with NO that include ROONO and $RONO_2$ dissociation to RO• + NO_2, we are not aware of any that focused on the kinetics of $\geq C_2$ RO• reacting with NO_2. The critical energy for disproportionation has not been reported with sufficient reliability to provide insight into the mechanism of production of $CH_2{=}O$ + HONO. Of the four theoretical studies of $CH_3O\bullet$ + NO_2, none reported a transition state for a direct

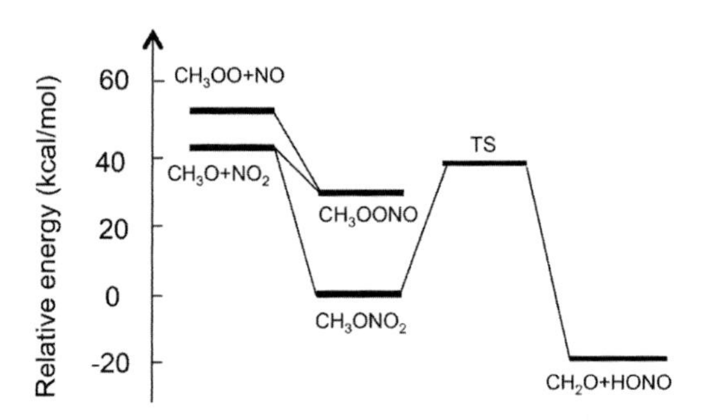

Fig. 3.12 Potential energy profile for the reaction $CH_3O\bullet + NO_2$. TS indicates the transition state.

hydrogen atom-abstraction mechanism for reaction (3.55); but only Pan *et al.*[148] reported trying to find such a transition state.

A roaming mechanism, mentioned in Sec. 3.2.2.3, is likely responsible for isomerization of CH_3OONO to CH_3ONO_2.[82,83] Most computational studies of this reaction used an inappropriate assumption about the nature of the transition state, as discussed in Ref. 83. In our view, the recent computational results of Arenas *et al.* are consistent with this conclusion, although that was not their interpretation.[149]

Theoretical work on the related CH_3ONO system has been carried out using various levels of theory.[78–81] The potential energy profile displayed in Fig. 3.13 is derived from most recent computational results.[78,79,81] Two types of reaction, disproportionation of CH_3ONO and its isomerization to CH_3NO_2, can occur both by an ordinary transition state and a roaming transition state (see end of Sec. 3.2.2.3). The potential energy profile is consistent with the observed activation energy for disproportionation and the absence of formation of CH_3NO_2. A study of the reaction $CH_3CH_2O\bullet + NO$ found that the major reaction channels were similar to those for $CH_3O\bullet + NO$.

The reactions of methoxy radical with NO_2 or NO are examples of radical-radical recombination reactions. The experimental pressure-dependent rate constant of this type of reaction can be used to constrain the RRKM/Master Equation (ME) simulations of the reaction.[171,172]

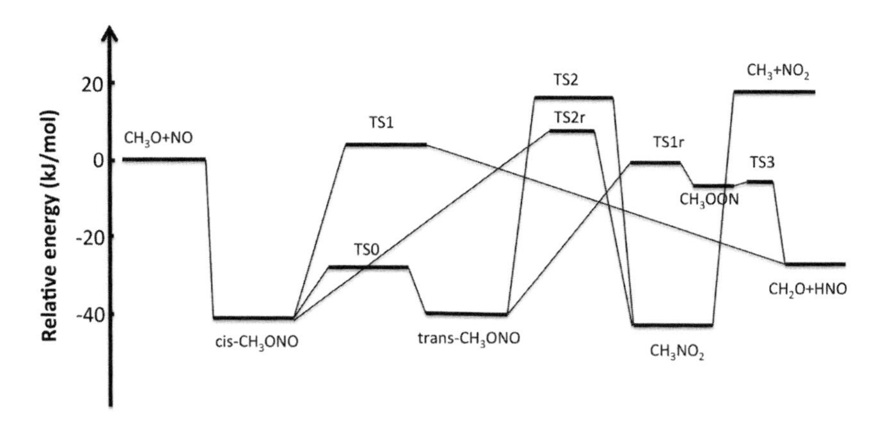

Fig. 3.13 Potential energy profile for $CH_3O\bullet + NO$, combining results discussed in the text. TS indicates transition state.

Barker *et al.*[172] fitted the experimental fall-off curve[158] at 297 K for the $CH_3O\bullet + NO_2$ reaction to estimate the collisional energy transfer parameter (α), which is implicitly defined by:

$$P(E', E) = \frac{1}{N(E)} \exp\left(\frac{-(E - E')}{a(E)}\right), \tag{3.59}$$

where $P(E', E)$ is the probability density for energy transfer from a high vibrational energy, E, to lower energy, E', in a deactivation step, $N(E)$ is a normalization factor. The energy transfer parameter $\alpha(E)$ is almost identical to the average energy transferred in deactivating collisions.

Although the fitted rate constants obtained by Barker *et al.*[172] agreed well with the experimental values, they pointed out that the fitted value of the collisional energy transfer parameter depends enormously on the assumptions used to treat the transition state, which is not a well-defined saddle point on the potential energy surface. Using a similar model, they found a value of α for formation of 2-$C_5H_{11}ONO_2$ that was 1/40 of that for formation of CH_3ONO_2. Unfortunately, this contrasted expectations that α for formation of 2-$C_5H_{11}ONO_2$ would be at least as large as that for formation of CH_3ONO_2. As a consequence, Barker *et al.* concluded that an erroneous assumption made in constructing models or errors in the experimental data caused the unusual results. Barker *et al.* suggested that experimental re-investigations of the kinetics of both reactions in an extended pressure range would enable validation or refinement of their fitted values of α.

Annesley *et al.*[79] fitted their experimental results for $CH_3ONO + M$ in Kr bath gas to an RRKM/ME model, and obtained α of 0.9 kcal mol^{-1} at 300 K. They also fitted the experimental results of Zaslonko *et al.*[173] for the same reaction in a lower temperature range in a different bath gas (argon), and obtained α equal to 0.6 kcal mol^{-1} at 300 K. Annesley *et al.*[79] noted that the value of α for similar reactions in other studies[174,175] showed little difference in α between bath gas Ar and Kr (less than 5%). Therefore, the discrepancy in α values for $CH_3ONO + M$ in Ar vs. Kr raises questions about the accuracy of some experimental results.

3.3.5. *Reactions Occurring in Alkoxy Radicals Formed from OVOC*

3.3.5.1. *α C–H Scission*

The C–H scission of an alkoxy radical produces a carbonyl compound, and is followed by the $H + O_2$ reaction:

$$RR'CHO\bullet \rightarrow RR'C = O + H\bullet, \tag{3.60}$$

$$H\bullet + O_2 \rightarrow HOO\bullet. \tag{3.61}$$

As a result, the products of C–H scission are indistinguishable from those of the direct reaction of the alkoxy radical with O_2. As such, its effects on ozone formation (terminating organic radical chemistry) and SOA formation are indistinguishable from those of the direct reaction with O_2. Under atmospheric conditions, or the conditions of most chamber experiments, the C–H scission and O_2 reactions will not be distinguished. However, that does not mean these two processes *cannot* be differentiated based on product yields; under conditions where C-H scission is occurring, lowering $[O_2]$ but not the total pressure should reduce k'_{O_2} but not the rate constant for C–H scission. Alternatively, reducing the total pressure without changing $[O_2]$ might reduce the rate constant for C–H scission.

There is ample evidence that C–H scission is of little importance for *alkane*-derived alkoxy radicals. Hippler *et al.* surveyed calculations on k_{C-H} for alkane-derived alkoxy radicals.[89] The review article by Curran constructed expressions for k_{C-H} based on rate data for the reverse reactions.[14] Unfortunately, there have been few studies of functionalized alkoxy radicals. Veyret *et al.* first suggested the importance of C-H scission, in $HOCH_2O\bullet$.[176] Jenkin *et al.* carried out relative rate studies for C–H scission and the O_2 reaction of $CH_3OCH_2O\bullet$, finding the two processes to be competitive at a total pressure of 25 Torr but only at very low $[O_2]$ (corresponding to 1-6 Torr).[177] Henon *et al.*[178] computed k_{C-H} for $HOCH_2O\bullet$, $CH_3OCH_2O\bullet$, and $CH_3OCH_2OCH_2O\bullet$, and Song *et al.* refined these calculations for $CH_3OCH_2O\bullet$.[179] Both these computational studies considered the pressure dependence of k_{C-H}.

Experiment and theory offer support for high rate constants for C–H scission in the three alkoxy radicals specified above under atmospheric conditions. At 1 atm and ambient temperature, k_{C-H} seems to be near the high-pressure limit for $CH_3OCH_2OCH_2O\bullet$[178] and far from the high-pressure limit for $HOCH_2O\bullet$;[178] there is disagreement on the value of k_{C-H} for $CH_3OCH_2O\bullet$ and its pressure-dependence.[178,179] The work of Henon *et al.* suggests that C-H scission for $CH_3OCH_2O\bullet$ is about ten times slower than reaction with O_2 in 1 atm of air at 298 K.[178] By contrast, Song *et al.* predicted C–H scission for $CH_3OCH_2O\bullet$ to be somewhat faster than the O_2 reaction under these conditions[179]; their conclusion changed if they adjusted their critical energy to fit the experimental data of Veyret *et al.*[176]

Since β-substitution by hydroxyl and ether functional groups lowers the barrier to C–H scission relative to that for analogous unsubstituted alkoxy radicals, one might expect that other β-substituents, such as chlorine or carbonyl groups, would enhance C–H scission.

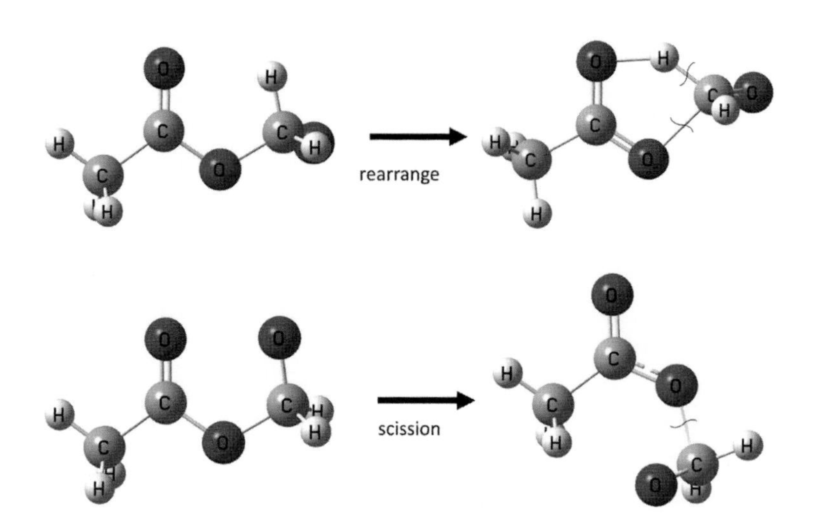

Fig. 3.14 Illustration of the reactant and transitions states for the α-ester rearrangement and C–C scission of $CH_3C(=O)OCH_2O\bullet$. The short curved lines indicate which bonds break as the transition states form products. Note that the two reactions start from different conformers. From Ref. 127.

3.3.5.2. *The α-ester Rearrangement*

The α-ester rearrangement is illustrated by the reaction below and in Fig. 3.14:

$$\mathrm{RC(=\hat{O})OCR'HO\bullet} \rightarrow \mathrm{RC(=O)\hat{O}H + R'C\bullet{=}O} \tag{3.62}$$

which competes with C–C scission:

$$\mathrm{RC(=O)OCHR'O\bullet} \rightarrow \mathrm{RC(=O)O\bullet + R'CH(=O)}, \tag{3.63}$$

where we have labeled the oxygen to which the H-atom is transferred as Ô. This reaction was first described by Tuazon *et al.*[180] Theory firmly indicates that the α-ester rearrangement occurs in a one-step process.[127,181] The α-ester rearrangement should be feasible for any ester of the general structure $RC(=O)OCHR'O\bullet$.

Since the review by Orlando *et al.*[11] there has been only a little experimental work involving this reaction.[182–184] The main conclusions of Orlando *et al.* are presented here. First, there is a barrier of 10–11 kcal mol^{-1} for the rearrangement of $CH_3C(=O)OCH_2O\bullet$ (generated from methyl acetate) and $HC(=O)OCH_2O\bullet$ (formed from methyl formate), and chemical activation can contribute significantly to occurrence of rearrangement in these two radicals. The critical energies suggested by theory for these two radicals are consistent with experiment.

It is noteworthy that Rayez *et al.*[181] computed a much lower barrier (6.5 kcal mol^{-1}) for the α-ester rearrangement of $CH_3C(=O)OCH(O\bullet)CH_3$ (from ethyl acetate) than the 10–11 kcal mol^{-1} noted above for the corresponding radicals from methyl formate and methyl acetate. This is consistent with experimental findings, cited by Orlando *et al*, that the larger the alkoxy group (R′ in reaction 3.63) the lower the barrier to reaction. Both C–C scission and the α-ester rearrangement are enhanced by substitution of the α carbon. Note that the *A*-factor for C–C scission ($2-5\times10^{13}$ sec^{-1}), exceeds that for the α-ester rearrangement ($5-9\times10^{12}$ sec^{-1}).[181,183] This fact may partially explain why the analogous rearrangement transferring alkyl groups:

$$RC(=O)OCHR'O\bullet \rightarrow RC(=O)OR' + HC\bullet(=O) \qquad (3.64)$$

($R' = CH_3$, etc.) has not been observed. In addition, tunneling can greatly increase the rate of the α-ester rearrangement but not reaction (3.64).[127,181]

The critical energy for the α-ester rearrangement computed by theory depends rather sensitively on the choice of quantum chemical approach.[127,181,183] Tunneling appears to be less important in the α-ester rearrangement than in isomerization, but probably non-negligible.[181] The rate constant is believed to be near the high-pressure limit at 1 atm and 298 K,[181] but there is limited information on the combined pressure and temperature-dependence that would allow one to make predictions for upper tropospheric conditions.

Ferenac *et al.*[127] indicated that as the $RC(=\hat{O})O\text{-}CHRO\bullet$ bond is stretched, both oxygen atoms in the incipient $RC(=\hat{O})O\bullet \leftrightarrow RC(\hat{O}\bullet)=O$ species gain radical character. When a hydrogen atom on the –CHRO fragment is oriented so as to be close to $\hat{O}$, as in the upper half of Fig. 3.14, there is a low barrier to transferring that hydrogen atom. When reaction occurs in a conformation in which no hydrogen atom is close to Ô, as in the lower half of Fig. 3.14, hydrogen atom transfer appears to be unlikely.

The corresponding β-ester rearrangement:

$$\begin{aligned} &\mathrm{RC(=\hat{O})OCH_2C\mathbf{H}R'O\bullet} \rightarrow \mathrm{RC(=O)\hat{O}\mathbf{H}} \\ &+ \mathrm{\bullet CH_2C(=O)R} \leftrightarrow \mathrm{CH_2{=}C(O\bullet)R} \end{aligned} \tag{3.65}$$

has not been observed. We suggest that the reason the β-ester rearrangement does not occur is the distance of the $\beta{-}H$ from the oxygen atom (Ô). This could also be viewed in entropic terms as a much lower entropy penalty to form the transition state for the β-ester compared to the α-ester rearrangement.

3.3.6. *Chemistry of Alkoxy Radicals Produced from Isoprene and Related Compounds*

3.3.6.1. *Reactions of First and Second Generation Alkoxy Radicals from Isoprene*

Approximately 500 Mtons of isoprene are emitted to the atmosphere each year,[185] so its atmospheric oxidation pathways and those of its stable degradation products have attracted much attention. Figure 3.15 depicts the first-generation alkoxy radicals produced from the OH-initiated oxidation of isoprene. Refs. 130 and 186 can be consulted for details.

Product yield studies and theoretical calculations agree that the first-generation β-hydroxyalkoxy radicals from isoprene all react via C–C scission to ultimately form formaldehyde plus methylvinylketone or methacrolein (depending on the structure of the radical).[186–189] The fate of the δ-hydroxyalkoxy radicals is somewhat more complicated.[130,186] As previously illustrated in Fig. 3.8 (Sec. 3.3.2), 1,5 H-shift is probably the major fate of all these δ-hydroxyalkoxy radicals except for (E)-**VI**.

Very recently, Nguyen and Peeters determined that δ-hydroxyalkoxy radicals **V** and **VI** can undergo fast (E) $\leftrightarrows$ (Z) isomerization via an oxirane intermediate.[190] This type of cyclization/decyclization is a classic reaction in organic chemistry.[191]

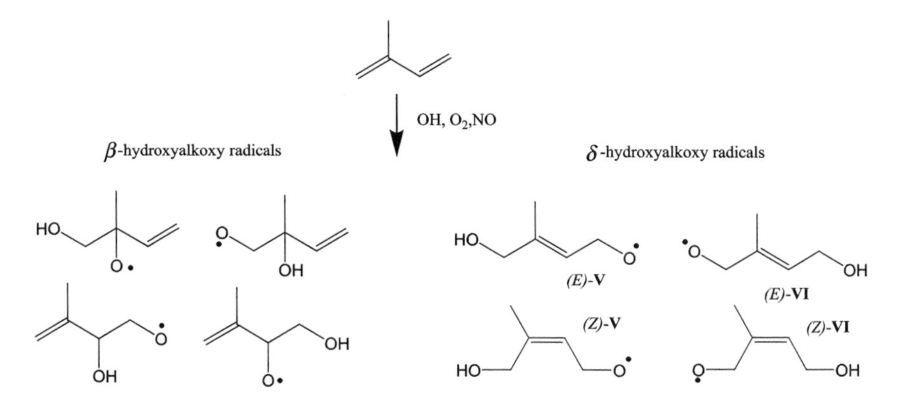

Fig. 3.15 Sketch of first-generation β-hydroxyalkoxy and δ-hydroxyalkoxy radicals formed from isoprene oxidation.

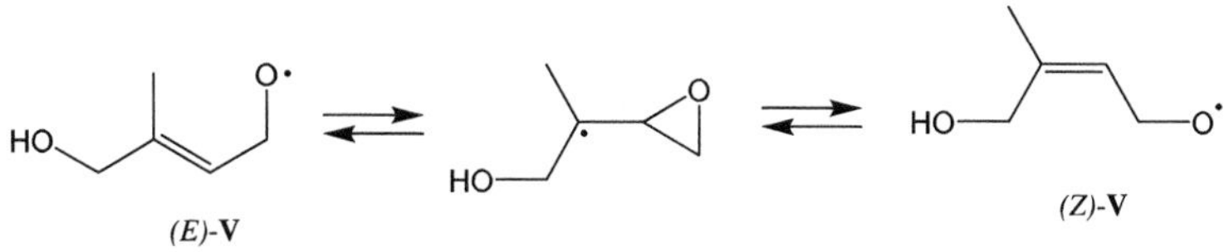

Fig. 3.16 Mechanism of the (E)–(Z) isomerization of δ-hydroxyalkoxy radicals from isoprene illustrated for **V**. The same chemistry occurs for **VI**.

Figure 3.16 depicts the mechanism of this $(E) \leftrightarrows (Z)$ isomerization for **V**. The rate constant for $(E) \leftrightarrows (Z)$ isomerization is computed to be $\sim 2 \times 10^9\,\mathrm{s}^{-1}$ at 298 K and 1 atm.[190] This is much faster than any other chemical reaction of these species. As a result, any (E)-**VI** produced in isoprene oxidation will probably be transformed to (Z)-**VI**, which will undergo 1,5 H-shift.

There have been disagreements in the literature[130,192,193] about the fraction of isoprene degradation leading to the (E) and (Z) isomers of radicals from isoprene (see Fig. 3.15). Interpretation of experimental product yields depends on the fate of first-generation alkoxy radicals and the production of second-generation alkyl, peroxy, and alkoxy radicals from isoprene.[192] The chemistry uncovered by Nguyen and Peeters may help explain why chamber experiments[192] were interpreted to mean that 85% of the first-generation radicals from isoprene possessed the (Z) configuration, as opposed to the ~50% predicted by theory[130] and inferred from another experiment.[193]

Dibble[194] discovered a new reaction type involving a pair of second-generation alkoxy radicals that had first been predicted to form by Paulson and Seinfeld.[186] The route for formation and subsequent reaction of one of these radicals, (Z)-**VIIO**, is shown in Fig. 3.17. These radicals possess two hydroxyl groups on the four-carbon backbone. The low energy conformer of each of these two species does not directly undergo the predicted[186] C–C scission

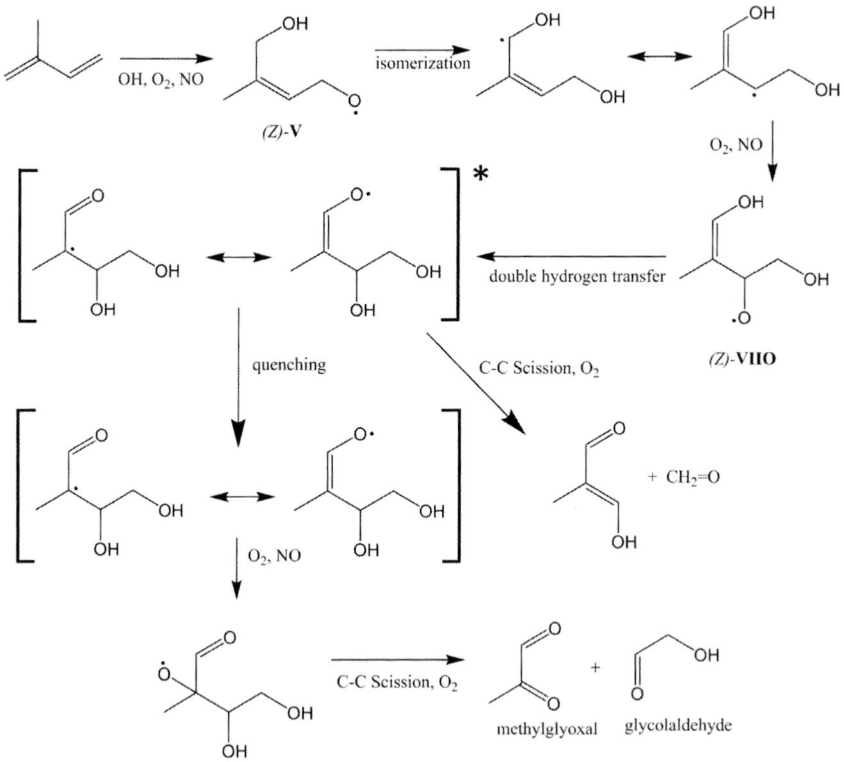

Fig. 3.17 Sequence of reactions leading from the first generation alkoxy radical (Z)-**V** through the second-generation alkoxy radical (Z)-**VIIO** to the double H-atom transfer and the competing reactions of the resonance-stabilized products of the double H-atom transfer. Analogous reactions occur for (Z)-**VI**. The notation "C–C scission, O_2" refers to the scission of parent radical followed by the reaction of the product RC• HOH radical with O_2.

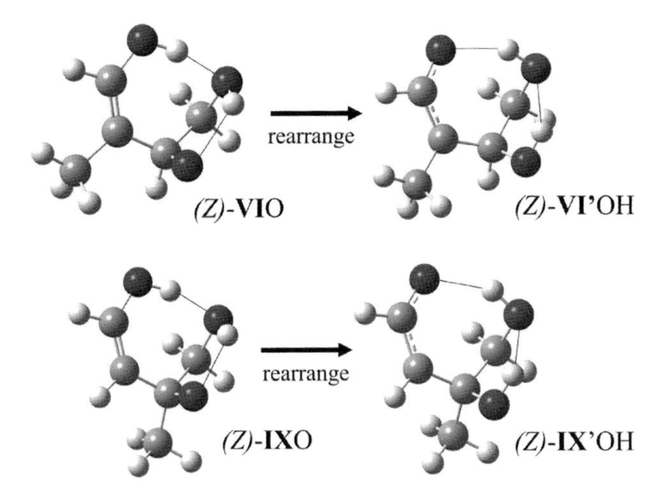

Fig. 3.18 Illustration of double hydrogen atom transfer reactions of (Z)-**VII**O and (Z)-**IX**O. Thin lines represent hydrogen bonds.

to produce C_4 carbonyl compounds plus $\bullet CH_2OH$. Dibble found, instead, a reaction in which hydrogen atoms of both hydroxyl groups were transferred in one step across the intramolecular hydrogen bonds (see Fig. 3.18). This leads to resonance stabilized vinoxy-type radicals: RR′C=CHO• ↔ RR′C•CH(=O). As a result of the conversion of a C=C bond to a C=O bond and the resonance stabilization, these reactions are exothermic by 15–20 kcal mol^{-1}. The fate of these radicals (illustrated in Fig. 3.17) depends on the competition between prompt C–C scission and quenching.[195]

Prompt C–C scission of the vinoxy-type radicals formed from the double hydrogen transfer should ultimately yield formaldehyde plus a stable $C_4H_6O_2$ compound.[195] By contrast, collisional quenching of these radicals would eventually lead (assuming sufficiently high [NO]) to rapid production of glyoxal + hydroxyacetone or methylglyoxal + glycolaldehyde. Paulot *et al.*[192] and Galloway *et al.*[196] found production of glycolaldehyde, hydroxyacetone, and gloyoxal that was too rapid to explain by chemistry following OH attack on stable (non-radical) degradation products of isoprene. They explained their observations by the double-hydrogen transfer proposed by Dibble, assuming that collisional quenching dominated the fate of the vinoxy-type radicals. Note that Peeters and Nguyen[197] have proposed that the peroxy radicals (Z)-**VIOO** and (Z)-**VIIOO** (precursors to (Z)-**VIO** and (Z)-**VIIO**) should undergo single hydrogen-atom transfer across intramolecular hydrogen bonds to ultimately yield the same products as proposed by Dibble. The mechanism of Peeters and Nguyen is consistent with the rapid production of products observed in experiments.[192,196]

3.3.6.2. *Reactions of Alkoxy Radicals from Methacrolein Methylvinylketone*

Methylvinylketone and methacrolein are major products of isoprene oxidation initiated by OH. Chamber studies and theory agree as to the fate of alkoxy radicals formed from methylvinylketone. As shown in Fig. 3.19, there are two competing scission reactions of the internal alkoxy radical, but the one leading to glycolaldehyde (GLYC) + acetyl ($CH_3C{\bullet}{=}O$) appears much faster than the one producing methylglyoxal (MGLY) + $\bullet CH_2OH$.[74,196,198] Quantum

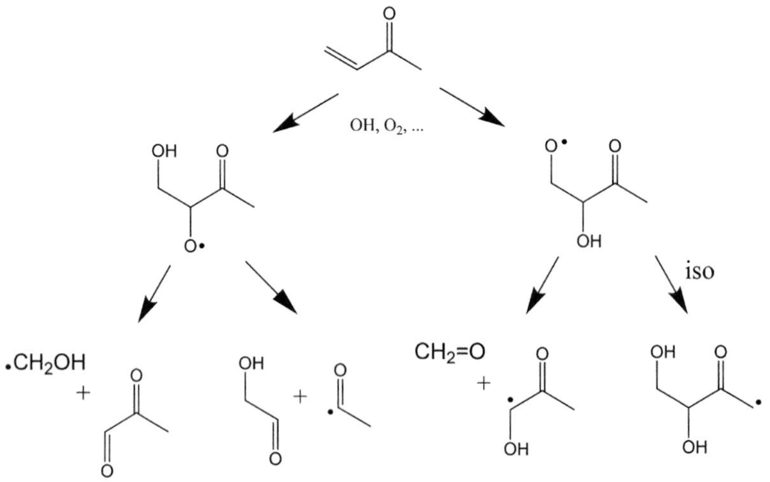

Fig. 3.19 Fate of the two alkoxy radicals produced in the OH-initiated oxidation of methylvinylketone. The internal alkoxy radical is on the left and the external alkoxy radical is on the right.

chemical calculations[74] obtained critical energies that favor the GLYC + acetyl channel over the MGLY + $\cdot CH_2OH$ channel by $\sim$7 kcal mol^{-1}. The SAR of Vereecken and Peeters[18] would suggest that the difference in activation energies is only 4 kcal mol^{-1}. Both the quantum calculations and the SAR indicate that scission is much faster than reaction of the internal alkoxy radical with O_2.

For the external radical, 1,5 H-shift appears feasible, but the SARs of Vereecken and Peeters for isomerization and scission would predict scission to have a much higher rate constant than isomerization.[17,18]

The structure and chemistry of alkoxy radicals produced following OH addition to methacrolein (shown in Fig. 3.20) parallels that of alkoxy radicals produced from OH addition to methylvinylketone (see Fig. 3.19). Specifically, the pathway forming a carbonyl radical (HC•=O, in the case of methacrolein) is favored over that forming $\cdot CH_2OH$.[199,200] Methacrolein, unlike methylvinylketone, can react

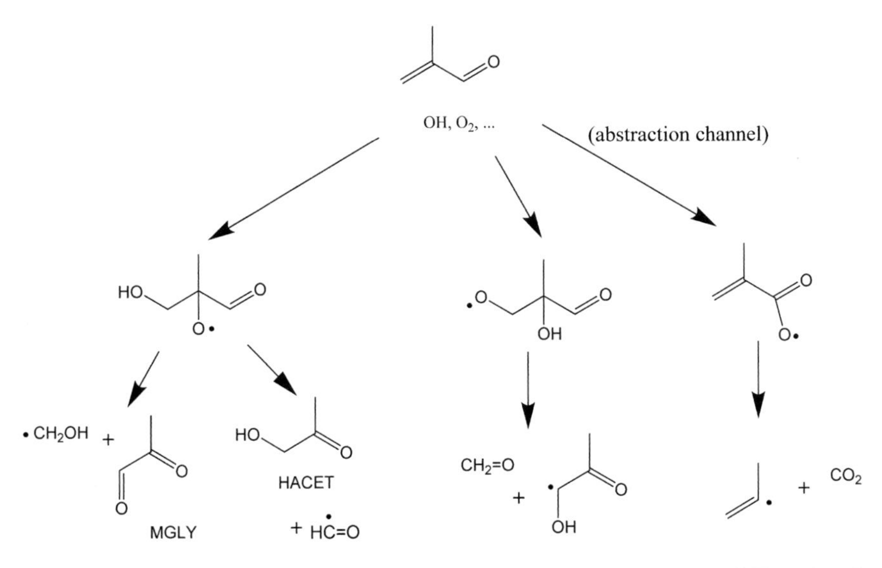

Fig. 3.20 Fate of the three alkoxy radicals produced in the OH-initiated oxidation of methacrolein. The internal alkoxy radical formed from OH addition is on the far left.

with OH radical via hydrogen abstraction, and does so to a significant extent (~45%).[199,200] This leads to formation of the radical $CH_2{=}C(CH_3)C({=}O)O\bullet$, whose fate is believed to be C–C scission.

3.3.6.3. *Reactions of Alkoxy Radicals from an Isoprene Hydrate (2-Methyl-3-Buten-2-ol)*

2-Methyl-3-buten-2-ol (MBO) is a hydrate of isoprene that is also emitted by vegetation, if only to the extent of ~2 Mtons $year^{-1}$.[185]

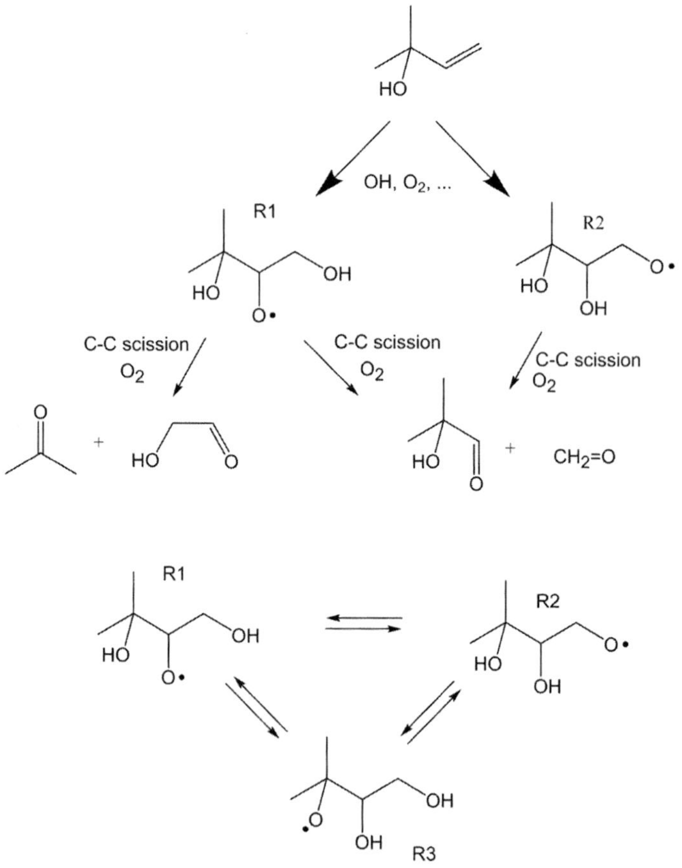

Fig. 3.21 Illustration of the chemistry of alkoxy radicals formed from 2-methyl-3-buten-2-ol. The lower portion shows the H-shift reactions considered by Dibble and Pham.[202]

The first generation alkoxy radicals from OH-initiated oxidation of MBO are indicated as R1 and R2 in Fig. 3.21. Chamber studies[201] identified stable products but were unable to determine which of the two reactions of R1 dominated. Dibble and Pham[202] carried out computations on the fate of R1 and R2. Their calculations indicated that scission of R1 leads (in two steps) to glycolaldehyde and acetone, while the competing path to make the same ultimate products as R2 is negligible. The authors were especially interested in the potential for H-atom migration across OH—O hydrogen bonds. They did not find transition states for double-hydrogen shifts across hydrogen bonds in these radicals. They did find relatively low-energy transition states (as low as 7 kcal mol^{-1}) for the 1,5 H-shift across an OH—O hydrogen bond from R2 to form a third alkoxy radical, R3. Critical energies of these H-atom shifts were higher than those for C–C scission reactions.

Kuwata *et al.* carried out high-level calculations of single hydrogen transfer across the hydrogen bond in $HOCH_2CH_2O\bullet$.[203] This reaction has no practical importance, because the reactants and products are identical, but it serves as a model system for similar 1,4 H-shift reactions across OH—O hydrogen bonds, such as H-shift reactions in R1 and R2 from MBO. The reaction barrier in $HOCH_2CH_2O\bullet$ was so high ($\sim$23 kcal mol^{-1}) that this reaction would not compete with C–C scission or reaction with O_2, despite strong enhancement of the rate constant by tunneling ($\kappa \approx 10^5$). On the basis of Refs. 202 and 203 it appears that single H-atom transfer across hydrogen bonds donated by hydroxyl groups to alkoxy radicals is not likely to be a general phenomenon. The double hydrogen atom transfer reactions found by Dibble in alkoxy radicals from isoprene (see Figs. 3.17 and 3.18) may only be feasible because of the reaction is significantly (15–20 kcal mol^{-1}) exothermic.[195]

3.4. Conclusions

There is no doubt that alkoxy radical chemistry strongly influences the production of ozone and SOA. The near-absence of experimental rate constants for isomerization, and the near-absence of experimental data on the effects of functional groups on reactivity, together represent enormous gaps in our knowledge of this chemistry. There is a clear need for additional methods to selectively generate and directly monitor the concentrations of alkoxy radicals. The CRDS approach of Sprague *et al.*,[25] published in 2012, does not monitor alkoxy radicals, but it represents the most recent advance in experimental methods for alkoxy radical kinetics in at least a decade.

The theoretical method used by Setokuchi and Sato[109] and validated by Hu and Dibble[108] for computing k_{O_2} should be applied to a variety of situations. Experimental and computational efforts are needed to test the results of Wu and Carr for chlorinated alkoxy radicals.[85–87] Theory should also be applied to alkoxy radicals from isoprene and its degradation products, to better understand the competition between isomerization and O_2 reaction of the δ-hydroxyalkoxy radicals. The many papers on product yields from OVOC provide an opportunity to jointly test theoretical approaches to computing k_{O_2} along with the SARs of Vereecken and Peeters[17,18] for isomeriation and C–C scission. Ultimately, we need a reliable SAR for k_{O_2}.

The SAR of Vereecken and Peeters for isomerization[18] should be extended to a wider range of functional groups, to match their SAR for scission. It would be valuable to investigate the apparent discrepancy between their SAR for scission and the results of Yeh *et al.*[139]

Finally, we note that there is a long history of studying alkoxy radicals in physical and synthetic organic chemistry, e.g., Refs. 123,151,191 and 204. Atmospheric chemists could benefit from the insights accumulated by organic chemists.

Acknowledgments

The authors wish to thank J. R. Barker, G. S. Tyndall, and T. J. Wallington for their very helpful feedback.

References

1. U.S. EPA. Air Quality Criteria for Ozone and Related Photochemical Oxidants (2006 Final). EPA/600/R-05/004aF-cF, 2006. Washington, D.C.,

2006. Available from: http://cfpub.epa.gov/ncea/cfm/recordisplay.cfm?deid=149923.
2. Q. Zhang, B. Yuan, M. Shao, X. Wang, S. Lu, K. Lu, *et al.* Variations of ground-level O_3 and its precursors in Beijing in summertime between 2005 and 2011. *Atmos. Chem. Phys.* **14**(12), 6089–6101 (2014).
3. A. P. Altshuller. Air pollution. *Ana. Chem.* **37**(5), 11R–20R (1965).
4. R. Atkinson and W. P. L. Carter. Reactions of alkoxy radicals under atmospheric conditions: The relative importance of decomposition versus reaction with O_2. *J. Atmos. Chem.* **13**(2), 195–210 (1991).
5. R. G. Derwent, M. E. Jenkin and S. M. Saunders. Photochemical ozone creation potentials for a large number of reactive hydrocarbons under European conditions. *Atmos. Environ.* **30**(2), 181–199 (1996).
6. R. G. Derwent, M. E. Jenkin, S. M. Saunders and M. J. Pilling. Photochemical ozone creation potentials for organic compounds in northwest Europe calculated with a master chemical mechanism. *Atmos. Environ.* **32**(14–15), 2429–2441 (1998).
7. Y. X. Wang, M. B. McElroy, D. J. Jacob and R. M. Yantosca. A nested grid formulation for chemical transport over Asia: Applications to CO. *J. Geophys. Res. Atmos.* **109**(D22), (2004).
8. G. Yarwood, S. Rao, M. Yocke and G. Z. Whitten. Updates to the carbon bond chemical mechanism: CB05. 2005.
9. D. R. Lynam and G. D. Pfeifer. Human health effects of highway-related pollutants. *Stud. Environ. Sci.* **44**, 259–244 (1991).
10. R. Atkinson. Atmospheric reactions of alkoxy and β-hydroxyalkoxy radicals. *Int. J. Chem. Kinet.* **29**(2), 99–111 (1997).
11. J. J. Orlando, G. S. Tyndall and T. J. Wallington. The atmospheric chemistry of alkoxy radicals. *Chem. Rev.* **103**(12), 4657–4689 (2003).
12. P. Devolder. Atmospheric fate of small alkoxy radicals: Recent experimental and theoretical advances. *J. Photochem. Photobiol. A Chem.* May **157**(2–3), 137–147 (2003).
13. R. Mereau, M.-T. Rayez and F. Caralp. Theoretical study of alkoxyl radical decomposition reactions: Structure–activity relationships. *Phys. Chem. Chem. Phys.* **2**(17), 3765–3772 (2000).
14. H. J. Curran. Rate constant estimation for C_1 to C_4 alkyl and alkoxyl radical decomposition. *Int. J. Chem. Kinet.* **38**(4), 250–275 (2006).
15. D. Johnson, P. Cassanelli and R. A. Cox. Correlation-type structure activity relationships for the kinetics of the decomposition of simple and β-substituted alkoxyl radicals. *Atmos. Environ.* **38**(12), 1755–1765 (2004).
16. R. Atkinson. Rate constants for the atmospheric reactions of alkoxy radicals: An updated estimation method. *Atmos. Environ.* **41**(38), 8468–8485 (2007).
17. L. Vereecken and J. Peeters. Decomposition of substituted alkoxy radicals-part I: A generalized structure-activity relationship for reaction barrier heights. *Phys. Chem. Chem. Phys.* **11**(40), 9062–9074 (2009).
18. L. Vereecken and J. Peeters. A structure-activity relationship for the rate coefficient of H-migration in substituted alkoxy radicals. *Phys. Chem. Chem. Phys.* **12**(39), 12608–12620 (2010).

19. S. P. Sander, J. P. D. Abbatt, J. R. Barker, J. B. Burkholder, R. R. Friedl, D. M. Golden, R. E. Huie, C. E. Kolb, M. J. Kurylo, G. K. Moortgat, V. L. Orkin and P. H. Wine. *Chemical Kinetics and Photochemical Data for Use in Atmospheric Studies.* (Jet Propulsion Laboratory. Pasadena: Jet Propulsion Laboratory, 2011).
20. H. Hein, A. Hoffman and R. Zellner. Direct investigation of reactions of 2-butoxy radicals using laser pulse initiated oxidation: reaction with O_2 and unimolecular decomposition at 293 K. *Berich. Bunsen. Phys. Chem.* **102**(12), 1840–1849 (1998).
21. P. L. Ross and M. V. Johnston. Excited state photochemistry of iodoalkanes. *J. Phys. Chem.* **99**(12), 4078–4085 (1995).
22. D. Schwartz-Lavi and S. Rosenwaks. Scalar and vectorial properties in the photodissociation of *tert*-butyl nitrite from the S_1 and S_2 states. *J. Chem. Phys.* **88**(11), (1988).
23. A. Untch, R. Schinke, R. Cotting and J. R. Huber. The vibrational predissociation of *cis*-methyl nitrite in the S_1 state: A comparison of exact quantum mechanical wave packet calculations with classical trajectory calculations and detailed experimental results. *J. Chem. Phys.* **99**(12), 9553 (1993).
24. C. S. Effenhauser, P. Felder and J. R. Huber. Photodissociation of alkyl nitrites in a molecular beam. Primary and secondary reactions. *J. Phys. Chem.* **94**(1), 296–302 (1990).
25. M. K. Sprague, E. R. Garland, A. K. Mollner, C. Bloss, B. D. Bean and M. L. Weichman. *et al.* Kinetics of *n*-butoxy and 2-pentoxy isomerization and detection of primary products by infrared cavity ringdown spectroscopy. *J. Phys. Chem. A* **116**(24), 6327–6340 (2012).
26. A. H. Blatt, *Organic Syntheses.* (Wiley, New York, 1966), pp. 108–109.
27. R. Chhantyal-Pun, M.-W. Chen, D. Sun and T. A. Miller. Detection and Characterization of Products from Photodissociation of XCH_2CH_2ONO (X = F, Cl, Br, OH). *J. Phys. Chem. A* **116**(49), 12032–12040 (2012).
28. W. Zhou, Y. Yuan and J. Zhang. Photodissociation dynamics of 1-propanol and 2-propanol at 193.3 nm. *J. Chem. Phys.* **119**(14) (2003).
29. J. G. Calvert, J. N. Jr. Pitts, *Photochemistry.* (John Wiley & Sons, Inc., New York, 1966). pp. 441–450.
30. J. R. Barker, S. W. Benson and D. M. Golden. The decomposition of dimethyl peroxide and the rate constant for $CH_3O + O_2 \rightarrow CH_2O + HO_2$. *Int. J. Chem. Kinet.* **9**, 31–53 (1977).
31. R. Atkinson, J. Arey and S. M. Aschmann. Atmospheric chemistry of alkanes: Review and recent developments. *Atmos. Environ.* **42**(23), 5859–5871 (2008).
32. J. Z. Yu, H. E. Jeffries and R. M. LeLacheur. Identifying airborne carbonyl compounds in the atmospheric oxidation products of isoprene by their PFBHA oximes using gas chromatography/ion trap mass spectrometry. Book of Abstracts, 210th ACS National Meeting, Chicago, IL, August 20–24 (1995).

33. C. Wang, W. Deng, L. G. Shemesh, M. D. Lilien, D. R. Katz and T. S. Dibble. Observation of fluorescence excitation spectra of *tert*-pentoxy and 3-pentoxy radicals. *J. Phys. Chem. A* **104**(45), 10368–10373 (2000).
34. C. C. Carter, S. Gopalakrishnan and J. R. Atwell. Laser Excitation Spectra of Large Alkoxy Radicals Containing 5−12 Carbon Atoms. *J. Phys. Chem. A* **105**(13), 2925–2928 (2001).
35. P. Cassanelli, D. Johnson and R. A. Cox. A temperature-dependent relative-rate study of the OH initiated oxidation of *n*-butane: The kinetics of the reactions of the 1- and 2-butoxy radicals. *Phys. Chem. Chem. Phys.* **7**(21), 3702–3710 (2005).
36. S. Gopalakrishnan, L. Zu and T. A. Miller. Rotationally resolved electronic spectra of the $\tilde{B}$-$\tilde{X}$ transition in multiple conformers of 1-butoxy and 1-pentoxy radicals. *J. Phys. Chem. A* **107**(26), 5189–5201 (2003).
37. H. Hein, A. Hoffmann, H. Somnitz and R. Zellner. A combined experimental and computational investigation of the reactions of 3-pentoxy radicals: Reaction with O_2 and unimolecular decomposition. Zeitschrift für Phys Chemie. **214**, 449–471 (2000).
38. H. Hein, A. Hoffmann and R. Zellner. Direct investigations of reactions of 1-butoxy and 1-pentoxy radicals using laser pulse initiated oxidation: Reaction with O_2 and isomerisation at 293 K and 50 mbar. *Phys. Chem. Chem. Phys.* **1**(16), 3743–3752 (1999).
39. D. Romanini, I. Ventrillard, G. Méjean, J. Morville and E. Kerstel. Cavity-Enhanced Spectroscopy and Sensing. In: G. Gagliardi, H.-P. Loock, Eds. (Springer Berlin Heidelberg, Berlin, Heidelberg, 2014), pp. 1–60.
40. G. Berden, R. Peeters and G. Meijer. Cavity ring-down spectroscopy: Experimental schemes and applications. *Int. Rev. Phys. Chem.* **19**(4), 565–607 (2010).
41. C. J. Cramer. Essentials of Computational Chemistry: Theories and Models. 2nd Ed. (Wiley, New York, 2004).
42. L. Vereecken, D. R. Glowacki and M. J. Pilling. Theoretical Chemical Kinetics in Tropospheric Chemistry: Methodologies and Applications. *Chem. Rev.* **115**(10), 4063–4114 (2015).
43. K. Raghavachari. Electron correlation techniques in quantum chemistry: Recent advances. *Annu. Rev. Phys. Chem.* **42**(1), 615–642 (1991).
44. S. F. Sousa, P. A. Fernandes and M. J. Ramos, General Performance of Density Functionals. *J. Phys. Chem. A* **111**, 10439–10111 (2007).
45. N. Oliphant and R. J. Bartlett. A systematic comparison of molecular properties obtained using Hartree-Fock, a hybrid Hartree-Fock density functional theory, and coupled-cluster methods. *J. Chem. Phys.* **100**(9), (1994).
46. K. Raghavachari, G. W. Trucks, J. A. Pople and M. Head-Gordon. A fifth-order perturbation comparison of electron correlation theories. *Chem. Phys. Lett.* **157**(6), 479–483 (1989).
47. G. E. Scuseria, C. L. Janssen and H. F. Schaefer. An efficient reformulation of the closed-shell coupled cluster single and double excitation (CCSD) equations. *J. Chem. Phys.* **89**(12), 7382 (1988).

48. W. Jiang and A. K. Wilson. *Ab Initio* composite approaches: Potential energy surfaces and excited electronic states. *Reports in Computational Chemistry*, R A Wheeler, Ed. 2012. pp. 29–51.
49. J. M. Simmie and K. P. Somers. Benchmarking compound methods (CBS-QB3, CBS-APNO, G3, G4, W1BD) against the active thermochemical tables: A litmus test for cost-effective molecular formation enthalpies. *J. Phys. Chem. A* **119**(28), 7235–7246 (2015).
50. K. Peterson, D. Feller and D. Dixon. Chemical accuracy in *ab initio* thermochemistry and spectroscopy: Current strategies and future challenges. *Theor. Chem. Acc.* **131**(1), 1–20 (2012).
51. G. da Silva. G3X-K theory: A composite theoretical method for thermochemical kinetics. *Chem. Phys. Lett.* **558**(12), 109–113 (2013).
52. D. P. Tew, W. Klopper, C. Neiss and C. Hattig. Quintuple-zeta quality coupled-cluster correlation energies with triple-zeta basis sets. *Phys. Chem. Chem. Phys.* **9**(16), 1921–1930 (2007).
53. E. Wigner. On the penetration of potential energy barriers in chemical reactions. *Z Phys. Chem. B* (19), 203 (1932).
54. C. Eckart. The penetration of a potential barrier by electrons. *Phys. Rev.* **35**, 13039, (1930).
55. H. S. Johnston and J. Heicklen. Tunneling corrections for unsymmetrical eckart potential energy barriers. *J. Phys. Chem.* **66**(3), 532–533 (1962).
56. D. R. Glowacki, C.-H. Liang, C. Morley, M. J. Pilling and S. H. Robertson, MESMER: An open-source master equation solver for multi-energy well reactions. *J. Phys. Chem. A* **116**(38), 9545–9560 (2012).
57. J. R. Barker. Multiple-Well, multiple-path unimolecular reaction systems. I. MultiWell computer program suite. *Int. J. Chem. Kinet.* **33**(4), 232–245 (2001).
58. J. R. Barker. Energy transfer in master equation simulations: A new approach. *Int. J. Chem. Kinet.* **41**(12), 748–763 (2009).
59. MultiWell-2014.1 Software, 2014, designed and maintained by R. John, Barker with contributors F. Nicholas, Ortiz, M. Jack, Preses, Lawrence L. Lohr, Andrea Maranzana, Philip J. Stimac, T. Lam Nguyen, and T. J. Dhilip Kumar; University of Michigan, *Ann Arbor.* http://clasp-research.engin.umich.edu/multiwell/?url=multiwell/
60. J. Zheng, S. Zhang, B. J. Lynch, J. C. Corchado, Y.-Y. Chuang, P. L. Fast, W.-P. Hu, Y.-P. Liu, G. C. Lynch, K. A. Nguyen, C. F. Jackels, A. F. Ramos, B. A. Ellingson, V. S. Melissas, J. Villà, I. Rossi, E. L. Coitiño, J. Pu, T. V. Albu, A. Ratkiewicz, R. Steckler, B. C. Garrett, B. C. Garrett, Alan D. Isaacson, and D. G. Truhlar. POLYRATE 2010-A: Computer Program for the Calculation of Chemical Reaction Rates for Polyatomics. 2010. Available from: http://comp.chem.umn.edu/polyrate/
61. L. Vereecken and J. Peeters. The 1,5-shift in 1-butoxy: A case study in the rigorous implementation of transition state theory for a multirotamer system. *J. Chem. Phys.* **119**(10), 5159–5170 (2003).
62. X. Xu, E. Papajak, J. Zheng and D. G. Truhlar. Multi-structural variational transition state theory: Kinetics of the 1,5-hydrogen shift isomerization of

the 1-butoxyl radical including all structures and torsional anharmonicity. *Phys. Chem. Chem. Phys.* **14**(12), 4204–4216 (2012).

63. J. Troe. Predictive possibilities of unimolecular rate theory. *J. Phys. Chem.* **83**(1), 114–126 (1979).
64. D. M. Golden. Evaluating data for atmospheric models, an example: $CH_3O_2 + NO_2 = CH_3O_2NO_2$. *Int. J. Chem. Kinet.* **37**(10), 625–632 (2005).
65. R. Atkinson, D. L. Baulch, R. A. Cox, J. N. Crowley, R. F. Hampson, R. G. Hynes, *et al.* Evaluated kinetic and photochemical data for atmospheric chemistry: Volume II – gas phase reactions of organic species. *Atmos. Chem. Phys.* **6**(11), 3625–4055 (2006).
66. D. M. Golden. Pressure dependent reactions for atmospheric and combustion models. *Chem. Soc. Rev.* **37**(4), 717–731 (2008).
67. S. C. Smith and R. G. Gilbert. *Theory of Unimolecular and Recombination Reactions.* (Blackwell Science, 1990).
68. P. Zhang and C. K. Law. A fitting formula for the falloff curves of unimolecular reactions, II: Tunneling effects. *Int. J. Chem. Kinet.* **43**(1), 31–42 (2011).
69. V. D. Knyazev and I. R. Slagle. Experimental and theoretical study of the $C_2H_3 \leftrightarrows H + C_2H_2$ reaction. Tunneling and the Shape of Falloff Curves. *J. Phys. Chem.* **100**(42), 16899–16911 (1996).
70. J. J. Orlando, G. S. Tyndall, M. Bilde, C. Ferronato, T. J. Wallington, L. Vereecken, *et al.* Laboratory and theoretical study of the oxy radicals in the OH- and Cl-initiated oxidation of ethene. *J. Phys. Chem. A* **102**(42), 8116–8123 (1998).
71. L. Vereecken, J. Peeters, J. J. Orlando, G. S. Tyndall and C. Ferronato. Decomposition of β-hydroxypropoxy radicals in the OH-initiated oxidation of propene. A theoretical and experimental study. *J. Phys. Chem. A* **103**(24), 4693–4702 (1999).
72. T. J. Wallington, M. D. Hurley, J. M. Fracheboud, J. J. Orlando, G. S. Tyndall, J. Sehested, *et al.* Role of excited CF_3CFHO radicals in the atmospheric chemistry of HFC-134a. *J. Phys. Chem.* **100**(46), 18116–18122 (1996).
73. L. Vereecken, and J. Peeters. Theoretical investigation of the role of intramolecular hydrogen bonding in β-hydroxyethoxy and β-hydroxyethylperoxy radicals in the tropospheric oxidation of ethene. *J. Phys. Chem. A* **103**(5), 1768–1775 (1999).
74. E. Praske, J. D. Crounse, K. H. Bates, T. Kurtén, H. G. Kjaergaard and P. O. Wennberg, Atmospheric fate of methyl vinyl ketone: Peroxy radical reactions with NO and HO_2. *J. Phys. Chem. A* **119**(19), 4562–4572 (2015).
75. A. S. Hasson, K. T. Kuwata, M. C. Arroyo and E. B. Petersen. Theoretical studies of the reaction of hydroperoxy radicals (HO_2) with ethyl peroxy ($CH_3CH_2O_2$), acetyl peroxy ($CH_3C(O)O_2$), and acetonyl peroxy ($CH_3C(O)CH_2O_2$. *J. Photochem. Photobiol. A Chem.* **176**(1–3), 218–230 (2005).
76. N. M. Donahue, G. T. Drozd, S. A. Epstein, A. A. Presto and J. H. Kroll. Adventures in ozoneland: Down the rabbit-hole. *Phys. Chem. Chem. Phys.* **13**(23), 10848–10857 (2011).

77. A. G. Suits. Roaming atoms and radicals: A new mechanism in molecular dissociation. *Acc. Chem. Res.* **41**(7), 873–881 (2008).
78. R. S. Zhu, P. Raghunath, and M. C. Lin. Effect of roaming transition states upon product branching in the thermal decomposition of CH_3NO_2. *J. Phys. Chem. A* **117**(32), 7308–7313 (2013).
79. C. J. Annesley, J. B. Randazzo, S. J. Klippenstein, L. B. Harding, A. W. Jasper, Y. Georgievskii, *et al.* Thermal dissociation and roaming isomerization of nitromethane: Experiment and theory. *J. Phys. Chem. A* **119**(28), 7872–7893 (2015).
80. Z. Homayoon and J. M. Bowman. Quasiclassical trajectory study of CH_3NO_2 decomposition via roaming mediated isomerization using a global potential energy surface. *J. Phys. Chem. A* **117**(46), 11665–11672 (2013).
81. M. Isegawa, F. Liu, S. Maeda and K. Morokuma. *Ab initio* reaction pathways for photodissociation and isomerization of nitromethane on four singlet potential energy surfaces with three roaming paths. *J. Chem. Phys.* **140**(24), 244–310 (2014).
82. J. M. O'Brien, E. Czuba, D. R. Hastie, J. S. Francisco and P. B. Shepson. Determination of the hydroxy nitrate yields from the reaction of C_2-C_6 alkenes with OH in the presence of NO. *J. Phys. Chem. A* **102**(45), 8903–8908 (1998).
83. T. S. Dibble. Failures and limitations of quantum chemistry for two key problems in the atmospheric chemistry of peroxy radicals. *Atmos. Environ.* **42**(23), 5837–5848 (2008).
84. S. J. Klippenstein, Y. Georgievskii, and L. B. Harding. Statistical theory for the kinetics and dynamics of roaming reactions. *J. Phys. Chem. A* **115**(50), 14370–14381 (2011).
85. F. Wu and R. W. Carr. The chloromethoxy radical: Kinetics of the reaction with O_2 and the unimolecular elimination of HCl at 306 K. *Chem. Phys. Lett.* **305**(1–2), 44–50 (1999).
86. F. Wu and R. W. Carr. Kinetic study of the reaction of the $CFCl_2CH2O$ Radical with O_2. *J. Phys. Chem.* **100**(22), 9352–9359 (1996).
87. F. Wu and R. W. Carr. Kinetics of CH_2ClO radical reactions with O_2 and NO, and the unimolecular elimination of HCl. *J. Phys. Chem. A* **105**(9), 1423–1432 (2001).
88. G. Falgayrac, F. Caralp, N. Sokolowski-Gomez, P. Devolder and C. Fittschen. Rate constants for the decomposition of 2-butoxy radicals and their reaction with NO and O_2. *Phys. Chem. Chem. Phys.* **6**(16), 4127 (2004).
89. H. Hippler, F. Striebel and B. Viskolcz. A detailed experimental and theoretical study on the decomposition of methoxy radicals. *Phys. Chem. Chem. Phys.* **3**(12), 2450–2458 (2001).
90. S. M. Aschmann and R. Atkinson. Products of the gas-phase reactions of the OH radical with n-butyl methyl ether and 2-isopropoxyethanol: Reactions of $ROC(\grave{O})<$ radicals. *Environ. Sci. Technol.* **31**(7), 501–513 (1999).

91. J. A. Montgomery, M. J. Frisch, J. W. Ochterski and G. A. Petersson. A complete basis set model chemistry. VII. Use of the minimum population localization method. *J. Chem. Phys.* **112**(15), 6532 (2000).
92. J. A. Montgomery, M. J. Frisch, J. W. Ochterski and G. A. Petersson. A complete basis set model chemistry. VI. Use of density functional geometries and frequencies. *J. Chem. Phys.* **110**(6), 2822 (1999).
93. W. Deng, C. Wang, D. R. Katz, G. R. Gawinski, A. J. Davis and T. S. Dibble. Direct kinetic studies of the reactions of 2-butoxy radicals with NO and O_2. *Chem. Phys. Lett.* **330**(5–6), 541–546 (2000).
94. W. Deng, A. J. Davis, L. Zhang, D. R. Katz and T. S. Dibble. Direct kinetic studies of reactions of 3-pentoxy radicals with NO and O_2. *J. Phys. Chem. A* **105**(39), 8985–8990 (2001).
95. L. Zhang, K. M. Callahan, D. Derbyshire and T. S. Dibble. Laser-induced fluorescence spectra of 4-methylcyclohexoxy radical and perdeuterated cyclohexoxy radical and direct kinetic studies of their reactions with O_2. *J. Phys. Chem. A* **109**(41), 9232–9240 (2005).
96. L. Zhang, K. A. Kitney, M. A. Ferenac, W. Deng and T. S. Dibble. LIF spectra of cyclohexoxy radical and direct kinetic studies of its reaction with O_2. *J. Phys. Chem. A* **108**(3), 447–454 (2004).
97. J. J. Orlando, L. T. Iraci and G. S. Tyndall. Chemistry of the cyclopentoxy and cyclohexoxy radicals at subambient temperatures. *J. Phys. Chem. A* **104**(21), 5072–5079 (2000).
98. O. Welz, F. Striebel and M. Olzmann. On the thermal unimolecular decomposition of the cyclohexoxy radical-an experimental and theoretical study. *Phys. Chem. Chem. Phys.* **10**(2), 320–329 (2008).
99. J. Chai, H. Hu, T. S. Dibble, G. S. Tyndall and J. J. Orlando. Rate constants and kinetic isotope effects for methoxy radical reacting with NO_2 and O_2. *J. Phys. Chem. A* **118**(20), 3552–3563 (2014).
100. P. J. Wantuck, R. C. Oldenborg, S. L. Baughcum and K. R. Winn. Removal rate constant measurements for methoxy radical by oxygen over the 298–973 K range. *J. Phys. Chem.* **91**(18), 4653–4655 (1987).
101. K. Lorenz, D. Rhasa, R. Zellner and B. Fritz. Laser photolysis — LIF kinetic studies of the reactions of CH_3O and CH_2CHO with O_2 between 300 and 500 K. *Berich. Bunsen. Phys. Chem.* **89**(3), 341–342 (1985).
102. D. Gutman, N. Sanders and J. E. Butler. Kinetics of the reactions of methoxy and ethoxy radicals with oxygen. *J. Phys. Chem.* **86**(1), 66–70 (1982).
103. H. A. Wiebe, A. Villa, T. M. Hellman and J. Heicklen. Photolysis of methyl nitrite in the presence of nitric oxide, nitrogen dioxide, and oxygen. *J. Am. Chem. Soc.* **95**(1), 7–13 (1973).
104. R. A. Cox, R. G. Derwent, S. V. Kearsey, L. Batt and K. G. Patrick. Photolysis of methyl nitrite: Kinetics of the reaction of the methoxy radical with O_2. *J. Photochem.* **13**(2), 149–163 (1980).

105. T. P. W. Jungkamp and J. H. Seinfeld. The mechanism of methoxy radical oxidation: Hydrogen abstraction versus trioxy radical formation. *Chem. Phys. Lett.* **263**(3–4), 371–378 (1996).
106. J. M. Bofill, S. Olivella, A. Solé and J. M. Anglada. The mechanism of methoxy radical oxidation by O_2 in the gas phase. Computational evidence for direct H atom transfer assisted by an intermolecular noncovalent O-O bonding interaction. *J. Am. Chem. Soc.* **121**(6), 1337–1347 (1999).
107. M. C. McCarthy, V. Lattanzi, D. Kokkin, O. Martinez and J. F. Stanton. On the molecular structure of HOOO. *J. Chem. Phys.* **136**(3), 034303 (2012).
108. H. Hu, and T. S. Dibble. Quantum chemistry, reaction kinetics, and tunneling effects in the reaction of methoxy radicals with O_2. *J. Phys. Chem. A* **117**, 14230–14242 (2013).
109. O. Setokuchi and M. Sato. Direct dynamics of an alkoxy radical (CH_3O, C_2H_5O, and *i*-C_3H_7O) reaction with an oxygen molecule. *J. Phys. Chem. A* **106**, 8124–8132 (2002).
110. H. Hu, T. S. Dibble, G. S. Tyndall and J. J. Orlando. Temperature-dependent branching ratios of deuterated methoxy radicals ($CH_2DO•$) reacting with O_2. *J. Phys. Chem. A* **116**(24), 6295–6302 (2012).
111. P. J. Wantuck, R. C. Oldenborg, S. L. Baughcum and K. R. Winn. Collisional quenching of methoxy ($\tilde{A}\ ^2A_1$) radical. *J. Phys. Chem.* **91**(12), 3253–3259 (1987).
112. W. P. L. Carter, K. R. Darnall, A. C. Lloyd, A. M. Winter and J. N. Jr. Pitts. Evidence for alkoxy radical isomerization in photooxidations of C_4–C_6 alkanes unders simulated atmospheric conditions. *Chem. Phys. Lett.* **42**(1), 22–27 (1976).
113. T. S. Dibble, Unpublished research. 2014.
114. P. Cassanelli, R. A. Cox, J. J. Orlando and G. S. Tyndall. An FT-IR study of the isomerization of 1-butoxy radicals under atmospheric conditions. *J. Photochem. Photobiol. A Chem.* **177**(2–3), 109–115 (2006).
115. D. Johnson, P. Cassanelli and R. A. Cox. Isomerization of simple alkoxyl radicals: New temperature-dependent rate data and structure activity relationship. *J. Phys. Chem. A* **108**(4), 519–523 (2004).
116. H. Somnitz and R. Zellner. Theoretical studies of unimolecular reactions of C_2–C_5 alkoxy radicals. Part II. RRKM dynamical calculations. *Phys. Chem. Chem. Phys.* **2**, 1907–1918 (2000).
117. S. M. Aschmann, J. Arey and R. Atkinson. Reactions of OH radicals with C_6–C_{10} cycloalkanes in the presence of NO: Isomerization of C_7–C_{10} cycloalkoxy radicals. *J. Phys. Chem. A* **115**(50), 14452–14461 (2011).
118. H. Somnitz and R. Zellner. Theoretical studies of unimolecular reactions of C_2–C_5 alkoxy radicals. Part III. A microscopic structure activity relationship (SAR). *Phys. Chem. Chem. Phys.* **2**, 4319–4325 (2000).
119. R. Mereau, M.-T. Rayez, F. Caralp and J.-C. Rayez. Isomerisation reactions of alkoxy radicals: Theoretical study and structure-activity relationships. *Phys. Chem. Chem. Phys.* **5**(21), 4828–4833 (2003).

120. H. Somnitz. The contribution of tunnelling to the 1,5 H-shift isomerisation reaction of alkoxyl radicals. *Phys. Chem. Chem. Phys.* **10**(7), 965–973 (2008).
121. A. C. Davis and J. S. Francisco. Reactivity trends within alkoxy radical reactions responsible for chain branching. *J. Am. Chem. Soc.* **133**(45), 18208–18219 (2011).
122. E. L. Eliel and H. Wilen. Stereochemistry of Organic Compounds. (John Wiley and Sons, New York, 1994).
123. A. E. Dorigo and K. N. Houk. The relationship between proximity and reactivity. An *ab initio* study of the flexibility of the OH + CH_4 hydrogen abstraction transition state and a force-field model for the transition states of intramolecular hydrogen abstractions. *J. Org. Chem.* **53**(8), 1650–1664 (1988).
124. G. Lendvay. *Ab Initio* studies of the isomerization and decomposition reactions of the 1-butoxy radical. *J. Phys. Chem. A* **102**, 10777–10786 (1998).
125. Standard Atmosphere 1976, U.S. Government Printing Office, Washington, D.C., 1976.
126. J. Zheng and D. G. Truhlar. Kinetics of hydrogen-transfer isomerizations of butoxyl radicals. *Phys. Chem. Chem. Phys.* **12**(28), 7782–7793 (2010).
127. M. A. Ferenac, A. J. Davis, A. S. Holloway and T. S. Dibble. Isomerization and decomposition reactions of primary alkoxy radicals derived from oxygenated solvents. *J. Phys. Chem. A* **107**(1), 63–72 (2003).
128. R. Mereau, M. T. Rayez, F. Caralp and J. C. Rayez. Theoretical study on the comparative fate of the 1-butoxy and β-hydroxy-1-butoxy radicals. *Phys. Chem. Chem. Phys.* **2**(9), 1919–1928 (2000).
129. A. C. Davis and J. S. Francisco. Hydroxyalkoxy radicals: Importance of intramolecular hydrogen bonding on chain branching reactions in the combustion and atmospheric decomposition of hydrocarbons. *J. Phys. Chem. A* **118**(52), 10982–11001 (2014).
130. T. S. Dibble. Isomerization of OH-isoprene adducts and hydroxyalkoxy isoprene radicals. *J. Phys. Chem. A* **106**(28), 6643–6650 (2002).
131. F. Caralp, P. Devolder, C. Fittschen, N. Gomez, H. Hippler, R. Mereau, *et al.* The thermal unimolecular decomposition rate constants of ethoxy radicals. *Phys. Chem. Chem. Phys.* **1**(12), 2935–2944 (1999).
132. C. Fittschen, H. Hippler and B. Viskolcz. The βC–C bond scission in alkoxy radicals: Thermal unimolecular decomposition of *t*-butoxy radicals. *Phys. Chem. Chem. Phys.* **2**(8), 1677–1683 (2000).
133. P. Devolder, C. Fittschen, A. Frenzel, H. Hippler, G. Poskrebyshev, F. Striebel, *et al.* Complete falloff curves for the unimolecular decomposition of *i*-propoxy radicals between 330 and 408 K. *Phys. Chem. Chem. Phys.* **1**(4), 675–681 (1999).
134. M. Blitz, M. J. Pilling, S. H. Robertson and P. W. Seakins. Direct studies on the decomposition of the *tert*-butoxy radical and its reaction with NO. *Phys. Chem. Chem. Phys.* **1**(1), 73–80 (1999).

135. D. Johnson, S. Carr and R. A. Cox. The kinetics of the gas-phase decomposition of the 2-methyl-2-butoxyl and 2-methyl-2-pentoxyl radicals. *Phys. Chem. Chem. Phys.* **7**(10), 2182–2190 (2005).
136. J. J. Orlando, G. S. Tyndall, L. Vereecken and J. Peeters. The atmospheric chemistry of the acetonoxy radical. *J. Phys. Chem. A* **104**(49), 11578–11588 (2000).
137. R. Mereau, M.-T. Rayez, J.-C. Rayez and P. C. Hiberty. Alkoxyl radical decomposition explained by a valence-bond model. *Phys. Chem. Chem. Phys.* **3**(17), 3656–3661 (2001).
138. J. Peeters, G. Fantechi. and L. Vereecken. A generalized Structure-Activity Relationship for the decomposition of (substituted) alkoxy radicals. *J. Atmos. Chem.* **48**(1), 59–80 (2004).
139. G. K. Yeh, M. S. Claflin and P. J. Ziemann, Products and mechanism of the reaction of 1-pentadecene with NO_3 radicals and the effect of a $-ONO_2$ group on alkoxy radical decomposition. *J. Phys. Chem. A* **119**(43), 10684–10696 (2015).
140. W. F. Schneider, T. J. Wallington, J. R. Barker and E. A. Stahlberg. $CF_3CFHO^\bullet$ radical: Decomposition vs. reaction with O_2. *Berich. Bunsen. Phys. Chem.* **102**(12), 1850–1856 (1998).
141. D. R. Glowacki, J. Lockhart, M. A. Blitz, S. J. Klippenstein, M. J. Pilling, S. H. Robertson, *et al.* Interception of excited vibrational quantum states by O_2 in atmospheric association reactions. *Science* **337**(6098), 1066–1069 (2012).
142. G. Fantechi, L. Vereecken and J. Peeters. The OH-initiated atmospheric oxidation of pinonaldehyde: Detailed theoretical study and mechanism construction. *Phys. Chem. Chem. Phys.* **4**(23), 5795–5805 (2002).
143. L. Vereecken and J. Peeters. A theoretical study of the OH-initiated gas-phase oxidation mechanism of β-pinene ($C_{10}H_{16}$): First generation products. *Phys. Chem. Chem. Phys.* **14**(11), 3802 (2012).
144. H.-J. Benkelberg, O. Boge, R. Seuwen and P. Warneck. Product distributions from the OH radical-induced oxidation of but-1-ene, methyl-substituted but-1-enes and isoprene in NO_x-free air. *Phys. Chem. Chem. Phys.* **2**(18), 4029–4039 (2000).
145. J. Zhao, R. Zhang and S. W. North. Oxidation mechanism of δ-hydroxyisoprene alkoxy radicals: Hydrogen abstraction versus 1,5 H-shift. *Chem. Phys. Lett.* **369**(1–2), 204–213 (2003).
146. R. J. Atkinson. Gas phase tropospheric chemistry of organic compounds. *Data, Monograph 2: Phys. Chem. Ref.* (1994).
147. A. Lesar, M. Hodoščcek, E. Drougas and A. M. Kosmas. Quantum mechanical investigation of the atmospheric reaction $CH_3O_2 + NO$. *J. Phys. Chem. A* **110**(25), 7898–7903 (2006).
148. X. M. Pan, Z. Fu, Z. S. Li, C. C. Sun, H. Sun, Z. M. Su, *et al.* Theoretical study on the mechanism of the gas-phase radical–radical reaction of CH_3O with NO_2. *Chem. Phys. Lett.* **409**(1–3), 98–104 (2005).

149. J. F. Arenas, F. J. Avila, J. C. Otero, D. Peláez and J. Soto. Approach to the atmospheric chemistry of methyl nitrate and methylperoxy nitrite. Chemical mechanisms of their formation and decomposition reactions in the gas phase. *J. Phys. Chem. A* **112**(2), 249–255 (2008).
150. L. L. Lohr, J. R. Barker and R. M. Shroll. Modeling the organic nitrate yields in the reaction of alkyl peroxy radicals with nitric oxide. 1. Electronic structure calculations and thermochemistry. *J. Phys. Chem. A* **107**(38), 7429–7433 (2003).
151. Y. Zhao, K. N. Houk and L. P. Olson. Mechanisms of peroxynitrous acid and methyl peroxynitrite, ROONO (R = H, Me), rearrangements: A conformation-dependent homolytic dissociation. *J. Phys. Chem. A* **108**(27), 5864–5871 (2004).
152. J. Zhang, T. Dransfield and N. M. Donahue. On the mechanism for nitrate formation via the peroxy radical + NO reaction. *J. Phys. Chem. A* **108**(42), 9082–9095 (2004).
153. J. Chai and T. S. Dibble. Pressure dependence and kinetic isotope effects in the absolute rate constant for methoxy radical reacting with NO_2. *Int. J. Chem. Kinet.* **46**(9), 501–511 (2014).
154. J. A. McCaulley, S. M. Anderson, J. B. Jeffries and F. Kaufman. Kinetics of the reaction of CH_3O with NO_2. *Chem. Phys. Lett.* **115**(2), 180–186 (1985).
155. P. Biggs, C. E. Canosa-Mas, J.-M. Fracheboud, A. D. Parr, D. E. Shallcross, R. Wayne, P. *et al.* Investigation into the pressure dependence between 1 and 10 Torr of the reactions of NO_2 with CH_3 and CH_3O. *J. Chem. Soc. Faraday Trans.* **89**(23), 4163–4169 (1993).
156. M. J. Frost and I. W. M. Smith. Rate constants for the reactions of CH_3O and C_2H_5O with NO over a range of temperature and total pressure. *J. Chem. Soc. Faraday Trans.* **86**(10), 1757–1762 (1990).
157. M. J. Frost and I. W. M. Smith. Corrigendum to Rate Constants for the Reactions of CH_3O and C_2H_5O with NO Over a range of temperature and total pressure. (Vol 86, pg 1751, 1990). *J. Chem. Soc. Trans.* **89**(23), 4251 (1993).
158. M. Wollenhaupt and J. N. Crowley. Kinetic studies of the reactions CH_3 + NO_2 → products, CH_3O + NO_2 → products, and OH + $CH_3C(O)CH_3$ → $CH_3C(O)OH$ + CH_3, over a range of temperature and pressure. *J. Phys. Chem. A* **104**(27), 6429–6438 (2000).
159. E. Martınez, J. Albaladejo, E. Jiménez, A. Notario and Y. D. de Mera. Temperature dependence of the limiting low-and high-pressure rate constants for the CH_3O + NO_2 + He reaction over the 250–390 K temperature range. *Chem. Phys. Lett.* **329**(3), 191–199 (2000).
160. A. Kukui, V. Bossoutrot, G. Laverdet and G. Le Bras. Mechanism of the reaction of CH_3SO with NO_2 in relation to atmospheric oxidation of dimethyl sulfide: Experimental and theoretical study. *J. Phys. Chem. A* **104**(5), 935–946 (2000).

161. C. Mund, C. Fockenberg and R. Zellner. LIF spectra of n-propoxy and i-propoxy radicals and kinetics of their reactions with O_2 and NO_2. *Berich. Bunsen. Phys. Chem.* **102**(5), 709–715 (1998).
162. R. J. Balla, H. H. Nelson and J. R. McDonald. Kinetics of the reactions of isopropoxy radicals with NO, NO_2, and O_2. *Chem. Phys.* **99**(2), 323–335 (1985).
163. C. Lotz and R. Zellner. Fluorescence excitation spectrum of the 2-butoxyl radical and kinetics of its reactions with NO and NO_2. *Phys. Chem. Chem. Phys.* **3**(13), 2607–2613 (2001).
164. C. Lotz and R. Zellner. Fluorescence excitation spectrum of the *tert*-butoxy radical and kinetics of its reactions with NO and NO_2. *Phys. Chem. Chem. Phys.* **2**(10), 2353–2360 (2000).
165. N. Sanders, J. E. Butler, L. R. Pasternack and J. R. McDonald. CH_3O ($^2\tilde{X}$ - $^2\tilde{E}$) production from 266 nm photolysis of methyl nitrite and reaction with NO. *Chem. Phys.* **48**(2), 203–208 (1980).
166. J. A. McCaulley, A. M. Moyle, M. F. Golde, S. M. Anderson and F. Kaufman. Kinetics of the reactions of CH_3O and CD_3O with NO. *J. Chem. Soc. Faraday Trans.* **86**(24), 4001–4009 (1990).
167. F. Caralp, M.-T. Rayez, W. Forst, N. Gomez, B. Delcroix, C. Fittschen, *et al.* Kinetic and mechanistic study of the pressure and temperature dependence of the reaction CH_3O + NO. *J. Chem. Soc. Faraday Trans.* **94**(22), 3321–3330 (1998).
168. S. Dobe, G. Lendvay, I. Szilagyi and T. Bérces. Kinetics and mechanism of the reaction of CH_3O with NO. *Int. J. Chem. Kinet.* **26**(9), 887–901 (1994).
169. V. Daële, A. Ray, I. Vassalli, G. Poulet and G. Le Bras. Kinetic study of reactions of $C_2H_5O_2$ with NO at 298 K and 0.55–2 torr. *Int. J. Chem. Kinet.* **27**(11), 1121–1133 (1995).
170. C. Fittschen, A. Frenzel, K. Imrik and P. Devolder. Rate constants for the reactions of C_2H_5O, i-C_3H_7O, and n-C_3H_7O with NO and O_2 as a function of temperature. *Int. J. Chem. Kinet.* **31**(12), 860–866 (1999).
171. D. M. Golden, J. R. Barker and L. L. Lohr. Master equation models for the pressure- and temperature-dependent reactions HO + NO_2 $\rightarrow$ $HONO_2$ and HO + NO_2 $\rightarrow$ HOONO. *J. Phys. Chem. A* **107**(50), 11057–11071 (2003).
172. J. R. Barker, L. L. Lohr, R. M. Shroll and R. Reading. Modeling the organic nitrate yields in the reaction of alkyl peroxy radicals with nitric oxide. 2. Reaction simulations. *J. Phys. Chem. A* **107**(38), 7434–7444 (2003).
173. I. S. Zaslonko, Y. P. Petrov and V. N. Smirnov, Thermal decomposition of nitromethane in shock waves: The effect of pressure and collision partners. *Kinet. Catal.* **38**(3), 321–324 (1997).
174. S. Saxena, J. H. Kiefer and S. J. Klippenstein. A shock-tube and theory study of the dissociation of acetone and subsequent recombination of methyl radicals. *Proc Combust Inst.* **32**(1), 123–130 (2009).

175. A. W. Jasper and J. A. Miller. Theoretical unimolecular kinetics for CH_4+ M $\leftrightarrows$ CH_3 + H + M in eight baths, M = He, Ne, Ar, Kr, H_2, N_2, CO, and CH_4. *J. Phys. Chem. A* **115**(24), 6438–6455 (2011).
176. B. Veyret, P. Roussel and R. Lesclaux. Mechanism of the chain process forming H_2 in the photooxidation of formaldehyde. *Int. J. Chem. Kinet.* **16**(12), 1599–1608 (1984).
177. M. E. Jenkin, G. D. Hayman, T. J. Wallington, M. D. Hurley, J. C. Ball, O. J. Nielsen, *et al.* Kinetic and mechanistic study of the self-reaction of methoxymethylperoxy radicals at room temperature. *J. Phys. Chem.* **97**(45), 11712–11723 (1993).
178. E. Henon, F. Bohr, N. Sokolowski-Gomez and F. Caralp. Degradation of three oxygenated alkoxy radicals of atmospheric interest: $HOCH_2O$, CH_3OCH_2O, $CH_3OCH_2OCH_2O$. RRKM theoretical study of the β C-H bond scission and the 1,6-isomerisation kinetics. *Phys. Chem. Chem. Phys.* **5**(24), 5431 (2003).
179. X. Song, H. Hou and B. Wang. Mechanistic and kinetic study of the O + CH_3OCH_2 reaction and the unimolecular decomposition of CH_3OCH_2O. *Phys. Chem. Chem. Phys.* **7**(23), 3980–3988 (2005).
180. E. C. Tuazon, S. M. Aschmann, R. Atkinson and W. P. L. Carter. The reactions of selected acetates with the OH radical in the presence of NO: Novel rearrangement of alkoxy radicals of structure $RC(O)OCH(\dot{O})\acute{R}$. *J. Phys. Chem. A* **102**(13), 2316–2321 (1998).
181. M.-T. Rayez, B. Picquet-Varrault, F. Caralp and J.-C. Rayez. $CH_3C(O)OCH(O)CH_3$ alkoxy radical derived from ethyl acetate: Novel rearrangement confirmed by computational chemistry. *Phys. Chem. Chem. Phys.* **4**(23), 5789–5794 (2002).
182. V. F. Andersen, T. A. Berhanu, E. J. K. Nilsson, S. Jørgensen, O. J. Nielsen, T. J. Wallington, *et al.* Atmospheric chemistry of two biodiesel model compounds: Methyl propionate and ethyl acetate. *J. Phys. Chem. A* **115**(32), 8906–8919 (2011).
183. A. S. Pimentel, G. S. Tyndall, J. J. Orlando, M. D. Hurley, T. J. Wallington, M. P. S. Andersen, *et al.* Atmospheric chemistry of isopropyl formate and *tert*-butyl formate. *Int. J. Chem. Kinet.* **42**(8), 479–498 (2010).
184. G. S. Tyndall, A. S. Pimentel and J. J. Orlando. Temperature dependence of the α-ester rearrangement reaction. *J. Phys. Chem. A* **108**(33), 6850–6856 (2004).
185. A. B. Guenther, X. Jiang, C. L. Heald. T. Sakulyanontvittaya, T. Duhl, L. K. Emmons, *et al.* The model of emissions of gases and aerosols from nature version 2.1 (MEGAN2.1): An extended and updated framework for modeling biogenic emissions. *Geosci. Model Dev.* **5**(6), 1471–1492 (2012).
186. S. E. Paulson and J. H. Seinfeld. Development and evaluation of a photooxidation mechanism for isoprene. *J. Geophys. Res. Atmos.* **97** (D18) 20703–20715 (1992).

187. E. C. Tuazon and R. Atkinson. A product study of the gas-phase reaction of Isoprene with the OH radical in the presence of NO_x. *Int. J. Chem. Kinet.* **22**(12), 1221–1236 (1990).
188. T. S. Dibble. A quantum chemical study of the C–C bond fission pathways of alkoxy radicals formed following OH addition to isoprene. *J. Phys. Chem. A* **103**(42), 8559–8565 (1999).
189. W. Lei and R. Zhang. Theoretical study of hydroxyisoprene alkoxy radicals and their decomposition pathways. *J. Phys. Chem. A* **105**(15), 3808–3815 (2001).
190. V. S. Nguyen and J. Peeters. Fast (E)–(Z) isomerization mechanisms of substituted allyloxy radicals in isoprene oxidation. *J. Phys. Chem. A* **119**(28), 7270–7276 (2015).
191. J. E. Baldwin. Rules for ring closure. *J. Chem. Soc. Chem. Commun.* (18), 734–736 (1976).
192. F. Paulot, J. D. Crounse, H. G. Kjaergaard, J. H. Kroll, J. H. Seinfeld and P. O. Wennberg. Isoprene photooxidation: New insights into the production of acids and organic nitrates. *Atmos. Chem. Phys.* **9**(2004), 1479–1501 (2009).
193. B. Ghosh, A. Bugarin, B. T. Connell and S. W. North. Isomer-selective study of the OH-initiated oxidation of isoprene in the presence of O_2 and NO: 2. The major OH addition channel. *J. Phys. Chem. A* **114**(7), 2553–2560 (2010).
194. T. S. Dibble. Intramolecular hydrogen bonding and double H-atom transfer in peroxy and alkoxy radicals from isoprene. *J. Phys. Chem. A* **108**(12), 2199–2207 (2004).
195. T. S. Dibble. Prompt chemistry of alkenoxy radical products of the double H-atom transfer of alkoxy radicals from isoprene. *J. Phys. Chem. A* **108**(12), 2208–2215 (2004).
196. M. M. Galloway, A. J. Huisman, L. D. Yee, A. W. H. Chan, C. L. Loza, J. H. Seinfeld, *et al.* Yields of oxidized volatile organic compounds during the OH radical initiated oxidation of isoprene, methyl vinyl ketone, and methacrolein under high–NO_x conditions. *Atmos. Chem. Phys.* **11**(4), 10779–10790 (2011).
197. J. Peeters. and T. L. Nguyen. Unusually fast 1,6-H shifts of enolic hydrogens in peroxy radicals: Formation of the first-generation C_2 and C_3 carbonyls in the oxidation of isoprene. *J. Phys. Chem. A* **116**(24), 6134–6141 (2012).
198. E. C. Tuazon and R. Atkinson. A product study of the gas-phase reaction of methyl vinyl ketone with the OH radical in the presence of NO_x. *Int. J. Chem. Kinet.* **21**(12), 1141–1152 (1989).
199. E. C. Tuazon and R. Atkinson. A product study of the gas-phase reaction of Methacrolein with the OH radical in the presence of NO_x. *Int. J. Chem. Kinet.* **22**(6), 591–602 (1990).
200. J. J. Orlando, G. S. Tyndall and S. E. Paulson. Mechanism of the OH-initiated oxidation of methacrolein. *Geophys. Res. Lett.* **26**(14), 2191–2194 (1999).

201. F. Reisen, S. M. Aschmann, R. Atkinson and J. Arey. Hydroxyaldehyde products from hydroxyl radical reactions of Z-3-Hexen-1-ol and 2-Methyl-3-buten-2-ol quantified by SPME and API-MS. *Environ. Sci. Technol.* **37**(20), 4664–4671 (2003).
202. T. S. Dibble and T. Pham. Peroxy and alkoxy radicals from 2-methyl-3-buten-2-ol. *Phys. Chem. Chem. Phys.* **8**(4), 456–463 (2006).
203. K. T. Kuwata, T. S. Dibble, E. Sliz and E. B. Petersen. Computational studies of intramolecular hydrogen atom transfers in the β-hydroxyethylperoxy and β-hydroxyethoxy radicals. *J. Phys. Chem. A* **111**, 5032–5042 (2007).
204. D. V. Avila, U. Ingold and J. Lusztyk. Solvent effects on the competitive β-scission and hydrogen atom abstraction reactions of the cumyloxyl radical. Resolution of a long-standing problem. *J. Am. Chem. Soc.* **115**(2), 466–470 (1993).

Chapter 4

Role of Nitric Acid Surface Photolysis on Tropospheric Cycling of Reactive Nitrogen Species

Xianliang Zhou and Lei Zhu*
New York State Department of Health, Wadsworth Center,
State University of New York,
Department of Environmental Health Sciences, Albany, NY 12201, USA
**lzhu@albany.edu*

A large number of field and laboratory studies have indicated that nitric acid (HNO_3) and nitrate (NO_3^-) on the surfaces are photochemically reactive and can be photolyzed at much higher rates than in the bulk gas and aqueous phases, producing oxides of nitrogen ($NO_x = NO + NO_2$) and nitrous acid (HONO). This review paper provides an historical perspective and summarizes the recent progress in this important research topic. Photolysis rates on various surfaces both in the absence and in the presence of water vapor, along with available results from the product studies are reviewed. Possible causes for the enhancement of HNO_3/NO_3^- photolysis rates on surfaces compared to those in bulk gas and aqueous phases are discussed. Photolysis mechanisms and the role of HNO_3 and NO_3^- photolysis/photochemistry in the cycling of reactive nitrogen species in the troposphere are discussed. Areas for further studies are suggested.

4.1. Introduction

Nitrogen oxides, NO_x ($NO_x = NO + NO_2$), are primary pollutants during combustion processes. Through a photochemical cycle involving NO_2 photolysis (R1) and NO oxidation by peroxyl radicals (R2), they play a central role[1,2] in the oxidation dynamics of odd hydrogen radicals ($HO_x = OH + HO_2$) and photochemical formation of ozone (O_3) in the troposphere:

$$NO_2 + hv \rightarrow NO + O, \quad \text{(R1)}$$

$$NO + HO_2 \rightarrow NO_2 + OH, \quad \text{(R2)}$$

$$O + O_2 \rightarrow O_3. \quad \text{(R3)}$$

Once emitted into the atmosphere, NO_x undergoes chemical transformations to form a series of products, including nitric acid (HNO_3), nitrous acid (HONO), nitrate radical (NO_3), dinitrogen pentoxide (N_2O_5), peroxynitric acid (HNO_4), alkyl nitrates ($RONO_2$) and peroxyacyl nitrates (PANs); these NO_x oxidation products along with NO_x are collectively defined as reactive nitrogen,[1,2] and denoted as NO_y. The reactive nitrogen species further impact the air quality in the down-wind regions.

The removal of NO_x from the atmosphere[1,2] is mainly through the reaction of NO_2 with OH to form nitric acid:

$$NO_2 + OH + M \rightarrow HNO_3 + M. \quad \text{(R4)}$$

The rate constant for the $NO_2 + OH$ reaction[3] in 760 Torr pressure at 298 K is about $1.0 \times 10^{-11}\ \text{cm}^3\ \text{molecule}^{-1}\ \text{s}^{-1}$. Assuming [OH] of $5 \times 10^6\ \text{cm}^{-3}$ gives a lifetime of NO_2 with respect to reaction with OH of ~6 h around the noontime. Several reactions also contribute indirectly to the conversion of NO_x to HNO_3, including gas phase reaction[1] of NO with hydroperoxyl (HO_2) radicals (Reaction (R2)), and hydrolysis of the oxidation products of NO_x, such as dinitrogen pentoxide[1] (N_2O_5), organic nitrates[4–6] ($RONO_2$), and halogen nitrates[7] (XNO_3, where X is primarily Br and I in the troposphere) on thin water films or in aerosol droplets.

In the stratosphere, gaseous HNO_3 is largely recycled back to NO_x by its gas-phase photolysis. Whereas in the troposphere, HNO_3 is relatively inert photochemically due to its small absorption cross sections in the actinic UV region. The photolysis rate constant of

HNO_3 is $\sim 7 \times 10^{-7}\,s^{-1}$ at ground level at 0° solar zenith angle, corresponding to a photolysis lifetime[1,8] of about 400 h under the full sun. As a highly 'sticky' and water-soluble species, HNO_3 is removed from the troposphere by dry or wet deposition. The dry deposition lifetime is on the order of several hours in the planetary boundary layer at a deposition velocity[2] of $1-5\,cm\,s^{-1}$. Accordingly, HNO_3 has been traditionally considered as a permanent sink[1,9,10] for NO_x.

While HNO_3 has been considered to be the end-product of tropospheric NO_x, comparisons of field and modeling work suggest that HNO_3 may be recycled back to photochemically reactive nitrogen forms.[11–13] The process is sometimes called "*renoxification*".[1] Several renoxification processes of HNO_3 have been proposed to bring the modeled values into closer agreement with the measurements. *Chatfield* proposed[11] HNO_3-to-NO_x recycling processes involving HCHO in aerosols and cloud droplets. *Underwood et al.* concluded[13] that heterogeneous reaction of HNO_3 on atmospheric mineral dust remobilized HNO_3 to NO_x. The heterogeneous reduction of HNO_3 into NO on black carbon aerosol and other carbonaceous aerosols,[12,14,15] heterogeneous reaction of CO and HNO_3 on sulfuric acid aerosols,[16] and heterogeneous reaction of surface adsorbed HNO_3 with gaseous NO on porous glass (silica) surfaces[17–19] were also proposed based on laboratory results. Observations in the remote troposphere suggest that unknown heterogeneous photochemical processes[20] are responsible for the conversion of HNO_3 to NO_x. Two-thirds of HNO_3 taken up on soot was observed to convert to NO_x in the laboratory.[21] These findings show that HNO_3 can indeed be reduced to photochemically active NO_x on various types of surfaces.

A large number of field and laboratory studies in the past 15 years have indicated that once deposited on surfaces, HNO_3 becomes more photochemically reactive, and the photolysis of deposited HNO_3 on ground surfaces can become an important renoxification pathway in the troposphere.[22–29] This review provides an historical perspective and summarizes the recent progress in this important research topic.

4.2. HONO Measurement in Rural Environments

The fast photolysis of nitric acid on surfaces was first proposed as a potential nitrous acid (HONO) source to sustain the higher than expected HONO concentrations in low-NO_x rural environments.[22,23]

HONO is a photochemically active reactive nitrogen species; its lifetime through photolysis (R5) is about 10 min at tropical noontime at ground level:

$$\mathrm{HONO} \underset{+\mathrm{M}}{\overset{hv}{\rightleftarrows}} \mathrm{NO} + \mathrm{OH}. \qquad \text{(R5, R-5)}$$

Since HONO was first unambiguously identified and measured in the troposphere over 35 years ago,[30] extensive field measurements and chamber studies have been conducted to unravel its atmospheric chemistry.[1,22,23,26,27,31–51] Most of the field studies were conducted in urban environments. Significant HONO concentrations have been observed in urban or semi-urban atmospheres, up to parts per billion by volume (ppbv) levels.[32,36,37,41,43,47,49] At the observed levels, the photolysis of HONO (R5) has been found to be a significant or even a major source of hydroxyl radical (OH) in urban atmosphere, accounting between 15% and 90% of the daily photolytic production[32,33,36,43,47] of HO_x.

It is generally recognized that NO_x is the dominant HONO precursor in urban atmosphere. Heterogeneous reactions of NO_2 with water and organic compounds on surfaces (R6) are known to produce[52–60] HONO:

$$\mathrm{NO_2} + \mathrm{H_2O},\ \text{organics (surface)} \rightarrow \mathrm{HONO}. \qquad \text{(R6)}$$

In contrast to the faster surface reaction, reaction of NO_2 with water in the bulk liquid phase ($2NO_2 + H_2O(aq) = HNO_2(aq) + HNO_3(aq)$) is slow due to the low solubility ($1.4 \times 10^{-2}\,\mathrm{M\,atm^{-1}}$ at 293 K) and slow hydrolysis rate ($3.0 \times 10^{7}\,\mathrm{M^{-1}\,s^{-1}}$) of NO_2.[61] The reaction between NO_2 and organics is more complex; it has been found to be greatly accelerated by sunlight through photosensitization,[53,55,58,59] and is likely the major daytime HONO source in urban environments.[32,43,47,50] The combination reaction of NO + OH (R-5) may also become an important HONO source in NO emission source areas.[47]

As more sensitive techniques were developed and became available, many HONO measurements have been conducted in the rural and remote environments.[22–24,26,27,32,38,62,63] Field measurement results demonstrate that significant levels of HONO consistently exist during the daytime in the low-NO_x environments, with median noontime values ranging from 50 pptv to 100 pptv at rural sites and from 10 pptv to ~40 pptv at remote site (Table 4.1). The observed HONO concentrations are much higher than the photo steady state concentrations ($[HONO]_{PSS}$), in the range of $<1 - 7$ pptv, calculated from the following Eq. (4.1) (Table 4.1):

$$[HONO]_{PSS} = \frac{k_{-5} \times [NO] \times [OH]}{J_{HONO} + k_7 \times [OH]}, \tag{4.1}$$

Table 4.1 Measured and calculated photo steady-state concentrations of HONO at noontime in low-NO_x rural and remote environments (average $NO_x \leq 1$ ppbv)

Site and time	$[HONO]_{noontime}$ pptv	$[HONO]_{PSS}$ pptv	Missing HONO source, pptv h^{-1}	References
Pinnacle State Park, NY Summer of 1998	60	≤10	170	23
Summit, Whiteface Mountain, NY Summer of 1999	55	≤5	250	26
Alert, NWT, Canada,	10 (at 5 m)	≤1	50	22
Spring of 2000	40 (20 cm)	≤3	220	
UMBS, Pellston, MI, Summer of 2008	55–70	≤10	180–290	27, 62

Note: The photo steady-state HONO concentration is estimated using Eq. (4.1) from the measured NO and J_{HONO} values and an assumed OH concentration of 5×10^6 molecule cm^{-3}. The "unknown" HONO source is the HONO source strength needed to sustain the observed noontime HONO concentration, based on Eq. (4.5).

where J_{HONO} is the HONO photolysis frequency, and k_{-5} and k_7 are the rate constants for the reactions (R-5) and (R7), respectively.

$$HONO + OH \rightarrow NO_2 + H_2O. \quad (R7)$$

During the day, the ground surface is mostly a net HONO source.[27,62] Tropospheric HONO is mostly removed by photolysis (R5) and by its reaction with OH radical (R7):

$$S^{HONO} = J_{HONO}[HONO] + k_7 \times [OH] \times [HONO]. \quad (4.2)$$

Photolysis is the dominant sink for HONO at noontime at the measurement sites listed in Table 4.1, with a lifetime of 10–15 min. The second term in Eq. (4.2) contributes less than 3% of the total HONO loss rate at noontime, assuming $[OH] = 5 \times 10^6$ molecule cm^{-3}. Total HONO loss rates ranged from 240 pptv h^{-1} to 450 pptv h^{-1}, as calculated for the rural sites listed in Table 4.1.

Two HONO production terms can be evaluated using measurement data, viz. contributions by the NO + OH recombination reaction (R-5) (P^{HONO}_{NO+OH}) and by the heterogeneous reactions of NO_2 (R6) ($P^{HONO}_{NO2\,het}$), using Eqs. (4.3) and (4.4), respectively:

$$P^{HONO}_{NO+OH} = k_{-5}[NO] \times [OH], \quad (4.3)$$

$$P^{HONO}_{NO2\,het} = \Delta\left(\frac{[HONO]_{night}}{[NO_2]_{night}}\right) \Big/ \Delta t \times [NO_2]. \quad (4.4)$$

Assuming $[OH] = 5 \times 10^6$ molecule cm^{-3}, P^{HONO}_{NO+OH} contributes less than 45 pptv hr^{-1}, or less than 10% of the total HONO production rate to sustain its photochemical loss calculated in Eq. (4.2). The $P^{HONO}_{NO2\,het}$ term is evaluated assuming the measured night-time increase rate of the $HONO/NO_2$ ratio

$$\left(\Delta\left(\frac{[HONO]_{night}}{[No_2]_{night}}\right) \Big/ \Delta t\right)$$

is solely contributed by the NO_2 heterogeneous reactions (R6). This pathway contributes less than 20 pptv h^{-1}, or less than 5% of the total HONO production rate. Therefore, NO_x is not a major daytime HONO precursor in the low-NO_x environments. To balance the HONO budget at noontime, an "unknown" HONO source term ($P^{\mathrm{HONO}}_{\mathrm{unknown}}$) is needed:

$$J_{\mathrm{HONO}}[\mathrm{HONO}] + k_7 \times [\mathrm{OH}] \times [\mathrm{HONO}] + \Delta[\mathrm{HONO}]/\Delta t$$

$$= k_{-5}[\mathrm{NO}] \times [\mathrm{OH}] + \frac{\Delta\left(\frac{[\mathrm{HONO}]_{\mathrm{night}}}{[\mathrm{NO_2}]_{\mathrm{night}}}\right)}{\Delta t} \times [\mathrm{NO_2}]_{\mathrm{noon}} + P^{\mathrm{HONO}}_{\mathrm{unknown}}. \tag{4.5}$$

The term $\Delta[\mathrm{HONO}]/\Delta t$ is the rate of change in concentration at noontime, which is usually small, less than 2 pptv h^{-1}. The "unknown" daytime HONO source has been reported in both urban and rural atmospheres during the day.[22–24,26,32,33,36,38,43,47,50] In low-NO_x environments, $P^{\mathrm{HONO}}_{\mathrm{unknown}}$ has been found to account for the majority of the total HONO production required to sustain the measured HONO concentrations[22–24,26,32] (Table 4.1).

Three processes may contribute to the "unknown" daytime HONO production; photolysis of gaseous ortho-nitrophenols and methyl-substituted nitroaromatics,[64] soil emission of nitrite produced by microbial activities,[65,66] and release of surface nitrite from nighttime deposition. The noontime HONO production from the photolysis of these specific nitroaromatics may be non-negligible for urban conditions.[55] However, these compounds have only been detected in polluted urban air, and they are not important HONO precursors in the remote and rural environments.

Soil emission of HONO is highly dependent on the soil nitrate and/or ammonium content (as precursors) and pH, microbial activity, moisture and permeability.[65–67] It may be a significant or even a major HONO source over the agriculture regions where nitrate and/or ammonium-containing fertilizers are regularly applied, because soils are often loosened and disturbed, and microbial organisms are active. The measurement sites listed in Table 4.1 are all in nutrient-poor forested environments, and thus soil emission of HONO is not expected to be important. The HONO gradient measurement within the forest canopy and over the forest floor indeed indicates that forest ground is a net HONO sink rather than a net HONO source during the day (Ref. [66]; unpublished data).

Ground surface, especially when wet, is a sink and temporary reservoir for tropospheric HONO, and the deposited nitrite may be released and become a HONO source for the overlying atmosphere.[23,24,26,62,68,69] However, this process is relatively rapid and may contribute to the HONO budget in the terrestrial boundary layer within 2-3 hours after sunrise[62] but is not expected to be an important HONO source at around noontime.

In searching for the precursors for daytime HONO in rural and remote environments, nitric acid appears to be the most likely candidate: HNO_3 is a major component of NO_y in the low-NO_x troposphere; it is highly sticky, can be deposited to the ground surfaces at a high rate[2] of 1–10 $\mathrm{cm\,s^{-1}}$, and thus may accumulate to a significant amount on ground surfaces, including surfaces of vegetation, rocks and soils. A hypothesis was then proposed: photolysis of HNO_3 deposited on ground surfaces is a major daytime HONO source in the low-NO_x rural atmospheric boundary layer[22–24,26,27]:

$$\mathrm{HNO_{3(g)}} + \text{ground surfaces} \Leftrightarrow \mathrm{HNO_{3(ads)}}\ (\text{or } \mathrm{NO^-_{3(ads)}}), \quad \text{(R8)}$$

$$\mathrm{HNO_{3(ads)}(NO^-_{3(ads)})} + \mathrm{hv} \rightarrow [\mathrm{HNO_3}]_{(ads)^*}\ (\text{or } \mathrm{NO^-_{3(ads)^*}}), \quad \text{(R9)}$$

$$[\mathrm{HNO_3}]_{(ads)^*}\ (\text{or } \mathrm{NO^-_{3(ads)^*}}) \rightarrow \mathrm{HNO_{2(ads)}} + \mathrm{O(^3P)_{(ads)}}, \quad \text{(R10)}$$

$$[\mathrm{HNO_3}]_{(ads)^*}\ (\text{or } \mathrm{NO^-_{3(ads)^*}}) \rightarrow \mathrm{NO_{2(ads)}} + \mathrm{OH_{(ads)}}. \quad \text{(R11)}$$

In the actinic region of solar radiation, NO_2 is likely the dominant primary product of HNO_3 photolysis on surfaces through reaction (R11), similar to the NO_2 quantum yield of near unity in the gas phase[70] and to the high NO_2:HNO_2 yield ratio of about 9:1 in the aqueous phase.[71] The produced $NO_{2(ads)}$ from reaction (R11) may react with adsorbed H_2O and organic to produce nitrous acid on the surface (R6′):

$$\mathrm{NO_{2(ads)}} + \mathrm{H_2O_{(ads)}},\ \text{organics}_{(ads)} \rightarrow \mathrm{HNO_{2(ads)}}. \quad \text{(R6′)}$$

The $HNO_{2(ads)}$ is then released from the surface into the air.

4.3. Supporting Evidence from Field Studies

There are several lines of evidence from field studies to support the hypothesis that the photolysis of HNO_3 on ground surfaces is a major daytime HONO source in the low-NO_x rural atmospheric boundary layer. No significant correlation between the noontime HONO (or noontime HONO production rate) and NO_x (or NO_2) has been found in all the field measurements at all the low-NO_x sites in Table 4.1 ($r^2 \leq 0.1$), suggesting that NO_x (or NO_2) is not a significant HONO precursor at noontime.[22–24,26,27,63] On the other hand, the noontime HONO concentration at the summit of Whiteface Mountain showed a moderate correlation with NO_y ($r^2 = 0.53$) and a strong correlation with the NO_y average over the prior 24-h period ($r^2 = 0.85$, Fig. 4.1).[26] Since HNO_3 is a major NO_y component at this site and is highly "sticky" toward vegetation surfaces, the 24-h NO_y average represents depositional accumulation of HNO_3 on the mountain slope surface. Indeed, when the data collected at the Pinnacle State Park site in 1998 are reanalyzed in the

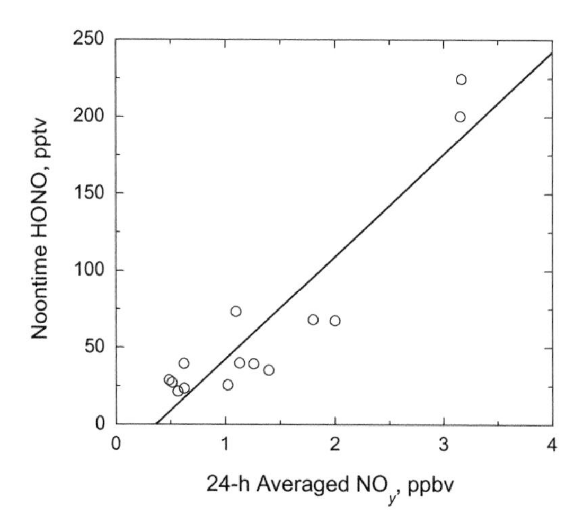

Fig. 4.1 Relationship between noontime HONO concentration (noon ± 1 h) and the NO_y concentration averaged over the previous 24-h period, at the summit of Whiteface Mountain in New York during the 1999 summer measurement intensive. The correlation coefficient $r^2 = 0.85$. The figure is reprinted from Ref. [26].

same way,[23,24] the noontime HONO concentration showed a stronger correlation with the previous 24-h averaged HNO_3 concentration ($r^2 = 0.74$) than with the noontime HNO_3 concentration ($r^2 = 0.62$). The improved correlation with the accumulative parameter is consistent with the proposal that the HNO_3 which has accumulated on vegetation and other ground surfaces is an important HONO precursor.

During the Polar Sunrise Experiment 2000 at Alert, Canada (ALERT 2000), strong diurnal variation in HONO concentrations were observed[22] in sync with the Eppley UV intensity (Fig. 4.2). Concentration gradient and flux measurements of HONO further indicated that photochemical processes in the snowpack emitted HONO into the overlying air and drove the observed diurnal variations in HONO concentration (Fig. 4.2). Snow chamber experiments confirmed that exposure of natural and artificial snow spiked with nitrate leads to significant productions of NO_x and HONO.[22,72–75] Interestingly, one of the supporting pieces of evidence is the finding of interference of sampling inlet manifold on HONO measurement. During the 2000 intensive summer measurement campaign at the PROPHET site (University of Michigan Biological Station in Pellston, MI), significant artifact HONO signals were detected

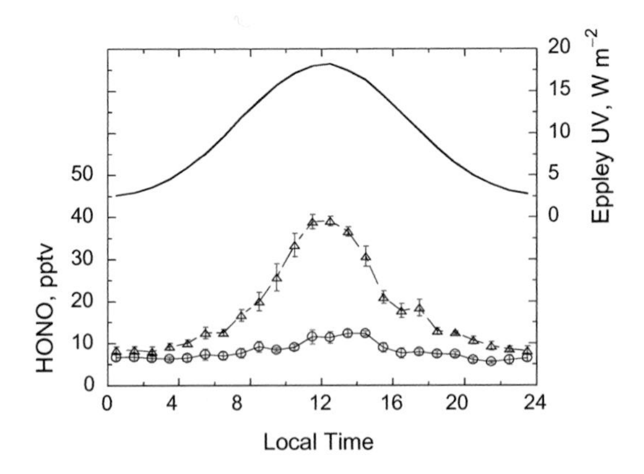

Fig. 4.2 Time series of hourly-average Eppley UV intensity (solid line), HONO mixing ratios at 5 m (○) and at 20 cm (Δ) above the snow surface for April 21–22, 2000 at Alert, Canada. The figure is reprinted from Ref. [22].

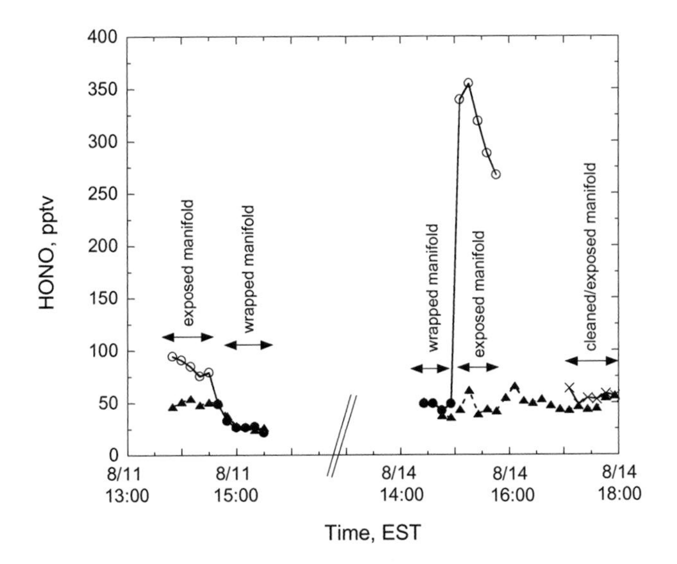

Fig. 4.3 HONO concentrations in air samples drawn from 34 m above ground via a dark Teflon PFA inlet (solid triangles) and via the Pyrex glass manifold (open circles — exposed to sunlight, solid circles — wrapped in aluminum foil, crosses — washed/exposed to sunlight) during the two experiments on August 11 and August 14, 2000 at the PROPHET site. The time periods and actions taken were indicated by the arrows. The figure is reprinted from Ref. [24].

during the daytime when the glass sample manifold walls were exposed to sunlight. The artifact HONO signals can be removed by shielding the manifold from the sunlight with Al foil or by washing the manifold wall surface with water (Fig. 4.3).[23,24] These results clearly indicate that the artifact HONO was produced via a photochemical process from something adsorbed on the inner wall surface of the glass inlet manifold. While several reactions may result in HONO formation, the HNO_3 photolysis seems to be the most likely candidate responsible for our observation. Nitric acid is a major NO_y component measured at this site,[76] and it is highly "sticky".[77,78] Fast adsorption of HNO_3 onto glass surfaces has been observed.[78] Since air is constantly pulled through the manifold, HNO_3 would adsorb and accumulate on the inlet wall surfaces over time, and once exposed to sunlight, the adsorbed HNO_3 absorbs the near UV portion of the sunlight transmitted through Pyrex glass

(≥290 nm) and undergoes photolysis to produce HONO through reactions (R9)–(R11) and (R6′). HONO flux measurement results have indicated that the forest canopy is a net HONO source.[22,27,63,67] The diurnal variation of HONO flux follows the solar UV intensity (Fig. 4.4), suggesting that photochemical processes on the canopy surface are responsible for HONO production. Furthermore, the daytime HONO flux at the PROPHET site was found to correlate well with the surface nitric acid photolysis potential (defined as the product of HNO_3 photolysis rate constant and leaf surface nitrate loading, $J_{HNO_3} \times L_{HNO_3(s)}$) at the top of forest canopy ($r^2 = 0.69$), which is in contrast with the lower correlation with J_{HNO_3} alone ($r^2 = 0.31$) or with $L_{HNO_3(s)}$ alone ($r^2 = 0.10$).[27] On the other hand, no significant correlation was found between

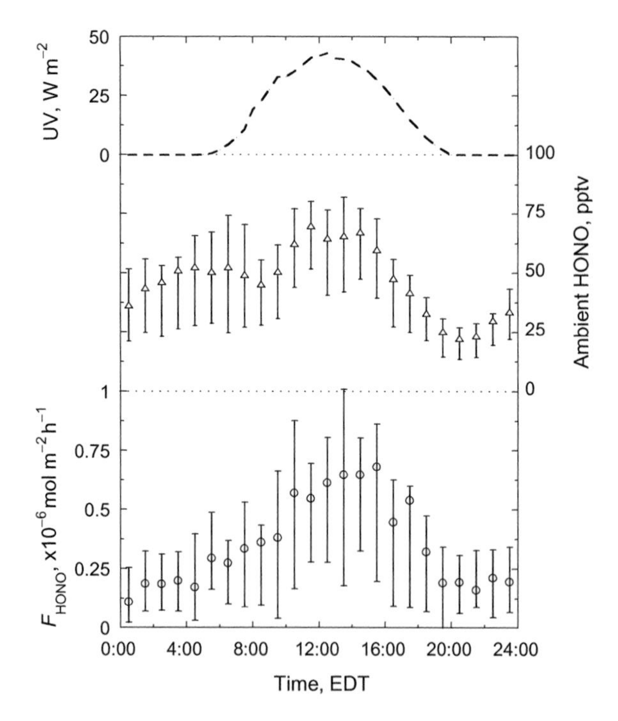

Fig. 4.4 Diurnal variations in hourly averaged median values of UV intensity, HONO mixing ratio, and HONO flux measured at the PRHOPHET site from July 15 to August 10, 2008. The vertical bars are the 25‰ and 75‰ ranges. The figure is reprinted from Ref. [22].

daytime HONO and NO_x ($r^2 = 0.1$), between daytime HONO flux and NO_x ($r^2 = 0.01$), and between daytime flux and $J_{NO_2} \times [NO_2]$ ($r^2 = 0.01$).[27,63] The prominent correlation between the daytime HONO flux and $J_{HNO_3} \times L_{HNO_3(s)}$ lends support to the hypothesis that the photolysis of HNO_3 on canopy surfaces is the major daytime HONO source in the low-NO_x rural boundary layer. Calculations based on field data of HONO concentration gradient/flux and surface HNO_3 loading at the Waldstein ecosystem research site in Germany also indicated that surface HNO_3 photolysis is the most important HONO source.[67] However, HONO flux measurements at Blodgett Forest during BEARPEX 2009 showed no significant upward HONO flux ($(-0.014 \pm 0.086) \times 10^{-6}$ mols m^{-2} h^{-1}) during the day. Low industrial SO_2 and high agricultural NH_3 emissions in California result in relatively elevated pH on canopy surfaces. Since HONO is a weak acid, with a pK_a of 3.5, the HONO-nitrite equilibrium would shift to nitrite at higher pH. At elevated pH, canopy and ground surfaces may behave more like a sink than a source for HONO at the Blodgett Forest site.[79,80] During the recent CalNex 2010 study, a very strong positive correlation ($r^2 = 0.985$) was observed between HONO flux and the product of NO_2 concentration and solar radiation at the Bakersfield site,[80] suggesting that photo-enhanced NO_2 reactions on ground surface was the dominant HONO source at elevated NO_2 concentrations.

4.4. Laboratory Light-Exposure Experiments

Laboratory experiments have been conducted[25] to test the hypothesis developed to explain field HONO observations, i.e., photolysis of HNO_3 deposited on ground surfaces is a major daytime HONO source in the low-NO_x rural atmospheric boundary layer.[22–24] Indeed, HONO and NO_x were produced from the photolysis of HNO_3 deposited on Pyrex glass surface when exposed to UV light (Fig. 4.5), at a rate of $(2-6) \times 10^{-5}$ s^{-1} under the normalized tropical noontime condition, which is $\sim$2 orders of magnitude higher than the rates in the aqueous and gas phases. Under dry conditions, NO_2 is the dominant product and accounts for 97% of the total production,

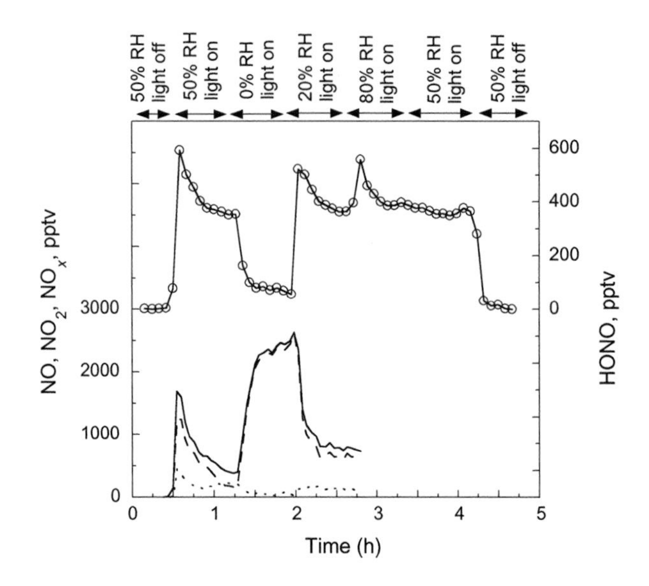

Fig. 4.5 Concentrations of HONO (open circles with solid line), NO (dotted line), NO_2 (dashed line) and NO_x (solid line) in the effluent gas from a flow reactor during a photochemical experiment. The reactor surface was coated with 1.2×10^{-7} moles of HNO_3. The gas flow rate through the reactor was $3.5\,L\,min^{-1}$. The figure is reprinted from Ref. [25].

confirming the proposal that reaction (R11) is the dominant pathway of HNO_3 photolysis on the surface. The relative yield of HONO was found to increase with relative humidity (RH), from ~2% at 0% RH to 35% at 20% RH and 55% at 50% RH, confirming the secondary production of HONO via reaction (R6′). It is interesting to note that the apparent HNO_3 photolysis rate on Pyrex surface ($J_{HNO_3(s)}$) was found to decrease with RH, from $6.0 \times 10^{-5}\,s^{-1}$ at 0% RH to $2.8 \times 10^{-5}\,s^{-1}$ at 20% RH and $2.2 \times 10^{-5}\,s^{-1}$ at 50% RH. As most environmental surfaces have several layers of water, $NO_{2(ads)}$ generated from the photolysis of nitric acid on the surfaces could react with water already adsorbed on surfaces to form HONO via the reaction (R6′) under ambient conditions.

Recently, Ye *et al.*[81] investigated the photolysis of HNO_3 on the surfaces of natural and artificial materials, including plant leaves, metal sheets, and construction materials. The photolysis rate constant of HNO_3 deposited on surfaces from outdoor ambient air ranged from $9 \times 10^{-6}\,s^{-1}$ to $3.7 \times 10^{-4}\,s^{-1}$, based on the total production rates of HONO and NO_x, 1–3 orders of magnitude higher than that of gaseous HNO_3. At 50% RH, HONO was found to be the major product from the photolysis of HNO_3 on most plant leaves, and NO_x was found to be the major product on the metal and non-porous granite surfaces. Although the leaf pH was not measured, the leaf surface is expected to be acidic; relatively high emissions of SO_2 and NO_x in the Northeastern US result in an acidic atmosphere.[82] In addition, the $J_{HNO_3(s)}$ value was found to decrease with the "apparent" HNO_3 surface density. Such results may appear to be contradictory, but can be understood if there are more than one layer of HNO_3 on the surfaces and the fastest HNO_3 surface photolysis occurs on the HNO_3 layer that directly interacts with the surface. Within a typical range of HNO_3 surface density of $(5 - 15) \times 10^{-6}\,mol\,m^{-2}$ in the low-NO_x forested areas,[27,67] photolysis of HNO_3 on the forest canopy is a major HONO source for the overlying atmosphere, with an upward HONO flux in the range of $4 \times 10^{-7} - 6 \times 10^{-7}\,mol\,m^{-2}h^{-1}$, comparable to the mid-day range of $2 \times 10^{-7} - 10 \times 10^{-7}\,mol\,m^{-2}\,h^{-1}$ and the median of $7 \times 10^{-7}\,mol\,m^{-2}\,h^{-1}$ from a forest canopy in Northern Michigan.[27,63]

In addition to investigating the interaction between HNO_3 and glass, leaf, metal, and construction material surfaces, gas phase nitric acid was exposed to hydrophobic organic film surfaces.[83] These organic films were found to take up gas phase nitric acid; a fraction of HNO_3 was found to dissociate in the film to form proton $+NO_3^-$. Photolysis of nitrated films by irradiation from a Xe lamp resulted in faster decrease of both proton and nitrate in samples exposed to irradiation than non-irradiated samples at time scales on the order of tens of minutes to several hours. Photochemistry of nitric acid and nitrate associated with urban films has been suggested to be important in the regeneration of gas phase HONO and/or NO_2. The photolysis of nitric acid deposited onto "real" or proxy urban grime was studied using attenuated total reflection (ATR) FTIR spectroscopy.[84] Urban grime samples were deposited onto the front surface of a ZnSe ATR crystal, the surface film thus formed was first exposed to nitric acid and subsequently illuminated with a solar simulator. Disappearance rates of nitrate were monitored using ATR-FTIR. The radiative flux was calibrated using an aqueous nitrate actinometer. A summer noontime photolysis rate of $1.2 \times 10^{-3}\,s^{-1}$ for HNO_3/NO_3^- on urban grime was reported in Toronto, which is about 4 orders of magnitude faster than the photolysis rate of $1.0 \times 10^{-7}\,s^{-1}$ for aqueous nitrate. Product channels of nitrate photolysis on urban grime were postulated to be similar to that in the aqueous phase photolysis, which can lead to the formation of NO_2, OH, and possibly HONO.

Apart from the macroscopic environmental surfaces, adsorption and photolysis of HNO_3 on microscopic dust particle surfaces were studied both in the absence and in the presence of water vapor.[85,86] A Knudsen cell reactor was used to study the uptake of HNO_3 on powdered samples of α-Al_2O_3, α-Fe_2O_3, SiO_2, MgO, and CaO, and of Gobi dust and Saharan sand.[85] Nitric acid was found to be deposited not only to the surface layers but also diffuse into the underlying layers of the particles. The initial sticking coefficients are in the range of $10^{-5} - 10^{-3}$ for HNO_3 concentration of $10^{11}-10^{12}$ molecule/cm^3. Irreversible uptake behavior was observed when HNO_3 was deposited on most of the oxide samples. Changes in the transmission FTIR spectrum of the oxide particles following exposure of particles such as SiO_2, α-Al_2O_3, TiO_2, γ-Fe_2O_3, CaO, and MgO to HNO_3 were monitored.[86] Nitric acid was found to molecularly and reversibly adsorb on SiO_2. On other particle surfaces, both molecular adsorption and dissociative adsorption occurred. In the presence of humidity, increased uptake of HNO_3 on α-Al_2O_3 and CaO particles was observed. For MgO and CaO, saturated solutions of $Mg(NO_3)_2$ and $Ca(NO_3)_2$ were formed upon exposure to HNO_3 in the presence of humidity. Gas phase products formed from broadband solar simulator irradiation of nitric acid/nitrate adsorbed on aluminum oxide particles were studied in an environmental aerosol chamber, and monitored with FTIR. Photolysis products include NO_2 and NO in the absence of humidity and NO_2 only in the presence of humidity.[87] An NO_x production rate constant of $5 \times 10^{-5}\,s^{-1}$ is estimated from the alumina surfaces under dry conditions. When a narrow-band (310 ± 10 nm) irradiation source was used, the only gas phase product observed in dry air was NO_2, whereas both NO_2 and N_2O were observed under humid conditions.[88] NO production was not detected under any of the experimental conditions.

4.5. Mechanistic Investigations

4.5.1. *Adsorbed HNO_3 Near UV Surface Absorption Cross Sections*

Rapidly accelerated photolysis rates of nitric acid on surfaces compared to that in the gas phase could be caused by a red-shift of the HNO_3 near UV absorption spectrum under the influence of the surface to which HNO_3 is adsorbed; the spectral shift allows more overlap with the solar actinic flux ($\lambda \geq 290$ nm). To understand the likely causes for the much faster HNO_3 surface photolysis rates, the near UV absorption cross sections of surface-adsorbed HNO_3 in the 290–365 nm region were measured using Brewster angle cavity ring-down spectroscopy.[28,29] A pair of super-clean, mutually compensating fused silica Brewster windows were placed in the path of the main optical axis inside the ring-down cavity. Nitric acid vapor was deposited onto window surfaces in vacuum. Nitric acid deposited on fused silica surfaces retained its molecular form. A previous literature study[86] also reported that HNO_3 molecularly and reversibly adsorbed on SiO_2 particle surfaces. Absorption of the linearly-polarized probe laser beam by a saturated monolayer of surface-adsorbed HNO_3 was measured as a function of near UV wavelength. To convert absorption by a saturated monolayer of surface adsorbed HNO_3 into HNO_3 near UV surface absorption cross sections, a maximum HNO_3 surface concentration on fused silica surfaces of $\sim 1.1 \times 10^{14}$ molecule/cm^2 was estimated with a van der Waals radius of 5.5 Å for HNO_3.[89] This calculated maximum HNO_3 surface concentration is in good agreement with the saturation monolayer coverage of HNO_3 on SiO_2 particle surfaces of $(7 \pm 3) \times 10^{13}$ molecule/cm^2, obtained with transmission FTIR spectroscopy.[86] Adsorbed HNO_3 near UV surface cross section data are tabulated in Table 4.2 and plotted in Fig. 4.6 along with nitric

Table 4.2 Surface and gas phase absorption cross section of HNO_3, as a function of wavelength, λ

λ (nm)	$\sigma_{surface}$ ($cm^2 molecule^{-1}$)[a]	σ_{vapor} ($cm^2 molecule^{-1}$)
290	$(1.85 \pm 0.21) \times 10^{-18}$	$(5.98 \pm 0.27) \times 10^{-21}$
295	$(1.59 \pm 0.18) \times 10^{-18}$	$(4.09 \pm 0.28) \times 10^{-21}$
300	$(1.48 \pm 0.15) \times 10^{-18}$	$(2.59 \pm 0.18) \times 10^{-21}$
305	$(1.09 \pm 0.17) \times 10^{-18}$	$(1.68 \pm 0.19) \times 10^{-21}$
310	$(1.31 \pm 0.05) \times 10^{-18}$	$(0.95 \pm 0.01) \times 10^{-21}$
315	$(1.30 \pm 0.12) \times 10^{-18}$	$(0.51 \pm 0.05) \times 10^{-21}$
320	$(1.00 \pm 0.23) \times 10^{-18}$	$(0.47 \pm 0.06) \times 10^{-21}$
325	$(0.47 \pm 0.05) \times 10^{-18}$	$(0.21 \pm 0.05) \times 10^{-21}$
330	$(0.45 \pm 0.05) \times 10^{-18}$	$(0.19 \pm 0.04) \times 10^{-21}$
335	$(0.31 \pm 0.05) \times 10^{-18}$	$(0.027 \pm 0.004) \times 10^{-21}$
340	$(0.27 \pm 0.03) \times 10^{-18}$	$(0.016 \pm 0.004) \times 10^{-21}$
345	$(0.20 \pm 0.03) \times 10^{-18}$	$(0.0080 \pm 0.0007) \times 10^{-21}$
350	$(0.17 \pm 0.04) \times 10^{-18}$	$(0.0053 \pm 0.0014) \times 10^{-21}$
355	$(0.13 \pm 0.01) \times 10^{-18}$	n.d.
360	$(0.10 \pm 0.01) \times 10^{-18}$	n.d.
365	$(0.044 \pm 0.004) \times 10^{-18}$	n.d.

[a] Cross section data in the 290–330 nm region were reported in Ref. [28]. Cross section data in the 335–365 nm region were reported in Ref. [29].

acid gas phase cross section values. Surface absorption cross sections of HNO_3 vary from $(1.85 \pm 0.21) \times 10^{-18}$ cm^2/molecule at 290 nm to $(0.044 \pm 0.004) \times 10^{-18}$ cm^2/molecule at 365 nm. Room temperature absorption cross sections for HNO_3 deposited on fused silica surfaces are two to four orders of magnitude larger than those in the gas phase, with the largest differences at the longer wavelengths. Results from the HNO_3 surface cross section measurements are consistent with the much enhanced HNO_3 photolysis rates on the surface observed from the field studies.

The nature of the near UV absorption band for HNO_3 is that of the electric dipole forbidden but vibronically allowed $n \rightarrow \pi^*$ transition between the NO_2 non-bonding orbital and the NO_2 π^* antibonding orbital in HNO_3.[90] Surface/HNO_3 interaction can affect the electronic transition probability of the surface-bound HNO_3, and its near UV cross section values. Quantum chemistry calculations[91] indicated that the equilibrium geometry of the electronically-excited state of HNO_3 is pyramidal whereas the ground state HNO_3 has

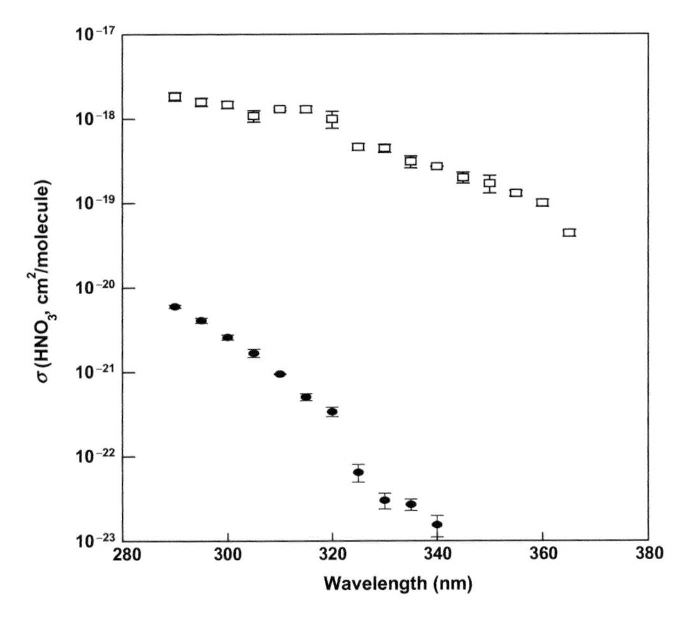

Fig. 4.6 Absorption cross sections of HNO_3 on fused silica surfaces (squares) and in the gas phase (circles) as a function of wavelength. Cross section data in the 290–330 nm region were reported in Ref. [28]. Cross section data in the 335–365 nm region were reported in Ref. [29].

a planar configuration. Intuitively, interaction between the ground state HNO_3 and the surfaces could distort ground state HNO_3 from planar towards pyramidal geometry and thus greatly increase the electronic transition probability between the ground state and the vibronically-excited state for surface-bound HNO_3, although theoretical calculation in support of such a conjecture has not been conducted. Also, the NO_2 π^* antibonding orbital can be stabilized more than the NO_2 nonbonding orbital by the multi-body interaction between the HNO_3 and the surfaces. Such interaction can also broaden both the NO_2 nonbonding orbital and the NO_2 π^* antibonding orbital in surface-bound HNO_3. Thus, the broadening and red-shifting of the HNO_3 near UV absorption spectrum on surfaces compared to that in the gas phase is a result of HNO_3/surface interactions.

4.5.2. *Nitric Acid Photolysis Product Channels in the Actinic UV Region*

Observation of HONO formation after exposing glass manifold surfaces coated with HNO_3 and H_2O with sunlight[24] appears to support the hypothesis that sunlight photolysis of nitric acid (HNO_3) deposited on ground surfaces is a photochemical HONO source. Reaction of H_2O with electronically-excited NO_2 (NO_2^*), produced from sunlight photolysis of adsorbed HNO_3, is a likely mechanism of HONO formation:

$$NO_2^* + H_2O \rightarrow HONO + OH. \qquad (R12)$$

To test if HONO formation is energetically feasible from the photolysis of HNO_3 and H_2O only system by sunlight, and if NO_2^* is a HNO_3 photolysis product, laboratory studies were conducted to investigate the 308 nm gas phase photolysis of HNO_3 in the presence and absence of water vapor.

Gas phase photolysis of HNO_3 in the actinic UV region ($\lambda \geq 290\,nm$) can proceed through the following pathways:

$$HNO_3 + h\nu \rightarrow OH + NO_2 \quad (\lambda \leq 604\,nm) \qquad (R13)$$

$$\rightarrow OH + NO_2^* \quad (\lambda \leq 381\,nm) \qquad (R14)$$

$$\rightarrow HONO + O(^3P) \quad (\lambda \leq 393\,nm). \qquad (R15)$$

where photochemical thresholds were estimated from the corresponding enthalpy changes.[92] Several studies[93–95] investigated the HNO_3 photolysis at around 300 nm. These studies mostly monitored the OH product. Laser-induced fluorescence (LIF) was used to obtain the OH quantum yield from the 308 nm photolysis of HNO_3 vapor, and an OH quantum yield of 1.05 ± 0.29 was reported.[94] Nitric acid photolysis channel (R15) was found to be insignificant with actinic UV irradiation. August *et al.*[93] applied Doppler-resolved LIF to probe the scalar and vector properties of the OH(X) fragments from 280 nm photodissociation of HNO_3. The near UV HNO_3 photolysis was proposed to occur through a vibronically-mediated transition to an electronic state of A'' character. 70% and 30% of the excess energy after the HNO_3 photolysis were distributed in the internal modes of NO_2, and in the OH and NO_2 translational mode, respectively. If similar partitioning of the excess energy from the 308 nm and 280 nm HNO_3 photolysis is assumed, the maximum internal energy of the NO_2 group following the 308 nm photolysis is about 30.7 kcal/mol. The lowest energy electronically-excited state for NO_2 is 1^2B_2. The calculated excess internal energy (30.7 kcal/mol) is only about 3 kcal/mol higher than the minimum energy difference between the ground electronic state and the first electronically-excited state (1^2B_2) of NO_2. Understanding the nature of the electronic states of NO_2 in the NO_x production channel from the HNO_3 photolysis in the actinic UV region is important toward understanding HONO formation in the troposphere. NO_2^* emission has not been reported from the gas phase HNO_3 photolysis in the actinic UV region. If the ground state NO_2 is the only NO_x product formed from the 308 nm HNO_3 photolysis, the proposed mechanism for the primary HONO formation via (R12) from sunlight photolysis of HNO_3 in the presence of H_2O may not hold. HONO can only be formed from the slower, gas-phase reaction:

$$2NO_2 + H_2O + M \rightarrow HNO_3 + HONO + M. \qquad \text{(R16)}$$

On the other hand, if NO_2^* is formed from the 308 nm HNO_3 photolysis in the gas phase, it may react with H_2O to form HONO+OH via the faster reaction (R12).[96]

The NO_x channels from the 308 nm gas phase photolysis of HNO_3 were investigated[70] by using excimer laser photolysis combined with cavity ring-down spectroscopy.[97,98] The photolysis products were measured in the 552–560 nm and 640–648 nm regions. Presented in Fig. 4.7 are HNO_3 absorption spectra in the 640–648 nm region in the absence and presence of the photolysis pulses (the photolysis/probe laser delay was set at 15 μs), and the difference spectrum. Also illustrated in the same figure for comparison is a bar graph of vibronic band origins and line intensities[99] of electronically-excited $NO_2(^2B_2)$. The HNO_3 photolysis product has a different spectral pattern from the vibronic band origins and line intensities in NO_2^*, which suggests that NO_2^* is not formed from HNO_3 photolysis at

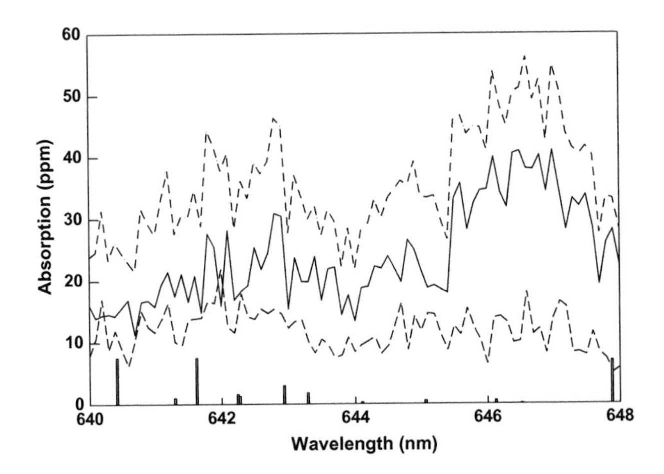

Fig. 4.7 Absorption spectra of 1.2 Torr HNO_3 in the 640–648 nm region, in the absence (solid line) and presence (dash-dot line) of photolysis at 308 nm. The difference spectrum is shown as medium dash line. The bar graph presents vibronic band origins and line intensities[99] in electronically-exited NO_2 ($^2B_2$). The figure is reprinted with permission from Ref. [70]. Copyright (2015) American Chemical Society.

308 nm. The probe wavelength range of 552–560 nm is at the longer-wavelength tail of the ground state NO_2 absorption spectrum.[100–102] It is also a spectral range where NO_2^* begins to absorb.[99,103–105] The similarity of the HNO_3 photolysis product spectrum in the 552–560 nm region to that of the NO_2 standard and the lack of dependence of absorption on photolysis/probe laser delay time indicate the NO_x product produced from the 308 nm HNO_3 photolysis is NO_2 in the ground electronic state. The NO_2 quantum yield was determined to be about unity. Nitric acid photolysis was also investigated in the presence of water vapor. Transient absorption profiles around 552 nm after the HNO_3 photolysis were acquired. Plotted in Fig. 4.8 is absorption at 552.80 nm as a function of photolysis/probe laser delay time, following the gas phase photolysis of a mixture containing 1.0 Torr HNO_3 and 0.2 Torr H_2O at 308 nm. Absorption at 552.80 nm was found to be invariant with delay time after the photolysis of equilibrated HNO_3/H_2O mixtures, implying that the HNO_3 photolysis product did not react with H_2O. If NO_2^* was produced from the photolysis of HNO_3 at 308 nm, and if NO_2^* was quenched by

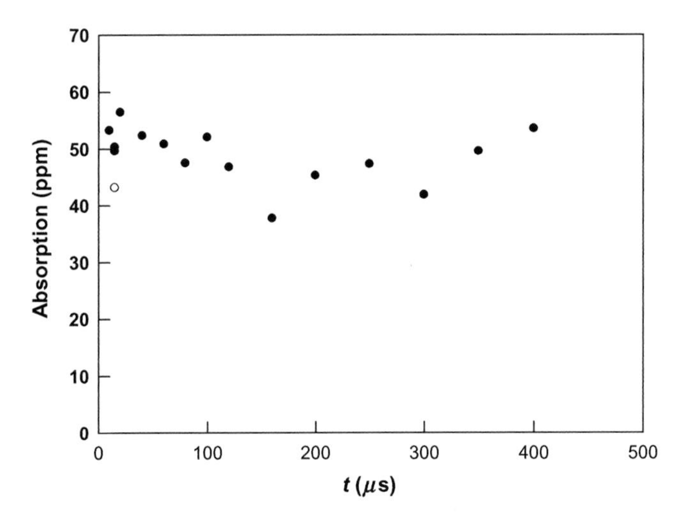

Fig. 4.8 Round-trip absorption at 552.80 nm as a function of photolysis/probe laser delay time after the 308 nm photolysis of equilibrated mixtures containing 1.0 Torr HNO_3/0.2 Torr H_2O. The figure is reprinted with permission from Ref. [70]. Copyright (2015) American Chemical Society.

H_2O, absorption at 552.80 nm would decay with time. The absence of time-dependent absorption from the gas phase photolysis of mixtures containing 1.0 Torr HNO_3/0.2 Torr H_2O or 1.1 Torr HNO_3/0.5 Torr H_2O is consistent with ground state NO_2 being the only NO_x product formed from the 308 nm photolysis of the HNO_3/H_2O mixture.

Potential HONO product following the 308 nm gas phase photolysis of HNO_3 in the presence of H_2O was examined by probing peak and valley of HONO absorption[106–110] at 342.0 nm and at 343.5 nm with and without photolysis pulses. Absorptions at 342.0 nm and 343.5 nm were found to be comparable in size, but did not exhibit temporal dependence following the photolysis of the HNO_3/H_2O mixtures. The study[70] excluded HONO formation at <1 ms after the 308 nm photolysis of the HNO_3/H_2O mixture. Such result is consistent with HONO not being a primary product formed from the 308 nm photolysis of the HNO_3/H_2O mixtures. The gas-phase reaction of the ground state NO_2 with H_2O (R16) is known to be extremely slow, and its heterogeneous analogue on the reactor wall surface is also expected to be too slow[111] to produce detectable HONO in the time scale of milliseconds.

Energetic constraints can explain why only the ground state NO_2 is produced from the 308 nm photolysis of HNO_3. Quantum chemistry calculations[90] indicated that nitric acid ground state and two lowest energy excited states are $1\,^1A'$, $1\,^1A''$, and $2\,^1A''$. These two lowest electronically-excited states are 3.40 eV and 4.54 eV higher in energy than the ground state HNO_3.[90,112] An HNO_3 molecule can absorb a 308 nm photon (3.73 eV), and be excited into the first electronically-excited $1^1A''$ state. $1^1A''$ quantum state of HNO_3 is correlated with X^2A_1 in NO_2, which is the ground electronic state of NO_2.[90,113] 308 nm photon energy is not energetic enough[90] to reach $2\,^1A''$ of HNO_3 (4.54 eV) which is correlated with the first electronically-excited state of NO_2.

Based upon the results of laboratory studies of nitric acid gas phase photolysis in the actinic UV region and the literature photodissociation potential energy surfaces of nitric acid, ground state NO_2 is also a NO_x product formed from adsorbed HNO_3 photolysis in the daytime troposphere. This is supported by the similar photolysis product spectra in the 552–560 nm region from the 308 nm HNO_3 photolysis in the gas phase and on the surfaces of Al, fresh ice film, and silica.[95,114] A recent careful study indicated the 552–560 nm product absorption from the 308 nm HNO_3 photolysis is that of ground state NO_2.[70] Therefore, the photolysis of HNO_3/H_2O only system on surface by actinic UV radiation is not expected to produce HONO via the fast $NO_{2(ads)^*} + H_2O_{(ads)} = HONO_{(ads)} + OH_{(ads)}$ reaction. Slower, secondary HONO formation may occur through the surface reaction (R6′).

The nitric acid photolysis quantum yield has been determined to be unity in the gas phase,[70,94] but only ~0.01 in the aqueous phase due to solvent cage effect.[71] The quantum yield for HNO_3 photolysis on solid surfaces is expected to be larger than in the aqueous phase, as the solvent cage effect is much reduced on surfaces. The photolysis rate is the sum of the products of solar flux, near UV absorption cross sections, and photolysis quantum yields for given wavelength intervals over the actinic UV region. Since near UV absorption cross sections of HNO_3 deposited on surfaces are two to four orders of magnitude larger than those in the gas phase, the sunlight photolysis of HNO_3 on surfaces is enhanced by 2–4 orders of magnitude. Nitric acid surface photolysis may be efficient in renoxification and in forming significant amount of OH radicals.

4.6. Significance and Future Work

Both field and laboratory studies have demonstrated that photolysis of HNO_3 deposited on the ground surfaces converts a NO_x reservoir to photochemically reactive NO_x and HONO. In addition to HONO being an important precursor for OH radical, NO_x plays a central role in the photochemical formation of oxidants including OH radical and O_3. The remobilized HONO and NO_x will render the overlying troposphere more photochemically reactive than generally predicted, especially in the rural and remote regions where direct anthropogenic emissions of NO_x are small. Therefore, this photolytic process should be incorporated into atmospheric models to more accurately predict the production of photooxidants and the degradation of atmospheric pollutants on the regional and global scales.

However, only limited numbers of studies have been conducted so far, many unknowns remain. To accurately parameterize this surface photolytic process for atmospheric models, we would need to acquire better information on the following three aspects:

(1) Temporal and spatial distributions of HNO_3 and nitrate surface density on various surfaces in the ambient environments.
(2) The near UV absorption cross sections, photolysis rate constants and product yields of HNO_3 and nitrate on various surfaces.
(3) Production of OH radicals from the photolysis of HNO_3/nitrate on various surfaces.

Although the photolysis of nitric acid on leaf and other surfaces was reported to be much faster than those in the aqueous phase, direct measurements of the near UV absorption cross sections and photolysis quantum yields of HNO_3/NO_3^- on aqueous and most common solid environmental surfaces have not been reported.

OH radical quantum yields from nitric acid photolysis on surfaces have not been directly measured. Further related laboratory study is needed.

End-products from the photolysis of nitric acid on proxy or real dust particles have been investigated. Few studies have been conducted to investigate the photolysis of nitric acid on aged ice particles. The contribution of the reaction of nitric acid with secondary organic aerosols and its subsequent photolysis to the recycling of reactive nitrogen species is not well understood.

In the real atmospheric environment, other pollutants such as organic compounds emitted from biogenic processes and anthropogenic activities can be deposited on the ground surfaces in addition to nitric acid. Complex mixture of reactants in addition to HNO_3 and water vapor, and complex surface matrices could act as photosensitizers in HONO formation. Nitric acid can also react with organics to form nitrated organic compounds on the surfaces. Such compounds can be photolyzed to form products which are different from the deposition of only HNO_3 on proxies of the ground surfaces. A comprehensive understanding of the synergistic interaction of nitric acid and other pollutants on the surfaces and their subsequent heterogeneous photochemistry is critically needed to assess the impacts of these processes on air quality.

Acknowledgments

We thank the book editors for their constructive comments. Funding is provided by the National Science Foundation under grants ATM-9615748, ATM-0122708, ATM-0632548 and AGS-1216166 for XZ and ATM-0653761, AGS0969985, and AGS1405610 for LZ.

References

1. B. J. Finlayson-Pitts and J. N. Pitts, Jr. *Chemistry of the Upper and Lower Atmosphere: Theory, Experiments, and Applications.* (Academic Press, San Diego, 2000).
2. J. H. Seinfeld and S. N. Pandis. *Atmospheric Chemistry and Physics, From Air Pollution to Climate Change*, 7th edn. (Wiley & Sons, Hoboken, NJ, 2006).

3. R. Atkinson, D. L. Baulch, R. A. Cox, J. N. Crowley, R. F. Hampson, R. G. Hynes, M. E. Jenkin, M. J. Rossi and J. Troe. IUPAC task group on atmospheric chemical kinetic data evaluation, *Atmos. Chem. Phys.* **4**, 1461 (2004), http://iupac.pole-ether.fr/htdocs/datasheets/pdf/NOx13_HO_NO2.pdf.
4. J. J. Zhang, T. Dransfield and N. M. Donahue. On the mechanism for nitrate formation via the peroxy radical + NO reaction. *J. Phys. Chem. A* **108**, 9082–9095 (2004).
5. E. C. Browne and R. C. Cohen. Effects of biogenic nitrate chemistry on the NO_x lifetime in remote. *Atmos. Chem. Phys.* **12**, 11917–11932 (2012).
6. E. C. Browne *et al.* Observations of total $RONO_2$ over the boreal forest: NO_x sinks and HNO_3 sources. *Atmos. Chem. Phys.* **13**, 4543–4562 (2013).
7. J. Savarino *et al.* Isotopic composition of atmospheric nitrate in a tropical marine boundary layer. *PNAS.* **110**, 17668–17673 (2012).
8. W. R. Stockwell *et al.* A new mechanism for regional atmospheric chemistry modeling. *J. Geophys. Res.* **102**, 25847–25879 (1997).
9. B. J. Johnson *et al.* Seasonal trends of nitric-acid, particulate nitrate, and particulate sulfate concentrations at a Southwestern United-States mountain site. *Atmos. Environ.* **28**, 1175–1179 (1994).
10. Y. Dubowski *et al.* Interactions of gaseous nitric acid with surfaces of environmental interest. *Phys. Chem. Chem. Phys.* **6**, 3879–3888 (2004).
11. R. B. Chatfield. Anomalous HNO_3/NO_x ratio of remote tropospheric air: Conversion of nitric acid to formic and NO_x? *Geophys. Res. Lett.* **21**, 2705–2708 (1994).
12. D. A. Hauglustaine *et al.* HNO_3/NOx ratio in the remote troposphere during MLOPEX 2: Evidence for nitric acid reduction on carbonaceous aerosols? *Geophys. Res. Lett.* **23**, 2609–2612 (1996).
13. G. M. Underwood *et al.* Heterogeneous reactions of NO_2 and HNO_3 on oxides and mineral dust: A combined laboratory and modeling study. *J. Geophys. Res.* **106**, 18055–18066 (2001).
14. D. J. Lary *et al.* Carbon aerosols and atmospheric photochemistry. *J. Geophys. Res.* **102**, 3671–3682 (1997).
15. D. J. Lary *et al.* Carbonaceous aerosols and their potential role in atmospheric chemistry. *J. Geophys. Res.* **104**, 15929–15940 (1999).
16. D. J. Lary and D. E. Shallcross. Potential importance of the reaction $CO+HNO_3$. *J. Geophys. Res.* **105**, 11617–11624 (2000).
17. M. Mochida and B. J. Finlayson-Pitts. FTIR studies of the reaction of gaseous NO with HNO_3 on porous glass: Implications for conversion of HNO_3 to photochemically active NO_x in the atmosphere. *J. Phys. Chem. A.* **104**, 9705–9711 (2000).
18. N. A. Saliba *et al.* Reaction of gaseous nitric oxide with nitric acid on silica surfaces in the presence of water at room temperature. *J. Phys. Chem.* **105**, 10339–10346 (2001).
19. A. M. Rivera-Figueroa, A. L. Sumner and B. J. Finlayson-Pitts. Laboratory studies of potential mechanisms of renoxification of tropospheric nitric acid. *Environ. Sci. Technol.* **37**, 548–554 (2003).

20. S. Sandholm *et al.* Summertime partitioning and budget of NOy compounds in the troposphere over Alaska and Canada: ABLE 3B. *J. Geophys. Res.* **99**, 1837–1861 (1994).
21. C. A. Rogaski, D. M. Golden and L. R. Williams. Reactive uptake and hydration experimentson amorphorous carbon treated with NO_2, SO_2, O_3, HNO_3 and H_2SO_4. *Geophys. Res. Lett.* **24**, 381–384 (1997).
22. X. L. Zhou, H. J. Beine, R. E. Honrath, J. D. Fuentes, W. Simpson, P. B. Shepson and J. W. Bottenheim. Snowpack photochemical production of HONO: A major source of OH in the Arctic boundary layer in springtime. *Geophys. Res. Lett.* **28**, 21 (2001), doi:10.1029/2001GL013531.
23. X. Zhou, K. Civerolo, H. Dai, G. Huang, J. Schwab and K. Demerjian. Summertime nitrous acid chemistry in the atmospheric boundary layer at a rural site in New York State. *J. Geophys. Res.* **107**, 4590 (2002), doi:10.1029/2001JD001539.
24. X. Zhou, Y. He, G. Huang, T. D. Thornberry, M. A. Carroll and S. B. Bertman. Photochemical production of nitrous acid on glass sample manifold surface. *Geophys. Res. Lett.* **29**, 14 (2002), doi:10.1029/2002GL 015080.
25. X. Zhou, H. Gao, Y. He, G. Huang, S. B. Bertman, K. Civerolo and J. Schwab. Nitric acid photolysis on surfaces in low-NOx environments: Significant atmospheric implications. *Geophys. Res. Lett* **30**, 2217 (2003), doi:10.1029/2003GL018620.
26. X. Zhou, G. Huang, K. Civerolo, U. Roychowdhury and K. L. Demerjian. Summertime observations of HONO, HCHO, and O_3 at the summit of Whiteface Mountain, New York. *J. Geophys. Res.* **112**, D08311 (2007), doi:10.1029/2006JD007256.
27. X. Zhou, N. Zhang, T. Michaela, D. Tang, J. Hou, S. B. Bertman, M. Alaghmand, P. B. Shepson, M. A. Carroll, S. Griffith, S. Dusanter and P. S. Stevens. Nitric acid photolysis on forest canopy surface as a tropospheric nitrous acid source. *Nature Geoscience.* **4**, (2011), doi:10.1038/ NGEO1164.
28. C. Zhu, B. Xiang, L. Zhu and R. Cole. Determination of absorption cross sections of surface-adsorbed HNO_3 in the 290–330 nm region by Brewster angle cavity ring-down spectroscopy. *Chem. Phys. Lett.* **458**, 373–377 (2008).
29. J. Du and L. Zhu. Quantification of the absorption cross sections of surface-adsorbed nitric acid in the 335–365 nm region by Brewster angle cavity ring-down spectroscopy. *Chem. Phys. Lett.* **511**, 213–218 (2011).
30. D. Perner and U. Platt. Detection of nitrous acid in the atmosphere by differential optical absorption. *Geophys. Res. Lett.* **6**, 917–920 (1979).
31. K. Acker, G. Spindler and E. Bruggemann. Nitrous and nitric acid measurements during the INTERCOMP2000 campaign in Melpitz. *Atmos. Environ.* **38**, 6497–6505 (2004).
32. K. Acker, D. Moller, W. Wieprecht, F. X. Meixner, B. Bohn, S. Gilge, C. Plass-Dulmer and H. Berresheim. Strong daytime production of OH from

HNO_2 at a rural mountain site. *Geophys. Res. Lett.* **33**, L02809 (2006), doi:10.1029/2005GL024643.

33. B. Alicke, U. Platt and J. Stutz. Impact of nitrous acid photolysis on the total hydroxyl radical budget during the Limitation of Oxidant Production/Pianura Padana Produzione di Ozono study in Milan. *J. Geophys. Res.* **107**(D22), 8196 (2002), doi:10.1029/2000JD000075.
34. B. Alicke, A. Geyer, A. Hofzumahaus, F. Holland, S. Konrad, H. W. Patz, J. Schafer, J. Stutz, A. Volz-Thomas and U. Platt. OH formation by HONO photolysis during the BERLIOZ experiment. *J. Geophys. Res.* **108**(D4), 8247 (2003), doi:10.1029/2001JD000579.
35. M. D. Andrés-Hernández, J. Notholt, J. Hjorth and O. Schrems. A DOAS study on the origin of nitrous acid at urban and non-urban sites. *Atmos. Environ.* **30**, 175–180 (1996).
36. Y. F. Elshorbany, J. Kleffmann, R. Kurtenbach, E. Lissi, M. Rubio, G. Villena, E. Gramsch, A. R Rickard, M. J. Pilling and P. Wiesen. Seasonal dependence of the oxidation capacity of the city of Santiago de Chile. *Atmos. Environ.* **44**, 5383–5394 (2010).
37. G. W. Harris, W. P. L. Carter, A. M. Winer, J. N. Pitts, Jr., U. Platt and D. Perner. Observations of nitrous acid in the Los Angeles atmosphere and the implications for the ozone-precursor relationships. *Environ. Sci. Technol.* **16**, 414–419 (1982).
38. J. Kleffmann, T. Gavriloaiei, A. Hofzumahaus, F. Holland, R Koppmann, L. Rupp, E. Schlosser, M. Siese and A. Wahner. Daytime formation of nitrous acid: A major source of OH radicals in a forest. *Geophys. Res. Lett.* **32**, L05818 (2005), doi:10.1029/2005GL022524.
39. G. Li, W. Lei, M. Zavala, R. Volkamer, S. Dusanter, P. Stevens and L. T. Molina. Impact of HONO sources on the photochemistry in Mexico City during the MCMA-2006/MILAGO Campaign. *Atmos. Chem. Phys.* **10**, 6551–6567 (2010).
40. A. Neftel, A. Blatter, R. Hesterberg and T. Staffelbach. Measurements of concentration gradients of HNO_2 and HNO_3 over a semi-natural ecosystem. *Atmos. Environ.* **30**, 3017–3025 (1996).
41. J. N. Pitts, Jr., H. W. Bierman, A. Atkinson and A. M. Winer. Atmospheric implications of simultaneous nighttime measurements of NO_3 radicals and HONO. *Geophys. Res. Lett.* **11**, 557–560 (1994).
42. A. R. Reisinger. Observations of HNO_2 in the polluted winter atmosphere: Possible heterogeneous production on aerosols. *Atmos. Environ.* **34**, 3865–3874 (2000).
43. X. Ren, W. H. Brune, A. Oliger, A. R. Metcalf, J. B. Simpas, T. Shirley, J. J. Schwab, C. Bai, U. Roychowdhury, Y. Li, C. Cai, K. L Demerjian, Y. He, X. Zhou, H. Gao and J. Hou. OH, HO_2, and OH reactivity during the PMTACS-NY Whiteface Mountain 2002 campaign: Observations and model comparison. *J. Geophys. Res.* **111**, D10S03 (2006), doi:10.1029/2005JD006126.
44. J. Stutz, B. Alicke and A. Neftel. Nitrous acid formation in the urban atmosphere: Gradient measurements of NO_2 and HONO over grass in

Milan, Italy. *J. Geophys. Res.* **107**, 8192 (2002), doi:10.1029/2001JD 000390.
45. J. Stutz, B. Alicke, R Ackermann, A. Geyer, S. Wang, A. B. White, E. J. Williams, C. W. Spicer and J. D. Fast. Relative humidity dependence of HONO chemistry in urban areas. *J. Geophys. Res.* **109**, D03307 (2004), doi:10.1029/2003JD004135.
46. Z. Vecera and P. K. Dasgupta. Measurement of ambient nitrous acid and a reliable calibration source for gaseous nitrous acid. *Environ. Sci. Technol.* **25**, 255–260 (1991).
47. G. Villena, J. Kleffmann, R. Kurtenbach, P. Wiesen, E. Lissi, M. A. Rubio, G. Croxatto and B. Rappengluck. Vertical gradients of HONO, NO_x and O_3 in Santiago de Chile. *Atmos. Environ.* **45**, 3867–3873 (2011).
48. S. H. Wang, R. Ackermann, C. W. Spicer, J. D. Fast, M. Schmeling and J. Stutz. Atmospheric observations of enhanced NO_2-HONO conversion on mineral dust particles. *Geophys. Res. Lett.* **30**, (2003), doi:10.1029/2003GL017014.
49. K. W. Wong, H.-J. Oh, B. Lefer, B. Rappenglück and J. Stutz. Vertical profiles of nitrous acid in the nocturnal urban atmosphere of Houston, TX. *Atmos. Chem. Phys.* **11**, 3595–3609 (2011), doi:10.5194/acp-11–3595-2011.
50. K. W. Wong, C. Tsai, B. Lefer, C. Haman, N. Grossberg, W. H. Brune, X. Ren, W. Luke and J. Stutz. Daytime HONO vertical gradients during SHARP 2009 in Houston, TX. *Atmos. Chem. Phys. Discuss.* **11**, 24365–24411 (2011).
51. N. Zhang, X. L. Zhou, P. B. Shepson, H. L. Gao, M. Alaghmand and B. Stirm. Aircraft measurement of HONO vertical profiles over a forested region. *Geophys. Res. Lett.* **36**, L15820 (2009), doi:10.1029/2009GL038999.
52. F. Arens, L. Gutzwiller, U. Baltensperger, H. W. Gaggeler and M. Ammann. Heterogeneous reaction of NO_2 on diesel soot particles. *Environ. Sci. Technol.* **35**, 2191–2199 (2001).
53. C. George, R. S. Strekowski, J. Kleffmann, K. Stemmler and M. Ammann. Photoenhanced uptake of gaseous NO_2 on solid organic compounds: A photochemical source of HONO? *Faraday Discuss.* **130**, 195–210 (2005).
54. M. E. Jenkin, R. A. Cox and D. J. Williams. Laboratory studies of the kinetics of formation of nitrous acid from the thermal reaction of nitrogen dioxide and water vapor. *Atmos. Environ.* **22**, 487–498 (1988).
55. J. Kleffmann. Daytime sources of nitrous acid (HONO) in the atmospheric boundary layer. *Chem. Phys. Chem.* **8**, 1137–1144 (2007).
56. G. Lammel and J. N Cape. Nitrous acid and nitrite in the atmosphere. *Chem. Soc. Rev.* **25**, 361–369 (1996).
57. F. Sakamaki, S. Hakakeyama and H. Akimoto. Formation of nitrous acid and nitric oxide in the heterogeneous dark reaction of nitrogen dioxide and water vapor in a smog chamber. *Int. J. Chem. Kinet.* **15**, 1013–1029 (1983).
58. K. Stemmler, M. Ammann, C. Donders, J. Kleffmann and C. George. Photosensitized reduction of nitrogen dioxide on humic acid as a source of nitrous acid. *Nature.* **440**, 195–198 (2006).

59. K. Stemmler, M. Ndour, Y. Elshorbany, J. Kleffmann, B. D'Anna, C. George, B. Bohn and M. Ammann. Light induced conversion of nitrogen dioxide into nitrous acid on submicron humic acid aerosol. *Atmos. Chem. Phys.* **7**, 4237–4248 (2007).
60. R. Svensson, E. Ljungsttom and O. Lindquist. Kinetics of the reaction between nitrogen dioxide and water vapor. *Atmos. Environ.* **21**, 1529–1539 (1987).
61. J. L. Cheung, Y. Q. Li, J. Boniface, Q. Shi, P. Davidovits, D. R. Worsnop, J. T. Jayne and C. E. Kolb. Heterogeneous interactions of NO_2 with aqueous surfaces. *J. Phys. Chem. A* **104**, 2655–2662 (2000).
62. Y. He, X. Zhou, J. Hou, H. Gao and S. B. Bertman. Importance of dew in controlling the air-surface exchange of HONO in rural forested environments. *Geophys. Res. Lett.* **33**, L02813 (2006), doi:10.1029/2005GL024348.
63. N. Zhang, *et al.* Measurements of ambient HONO concentrations and vertical HONO flux above a northern Michigan forest canopy. *Atmos. Chem. Phys.* **12**, 8285–8296 (2012).
64. I. Bejan, Y. Abd, E. Aal, I. Barnes, T. Benter, B. Bohn, P. Wiesen and J. Kleffmann. The photolysis of ortho-nitrophenols: A new gas phase source of HONO. *Phys. Chem. Chem. Phys.* **8**, 2028–2035 (2006).
65. H. Su *et al.* Soil nitrite as a source of atmospheric HONO and OH radicals. *Science.* **333**, 1616 (2011), doi:10.1126/science.1207687.
66. R. Oswald *et al.* HONO Emissions from soil bacteria as a major source of atmospheric reactive nitrogen. *Science.* **341**, 1233–1235 (2013).
67. M. Sörgel, I. Trebs, D. Wu and A. Held. A comparison of measured HONO uptake and release with calculated source strengths in a heterogeneous forest environment. *Atmos. Chem. Phys.* **15**, 9237–9251 (2015).
68. T. C. VandenBoer *et al.* Evidence for a nitrous acid (HONO) reservoir at the ground surface in Bakersfield, CA, during CalNex 2010. *J. Geophys. Res. Atmos.* **119**, 9093–9106 (2014).
69. T. C. VandenBoer *et al.* Nocturnal loss and daytime source of nitrous acid through reactive uptake and displacement. *Nat. Geosci.* **8**, 55–60 (2015).
70. L. Zhu, M. Sangwan, L. Huang, J. Du and L. T. Chu. Photolysis of nitric acid at 308 nm in the absence and in the presence of water vapor. *J. Phys. Chem. A*, **119**, 4907–4914 (2015).
71. J. Mack and J. R. Bolton. Photochemistry of nitrite and nitrate in aqueous solution: A review. *J. Photochem. Photobio. A: Chem.* **128**, 1–13 (1999).
72. H. J. Beine *et al.* Snow-pile and chamber experiments during the Polar Sunrise Experiment 'Alert 2000': Exploration of nitrogen chemistry. *Atmos. Environ.* **36**, 2707–2719 (2002).
73. H. J. Beine, A. Amoroso, F. Domine, M. D. King, M. Nardino, A. Ianniello and J. L France. Surprisingly small HONO emissions from snow surfaces at Browning Pass, Antarctica. *Atmos. Chem. Phys.* **6**, 2569–2580 (2006).
74. J. E. Dibb, M. Arsenault, M. C. Peterson and R. E. Honrath. Fast nitrogen oxide photochemistry in Summit, Greenland snow. *Atmos. Environ.* **36**, 2501–2511 (2002).

75. R. E. Honrath, Y. Lu, M. C. Peterson, J. E. Dibb, M. A. Arsenault, N. J. Cullen and K. Steffen. Vertical fluxes of NO_x, HONO, and HNO_3 above the snowpack at Summit, Greenland. *Atmos. Environ.* **36**, 2629–2640 (2002).
76. T. Thornberry *et al.* Observation of reactive oxidized nitrogen and speciation of NO_y during the PROPHET summer 1998 intensive. *J. Geophys. Res.* **106**, 24359–24386 (2001).
77. B. J. Huebert and C. H. Robert. The dry deposition of nitric acid to grass. *J. Geophys. Res.* **90**, 2085–2090 (1985).
78. J. A. Neuman, L. G. Huey, T. B. Ryerson and D. W. Fahey. Study of inlet materials for sampling atmospheric nitric acid. *Environ. Sci. Technol.* **33**, 1133–1136 (1999).
79. X. Ren, H. Gao, X. Zhou, J. D. Crounse, P. O. Wennberg, E. C. Browne, B. W. LaFranchi, R. C. Cohen, M. McKay, A. H. Goldstein and J. Mao. Measurement of atmospheric nitrous acid at Bodgett Forest during BEARPEX2007. *Atmos. Chem. Phys.* **10**, 6283–6294 (2010), doi:10.5194/acp-10-6283-2010.
80. X. Ren, J. E. Sanders, A. Rajendran, R. J. Weber, A. H. Goldstein, S. E. Pusede, E. C. Browne, K.-E. Min and R. C. Cohen. A relaxed eddy accumulation system for measuring vertical fluxes of nitrous acid. *Atmos. Meas. Tech.* **4**, 2093–2103 (2011), doi:10.5194/amt-4-2093-2011.
81. C. Ye, H. Gao, N. Zhang and X. Zhou. Photolysis of nitric acid on natural and artificial surfaces. *Environ. Sci. Technol.* Submitted (2015).
82. V Rao, T. Lee and J. Drukenbrod. *2008 National Emissions Inventory: Review, Analysis and Highlight*, USEPA Office of Air Quality Planning and Standards, Research Triangle Park, NC (2013), http://www.epa.gov/ttn/chief/eiinformation.html.
83. S. R Handley, D. Clifford and D. J. Donaldson. Photochemical loss of nitric acid on organic films: A possible recycling mechanism for NO_x. *Environ. Sci. Technol.* **41**, 3898–3903 (2007).
84. A. M. Baergen and D. J. Donaldson. Photochemcial renoxification of nitric acid on real urban grime. *Environ. Sci. Tech.* **47**, 815–820 (2013).
85. G. M. Underwood, P. Li, H. Al-Abadleh and V. H. Grassian. A Knudsen cell study of the heterogeneous reactivity of nitric acid on oxide and mineral dust particles. *J. Phys. Chem. A.* **105**, 6609–6620 (2001).
86. A. L. Goodman, E. T. Bernard and V. H. Grassian. Spectroscopic study of nitric acid and water adsorption on oxide particles: Enhanced nitric acid uptake kinetics in the presence of adsorbed water. *J. Phys. Chem. A.* **105**, 6443–6457 (2001).
87. H. Chen, J. G. Navea, M. A. Young and V. H. Grassian. Heterogeneous photochemistry of trace atmospheric gases with components of mineral dust aerosol. *J Phys. Chem. A.* **115**, 490–499 (2011).
88. G. Rubasinghege, S. Elzey, J. Baltrusaitis, P. M. Jayaweera and V. H. Grassian. Reactions on atmospheric dust particles: Surface photochemistry and size-dependent nanoscale redox chemistry. *J. Phys. Chem. Lett.* **1**, 1729–1737 (2010).

89. A. Bondi. van der Waals volumes and radii. *J. Phys. Chem.* **68**, 441–451 (1964).
90. J. R. Huber. Photochemistry of molecules relevant to the atmosphere: Photodissociation of nitric acid in the gas phase. *Chem. Phys. Chem.* **5**, 1663–1669 (2004).
91. Y. Y. Bai and G. A. Segal. Features of the electronic potential energy surfaces of nitric acid below 7 eV. *J. Chem. Phys.* **92**, 7479–7484 (1990).
92. D. R. Lide. *CRC Handbook of Chemistry and Physics.* (CRC Press, Boca Raton, FL, 2008).
93. J. August, M. Brouard and J. P. Simons. Photofragment vector correlations and dissociation dynamics in $HONO_2$. *J. Chem. Soc. Faraday Trans.* **2**, 84,587–84,598 (1988).
94. V. Riffault, T. Gierczak, J. B. Burkholder and A. R. Ravishankara. Quantum yields for OH production in the dissociation of HNO_3 at 248 and 308 nm and H_2O_2 at 308 and 320 nm. *Phys. Chem. Chem. Phys.* **8**, 1079–1085 (2006).
95. C. Zhu, B. Xiang, L. T. Chu and L. Zhu. 308 nm Photolysis of nitric acid in the gas phase, on aluminum surfaces and on ice films. *J. Phys. Chem. A.* **114**, 2561–2568 (2010).
96. S. Li, J. Matthews and A. Sinha. Atmospheric hydroxyl radical production from electronically excited NO_2 and H_2O. *Science.* **319**, 1657–1660 (2008).
97. A. O'Keefe and D. A. G. Deacon. Cavity ring-down optical spectrometer for absorption measurements using pulsed laser sources. *Rev. Sci. Instrum.* **59**, 2544–2551 (1988).
98. A. O'Keefe, J. J. Scherer, A. L. Cooksy, R. Sheeks, J. Heath and R. J. Saykally. Cavity ring down dye laser spectroscopy of jet-cooled metal clusters: Cu_2 and Cu_3. *Chem. Phys. Lett.* **172**, 214–218 (1990).
99. G. Persch, E. Mehdizadeh, W. Demtröder, T. H. Zimmermann, H. Köppel and L. S. Cederbaum. Vibronic level density of excited NO_2-states and its statistical analysis. *Ber. Bunsenges. Phys. Chem.* **92**, 312–318 (1988).
100. J. A. Davidson, C. A. Cantrell, A. H. McDaniel, R. E. Shetter, S. Madronich and J. G. Calvert. Visible-ultraviolet absorption cross sections for NO_2 as a function of temperature. *J. Geophys. Res. — Atmos.* **93**(D6), 7105–7112 (1988).
101. A. C. Vandaele, C. Hermans, P. C. Simon, M. Carleer, R. Colin, S. Fally, M. F. Merienne, A. Jenouvrier and B. Coquart. Measurements of the NO_2 absorption cross-section from 42,000 cm^{-1} to 10,000 cm^{-1} (238–1000 nm) at 220 K and 294 K. *J. Quant. Spec. Rad. Trans.* **59**, 171–184 (1998).
102. W. Schneider, G. K. Moortgat, G. S. Tyndall and J. P. Burrows. Absorption cross-sections of NO_2 in the uv and visible region (200–700 nm) at 298 K. *J. Photochem. Photobiol. A: Chem.* **40**, 195–217 (1987).
103. R. Georges, A. Delon and R. Jost. The visible excitation spectrum of jet cooled NO_2: The chaotic behavior of a set of $^2B_2$ vibronic levels. *J. Chem. Phys.* **103**, 1732–1747 (1995).

104. A. Delon and R. Jost. NO_2 jet cooled visible excitation spectrum: Vibronic chaos induced by the X^2A_1-A^2B_2 interaction. *J. Chem. Phys.* **95**, 5701–5718 (1991).
105. A. Delon, R. Georges and R. Jost. The visible excitation spectrum of jet cooled NO_2: Statistical analysis of rovibronic interactions. *J. Chem. Phys.* **103**, 7740–7772 (1995).
106. A. Bongartz, J. Kames, F. Welter and U. Schurath. Near-UV absorption cross sections and trans/cis equilibrium of nitrous acid. *J. Phys. Chem.* **95**, 1076–1082 (1991).
107. P. Pagsberg, E. Bjergbakke, E. Ratajczak and A. Sillesen. Kinetics of the gas phase reaction OH + NO(+ M) $\rightarrow$ HONO(+ M) and the determination of the UV absorption cross sections of HONO. *Chem. Phys. Lett.* **272**, 383–390 (1997).
108. W. R. Stockwell and J. G. Calvert. The near ultraviolet absorption spectrum of gaseous HONO and N_2O_3. *J. Photochem.* **8**, 193–203 (1978).
109. J. Stutz, E. S. Kim, U. Platt, P. Bruno, C. Perrino and A. Febo. UV-visible absorption cross sections of nitrous acid. *J. Geophys. Res.* **105**, 14585–14592 (2000).
110. R. Vasudev. Absorption spectrum and solar photodissociation of gaseous nitrous acid in the actinic wavelength region. *Geophys. Res. Lett.* **17**, 2153–2155 (1990).
111. K. A. Ramazan, D. Syomin and B. J. Finlayson-Pitts. The photochemical production of HONO during the heterogeneous hydrolysis of NO_2. *Phys. Chem. Chem. Phys.* **6**, 3836–3843 (2004).
112. M. Nonella, H. U. Suter and J. R. Huber. An *ab initio* and dynamics study of the photodissociation of nitric acid HNO_3. *Chem. Phys. Lett.* **487**, 28–31 (2010).
113. T. L. Myers, N. R. Forde, B. Hu, D. C. Kitchen and L. J. Butler. The influence of local electronic character and nonadiabaticity in the photodissociation of nitric acid at 193 nm. *J. Chem. Phys.* **107**, 5361–5373 (1997).
114. O. Abida, J. Du and L. Zhu. Investigation of the photolysis of the surface-adsorbed HNO_3 by combining laser photolysis with Brewster angle cavity ring-down spectroscopy. *Chem. Phys. Lett.* **534**, 77–82 (2012).

Chapter 5

Atmospheric Chemistry of Halogenated Organic Compounds

Timothy J. Wallington[*,§] Mads P. Sulbaek Andersen[†,‡]
and Ole John Nielsen[‡]

[*] *Research and Advanced Engineering, Ford Motor Company, Dearborn, Michigan 48121-2053, USA*

[†] *Department of Chemistry and Biochemistry, California State University, Northridge, California 91330, USA*

[‡] *Copenhagen Center for Atmospheric Research, Department of Chemistry, University of Copenhagen, Universitetsparken 5, DK-2100 Copenhagen Ø, Denmark*

[§] *twalling@ford.com*

Halogenated organic compounds play an important role in atmospheric chemistry. We provide an overview of the atmospheric chemistry of halogenated organic compounds starting with a discussion of sources, emissions, and atmospheric concentrations. The chemistry associated with formation and loss of stratospheric ozone and processes related to halogenated organics is described. A discussion of the atmospheric chemistry of halogenated alkanes, alkenes, oxygenates, sulfur-, nitrogen-, and phosphorus-containing organics is provided. The contribution of halogenated organics to radiative forcing of climate change and the environmental impact of halogenated organic compounds is described.

5.1. Introduction

A wide variety of halogenated organic compounds are emitted into the atmosphere from a range of sources, both natural and man-made. Halogenated organic compounds have a rich atmospheric chemistry. Their atmospheric lifetime range from minutes for compounds which undergo rapid photolysis in the sunlit troposphere such as iodinated organics (e.g., CH_2I_2)[1] to millennia for compounds which are unreactive towards gas-phase reactions in the lower atmosphere such as perfluorocarbons (PFCs) (e.g., CF_4).[2] Recognition of ozone depletion caused by chlorine-based catalytic chain reactions initiated by photolysis of chlorofluorocarbons (CFCs) in the stratosphere[3] prompted a large research effort focused on understanding the atmospheric chemistry of halocarbons in the 1980s and 1990s. In particular, the observation of seasonal loss of ozone in the Southern Hemisphere,[4] which became known as the Antarctic Ozone Hole, prompted international action to limit the emissions of ozone depleting substances such as CFCs and bromofluorocarbons. A large amount of research was performed in the 1990s and 2000s to determine the atmospheric chemistry and environmental impact of a range of halogenated organic compounds which were proposed as substitutes for ozone depleting substances. As a result of this research effort the halogenated organics are the class of organic compounds whose atmospheric chemistry is perhaps the best understood. In this chapter, we review the emissions, atmospheric concentrations, atmospheric lifetimes, and environmental impacts of halogenated organics. We start with a description of the naming systems in common use for these compounds beginning with halocarbons and then expanding to other halogenated organic compounds.

There is some ambiguity in the literature as to the definition of a halocarbon. The Intergovernmental Panel on Climate Change (IPCC) and the World Meteorological Organization (WMO) use the term halocarbon to cover chemical compounds containing carbon and one, or more, halogen atoms.[5,6] In IPCC and WMO reports both CCl_4 and CH_3Cl are considered halocarbons. In contrast, the International Union of Pure and Applied Chemistry (IUPAC) definition of a halocarbon is a "hydrocarbon in which *all* of the hydrogen atoms are replaced by halogens",[7] i.e., a compound containing only carbon and halogen(s). If a molecule contains any element other than carbon and a halogen(s) then under the IUPAC definition it is not a halocarbon; CCl_4 is a halocarbon, CH_3Cl is not. We adopt the IUPAC definition in the following discussion. The present review of halogenated organic compounds encompasses halocarbons and compounds such as CH_3Cl which are often, and incorrectly, classified as halocarbons.

The IUPAC has developed systematic rules for naming all organic compounds including halogenated compounds.[8] In the IUPAC convention the substituents in a molecule are assigned numbers (unless the compound can be defined without the need for a number). Thus, CH_3CH_2Cl is chloroethane, while $CH_3CH_2CH_2Cl$ and $CH_3CHClCH_3$ are 1- and 2-chloropropane, respectively. The prefixes di-, tri-, tetra-, penta-, hexa-, etc. are used to name compounds containing more than one given substituent atom. Numbers are assigned to give the lowest sum, thus CH_3CCl_3 is 1,1,1-trichloroethane, and $CH_2BrCH_2CHBr_2$ is 1,1,3-tribromopropane. If a compound can be described using the same set of numbers, the substituent first in the alphabet is placed first; thus $CH_2ICH_2CH_2Cl$ is 1-chloro-3-iodopropane. Prefixes (di-, tri-, tetra-, etc.) are ignored in the alphabetization, so $CF_3CH_2CCl_3$ is 1,1,1-trichloro-3,3,3-trifluoropropane. If there is a conflict in priority between the lowest sum in the numbering system and alphabetization of the substituent names, the numbering takes precedence; thus CF_3CCl_2H is 2,2-dichloro-1,1,1-trifluoroethane. Finally, if a given substituent occupies all possible sites in the molecule the prefix "per-" is used, hence $CF_3CF_2CF_2CF_3$ is perfluorobutane (rather than decafluoropropane).

The IUPAC naming system for the halogenated compounds, while logical and clear to the chemical community, is considered by some non-chemists to be rather cumbersome. Systems of naming halogenated compounds have developed by industrial users and are in popular use. The American Society of Heating Refrigeration and Air Conditioning Engineers (ASHRAE) and the American National Standards Institute (ANSI) have developed an alternative haloalkane numbering system (ANSI/ASHRAE Standard 34-1992)[9] which is in widespread use. In this numbering system a compound is denoted by several uppercase letters ("composition denoting prefix") which define the elements present in the molecule, followed by several numbers which describe how many of each substituents are present, followed by one, or more, lower case letters to differentiate different isomers (if necessary). The composition denoting prefix always ends in "C" to indicate the presence of carbon. The presence of hydrogen, fluorine, chlorine, bromine, and iodine is indicated by H, F, C, B, and I. Hydrogen is always placed first, carbon always last, the other elements are placed alphabetically. Hence, CF_3CCl_2H is a hydrochlorofluorocarbon and is given the prefix HCFC, CF_3CHF_2 and CF_3H are hydrofluorocarbons and have the prefix HFC, and CF_2Cl_2 is a chlorofluorocarbon with the prefix CFC.

In the ANSI/ASHRAE haloalkane numbering system the numbers following the prefix have the following meanings: first number is number of carbon atoms minus 1, second number is the number of hydrogen atoms plus 1, and the third number is the number of fluorine atoms. For substituted methanes, the first number is zero and is omitted; CF_3H is HFC-23 while CF_2Cl_2 is CFC-12. For substituted ethanes, a lower case letter suffix is sometimes added to distinguish between different isomers. The isomer with the smallest difference in the sum of the masses of the substituents on each carbon has no letter, the compound with next smallest difference is labeled "a", the compound with the next smallest difference is labeled "b", etc. Examples include CF_2HCH_2Cl (HCFC-142), $CFClHCH_2F$ (HCFC-142a), and CF_2ClCH_3 (HCFC-142b). For chlorine and fluorine substituted propanes two lower case letters are added as a suffix to distinguish different isomers. The first letter defines the substituents on the central carbon atom, "a" for $-CCl_2-$, "b" for $-CClF-$, "c" for $-CF_2-$, "d" for $-CHCl-$, "e" for $-CHF-$, and "f" for $-CH_2-$ (ordered by decreasing mass). The second letter reflects the difference in mass of the substituents on the two terminal carbon atoms with "a" used for the isomer with the smallest difference, "b" for the next smallest, and so on. For example $CHF_2CHFCHF_2$ is HFC-245ea, CF_3CHFCH_2F is HFC-245eb, $CF_3CF_2CCl_2H$ is HCFC-225ca, and CF_2ClCF_2 CHFCl is HCFC-225cb. For linear compounds with four or more carbons, letters are added using the "a" to "f" system noted above for the methylene carbons and the following for the methyl groups: "j" for $-CCl_3$, "k" for $-CCl_2F$, "l" for $-CCF_2$, "m" for $-CF_3$, "n" for $-CHCl_2$, "o" for $-CH_2Cl$, "p" for $-CHF_2$, "q" for $-CH_2F$, "r" for $-CHClF$, and "s" for $-CH_3$. Letters begin at one end of the molecule and address each carbon in turn in a fashion which first minimizes the number of letters and second uses a combination of letters which appear as early as possible in the alphabet, e.g., $CF_3CF_2CH_2CH_2F$ is HFC-356mcf. Hyphens are added to avoid confusion when double digits are required, e.g., $CF_3CHFCHFCF_2CF_3$ has ten fluorine atoms and is HFC 43–10 mee.

Short chain haloalkenes are a class of compounds that are being developed as replacements for CFCs and saturated HFCs. Haloalkenes have one, or more, reactive >C=C< bonds and consequently have atmospheric lifetimes which are much shorter than for the corresponding haloalkanes. As an example, $CF_3CF{=}CH_2$ (HFO-1234yf) has an atmospheric lifetime of $\sim$ 11 days[10] and is being used as a refrigerant in vehicle air conditioning systems in place of CF_3CFH_2 (HFC-134a) which has an atmospheric lifetime of $\sim$ 13.4 years.[11] In the ANSI/ASHRAE naming system haloalkenes[12] are named using letters to describe the elemental composition following the rules for haloalkanes followed by "O" for "olefin". The letters are followed by four numbers and letters if needed. The numbers, in order, are the number of: (i) >C=C< bonds, (ii) carbon atoms −1, (iii) hydrogen atoms +1, and (iv) the number of fluorine atoms. Halopropenes have found commercial applications. For the halopropenes two letters are added. The first letter after the numbers signifies the substituent on the central carbon atom: "x" for –Cl, "y" for –F, and "z" for –H. The second letter signifies the substituents on the terminal carbon atom: "a" for $-CCl_2$, "b" for –CClF, "c" for $-CF_2$, "d" for –CHCl, "e" for –CHF, and "f" for $-CH_2$. Thus, $CF_3CF{=}CH_2$ is HFO-1234yf while $CF_3CH{=}CH_2$ is HFO-1243zf. Where needed "E" or "Z" are added in parentheses to denote isomers with cis (Z, zussammen) or trans (E, entgegen) substituents. Thus, cis-$CF_3CH{=}CHCl$ is HCFO-1233zd(Z) and trans-$CF_3CH{=}CHCl$ is HCFO-1233zd(E). Many of the haloalkanes and haloalkenes are used as the working fluid in refrigeration systems. As shorthand the letters in many of the names are often replaced with "R" for "refrigerant". Thus, HFC-134a becomes R-134a, HCFC-22 becomes R-22, and HFO-1234yf becomes R-1234yf.

Bromofluorocarbons and bromochlorofluorocarbons are effective fire suppression agents useful in military and other applications. The US Army Corps of Engineers has developed a separate naming system for such compounds. The class of compounds are known as Halons; individual compounds are designated "Halon-xyza" where "x" is the number of carbon atoms, "y" is the number of fluorine atoms, "z" is the number of chlorine atoms, and "a" is the number of bromine atoms in the molecule. Examples include Halon-1211 ($CBrClF_2$) and Halon-1301 ($CBrF_3$). The Halon naming system has the advantage of simplicity, but the disadvantage that it does not distinguish between isomers. Thus, Halon-2402 could, in principle, refer to either $CBrF_2CBrF_2$, or CBr_2FCF_3. For historical reasons associated with its commercial use under the name "Halon-2402", only the $CBrF_2CBrF_2$ isomer is named Halon-2402. In the present chapter, we use chemical formulae and where the chemical is of commercial significance the ANSI/ASHRAE, Halon nomenclature, or informal names such as "methyl chloroform".

The present chapter provides an overview of the atmospheric chemistry of halogenated organic compounds. We start with a discussion of their sources, emissions, and atmospheric concentrations. The chemistry associated with formation and loss of stratospheric ozone and processes related to halogenated organics is then described. This is followed by a discussion of the atmospheric chemistry of halogenated alkanes, alkenes, oxygenates, sulfur-, nitrogen-, and phosphorus-containing organics. The chemistry of halogenated alkanes has been reviewed in detail by Burkholder *et al.*[13] and Calvert *et al.*[14] and hence we focus here on the atmospheric chemistry of halogenated alkenes, oxygenates, sulfur-, nitrogen-, and phosphorus-containing organics. We discuss the contribution of halogenated organics to radiative forcing of climate change and draw some general conclusions regarding the atmospheric chemistry and environmental impact of halogenated organic compounds.

5.2. Halogenated Organics in the Atmosphere: Sources, Emissions, and Concentrations

Halogenated organic compounds are emitted into the atmosphere from both natural and anthropogenic sources. Anthropogenic sources account for $\sim 85\%$ of the chlorine and 50% of the bromine contained in halogenated organic compounds currently in the atmosphere.[15] Interest in the atmospheric chemistry of halogenated organic compounds stems from the stratospheric ozone depletion associated with the release of CFCs. CFCs, were first synthesized in the 1890s using what is now known as the Swarts reaction using antimony trifluoride[16] in the presence of antimony pentafluoride catalyst to replace chlorine with fluorine substituents in organic molecules. CFCs are non-toxic, non-corrosive, and non-flammable. The suitability of CFCs as refrigerants to replace the toxic refrigerants NH_3 and SO_2 was recognized by Midgley[17] in the late 1920s and led to the commercialization of CFCs as refrigerants in the 1930s. By the 1960s and 1970s CFCs have been in widespread commercial use as refrigerants, aerosol propellants, foam blowing agents, and solvents and their production was increasing dramatically as shown in Fig. 5.1. CFC-11 was used in foam blowing, CFC-12 was used in refrigeration, a mix of CFC-11 and CFC-12 was used in aerosol propellants, and CFC-113 was used as a solvent in degreasing applications for electronic components.

Recognition of the adverse effect of CFCs on stratospheric ozone led to the formulation and international adoption of the "Montreal Protocol on Substances which Deplete Stratospheric Ozone" which called for large reductions in the production and use of CFCs and entered into force in 1989. Increasing recognition of the seriousness of ozone depletion and the need for control of CFC emissions led to the London (1990), Copenhagen (1992), Montreal (1997), and Beijing (1999) Amendments of the Protocol which further limited the production of CFCs. The results of this international action are seen

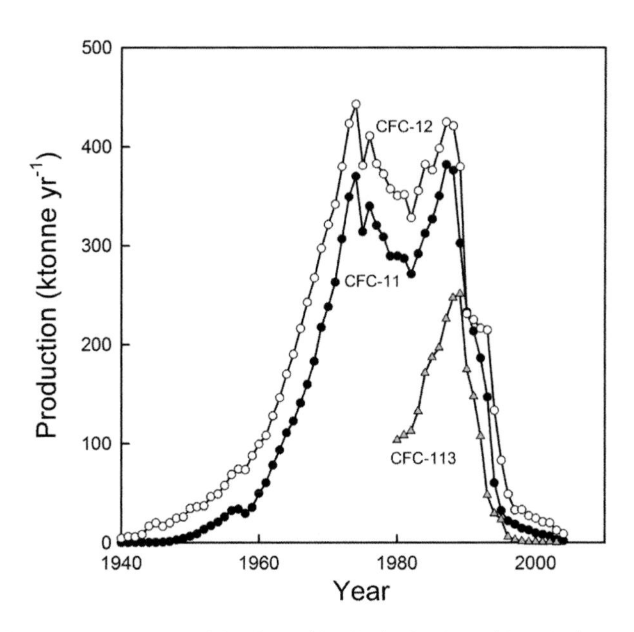

Fig. 5.1 Global production of CFC-11 ($CFCl_3$), CFC-12 (CF_2Cl_2), and CFC-113 ($CF_2ClCFCl_2$); data were taken from AFEAS[18] and WMO.[15]

in Fig. 5.1 and 5.2 which show the dramatic progress in phase-out of production and emissions of CFCs and CH_3CCl_3. Replacement compounds for CFCs have been developed. The two most important classes of replacement compounds are hydrochlorofluorocarbons (HCFCs) and hydrofluorocarbons (HFCs). Fig. 5.3 shows the increase in global emissions of HCFC-22 ($CFCl_2H$), HCFC-141b (CH_3CFCl_2), HFC-134a (CF_3CFH_2), HFC-125 (CF_3CF_2H), HFC-143a (CH_3CF_3), HFC-152a (CH_3CHF_2), and HCFC-32 (CH_2F_2) which have found substantial use as CFC replacements.

There are no known significant natural sources of CFCs and their atmospheric burden is attributable to human activities. Despite the cessation in production, some emission of CFCs into the atmosphere still occurs from legacy sources (e.g., from old foam, old refrigeration units). However, over time the emissions from such sources are becoming less important. As seen from Fig. 5.4, the atmospheric concentrations of CFCs have peaked and are now decreasing. CFCs have atmospheric lifetimes which are typically of the order of

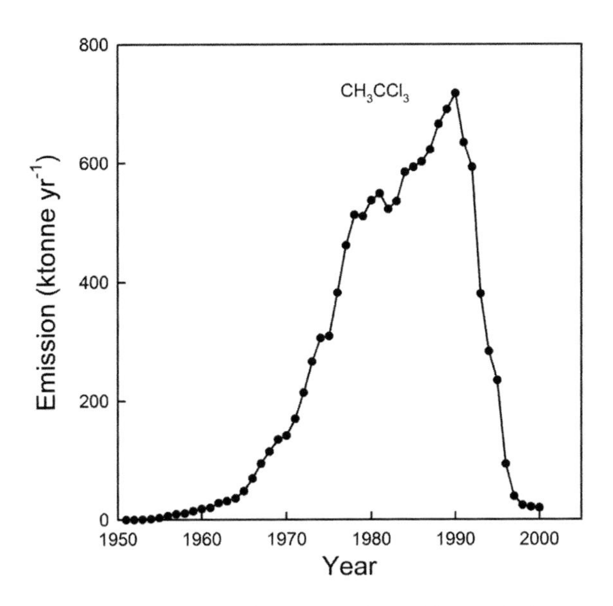

Fig. 5.2 Global emissions of methyl chloroform (CH_3CCl_3); data were taken from Ref. [19].

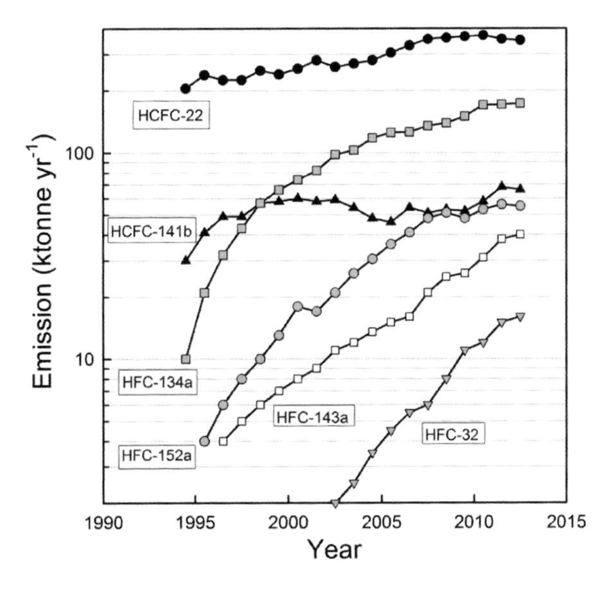

Fig. 5.3 Global emissions of HCFC-22 ($CHClF_2$), HCFC-141b (CH_3CFCl_2), HFC-134a (CF_3CFH_2), HFC-152a (CH_3CHF_2), HFC-143a (CH_3CF_3), and HFC-32 (CH_2F_2); data were taken from Ref. [20].

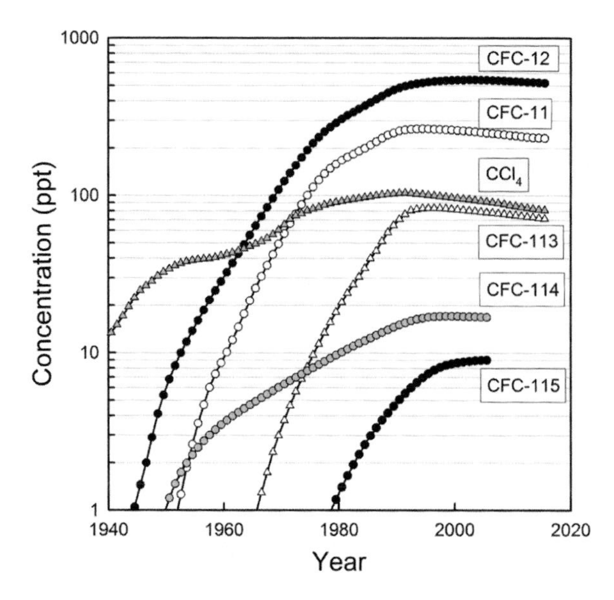

Fig. 5.4 Global average surface concentrations of CFC-11 ($CFCl_3$), CFC-12 (CF_2Cl_2), CFC-113 ($CF_2ClCFCl_2$), CFC-114 (CF_2ClCF_2Cl), CFC-115 (CF_3CF_2Cl), and carbon tetrachloride (CCl_4); data were taken from Refs. [21,22].

100 years. The decrease in atmospheric concentrations shown in Fig. 5.4 is expected to continue with the concentrations of the CFCs decaying with a time constant of ~100 years. CFCs are persistent and will be present in the Earth's atmosphere at appreciable concentrations (>10% of atmospheric chlorine) for centuries to come.

Carbon tetrachloride was used in large quantities as a precursor in the production of CFC-11 and CFC-12, as a solvent, and as a fire fighting agent. It is still used as a chemical feedstock. The atmospheric lifetime of CCl_4 is relatively short at 26 years. Interestingly there is a large mismatch between the ~ 60 ktonne yr^{-1} global emission source inferred from the atmospheric concentration and lifetime and the ~6 ktonne yr^{-1} from emission inventories.[15] It has been suggested that the discrepancy is explained by emissions from contaminated soils and industrial wastes[15,24,25] not included in the emission inventory. It is also possible that the emissions in the inventory are underreported. Further research is needed to better understand the emission source of CCl_4.

Figs. 5.5 and 5.6 show the concentrations of CF_2ClH (HCFC-22), CH_3CCl_3 (methylchloroform), CH_3Br (methyl bromide), Halon-1211 ($CBrClF_2$), HFC-365mfc ($CF_3CH_2CF_2CH_3$), HFC-134a (CF_3CFH_2)

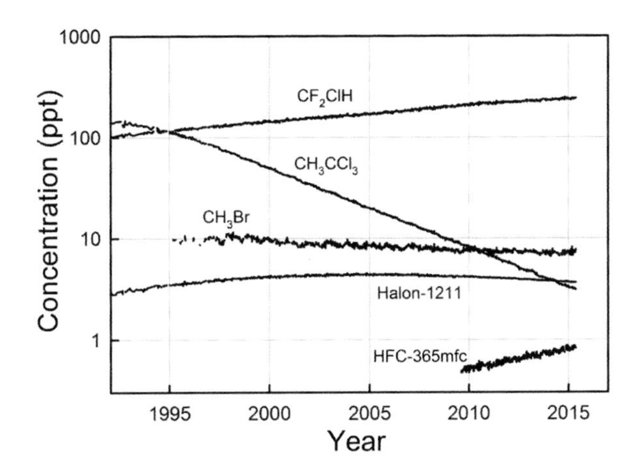

Fig. 5.5 Concentrations of CF_2ClH (HCFC-22), CH_3CCl_3 (methyl chloroform), CH_3Br (methyl bromide), Halon-1211 ($CBrClF_2$), and HFC-365mfc ($CF_3CH_2CF_2CH_3$) measured at Mauna Loa, Hawaii (see Ref [23]).

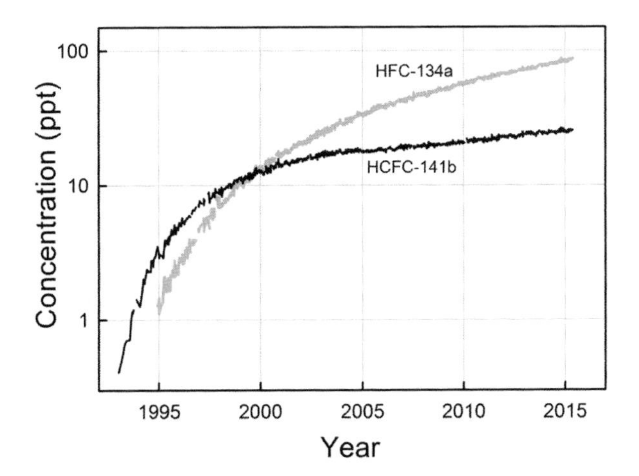

Fig. 5.6 Concentrations of HFC-134a (CF_3CFH_2) and HCFC-141b ($CFCl_2CH_3$) measured at Mauna Loa, Hawaii (see Ref. [23]).

and HCFC-141b ($CFCl_2CH_3$) measured at Mauna Loa, Hawaii. The increase in atmospheric levels of HCFC-22, HCFC-141b, HFC-134a, and HFC-365mfc reflects their use as CFC replacements. The levels of CH_3CCl_3 are decreasing rapidly reflecting its phase out (see Fig. 5.2) and relatively short atmospheric lifetime (5 years). Use of methyl bromide as a fumigant in agricultural activities is now controlled and decreasing and as a result the atmospheric levels of this short lived (0.8 years) compound are decreasing. Global production of Halons was phased out in 2010. Estimates from inventories and from the trend of atmospheric concentrations indicate that global emissions of Halons have decreased by ~50% since their peak in the 1990s.[15] Halons are used as fire extinguishing agents and their emission into the atmosphere is expected to continue for decades, albeit at decreasing rates, as these banks are used up. The atmospheric concentrations of the shorter lived Halon-1202 (CBr_2F_2), Halon-1211 ($CBrClF_2$), and Halon-2402 ($CBrCF_2CBrF_2$) are decreasing while the atmospheric level of the longer lived Halon 1301 ($CBrF_3$) is increasing. As a group the atmospheric concentration of Halons is decreasing at a rate of 0.06 ppt Br yr^{-1}.[15]

Figure 5.7 shows the atmospheric levels of CH_3Cl, CH_2Cl_2, and $CCl_2{=}CCl_2$ measured at Mauna Loa, Hawaii. These compounds react relatively rapidly with OH radicals and have short atmospheric lifetimes (1.0 year, 0.4 years, and 90 days, respectively). As seen from Fig. 5.7, the atmospheric concentrations of these compounds display a seasonal variation. This variation reflects the annual cycle of OH radical concentration which peaks in the summer with increased solar flux driving the photochemical production of OH radicals. Methyl chloride is the most abundant halogen containing organic in the atmosphere and is present at a level of ~540 ppt just slightly ahead of CFC-12 (520 ppt),[15] however when expressed in terms of the chlorine content, CFC-12 is the most abundant halogen containing organic in the atmosphere. Natural sources of methyl chloride including tropical forests, oceans, fungi, salt marshes, wetlands, rice paddies, and mangroves account for ~80% of emissions. Anthropogenic sources account for ~20% of methyl

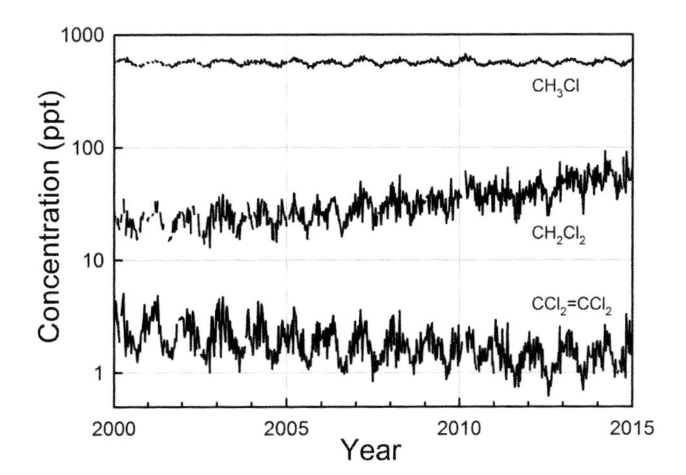

Fig. 5.7 Concentrations of CH_3Cl (methyl chloride), CH_2Cl_2 (methylene chloride), and C_2Cl_4 (tetrachloroethylene) measured at Mauna Loa, Hawaii (see Ref. [23]).

chloride emissions with biomass burning making the largest single contribution.[15]

Methylene chloride (CH_2Cl_2) is used as a solvent and has mainly anthropogenic sources. CH_2Cl_2 is not controlled under the Montreal Protocol and its emissions have increased significantly recently. With its short lifetime (0.4 years) the atmospheric concentrations are low and respond quickly to emissions. The atmospheric concentration of CH_2Cl_2 has increased by ~60% over the last decade.[15] Perchloroethylene ($CCl_2{=}CCl_2$) has a lifetime of just 90 days and as evident in Fig. 5.7 the atmospheric concentrations of this compound measured at Mauna Loa are low and have a pronounced seasonal variation. As with all short-lived compounds, the atmospheric concentrations of CH_3Cl, CH_2Cl_2, and $CCl_2{=}CCl_2$ vary by location reflecting the importance of local sources.

5.3. Stratospheric Ozone Chemistry

5.3.1. *Stratospheric Ozone Layer*

About 90% of atmospheric ozone resides in the stratosphere where it constitutes the so-called "ozone layer" located at approximated

15–35 km altitude. At room temperature and pressure the amount of ozone in the stratosphere would have a thickness of ~3 mm. The column ozone is typically reported in Dobson Units (DU). One Dobson Unit is a layer of ozone which at standard temperature and pressure would be 0.01 mm thick (this corresponds to $\sim 2.7 \times 10^{16}$ molecule cm^{-2}). Typical levels are ~300 DU. Ozone absorbs strongly in the UV and is very effective at absorbing harmful ultraviolet solar radiation. Depletion of the stratospheric ozone layer is a serious issue because it permits penetration of UV-B radiation (280–320 nm) to the Earth's surface. Nucleic acids and other biological macromolecules absorb strongly in the UV. Exposure to UV-B is damaging to biological systems and is associated with human skin cancer and suppression of the human immune system. Adverse effects of increased UV-B exposure are expected for other biological species and for overall ecosystem health.

Chapman[26] was the first to propose a quantitative chemical mechanism to account for stratospheric ozone. He proposed the following reactions:

$$O_2 + h\nu \rightarrow O + O, \tag{5.1}$$

$$O + O + M \rightarrow O_2 + M, \tag{5.2}$$

$$O + O_2 + M \rightarrow O_3 + M, \tag{5.3}$$

$$O + O_3 \rightarrow 2O_2, \tag{5.4}$$

$$O_3 + h\nu \rightarrow O_2 + O. \tag{5.5}$$

M represents a third body such as N_2 which can remove the excess energy released during the formation of the chemical bond. Reaction (5.1) followed by (5.3) is the source of stratospheric ozone. Measurements of the rate coefficient for reaction (5.2) have shown that it is too slow to be of significance in the stratosphere. Reactions (5.3) and (5.5) proceed rapidly and interconvert O and O_3 on a time scale of the order of minutes in the sunlit stratosphere and it is convenient to lump O and O_3 together as "odd oxygen", O_x. Applying a steady-state analysis to O_3 gives:

$$d[O_3]/dt = k_3[O][O_2][M] - k_4[O][O_3] - k_5[O_3] = 0$$

$[O_2] \gg [O_3]$ and $k_3[O_2][M] \gg k_4[O_3]$, hence we can derive

$$[O] = k_5[O_3]/(k_3[O_2][M]).$$

Applying a steady-state analysis for odd oxygen,

$$d[O_x]/dt = 2k_1[O_2] - 2k_4[O][O_3] = 0$$

substituting for [O], and rearranging gives

$$[O_3] = (k_1 k_3[M]/k_4 k_5)^{0.5}[O_2].$$

The absorption spectrum and photolysis quantum yields of O_2 and O_3 are known so the values of k_1 and k_5 can be calculated, the rate coefficients k_3 and k_4 have been measured, and the altitude dependence of [M] and $[O_2]$ are known. The ozone profile from Chapman chemistry can be calculated numerically using a model starting at the top of the atmosphere using the solar flux and working downwards in incremental steps, calculating the ozone concentrations for each step and making allowance for UV attenuation. The simple Chapman mechanism oxygen-only chemistry correctly predicts the formation of a layer of ozone at an altitude of ~25 km. However, while the shape of the ozone layer matches well with observations, the magnitude of the ozone concentration predicted by the Chapman chemistry is approximately a factor of 2 greater than the observations.[14] Clearly there are ozone loss processes missing from the Chapman scheme which need to be considered.

With regard to physical loss processes, photolysis is included in the Chapman scheme, and it difficult to imagine other physical loss mechanisms in the stratosphere. With regard to chemical loss processes, given the large concentration of ozone relative to other species in the stratosphere the processes cannot be simple reactions as ozone would easily titrate out any other reactant, but must be catalytic in nature. Bates and Nicolet[27] were the first to suggest that catalytic cycles involving radicals produced by photolysis of water vapor could be important in stratospheric ozone chemistry.

5.3.2. HO_x, NO_x and ClO_x catalytic ozone destruction cycles

In catalytic ozone destruction cycles a catalyst species X reacts with ozone to give XO which can react with O atoms to regenerate X. The important catalyst species are X = H, OH, NO, NO_2, Cl, and Br. The cycles can be written in general form as:

$$X + O_3 \rightarrow XO + O_2, \qquad (5.7)$$

$$\underline{XO + O \rightarrow X + O_2,} \qquad (5.8)$$

$$\text{Net}: O + O_3 \rightarrow 2O_2$$

The net effect of reactions (5.7) and (5.8) is loss of odd oxygen without any loss in X and is equivalent to speeding up the rate of Reaction (5.4) in the Chapman mechanism. Bates and Nicolet[27] recognized that HO_x ($x = 0, 1, 2$) radicals produced by photolysis of water vapor in the stratosphere and mesosphere could participate in such catalytic cycles. The "HO_x" cycles include:

$$\begin{array}{l} H + O_3 \rightarrow OH + O_2 \\ \underline{OH + O \rightarrow H + O_2} \\ \text{Net}: O_3 + O \rightarrow O_2 + O_2 \end{array},$$

$$\begin{array}{l} OH + O_3 \rightarrow HO_2 + O_2 \\ \underline{HO_2 + O \rightarrow OH + O_2} \\ \text{Net}: O_3 + O \rightarrow O_2 + O_2 \end{array},$$

$$\begin{array}{l} OH + O \rightarrow H + O_2 \\ H + O_2 + M \rightarrow HO_2 + M \\ \underline{HO_2 + O \rightarrow OH + O_2} \\ \text{Net}: O + O + M \rightarrow O_2 + M \end{array},$$

and

$$\begin{array}{l} OH + O_3 \rightarrow HO_2 + O_2 \\ \underline{HO_2 + O_3 \rightarrow OH + O_2 + O_2} \\ \text{Net}: O_3 + O_3 \rightarrow O_2 + O_2 + O_2 \end{array}.$$

The result of these cycles is conversion of odd oxygen ($O + O_3$) into O_2. The concentrations of H, OH, and HO_2 vary with height and so the importance of the different HO_x cycles varies with altitude.

Crutzen[28] and Johnston[29] recognized the importance of NO formation in the stratosphere by reaction of $O(^1D)$ atoms with N_2O and the potential for significant aircraft NO emissions into the stratosphere and suggested that catalytic cycles reactions involving nitrogen oxides could deplete stratospheric ozone

$$\begin{array}{l} NO + O_3 \rightarrow NO_2 + O_2 \\ \underline{NO_2 + O \rightarrow NO + O_2} \\ \text{Net}: O + O_3 \rightarrow 2O_2 \end{array}$$

and

$$\begin{array}{l} NO + O_3 \rightarrow NO_2 + O_2 \\ NO_2 + O_3 \rightarrow NO_3 + O_2 \\ \underline{NO_3 + h\nu \rightarrow NO + O_2} \; . \\ \text{Net}: O + O_3 \rightarrow 2O_2 \end{array}$$

Molina and Rowland[3] and Stolarski and Cicerone[30] suggested that a catalytic cycle based on chlorine oxides could also contribute. Furthermore, Molina and Rowland[3] recognized that photolysis of CFCs in the stratosphere could become an importance source of chlorine atoms and that CFC emissions threatened the ozone layer.

$$\begin{array}{l} CFCl_3(\text{CFC-11}) + h\nu \rightarrow CFCl_2 + Cl \\ Cl + O_3 \rightarrow ClO + O_2 \\ \underline{ClO + O \rightarrow Cl + O_2} \; . \\ \text{Net}: O + O_3 \rightarrow 2O_2 \end{array}$$

The chlorine-based catalytic cycle is efficient with each chlorine atom capable of destroying thousands of molecules of ozone before it reacts with methane and is converted into relatively unreactive HCl. Concerns were also raised that a catalytic BrO_x cycle initiated by bromine released from photolysis of Halons (*e.g.*, CF_2ClBr, Halon-1211, CF_3Br, Halon-1301; and CF_2BrCF_2Br, Halon-2402) could also contribute to ozone loss.

$$CF_2Br_2(\text{Halon-1202}) + h\nu \rightarrow CF_2Br + Br$$
$$Br + O_3 \rightarrow BrO + O_2$$
$$\underline{BrO + O \rightarrow Br + O_2.}$$
$$\text{Net}: O + O_3 \rightarrow 2O_2$$

The sinks for bromine atoms are less effective than for chlorine atoms and as a result on a per-atom basis bromine atoms are ~60–65 times more efficient in removing ozone than chlorine atoms.[15]

The HO_x, NO_x, ClO_x, and BrO_x catalytic ozone destruction cycles do not operate in isolation but are interconnected by coupling reactions and common reservoir species. An example of a reservoir species is chlorine nitrate, $ClONO_2$, which is formed by reaction of ClO radicals with NO_2:

$$ClO + NO_2 + M \rightarrow ClONO_2 + M.$$

The formation of chlorine nitrate removes two active species and slows down the loss of ozone. However, this slowdown is temporary as photolysis of chlorine nitrate reforms the active species:

$$ClONO_2 + h\nu \rightarrow Cl + NO_3.$$

The reaction of ClO with BrO radicals is an example of an important coupling cycle:

$$ClO + BrO \rightarrow Cl + Br + O_2$$
$$Cl + O_3 \rightarrow ClO + O_2$$
$$\underline{Br + O_3 \rightarrow BrO + O_2}\ .$$
$$\text{Net}: 2O_3 \rightarrow 3O_2$$

5.3.3. *Stratospheric Ozone Depletion and the Antarctic Ozone Hole*

The finding by Molina and Rowland[3] that large scale release of CFCs would lead to significant stratospheric ozone loss led to legislation which banned CFC use in aerosol cans in the U.S., Canada, Norway and Sweden. The global production of CFC-11 and

CFC-12 decreased for several years following their peak production in 1974, but then started to increase again (see Fig. 5.1). Results from photochemical models in the 1980s suggested that the release of CFCs was responsible for a small (few percent) loss in stratospheric ozone. However, it was difficult to discern such a small trend in the observational record. The picture became much clearer in 1985 with the report from the British Antarctic Survey station of a major drop in atmospheric ozone overhead at Halley Bay.[4] As shown in Fig. 5.8, the total ozone column over Halley Bay had decreased from 320 Dobson Units to 200 DU in the Antarctic spring (October). The ozone loss detected by the ground based instruments at Halley Bay had also been recorded in NASA satellite data, but the low values in the latter data set were assumed to be incorrect and were not reported. Reanalysis of the satellite data confirmed the existence of the seasonal decrease in ozone above Halley Bay and revealed this was a continent wide phenomenon. The spring time loss of ozone over Antarctica has become known as the "ozone hole". The ozone hole is deep with almost total ozone loss at some altitudes, covers an area of ~20–25 million km^2, and has occurred every September–October

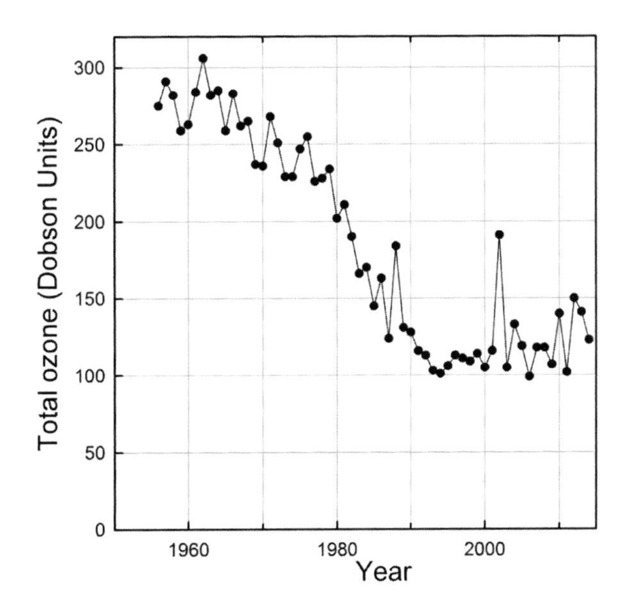

Fig. 5.8 Minimum overhead ozone observed in October at Halley Bay, Ref. [35]

for the past 30 years.[31] The ozone hole came as a complete surprize to the atmospheric chemistry community and was a very dramatic development which captured the public attention.

The ozone hole is the result of the unique conditions in the Antarctic stratosphere. There is a strong circumpolar flow which isolates the air over Antarctica. This isolation combined with the absence of sunlight leads to extremely low temperatures in the Antarctic stratosphere during winter. The low temperatures enable the formation of polar stratospheric clouds consisting of ice and sulfuric acid onto which HCl and HNO_3 can adsorb. Heterogeneous reactions on polar stratospheric clouds convert unreactive chlorine reservoir species such as $ClONO_2$ and HCl into more active forms such as Cl_2 and $ClNO_2$[32]:

$$\mathrm{ClONO_2(g) + HCl(s) \rightarrow HNO_3(s) + Cl_2(g),}$$
$$\mathrm{N_2O_5(g) + HCl(s) \rightarrow ClNO_2(g) + HNO_3(s).}$$

During the winter the concentrations of the more active forms such as Cl_2 and $ClNO_2$ build up and with the arrival of the spring these species are photolyzed to provide a large flux of chlorine atoms which react with ozone.

$$\mathrm{Cl_2 + h\nu \rightarrow Cl + Cl,}$$
$$\mathrm{ClNO_2 + h\nu \rightarrow Cl + NO_2,}$$
$$\mathrm{Cl + O_3 \rightarrow ClO + O_2.}$$

Under the unusual conditions in the springtime Antarctic stratosphere with high concentrations of ClO radicals the self-reaction of ClO radicals becomes important and leads to yet another catalytic ozone destruction cycle.[33]

$$\mathrm{ClO + ClO + M \rightarrow ClOOCl + M}$$
$$\mathrm{ClOOCl + h\nu \rightarrow Cl + OOCl}$$
$$\mathrm{ClOO + M \rightarrow Cl + O_2}$$
$$\underline{\mathrm{2Cl + O_3 \rightarrow 2ClO + O_2}}$$
$$\mathrm{Net: 2O_3 \rightarrow 3O_2.}$$

The observation of the dramatic ozone loss in the Antarctic ozone hole coupled with experimental, field, and modeling research showing that man-made halogenated organic compounds were to blame led to the 1987 Montreal Protocol which has limited the emissions of ozone depleting substances. The Montreal Protocol has been remarkable effective in curbing emissions and protecting stratospheric ozone. The combined tropospheric abundance of anthropogenic ozone depleting substances peaked in 1994 and by 2012 had decreased by ~10% from its peak value.[15] While the Antarctic ozone hole has returned every September–October, a small increase in springtime Antarctic total ozone has been observed since 2000[3] and there has been an ~2.5–5.0% per decade increase in upper stratospheric ozone (35–45 km altitude) in middle latitudes and the tropics since 2000.[15] Absent the Montreal Protocol, it has been predicted that the Antarctic ozone hole would have grown 40% by 2013, that a deep Arctic ozone hole would have developed in 2011, and that the ozone layer worldwide would have thinned by ~15%.[34] Such large global losses of stratospheric ozone would likely have had extremely negative impacts on human and ecosystem health. The Montreal Protocol is widely recognized as perhaps the most effective environmental regulation in history.

5.4. Climate Impacts of Halogenated Organics

Shortly after the pioneering work of Rowland and Molina[3] showing that CFCs posed a threat to stratospheric ozone, Ramanathan[36] and Wang *et al.*[37] pointed out that CFCs were powerful greenhouse gases and also posed a threat to global climate. The effectiveness of CFCs as greenhouse gases stems from two factors. First, C–F and C–Cl bonds absorb strongly in what is termed the "atmospheric window" region of the IR at ~800–1200 cm^{-1}. The mechanism by which the Earth cools itself is via loss of IR radiation into space. Gases in the atmosphere, principally H_2O vapor, CO_2, ozone, and methane absorb IR radiation and hinder the transfer of terrestrial radiation out into space. There is a "window" at approximately at 800–1200 cm^{-1} where there is no absorption by H_2O vapor, CO_2, ozone, or methane and where blackbody radiation from the Earth's surface (terrestrial radiation) can pass unhindered through the atmosphere and into space. CFCs absorb in this window and trap heat which would otherwise escape into space. C-Br and C-I bonds do not absorb strongly in the atmospheric window region and hence halogenated compounds containing only bromine and/or iodine are not strong greenhouse gases. Second, the long atmospheric lifetime of CFCs enables their heat trapping effect to accumulate over time making them potent greenhouse gases.

The first step in estimating the climate impact of a halogenated organic is to measure its IR absorption spectrum. This can be achieved experimentally typically by using Fourier transform infrared (FTIR) spectroscopy or it can be achieved computationally via *ab initio* quantum mechanical calculations. Ideally, both experimental and theoretical approaches are employed which provide a cross check of the results. The ability of computational methods to calculate IR spectra has increased substantially over the past decade to a level where, for the purposes of estimating contributions to radiative forcing of climate change, the accuracy of computational methods is now approaching that of experimental techniques. Computational methods have the substantial advantage that authentic samples of the compounds are not required which avoids the considerable expense and time associated with chemical synthesis and purification. Computational methods are ideal for screening the IR spectra of large numbers of compounds to identify candidates with suitable properties which can be further investigated using experimental measurements. Illustrative spectra recorded experimentally for CCl_4, CF_2Cl_2, and CF_3CFH_2 are shown in Fig. 5.9.

The second step is to use the IR absorption spectrum to compute the impact that the addition of a particular halogenated organic to the atmosphere has on the radiation budget. A change in the radiation budget is generally quantified in terms of the change in the net, downward minus upward, radiative flux (expressed in $\mathrm{W\,m^{-2}}$) at the tropopause, or top of atmosphere, and is defined as "radiative forcing".[11] To compare the effects of different greenhouse gases the radiative forcing per unit change in concentration is computed.

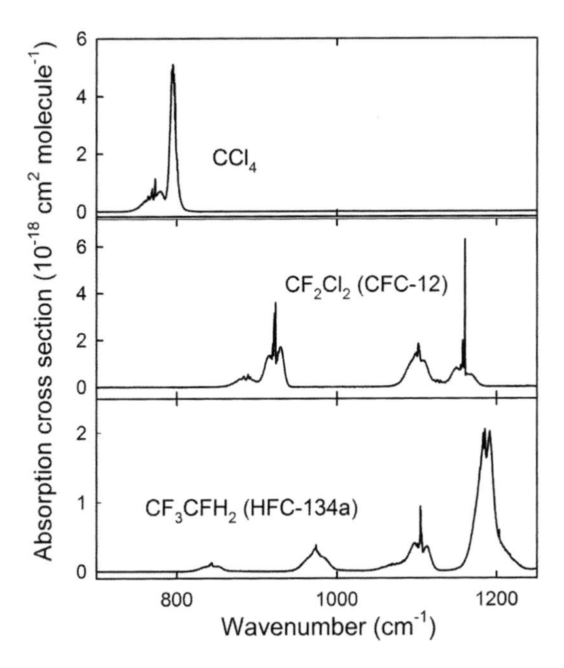

Fig. 5.9 IR spectra of CCl_4, CF_2Cl_2, and CF_3CFH_2 in 700 Torr of air at 296 K.

This is referred to as "radiative efficiency" (RE) and has units of $W\,m^{-2}\,ppb^{-1}$. Radiative efficiencies can be computed using models of radiative transfer in the atmosphere. Running radiative transfer models is a complex task. Pinnock *et al.*[38] used radiative transfer models to produce simple techniques to compute the radiative forcing directly from the absorption cross-sections. Pinnock *et al.*[38] used a narrow-band model to compute the RE per unit absorption cross-section as a function of wavenumber using a global and annual mean atmosphere. This provides a quantitative assessment of the spectral regions where halogenated organic radiative forcing is most effective. Pinnock *et al.*[38] provided values for the average RE per unit IR absorption for $10\,cm^{-1}$ intervals from $0\,cm^{-1}$ to $2500\,cm^{-1}$. Dividing the IR absorption spectrum into $10\,cm^{-1}$ bins, calculating the average absorption cross section for each bin, multiplying by the appropriate Pinnock factors, and summing the result provides an estimate of the radiative efficiency of halogenated organic compounds.

The method outlined by Pinnock *et al.*[38] is simple and effective and has found widespread use. In the method it is assumed that the compound is well mixed throughout the atmosphere. For compounds with atmospheric lifetimes greater than 10 years (e.g., CFCs) this is a good approximation, however compounds with lifetimes less than ~ 1 year are not well mixed. For short lived compounds (e.g., halogenated alkenes) a correction is needed to account for non-homogeneous mixing. Greenhouse gases are most effective in affecting radiative forcing when they are present near the tropopause. Short lived compounds generally do not survive transport to the tropopause and as a result their radiative efficiency is lower than that computed assuming uniform mixing. Hodnebrog *et al.*[44] updated the method of Pinnock *et al.*[38] to account for developments in radiative transfer models and the non-uniform mixing of short lived compounds. The correction to account for non-uniform distribution can be very significant for short-lived compounds. Hodenebrog *et al.*[44] derived the following correction factor, f, as a function of atmospheric lifetime, τ (years):

$$f(\tau) = \frac{a\tau^b}{1 + c\tau^d},$$

where a, b, c and d are constants with values of 2.962, 0.9312, 2.994, and 0.9302, respectively. For compounds with lifetimes of 0.01, 0.1, and 1 year the correction factors are 0.036, 0.238, and 0.688, respectively. The radiative efficiencies of selected halogenated organic compounds computed by Hodnebrog *et al.*[44] are given in Table 5.1.

The third step is to use the radiative efficiency and atmospheric lifetime of the compound to compute its global warming potential (GWP). The GWP metric was developed to compare the contributions of different molecules to radiative forcing of climate change.[11] GWP values measure the time-integrated radiative forcing due to a pulse emission of a unit mass of gas. They can be given on an absolute scale where the absolute GWP for gas i ($AGWP_i$) would have units of $W\,m^{-2}\,kg^{-1}$ year, or on a relative scale as a dimensionless value by dividing the $AGWP_i$ by the AGWP of a reference gas, normally

Table 5.1 Atmospheric lifetimes, REs in units of $W\,m^{-2}\,ppb^{-1}$, POCPs, GWPs, and ODPs of selected organic and halogenated organic compounds.[1,11,15,40–49]

Compound	Atmospheric lifetime	Radiative efficiency	POCP	GWP	ODP
Alkanes-Alkenes					
CH_4 (methane)	12.4 years	0.000363	0.6	30	0
C_2H_6 (ethane)	45 days	0.0032	8.8	<1	0
C_3H_8 (propane)	10 days	0.0031	18.3	<1	0
$CH_2{=}CH_2$ (ethene, ethylene)	1.4 days	0.035	100	<1	0
$CH_3CH{=}CH_2$ (propene, propylene)	0.44 days	0.035	105.3	<1	0
$CH_3CH_2CH{=}CH_2$ (1-butene)	0.37 days	0.035	95.2	<1	0
CFCs					
CF_3Cl (CFC-11)	45 years	0.26	0.0	4660	1.0
CF_2Cl_2 (CFC-12)	100 years	0.32	0.0	10200	0.73
$CFCl_2CF_2Cl$ (CFC-113)	85 years	0.30	0.0	5820	0.81
$CClF_2CF_3$ (CFC-115)	540 years	0.20	0.0	7310	0.26
HCFCs					
$CHClF_2$ (HCFC-22)	11.9 years	0.21	0.1	1760	0.034
CF_3CCl_2H (HCFC-123)	1.3 years	0.15	0.3	79	0.01
CH_3CCl_2F (HCFC-141b)	9.2 years	0.16	0.1	782	0.102
CH_3CClF_2 (HCFC-142b)	17.2 years	0.19	0.1	1980	0.057
HFCs					
CH_2F_2 (HFC-32)	5.2 years	0.11	0.2	677	0
CF_3CHF_2 (HFC-125)	28.2 years	0.23	0.0	3170	0
CF_3CH_3 (HFC-143a)	47.1 years	0.16	0.0	4800	0

(Continued)

Table 5.1 (*Continued*)

Compound	Atmospheric lifetime	Radiative efficiency	POCP	GWP	ODP
CF_3CH_2F (HFC-134a)	13.4 years	0.16	0.1	1300	0
CHF_2CH_3 (HFC-152a)	1.5 years	0.10	1.0	124	0
CF_3CHFCF_3 (HFC-227ea)	38.9 years	0.27	0.0	3350	0
CF_3CH_2FCHF (HFC-245eb)	3.1 years	0.20	0.2	290	0
HCCs and CCl_4					
CH_3Cl (methyl chloride)	1.0 years	0.01	0.4	12	0.015
CH_2Cl_2 (methylene chloride)	0.4 years	0.03	3.5	9	
$CHCl_3$ (chloroform)	0.4 years	0.08	2.4	16	
CCl_4 (carbon tetrachloride)	26 years	0.17	0	1730	0.72
CH_3CCl_3 (methyl chloroform)	5.0 years	0.07	0.2	160	0.14
HBCs					
CH_3Br	0.8 years	0.00	n.a.	2	0.57
CH_2Br_2	0.3 years	0.01	n.a.	1	3.5
HICs					
CH_3I (methyl iodide)	4.1 days	0.00	n.a.	<1	0.017
CH_2I_2 (dibromomethane)	4.9 minutes	0.00	n.a.	<1	n.a.
Halons					
$CBrF_3$ (Halon-1301)	65.0 years	0.30	0.0	6290	15.2
$CBrClF_2$ (Halon-1211)	16.0 years	0.29	0.0	1750	6.9
CHFBrCF3 (Halon-2401)	2.9 years	0.19	0.0	184	
$CBrF_2$CBrF2 (Halon-2402)	20 years	0.31	0.0	1470	15.7
Perfluorocarbons					
CF_4 (perfluoromethane)	50,000 years[g]	0.09	0.0	6,630	0

C_2F_6 (perfluoroethane)	10,000 years[g]	0.25	0.0	11,100	0
C_3F_8 (perfluoropropane)	2,600 years[g]	0.28	0.0	8,900	0
Haloalkenes					
$CF_2{=}CH_2$ (HFO-1132a)	4.0 days	0.00	18.0	<1	0
$CF_2{=}CF_2$ (HFO-1114)	1.1 days	0.00	12.5	<1	0
$CF_3CH{=}CH_2$ (HFO-1243zf)	7.0 days	0.01	10.7	<1	0
$CF_3CFH{=}CH_2$ (HFO-1234yf)	10.5 days	0.02	7.0[i]	<1	0
$CF_3CF{=}CF_2$ (HFO-1216)	4.9 days	0.01	5.4	<1	0
Z-$CF_3CF{=}CHF$ (HFO-1225ye(Z))	8.5 days	0.02	5.6	<1	0
E-$CF_3CF{=}CHF$ (HFO-1225ye(E))	4.9 days	0.01	7.3	<1	0
$CF_3CF_2CH{=}CH_2$ (HFO-1345zfc)	7.6 days	0.01	6.6	<1	0
E-$CF_3CH{=}CHF$ (HFO-1234ze(E))	16.4 days	0.04	6.4	<1	0
Z-$CF_3CH{=}CHCF_3$(HFO-1336(Z))	22.0 days	0.07	3.1	2	0
E-$CF_3CH{=}CHCl$ (HCFO-1233zd(E))	26 days	0.04	3.9	1	0.00034
Z-$CF_3CH{=}CHCl$ (HCFO-1233zd(Z))	12 days	0.02	6.4	<1	<0.00034
$CH_2{=}CHCl$ (vinyl chloride)	1.5 days	n.a.	25.6	n.a.	
$CHCl{=}CCl_2$ (trichloroethylene)	4.9 days	n.a.	31.6	n.a.	0.00037
$CCl_2{=}CCl_2$ (tetrachloroethylene)	90 days	n.a.	2.7	n.a.	0.0050
Alcohols					
CH_3OH (methanol)	13 days	0.004	16.5	1	0
C_2H_5OH (ethanol)	4 days	n.a.	39.7	n.a.	0
Halogenated alcohols					
CF_3OH (trifluoromethanol)	10 days	n.a.	n.a	n.a.	0
$(CF_3)_2CHOH$ (1,1,1,3,3,3-hexafluoroisopropanol)	1.9 years	0.26	n.a.	182	0

(*Continued*)

Table 5.1 (*Continued*)

Compound	Atmospheric lifetime	Radiative efficiency	POCP	GWP	ODP
$CF_3(CF_2)_nCH_2CH_2OH$ ($n = 3, 5, 7$), 4:2, 6:2, 8:2-fluorotelomer alcohols	2 weeks	0.06	n.a.	<1	0
Aldehydes					
HCHO (formaldehyde)	0.5 days	0.00	47.1	0	0
CH_3CHO (acetaldehyde)	0.3 days	0.00	55.0	0	0
C_2H_5CHO (propionaldehyde)	0.2 days	0.00	61.2	0	0
Halogenated aldehydes					
$COCl_2$ (phosgene)	5 days	n.a.	n.a.	n.a.	0
CF_3CHO (fluoral)	2 days	n.a.	n.a.	n.a.	0
Ketones					
$CH_3C(O)CH_3$ (acetone)	18 days	0.003	7.5	0	0
$C_2H_5C(O)CH_3$(butanone)	2 days	0.00	35.3	0	0
Halogenated ketones					
$C_2F_5C(O)CF(CF_3)_2$	7 days	0.03	n.a	<1	0
Ethers					
CH_3OCH_3 (dimethyl ether)	4 days	0.0024[a]	19.8	0[a]	0
$C_2H_5OC_2H_5$ (diethyl ether)	8 hours	<0.0024[b]	46.4	0[b]	0
Halogenated ethers					
CH_3OCF_3 (HFE-143a)	4.8 years	0.18	n.a.	523	0
CHF_2OCF_3 (HFE-125)	119 years	0.41	n.a.	12,400	0

CHF_2OCF_2CHFCl (HCFE-235ca2, enflurane)	4.3 years	0.41	n.a.	583	n.a .
$CHF_2OCHFCF_3$ (HCFE-236ea2, desflurane)	10.8 years	0.45	n.a .	1790	0
$CHF_2OCHClCF_3$ (HCFE-235da2, isoflurane)	3.5 years	0.42	n.a .	491	n.a.
$C_4F_9OCH_3$ (HFE-449s1)	4.7 years	0.36	n.a.	421	0
$C_4F_9OC_2H_5$ (HFE-569sf2)	0.8 years	0.30	n.a.	57	0
$CF_3CH_2OCF_2CHF_2$ (HFE-347pcf2)	6.0 years	0.48	n.a.	889	0
$CHF_2OCF_2OCHF_2$ (HFE-236ca12)	25.0 years	0.65	n.a.	5350	0
$CHF_2OCF_2CF_2OCHF_2$ (HFE-338pcc13)	12.9 years	0.86	n.a.	2910	0
$CHF_2OCF_2OCF_2CF_2OCHF_2$ (HFE-43-10pccc)	13.5 years	1.02	n.a.	2820	0

Note: n.a. = not available; [a]calculated using an IR spectrum measured at Ford, [b]inferred from result for dimethyl ether.

CO_2. Thus, the GWP is defined as[11]:

$$\mathrm{GWP}_i(H) = \frac{\int_0^H \mathrm{RF}_i(t)dt}{\int_0^H \mathrm{RF}_{\mathrm{CO_2}}(t)dt} = \frac{\mathrm{AGWP}_i(H)}{\mathrm{AGWP}_{\mathrm{CO_2}}(H)}.$$

The time horizon (H) over which the integration is performed is a value judgement. The IPCC in its reports usually presents GWPs evaluated over 20, 100, and 500 years to provide a range of GWPs for short-, medium-, and long-time horizons. National and international agreements such as the Kyoto Protocol typically adopt a time horizon of 100 years for GWPs. We provide GWPs evaluted over 100 years in Table 5.1.

For a gas i, if A_i is the RE, τ_i is the lifetime (and assuming its removal from the atmosphere can be represented by exponential decay), and H is the time horizon, then the integrated RF up to year H is given by[44]:

$$\mathrm{AGWP}_x(H) = A_x\tau\left(1 - \exp\left(-\frac{H}{\tau}\right)\right).$$

This is an approximation that holds for long-lived gases but is less accurate for shorter lived gases whose lifetimes depend on location of emissions and physical and chemical conditions of the atmosphere.

The AGWP for CO_2 is complicated, because its atmospheric response time (or lifetime of a perturbation) cannot be represented by a simple exponential decay. This situation arises because CO_2 is absorbed into the various regions of the oceans (surface water, thermocline, deep ocean) on a range of different timescales. The decay of atmospheric CO_2 following a pulse emission at time t (years) is usually approximated as[39]:

$$\mathrm{IRF}(t) = a_0 + \sum_{i=1}^{3} a_i \exp\left(-\frac{t}{\alpha_i}\right),$$

where the parameter values are $a_0 = 0.2173$, $a_1 = 0.2240$, $a_2 = 0.2824$, $a_3 = 0.2763$, $\alpha_1 = 394.4$, $\alpha_2 = 36.54$ and $\alpha_3 = 4.304$. The RE of CO_2 expressed per mass is $1.75 \times 10^{-15}\ \mathrm{W\,m^{-2}\,kg^{-1}}$, and the AGWPs for CO_2 are 2.495×10^{-14}, 9.171×10^{-14} and 32.17×10^{-14}

$W\,m^{-2}$ yr $(kgCO_2)^{-1}$ for time horizons of 20, 100, and 500 years, respectively. Dividing the 100 year horizon $AGWP_i$ for species i by $AGWP_{CO2} = 9.171 \times 10^{-14}\,W\,m^{-2}$ yr $(kgCO_2)^{-1}$ gives the GWP of species i.

GWPs for selected halogenated organic compounds are listed in Table 5.1 and range from 12,400 for CHF_2OCF_3 to <1 for $C_2F_5C(O)CF(CF_3)_2$. The large range in the ability of halogenated organic compounds to contribute to radiative forcing of climate change mainly reflects the large range in atmospheric lifetimes of the halogenated organic compounds. CHF_2OCF_3 has a lifetime of 119 years while $C_2F_5C(O)CF(CF_3)_2$ has a lifetime of only 7 days. Lifetime is the most important determinant of GWP because it enters in the calculation twice; correcting the radiative efficiency to account for the non-uniform distribution, and calculation of the integrated radiative forcing which is limited by the rapid decay of the atmospheric concentration of the compound.

To place the contribution to radiative forcing of climate change of halogenated organic compounds into a broader perspective we need to compare with the major long-lived greenhouse gases CO_2, N_2O, and CH_4. Taking the increase in atmospheric concentrations since preindustrial times (1750) of CO_2, N_2O, CH_4, CFCs, HFCs, HCFCs, and halons [11,15,22] and multiplying by the radiative efficiencies of $CO_2(1.37 \times 10^{-2}\ W\,m^{-2}\ ppm^{-1}$[11]), $CH_4(3.63 \times 10^{-4}\ W\,m^{-2}\ ppb^{-1}$[11]), $N_2O(3.00 \times 10^{-3}\ W\,m^{-2}\ ppb^{-1}$[11]), halogenated organic compounds (from Table 5.1) gives the evolution of instantaneous radiative forcing over the last century shown in Figs. 5.10–5.15 from these species. As seen from Figs. 5.10 and 5.11 the radiative forcing from CFCs increased rapidly from the 1940s to the 1980s, peaked in about 2000 and is now in decline. A similar trend is evident for the contribution from the halons although they were introduced later and in lower amounts than the CFCs and the absolute magnitude of their contribution is about 100 times lower than that of the more abundant CFCs. In contrast to the CFCs and halons, the radiative forcing by HCFCs and HFCs is growing although the absolute contribution of these compounds is only about 3% and 1% of that of CO_2. The growth in radiative forcing by HCFCs and HFCs reflects the fact

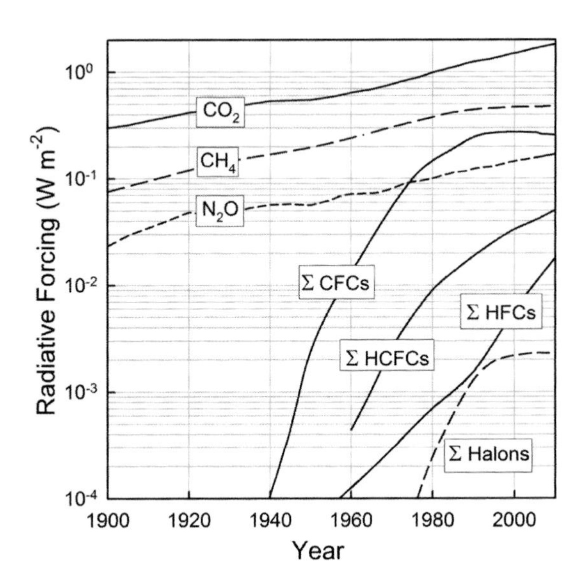

Fig. 5.10 Radiative forcing from CO_2, CH_4, N_2O, CFCs, HCFCs, HFCs, and halons.[11]

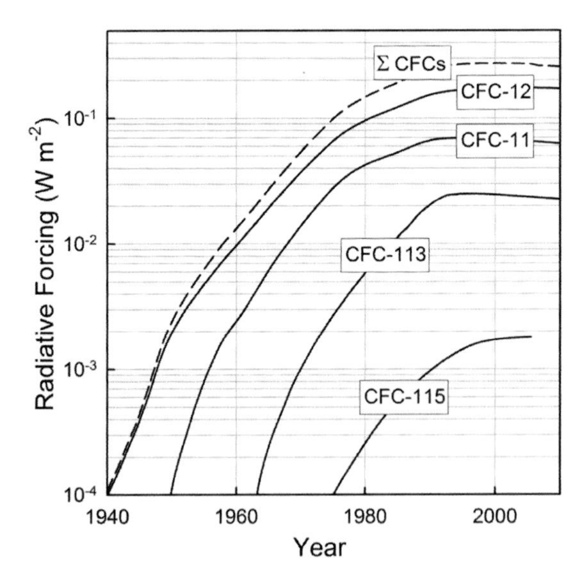

Fig. 5.11 Radiative forcing from the CFCs.[11]

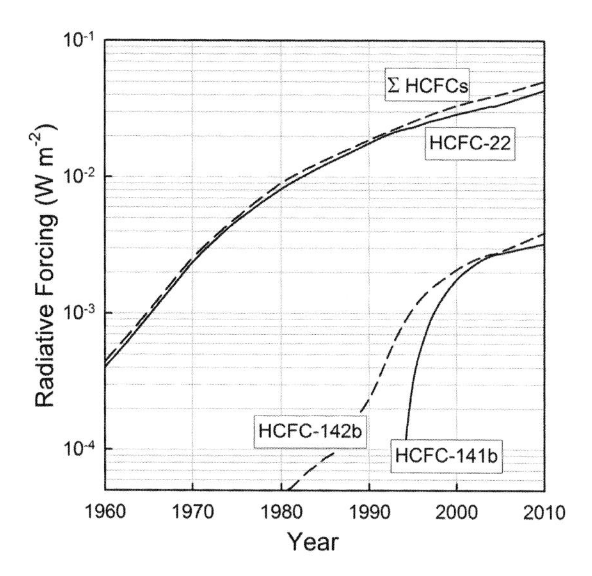

Fig. 5.12 Radiative forcing from the HCFCs.[11]

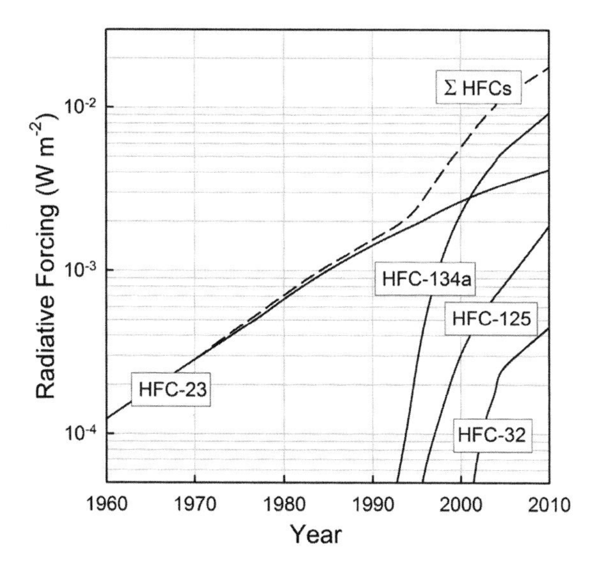

Fig. 5.13 Radiative forcing from the HFCs.[11]

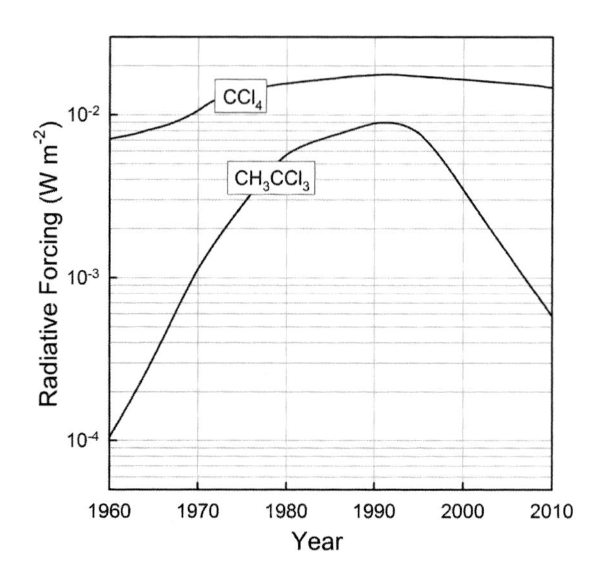

Fig. 5.14 Radiative forcing from CCl_4 and CH_3CCl_3.[11]

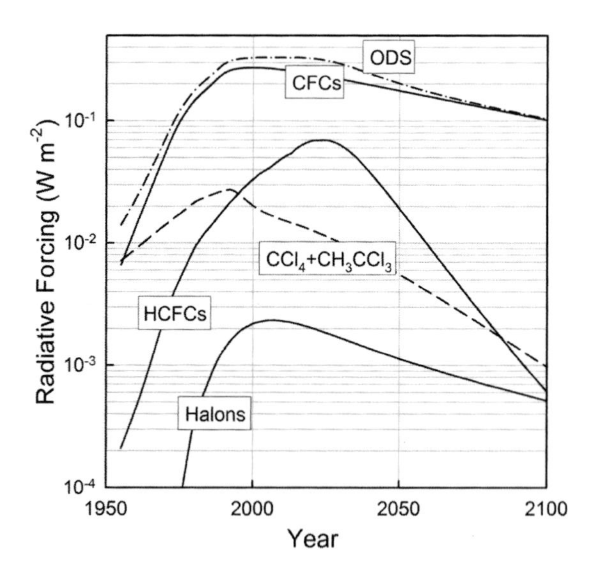

Fig. 5.15 Radiative forcing from the sum of ODSs and the individual contributions from CFCs, HCFCs, CCl_4 and CH_3CCl_3, and halons through 2100 in the WMO Ozone Assessment baseline (A1) scenario.[15]

that these compounds are being used as replacements for the CFCs. HCFCs contain chlorine and their production and use is controlled under the Montreal Protocol. As seen in Fig. 5.12 the rate of increase in radiative forcing from HCFCs has decreased over the past decade and as illustrated in Fig. 5.15 their radiative forcing is projected to decrease substantially in the coming decades. As shown in Fig. 5.10, the radiative forcing by CFCs exceeded that of N_2O in around 1975 and is currently approximately half of that from CH_4 and one tenth of that from CO_2. Clearly halogenated organic compounds, and in particular the CFCs, make a significant contribution to radiative forcing of climate change.

Figures 5.11–5.14 show the radiative forcing contributions from individual CFCs, HCFCs, HFCs, HCCs, and ozone depleting substances (ODS). Fig. 5.11 illustrates the major climate benefits of the global phase-out of the production and use of CFCs under the Montreal Protocol (see Fig. 5.1) which led to a peak in the atmospheric concentrations of CFC-11, CFC-12, CFC-113, and CFC-115 in ~2000 (see Fig. 5.4) and a slight decline over the past decade. As seen from Figs. 5.11 and 5.15, the radiative forcing by CFCs and ozone depleting substances (ODS) peaked at $\sim 0.3\,W\,m^{-2}$ in 2000–2010. It has been estimated that in the absence of the Montreal Protocol the radiative forcing by ODS would have reached 0.60–0.65 $W\,m^{-2}$ by 2010.[50] Total avoided net annual ODS emissions by 2010 have been estimated to be equivalent to ~10 Gt CO_2/year.[50] This is approximately one third of the fossil fuel CO_2 emissions in 2010 and approximately five times the annual reduction called for using the Kyoto Protocol for 2008–2012. In addition to protecting the stratospheric ozone layer, the Montreal Protocol has been very successful in protecting climate.

HFCs are the only class of halogenated organic compounds which are currently showing substantial growth in atmospheric concentration and radiative forcing and may continue to grow in the coming decades. The current radiative forcing from HFCs is rather minor at about 1% of that of CO_2. However, it has been argued that the potential future growth in emissions of HFCs, particularly those with long lifetimes and high GWPs such as HFC-125 and HFC-143a, could negate some of the climate benefits of the Montreal Protocol.[51] It has been proposed that HFC emissions should be controlled[52] under the Montreal Protocol although the cost-effectiveness of such control has been questioned.[53]

5.5. Halogenated Alkanes

5.5.1. *CFCs, Halons, CCl_4, and PFCs*

Perhalogenated compounds of atmospheric importance include CFCs, Halons, CCl_4, and PFCs. These perhalogenated compounds do not react with OH, O_3, Cl, or NO_3 and do not absorb at wavelengths relevant to solar radiation in the troposphere. They are transported through the troposphere to the stratosphere where they undergo photolysis and reaction with $O(^1D)$ atoms. The photolysis of CFCs releases chlorine atoms which then participate in a series of reactions, the ClO_x catalytic cycles (see above), which lead to destruction of stratospheric ozone. Halons such as Halon-1301 ($CBrF_3$) and Halon-1211 ($CBrClF_2$), mainly used as fire suppression agents, also undergo photolysis in the stratosphere. They release bromine atoms, which participate in BrO_x catalytic cycles and destroy stratospheric ozone. The main fate of CCl_4 is photolysis in the stratosphere releasing chlorine atoms, minor fates are loss via uptake by oceans and soils. C_2F_6 and larger PFCs are photolysed by Lyman-α radiation (121.6 nm) in the mesosphere and have extremely long atmospheric lifetimes determined by the very long time scales (millenia) for the atmosphere to circulate through the mesosphere. Photolysis releases F atoms which react rapidly with hydrogen containing species to form HF which is a sink for F atoms; there is no catalytic ozone destruction associated with F atoms.[54,55] CF_4 is an interesting special case. Photolysis of CF_4 even in the mesosphere is exceedingly slow. The dominant atmospheric fate of CF_4 is thought to be pyrolysis when air is combusted. Loss via reaction with $O(^1D)$ atoms and via ion-molecule reactions also contribute as fates of CF_4. The fraction of the global atmosphere which participates in combustion each year is very small and hence CF_4 has a very long lifetime which is estimated to be >50,000 years.[56] PFCs are by-products of the aluminum industry and they are used in the semiconductor industry.

CFCs, Halons, and CCl_4 are ozone depleting substances. The relative impact of these compounds on stratospheric ozone is gauged using the concept of ODP. Steady-state ODPs are defined as the change in global ozone for a sustained unit mass emission of a specific compound relative to the change in global ozone calculated for the sustained unit mass emission of CFC-11 ($CFCl_3$).[15] By definition the ODP for CFC-11 is 1.0. ODPs for CCl_4 and selected CFCs and Halons are given in Table 5.1.

CFCs, Halons, PFCs, and CCl_4 contain carbon–halogen bonds which absorb in the atmospheric window region and have the potential to contribute to radiative forcing of climate change. The radiative efficiency of a molecule reflects its heat trapping ability in the atmosphere, has units of $W\,m^{-2}\,ppb^{-1}$, and can be calculated using a radiative transfer model. The GWP is the integrated radiative forcing over a given time horizon following the pulsed emission of 1 kg of gas compared to the integrated radiative forcing for release of 1 kg of CO_2. Radiative efficiencies and GWPs for CFCs, Halons, PFCs, and CCl_4 are given in Table 5.1. As seen from Table 5.1 CFCs and PFCs are long-lived compounds with high GWPs.

Recognition of the adverse impact of CFCs, Halons, and CCl_4 on stratospheric ozone and the high GWPs of PFCs has led to an international effort to develop substitutes for these compounds and to limit their emissions. There are two general strategies to design compounds which will have minimal, or no, impact on stratospheric ozone: (i) do not include the ozone destructive elements chlorine and bromine in the compounds and (ii) introduce chemical groups into the molecule which give the compound a short enough lifetime in the troposphere that insignificant amounts survive to reach the stratosphere. As we will describe below, shortening the tropospheric lifetime can be achieved by several means. Options include adding one or more C–H bonds or >C=C< double bonds making the molecule reactive towards OH radicals, or adding a carbonyl group to the molecule which makes it susceptible to photolysis in the troposphere.

5.5.2. *HFCs and HCFCs*

As implied by the emission and atmospheric concentration estimates in Figs. 5.3, 5.5, and 5.6, HFCs and HCFCs have found widespread use as CFC replacements. There are no significant natural sources of HFCs and HCFCs. HCFCs differ from CFCs in one critical aspect; HCFC molecules contain one or more C–H bonds. The presence of a C–H bond provides an Achilles heel making the molecule susceptible to reaction with OH radicals in the troposphere and decreasing the fraction of emissions that survive transport to the stratosphere. The atmospheric lifetimes of HCFCs are determined mostly by their reaction with OH radicals in the troposphere and are typically of the order of 1–10 years as opposed to the ~100 year lifetime for CFCs. There are no significant loss mechanisms for CFCs in the troposphere and as a result 100% of CFC emissions survive transport through the troposphere to reach the stratosphere. In contrast, only ~1–10% of HCFC emissions survive to reach the stratosphere and consequently HCFCs have ~10–100 times lower impact on stratospheric ozone than CFCs (see Table 5.1). Hydrofluorocarbons do not contain chlorine or bromine and do not contribute to the well-established Cl- and Br-based catalytic ozone destruction cycles. In contrast to chlorine and bromine, fluorine does not participate in catalytic ozone destruction cycles. While HFCs do not themselves participate in ozone destruction chemistry,[54,55] HFCs can modify atmospheric temperatures and circulation which might affect stratospheric ozone.[57] Increased temperatures in the stratosphere lead to increased ozone loss via the gas-phase catalytic cycles and decreased ozone loss via the decreased importance of heterogeneous chemistry with lower amounts of polar stratospheric clouds. Changes in circulation lead to decreased ozone in some locations and increased ozone in others.[57] The overall effect is very small and rather uncertain reflecting a near cancellation of the positive and negative effects and ODPs of 0.00039–0.030 were estimated for different HFCs.[57] For practical purposes the ODPs of the HFCs can be considered to be essentially zero.

The atmospheric oxidation mechanisms of HFCs and HCFCs are similar and can be conveniently considered together. HFC-134a (CF_3CFH_2) is the most widely used HFC and we will consider its atmospheric oxidation in detail as a representative example for HFCs and HCFCs. The atmospheric chemistry of HFC-134a has been the subject of studies by several research groups and is well established.[58] As illustrated in Fig. 5.16, the atmospheric oxidation of HFC-134 a is initiated by reaction with OH radicals. This reaction takes place with a rate constant described by the Arrhenius expression $k_{OH} = 4.9 \times 10^{-13} \exp(-1395/T)$ cm^{-3} $molecule^{-1}$ s^{-1}.[59] Compounds such as HFCs and HCFCs are lost mainly via reaction with OH radicals. The rate coefficients for reactions of OH radicals with many HFCs and HCFCs have been measured and critically evaluated values are provided by the IUPAC[59] and JPL[60] data evaluation groups. Atmospheric lifetimes of HFCs and HCFCs can be estimated using the following methodology from Calvert *et al.*[49] For compounds whose lifetime is <1 day, an average daytime [OH] of 2.5×10^6 cm^{-3} is used in conjunction with the rate coefficient for reaction with OH, evaluated at 298 K. For species with lifetimes between about 1 day and 1 year, a diurnally-averaged [OH] of 1.0×10^6 cm^{-3} is used, again in conjunction with the OH rate coefficient at 298 K. Species with lifetimes >1 year are mixed throughout the troposphere and for these cases the [OH] weighted tropospheric average temperature of 272 K is used, with an assumed tropospheric average $[OH] \approx 10^6$ cm^{-3}. At 272 K the rate coefficient for reaction of HFC-134a with OH radicals is $k_{OH} = 2.9 \times 10^{-15}$ cm^{-3} $molecule^{-1}$ s^{-1} which combined with [OH] $\approx 10^6$ cm^{-3} gives a lifetime of ≈11 years. This estimate is consistent with the WMO estimate of 13.4 years (listed in Table 5.1) which was derived from the ratio of rate coefficients for reactions of OH radicals with HFC-134a and CH_3CCl_3 assuming a tropospheric lifetime of CH_3CCl_3 towards reaction with OH of 6.1 years and accounting for minor losses of HFC-134a by uptake into the oceans and reactions in the stratosphere.

The lifetime of HFC-134a with respect to reaction with OH is 13.4 years.[2] The fluorinated alkyl radical CF_3CFH adds O_2 rapidly (within $1\mu s$) to give the corresponding peroxy radical CF_3CFHO_2. As with other peroxy radicals the fate of the CF_3CFHO_2 radical is reaction with either NO, NO_2, or HO_2 radicals. Peroxy radicals react rapidly with NO_2 to give alkyl peroxynitrates (RO_2NO_2). Using the laboratory kinetic data for the reaction of $CF_3CFH_2O_2$ radicals with NO_2 the lifetime of $CF_3CFH_2O_2$ radicals with respect to reaction with NO_2 is estimated to be ~10–100 min for the representative tropospheric background NO_2 concentration range of 1–10 ppt. Alkyl peroxynitrates are thermally unstable and decompose rapidly to regenerate RO_2 radicals and NO_2. At room temperature in one atmosphere of air the peroxynitrate derived from HFC-134a has a lifetime of <90 sec. The lifetime of CF_3CFHO_2 radicals with respect to reaction with HO_2 has been estimated to be ~5 min.[61] The reaction of peroxy radicals with HO_2 radicals gives hydroperoxides and, in some cases, carbonyl products. Product data are available for two HFC derived peroxy radicals: CH_2FO_2 and CF_3CFHO_2. Reaction of CH_2FO_2 radicals with HO_2 gives a 30% yield of the hydroperoxide, CH_2FOOH, and 70% yield of the carbonyl product, HCOF. In the reaction of CF_3CFHO_2 with HO_2 radicals less than 5% of the products appear as the carbonyl CF_3COF. The factors which determine the relative importance of the hydroperoxide and carbonyl forming channels are unclear at present and more work is needed in this area. The hydroperoxide $CF_3CFHOOH$ is expected to be converted back into CF_3CFHO_2 radicals via photolysis and reaction with OH. Alkyl peroxy radicals react rapidly with NO to give alkoxy radicals and NO_2 as major products and alkyl nitrates as minor products. The experimental data suggest that nitrate formation from reactions of fluorinated alkyl peroxy radicals with NO is of little, or no, significance. The reaction of CF_3CFHO_2 radicals gives CF_3CFHO radicals and NO_2. This reaction produces a substantial fraction (~60%) of chemically activated (i.e., vibrationally excited) CF_3CFHO radicals (denoted as CF_3CFHO^* in Fig. 5.16).[62] The excited alkoxy radicals possess sufficient internal energy to overcome the ~8 kcal mol^{-1} barrier to C–C bond scission and decompose on

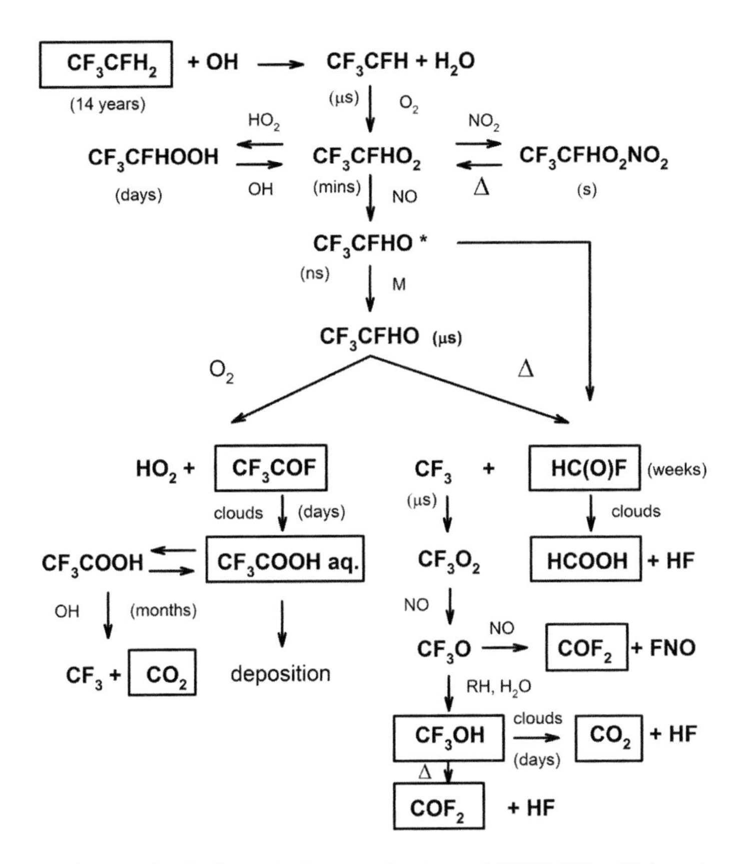

Fig. 5.16 Atmospheric degradation mechanism of HFC-134a. Values in parentheses are order-of-magnitude lifetime estimates; species in boxes are observed products. *The Handbook of Environmental Chemistry*, Vol. 3, Part N, Organofluorines (ed. by A.H. Neilson, 2002). With permission of Springer.

a timescale of the order of 10^{-10} seconds to give CF_3 radicals and HCOF. The CF_3CFHO radicals which become thermalized undergo both reaction with O_2 and thermal decomposition via C–C bond scission.

$$CF_3CFHO + O_2 \rightarrow CF_3COF + HO_2,$$
$$CF_3CFHO + M \rightarrow CF_3 + HCOF + M.$$

In one atmosphere of air (1 atmosphere = 1013 mbar = 760 Torr) diluent the competition between reaction with O_2 and dissociation as loss processes for the thermalized CF_3CFHO radicals is described by the rate constant ratio $k_{O2}/k_{diss} = 2.4 \times 10^{-25} \exp(3590/T)$ cm^3 molecule^{-1}.[1] The relative importance of the two channels is given by $k_{O2}[O_2]/k_{diss}$. In one atmosphere of air at 288 K (0 km altitude in the U.S. Standard Atmosphere) $k_{O2}[O_2]/k_{diss} = 6.22 \times 10^{-20} \times 5.34 \times 10^{18} = 0.33$, hence 25% of CF_3CFHO radicals undergo reaction with O_2 while the remaining 75% decompose. With increasing altitude both temperature and pressure fall substantially. Decreasing temperature slows down the rates of both processes but the unimolecular decomposition process has an activation energy which is substantially larger and slows down more than that of the bimolecular reaction with oxygen. Hence, the rate of reaction with O_2 becomes more important with increasing altitude (even after accounting for decreased $[O_2]$ at higher altitudes). Ignoring for the moment the pressure dependence of the O_2 reaction, we can calculate that at an altitude of 10 km ($T = 223$ K, total pressure = 265 mbar) that $k_{O2}[O_2]/k_{diss} = 2.35 \times 10^{-18} \times 1.81 \times 10^{18} = 4.25$, hence 81% of CF_3CFHO radicals undergo reaction with O_2 while the remaining 19% decompose. Accounting for the distribution of OH radicals in the troposphere, the temperature and pressure profiles, and the chemical activation effect it has been estimated that 7–20% of the atmospheric degradation of HFC-134a gives CF_3COF.[62]

CF_3COF does not take part in any gas-phase chemistry in the troposphere and is removed on a time scale of ~5–15 days by uptake and hydrolysis within cloud droplets to give trifluoroacetic acid, $CF_3C(O)OH$. CF_3 radicals add O_2 to give CF_3O_2 radicals which react with NO to give CF_3O radicals. The usual modes of alkoxy radical loss are not possible for the CF_3O radical. Reaction with O_2 and decomposition via F atom elimination are both thermodynamically unfavorable under atmospheric conditions. Instead, CF_3O radicals react mainly with NO and hydrocarbons.

$$CF_3O + NO \rightarrow COF_2 + FNO,$$

$$CF_3O + CH_4 \rightarrow CF_3OH + CH_3.$$

Reaction with NO yields COF_2. COF_2 does not react with any gas phase trace atmospheric species and its photolysis is slow. COF_2 is removed from the atmosphere by incorporation into water droplets and hydrolysis to give CO_2 and HF and by photolysis in the upper stratosphere to give FCO radicals and F atoms. FNO photolyzes to give NO and a F atom. F atoms reversibly form FO_2 radicals by combining with O_2, and also react with CH_4 and H_2O to give HF which will be rained out of the atmosphere. The reaction of CF_3O radicals with hydrocarbons such as CH_4 produces CF_3OH. The CF_3O–H bond is unusually strong for an alcohol at 120 kcal mol^{-1} (1 kcal = 4.18 kJ). CF_3OH does not participate in any gas-phase chemical reactions and is not photolyzed in the lower atmosphere. CF_3OH is thermodynamically unstable with respect to decomposition into COF_2 and HF and undergoes heterogeneous decomposition on atmospheric aerosols. COF_2 undergoes surface-mediated hydrolysis to give CO_2 and HF.

To summarize, the gas phase atmospheric degradation of HFC-134a is initiated by reaction with OH radicals which occurs relatively slowly giving HFC-134a an atmospheric lifetime of ~13.4 years. Following reaction with OH radicals a series of rapid reactions involving free radicals occur on a time scale of minutes, or less, leading to the formation of four primary products: CF_3COF, HCOF, COF_2, and CF_3OH. These primary products do not participate further in gas-phase chemistry but are removed from the atmosphere by heterogeneous processes and hydrolysis to give $CF_3C(O)OH$, CO_2, and HF on a time scale of ~5–15 days.[58]

As illustrated in Fig. 5.16 for HFC-134a, there is a wealth of data from experimental and theoretical studies and the atmospheric oxidation mechanisms of commercially important HFCs and HCFCs are in general well understood. The atmospheric degradation of HFCs and HCFCs is initiated by reaction with OH radicals. The atmospheric lifetimes of five of the top six commercially significant HFCs and HCFCs shown in Fig. 5.3 are in the range 1.5–15 years, the outlier is HFC-143a with a lifetime of 47.1 years (see Table 5.1). HFCs and HCFCs have C–F and C–Cl bonds which absorb in the atmospheric window region of the infrared and can contribute to radiative forcing of climate change. Table 5.1 lists GWPs for selected HFCs and HCFCs which shows that in general these compounds have GWPs which are approximately an order of magnitude lower than the CFCs they have replaced. The reactions of HFCs and HCFCs with OH radicals give halogenated alkyl radicals which add O_2 to form the corresponding peroxy radicals. Reactions of these peroxy radicals are similar to those discussed above for HFC-134a and lead to the halogenated carbonyl oxidation products listed in Table 5.2.[58] The halogenated carbonyl compounds are intermediate oxidation products and provide a convenient break point in our discussion. The oxidation of HCFCs and HFCs to the intermediate products in Table 5.2 occurs purely by gas-phase reactions. The rate determining step is the initial reaction with OH radicals which occurs on a time scale of years and can be contrasted with the subsequent gas-phase reactions which occur on a time scale of minutes or seconds. With the exception of aldehydes such as CF_3CHO and $CFCl_2CHO$, the intermediate halogenated carbonyl products are unreactive toward further gas-phase reactions and are removed by heterogeneous processes and hydrolysis on a time scale of 5–15 days leading to the final products given in Table 5.2. CF_3CHO and $CFCl_2CHO$ are reactive towards OH radicals and also undergo photolysis leading to the final products given in Table 5.2.

The global emissions of HFCs and HCFCs are of the order of 100 kt per year. Assuming an empirical formula of –CHX– (X = F or Cl) for these compounds, uniform distribution in the atmosphere, and annual global precipitation of 4.9×10^{17} L the level of HF/HCl expected from degradation of HFCs and HCFCs is of the order of $10^{-9} - 10^{-8}$ molar.[58] HFCs and HCFCs are not uniformly distributed in the atmosphere and hence some regional concentrations of HF/HCl will be greater than the global average by perhaps an order of magnitude (i.e., $10^{-8} - 10^{-7}$molar). The concentration of fluoride/chloride and the additional acidity in precipitation resulting from the atmospheric oxidation of HFCs and HCFCs is negligible. Trifluoroacetic acid is a persistent degradation product of some HFCs and HCFCs (see Table 5.2). The sources (natural and anthropogenic), sinks, and potential environmental

Table 5.2 Atmospheric degradation products for selected HFCs and HCFCs.[58]

Compound	Intermediate products	Final products
HFC-23 (CF_3H)	COF_2,CF_3OH	CO_2, HF
HFC-32 (CH_2F_2)	COF_2	CO_2, HF
HFC-41 (CH_3F)	HCOF	HC(O)OH, HF
HFC-125 (CF_3CF_2H)	COF_2, CF_3OH	CO_2, HF
HFC-134a (CF_3CFH_2)	HCOF, CF_3OH, COF_2, CF_3COF	CO_2, HF, $CF_3C(O)OH$
HFC-143a (CF_3CH_3)	CF_3CHO, CF_3OH, COF_2, CO_2	CO_2, HF, $CF_3C(O)OH$
HFC-152a (CF_2HCH_3)	COF_2	CO_2, HF
HFC-227ea (CF_3CHFCF_3)	COF_2, CF_3OH, CF_3COF	CO_2, HF, $CF_3C(O)OH$
HFC-236cb ($CF_3CF_2CH_2F$)	HCOF, CF_3OH, COF_2, C_2F_5COF	CO_2, HF, $C_2F_5C(O)OH$
HCFC-22 ($CHClF_2$)	COF_2	CO_2, HF
HCFC-123 (CF_3CCl_2H)	CF_3COCl, CF_3OH, COF_2	CO_2, HF, $CF_3C(O)OH$
HCFC-141b (CH_3CCl_2F)	CCl_2FCHO, COFCl	CO_2, HF, HCl
HCFC-142b (CH_3CClF_2)	CF_2ClCHO, COF_2	CO_2, HF, HCl

effects of trifluoroacetic acid have been reviewed elsewhere.[63–65] It is well established that trifluoroacetic acid is ubiquitous in precipitation and ocean water even in remote areas.[66–71] The oceans contain ∼300 million tonnes of trifluoroacetic acid.[68] The natural global environmental loading of trifluoroacetic acid greatly exceeds that expected from human activities including from the atmospheric degradation of HFCs and HCFCs.

Tromp *et al.*[72] argued that trifluoroacetic acid would accumulate in seasonal wetlands, however Boutonnet *et al.*[66] argue that the assumptions made by Tromp *et al.* were highly improbable. Benesch *et al.* [73] showed that trifluoroacetic acid does not adversely affect the development of soil microbial communities and pool plant species in vernal ponds. With respect to trifluoroacetic acid formation from the atmospheric degradation of HCFCs and HFCs it was concluded in the 2007 WMO Ozone Assessment report[65] that "trifluoroacetic acid from the degradation of HCFCs and HFCs will not result in environmental concentrations capable of significant ecosystem damage". In the concentrations expected from HFC and HCFC degradation, none of the products listed in Table 5.2 are toxic or have adverse ecosystem health impacts.[58]

5.5.3. *Hydrochlorocarbons*

The atmospheric oxidation mechanism of hydrochlorocarbons is similar to that of the HFCs and HCFCs. However, unlike for the HFCs and HCFCs there are substantial natural sources of hydrochlorocarbons. Methyl chloride is the most abundant halogenated organic in the atmosphere (540 ppt) and the flux of this predominately natural compound into the atmosphere (3700 Gg yr^{-1} [15]) is the largest of any halogenated organic (1000 Gg = 1000 ktonne = 1 Mtonne). The current global emissions of CH_2Cl_2, $CHCl_3$, and $CCl_2 = CCl_2$ are estimated to be 800, 300, and 150 Gg yr^{-1}, respectively.[15] As shown in Fig. 5.2, the emissions of CH_3CCl_3 peaked at ∼700 Gg yr^{-1} (700 ktonne yr^{-1}) in 1990 and following controls under the Montreal Protocol have declined to ∼2 Gg yr^{-1} in 2012.[15] The global average atmospheric concentration of CH_3CCl_3 peaked at ~140 ppt in the early 1990s and has been declining steadily since and in 2015 was ~3 ppt (see Fig. 5.5). Atmospheric concentrations of CH_3CCl_3 can be monitored conveniently using gas chromatography with electron capture detection and CH_3CCl_3 is an unusually important molecule in atmospheric chemistry research for the following reasons. CH_3CCl_3 has essentially no natural sources; its anthropogenic emissions were reasonably well characterized and were largely eliminated over the decade 1990–2000. The atmospheric loss mechanism for CH_3CCl_3 is dominated by reaction with OH radicals and its reaction kinetics with OH are well established; $k_{OH} = 1.2 \times 10^{-13} \exp(-1440/T)$ $cm^{-3}molecule^{-1}\ s^{-1}$.[59] The rate of decay of CH_3CCl_3 in the atmosphere provides a convenient and sensitive measurement of the average tropospheric OH concentration.[74] For compounds which are well mixed in the troposphere, scaling the OH reactivity at 272 K to that of CH_3CCl_3 and multiplying by the tropospheric lifetime of CH_3CCl_3 provides an estimate of the tropospheric lifetime with respect to reaction with OH.[75] As indicated in Table 5.1, the atmospheric lifetimes of the major hydrochlorocarbons, CH_3Cl, CH_2Cl_2, $CHCl_3$, and CH_3CCl_3 are relatively short and the GWPs are small.

The oxidation of CH_3Cl and CH_2Cl_2 produces HC(O)Cl in essentially 100% yield, $CHCl_3$ gives $COCl_2$, and CH_3CCl_3 gives CCl_3CHO and $COCl_2$.[1] As with the halocarbonyl intermediate oxidation products from HFCs and HCFCs, the fate of HC(O)Cl and $COCl_2$ is heterogeneous uptake and decomposition or hydrolysis to give HCl and CO_2 on a time scale of ~5–15 days. CCl_3CHO undergoes photolysis on a time scale of ~1 day leading to the formation of $COCl_2$ which in turn will undergo hydrolysis to HCl and CO_2.[49]

5.5.4. *Hydrobromocarbons and Hydroiodocarbons*

The use of methyl bromide as a fumigant in agricultural activities has been controlled under the Montreal Protocol and its emissions following use as a fumigant dropped by a factor of ~5 from 48 Gg

yr^{-1} in 1995–1998 to 10 Gg yr^{-1} in 2012. The total global emissions of CH_3Br are estimated to be 84 Tg yr^{-1} in 2012 with the biological activity in the oceans accounting for 40% of this total. CH_2Br_2, $CHBr_3$, and CH_3I are produced by microalgae and macroalgae (seaweed) in open ocean and coastal marine areas. The emissions vary substantially both spatially and temporally depending on the types and amounts of algae which are present and the local meteorological conditions. There are significant ranges in the published estimates of global emissions of hydrobromocarbons and hydroiodocarbons into the atmosphere; CH_2Br_2, 60–280 Gg Br yr^{-1}; $CHBr_3$, 360–820 Gg Br yr^{-1}; and CH_3I, 270–550 Gg I yr^{-1}.[15] Atmospheric concentrations of CH_3Br have been declining steadily since 2000 as shown in Fig. 5.5 as a result of the decreased anthropogenic emissions associated with its use as a fumigant. CH_2Br_2, $CHBr_3$, and CH_3I are emitted from natural sources and have atmospheric concentrations that vary greatly by location reflecting local emissions and their short atmospheric lifetimes.

In contrast to CFCs, HFCs, HCCs, and HCFCs, photolysis is an important tropospheric loss mechanism for hydrobromocarbons and hydroiodocarbons. The mechanism of photolysis of halogenated alkanes has been reviewed by Calvert *et al.*[1] Absorption of UV occurs via an $n \rightarrow \sigma^*$ transition with a non-bonding electron of the halogen promoted into an anti-bonding sigma orbital of the C–X bond. The ultra-violet absorption spectra of halogenated alkanes (RX) with similar carbon backbones shift to longer wavelength (lower energy) along the series X = F, Cl, Br, to I, reflecting progressively lower electron affinity of the halogen atom.[1] Absorption by C–F and C–Cl bonds only occurs at wavelengths below ~240 nm. CFCs, HFCs, HCFCs, chlorocarbons, and hydrochlorocarbons do not photolyze in the troposphere as sufficiently short wavelength UV radiation does not reach the troposphere. In contrast to C–F and C–Cl bonds, C–I bonds absorb strongly at tropospherically relevant wavelengths and iodine-containing haloalkanes have atmospheric lifetime of days, or less, with respect to photolysis in the troposphere (e.g., 4.9 days for CH_3I, 4.9 hours for CH_2ICl, and 4.9 min for CH_2I_2).[1] Absorption by C–Br bonds at tropospherically relevant wavelengths is generally weak but photolysis is a significant atmospheric loss mechanism for some hydrobromocarbons such as $CHBr_3$ which has a photolysis lifetime of 27 days.[1]

Degradation of hydrobromocarbons and hydroiodocarbons is initiated by photolysis and OH reaction. Reaction with OH radicals is generally the dominant loss mechanism for hydrobromocarbons, while photolysis is dominant for the hydroiodocarbons. Both classes of compounds have short atmospheric lifetimes and release free Br and I atoms during their oxidation. Reactions involving Br and I are important in tropospheric ozone chemistry particularly in marine environments. It has been shown that halogen chemistry associated with emissions of hydrobromocarbons and hydroiodocarbons from open-ocean sources is responsible for significant ozone loss in the tropical marine boundary layer.[76] In model studies the inclusion of halogen chemistry (primarily bromine and iodine) associated with emissions of $CHBr_3$, CH_2Br_2, CH_2BrCl, $CHBr_2Cl$, and $CHBrCl_2$, CH_2I_2, CH_2IBr and CH_2ICl leads to an annually averaged decrease of $\sim$10% in the average tropical tropospheric ozone column.[77]

5.6. Halogenated Alkenes

Halogenated alkenes are a class of industrially important compounds. Vinyl chloride ($CH_2{=}CHCl$) is the monomer of polyvinylchloride (PVC) which is a widely used plastic in construction materials and in consumer plastics. Tetrafluoroethene is the monomer of polytetrafluoroethene (PFTE). PTFE has a wide range of uses in industrial and consumer applications. PTFE is used as an insulating material in electrical systems, in applications requiring chemically inert or low friction surfaces, and to provide non-stick surfaces in cook ware. Recently there has been interest in the use of halogenated propenes as substitutes for long lived HFCs which themselves are substitutes for CFCs. As discussed above, CFCs are unreactive in the lower atmosphere (troposphere) and survive transport to the stratosphere where they undergo photolysis releasing chlorine which participates in catalytic ozone destruction.

HFCs are a class of compounds that have been developed to replace CFCs. Hydrofluorocarbons do not contain chlorine and hence do not contribute to chlorine-based stratospheric ozone destruction.

However, HFCs by virtue of their highly polar C–F bonds absorb strongly in the infrared. HFCs with long atmospheric lifetimes have high GWPs. Concerns over radiative forcing of climate change have led to interest in replacing longer lived HFCs with shorter lived compounds. The $>C=C<$ double bond in alkenes is highly reactive towards OH radicals and as a result alkenes tend to have short atmospheric lifetimes. Compounds with short atmospheric lifetimes have low GWPs. As an example HFC-134a (CF_3CFH_2) which has a GWP of 1300 is being replaced by HFO-1234yf ($CF_3CF=CH_2$) which has a GWP of <1.

Short-chain halogenated alkenes are finding use in several commercial applications as listed in Table 5.3.[46] For example, *trans*-$CF_3CH=CHF$ (HFO-1234ze(E)), $CF_3CF=CF_2$ (FO-1216), *cis*-$CF_3CH=CHCl$ (HCFO-1233zd(Z)), and *trans*-CF_3-$CH=CHCl$ (HCFO-1233zd(E)) are being used for degreasing of mechanical parts, dry cleaning, and foam blowing.[46] 2,3,3,3-tetrafluoropropene, $CF_3CF=CH_2$ (HFO-1234yf) is being used as a replacement for 1,1,1,2-tetrafluoroethane, CF_3CH_2F (HFC-134a), in vehicle air conditioning units.[46,78] Haloalkenes are hydrophobic and volatile and following release into the environment will partition into the atmosphere.

The reaction of OH radicals with $>C=C<$ double bonds is very rapid and as a result dominates the atmospheric removal mechanism for haloalkenes. The inclusion of one, or more, double bonds is a particularly effective method to increase the reactivity of organic molecules towards OH radicals. Table 5.4 lists rate coefficients at 298 K in one atmosphere of air for selected halogenated alkenes and corresponding halogenated alkanes. Fig. 5.17 shows a plot of the haloalkene rate coefficients versus those for the corresponding haloalkanes. As seen from Fig. 5.17, in general there is a 1–2 order of magnitude difference between the reactivity of the alkene compared to the alkane. The exception to this general trend is $CCl_2=CCl_2$ which interestingly has the same rate coefficient for reaction with OH as does $CHCl_2CHCl_2$, but the mechanism is different. The mechanism of reaction of OH radicals with alkenes is dominated by

Table 5.3 Commercial applications of haloalkenes.[46]

Haloalkene	Applications
$CF_2{=}CF_2$	Monomer for PTFE used as insulator in electrical devices, in applications where inertness is required, non-stick coating
$CH_2{=}CHCl$	Monomer for polyvinyl chloride (PVC) used as a construction material, used in consumer plastics
$CHCl{=}CCl_2$	Industrial solvent, degreasing agent
$CCl_2{=}CCl_2$	Industrial solvent, dry cleaning
CF_3-$CF{=}CH_2$ (HFO-1234yf)	Refrigerant in vehicle AC systems
trans-CF_3-$CH{=}CHF$ (HFO-1234ze(E))	Aerosol propellant, in expanded polystyrene (styrofoam) insulation industry, refrigerant
CF_3-$CF{=}CF_2$ (HFO-1216)	Co-monomer for FEP polymer, raw material for certain routes to make 1234yf
Z-CF_3-$CH{=}CHCl$ (HCFO-1233zd(Z))	Degreasing of mechanical parts and dry cleaning
E-CF_3-$CH{=}CHCl$ (HCFO-1233zd(E))	Polyurethane foam blowing agent gas when the insulation performance is critical.

electrophilic addition to the $>C{=}C<$ double bond while the reaction of OH radicals with alkanes proceeds via H-abstraction from one of the C–H bonds. Hydrogen atom abstraction from alkanes is generally a rather slow process, whereas addition of OH to double bonds generally proceeds rapidly and this is reflected in the fact that all the data in Fig. 5.17 lie on, or above, the dotted line which marks equal reactivities.

Table 5.5 lists the available rate coefficients at 298 K in one atmosphere of air for haloalkenes. Figures 5.18–5.21 show plots of the rate coefficients for reaction with OH radicals with chlorinated ethenes, halogenated ethenes, fluorinated propenes, and fluorinated butenes as a function of the number of halogen atoms in the molecules. In general there is a decrease in reactivity with increasing number of

Table 5.4 Rate coefficients (units of cm^3 $molecule^{-1}$ s^{-1}) for reactions of OH radicals with selected halogenated alkenes and the corresponding halogenated alkanes[1] in one atmosphere of air at 298 K.

Alkene	k_{OH} (298 K)	Alkane	k_{OH} (298 K)
$CH_2=CH_2$	8.5×10^{-12}	C_2H_6	2.57×10^{-13}
$CH_2=CHCH_3$	2.63×10^{-11}	C_3H_6	1.10×10^{-12}
$CH_2=CHCH_2CH_3$	3.14×10^{-11}	n-C_4H_9	2.45×10^{-12}
$CH_2=CHF$	5.0×10^{-12}	CH_3CH_2F	2.18×10^{-13}
$CF_2=CH_2$	2.8×10^{-12}	CF_2HCH_3	3.45×10^{-14}
$CF_2=CF_2$	1.00×10^{-11}	CF_2HCF_2H	5.96×10^{-15}
$CF_3CH=CH_2$	1.45×10^{-12}	$CF_3CH_2CH_3$	5.48×10^{-14}
E-$CF_3CF=CHF$	2.3×10^{-12}	CF_3CHFCH_2F	1.81×10^{-14}
$CF_3CF=CF_2$	2.2×10^{-12}	$CF_3CHFCHF_2$	5.14×10^{-15}
Z-$CF_3CH=CHCF_3$	4.90×10^{-13}	$CF_3CH_2CH_2CF_3$	8.4×10^{-15}
$CH_2=CHCl$	6.9×10^{-12}	CH_3CH_2Cl	3.28×10^{-13}
cis-$CHCl=CHCl$	2.28×10^{-12}	CH_2ClCH_2Cl	2.42×10^{-13}
trans-$CHCl=CHCl$	1.80×10^{-12}	$CHCl_2CHCl_2$	2.74×10^{-13}
$CCl_2=CCl_2$	1.70×10^{-13}	$CHCl_2CHCl_2$	1.7×10^{-13}
$CH_2=CHBr$	6.8×10^{-12}	CH_3CH_2Br	3.3×10^{-13}

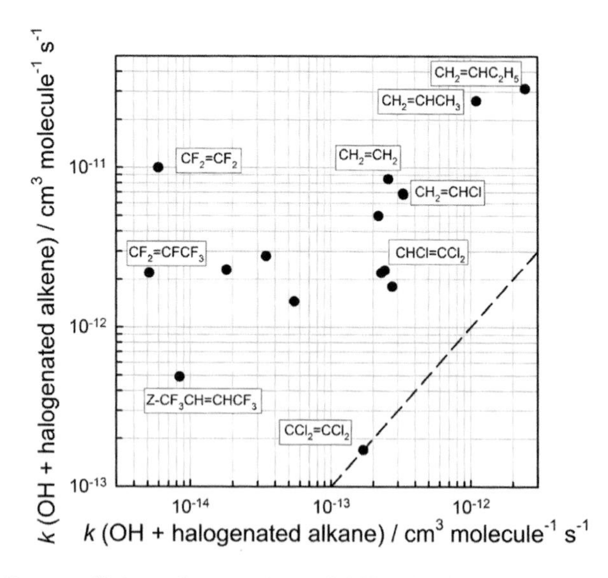

Fig. 5.17 Rate coefficients for reactions of OH radicals with halogenated alkenes (e.g., $CCl_2=CCl_2$) at 298 K in one atmosphere of air vs. rate coefficients for reaction of the corresponding halogenated alkanes (e.g., $CHCl_2CHCl_2$).

Table 5.5 Rate coefficients (units of cm^3 $molecule^{-1}$ s^{-1}) for reaction of OH radicals with alkenes and haloalkenes in one atmosphere of air at 298 K.[14,60,79,91]

Alkene/Haloalkene	k_{OH} (298 K)	Alkene/Haloalkene	k_{OH} (298 K)
C_2H_4	8.52×10^{-12}	$CH_2{=}CHCH_2Cl$	1.69×10^{-11}
$CH_2{=}CHF$	5.0×10^{-12}	E-$CHCl{=}CHCH_2Cl$	1.44×10^{-11}
$CH_2{=}CF_2$	2.8×10^{-12}	Z-$CHCl{=}CHCH_2Cl$	8.45×10^{-12}
$CF_2{=}CF_2$	1.0×10^{-11}		
$CH_2{=}CHCl$	6.9×10^{-12}	$CH_2{=}CHCH_2Br$	1.75×10^{-11}
$CH_2{=}CCl_2$	1.1×10^{-11}		
cis-$CHCl{=}CHCl$	2.28×10^{-12}	E-$CHCl{=}CHCF_3$	4.4×10^{-13}
trans-$CHCl{=}CHCl$	1.80×10^{-12}	$CH_2{=}CBrCF_3$	3.9×10^{-12}
$CHCl{=}CCl_2$	2.2×10^{-12}		
$CCl_2{=}CCl_2$	1.7×10^{-13}	$CH_2{=}CHC_2H_5$	3.14×10^{-11}
		$CH_2{=}CHCF_2CF_3$	1.4×10^{-12}
$CH_2{=}CHBr$	6.8×10^{-12}	$CF_2{=}CF\text{-}CF{=}CF_2$	9.64×10^{-12}
$CFCl{=}CF_2$	7.57×10^{-12}	(Z)-$CF_3CH{=}CHCF_3$	4.9×10^{-13}
$CCl_2{=}CF_2$	7.44×10^{-12}	$CH_2{=}C(CF_3)_2$	6.58×10^{-13}
$CHBr{=}CF_2$	4.5×10^{-12}	*cis*-$CF_3CF{=}CFCF_3$	3.8×10^{-13}
$CFBr{=}CF_2$	7.6×10^{-12}	*trans*-$CF_3CF{=}CFCF_3$	5.8×10^{-13}
		cyclo-$CF{=}CFCF_2CF_2$–	3.79×10^{-14}
C_3H_6	2.63×10^{-11}	cyclo-$CH{=}CFCF_2CF_2$–	6.15×10^{-14}
$CH_2{=}CHCH_2F$	1.6×10^{-11}	cyclo-$CH{=}CHCF_2CF_2$–	1.73×10^{-13}
$CH_2{=}CFCH_3$	1.5×10^{-11}		
$CH_2{=}CHCF_3$	1.45×10^{-12}	(Z/E)-$CF_3CF{=}CClCF_3$	3.3×10^{-13}
$CH_2{=}CFCF_3$	1.1×10^{-12}		
E-$CHF{=}CHCF_3$	7.0×10^{-13}	$CH_2{=}CBrCF_2CF_3$	3.4×10^{-12}
E-$CHF{=}CFCF_3$	2.3×10^{-12}	$CH_2{=}CHCF_2CF_2Br$	1.7×10^{-12}
Z-$CHF{=}CFCF_3$	1.3×10^{-12}		
$CF_2{=}CFCF_3$	2.2×10^{-12}	$(CF_3)_2C{=}CFC_2F_5$	7.2×10^{-14}

halogen substituents in the molecule. This general trend can be rationalized on the basis of the halogen substituents withdrawing electron density from the $>C{=}C<$ double bond and hence decreasing the rate of electrophilic addition. However, there are some quite striking exceptions to this simple picture of reactivity trends.

Figure 5.18 shows that there is little, or no, substantial trend in OH reactivity on adding fluorine substituents to ethene with the rate coefficients for reaction of OH with $CF_2{=}CF_2$, $CFBr{=}CF_2$, $CFCl{=}CF_2$, and $CCl_2{=}CF_2$ being indistinguishable from that of

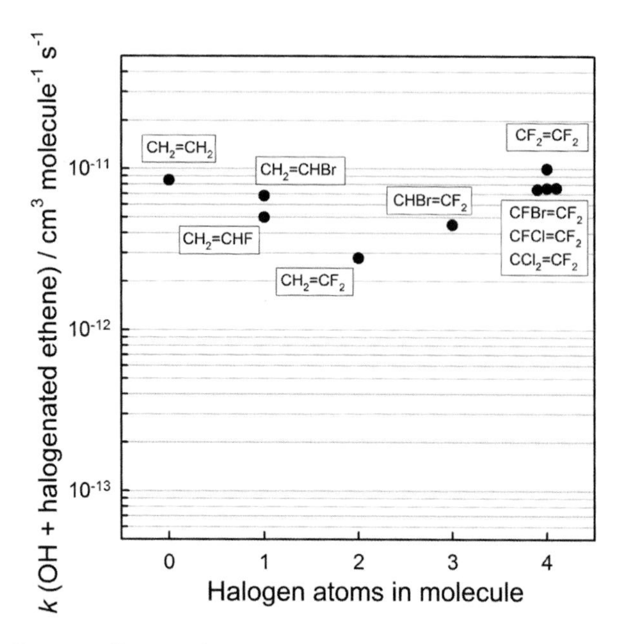

Fig. 5.18 Rate coefficients for the reaction of OH radicals with halogenated ethenes at 298 K in one atmosphere of air vs. the number of halogen atoms in the molecule.

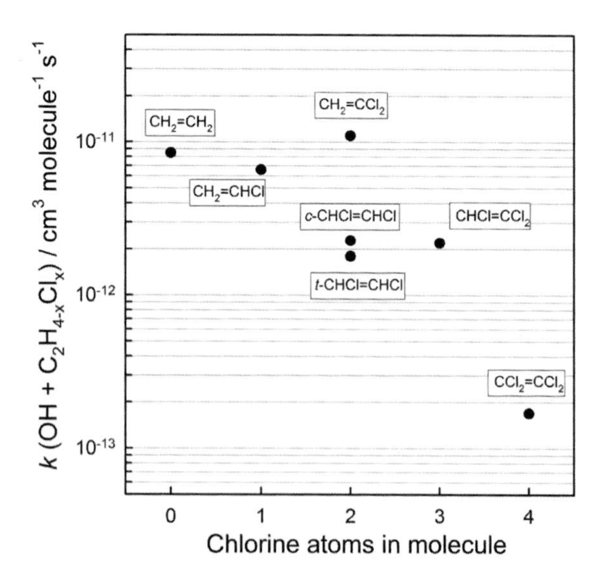

Fig. 5.19 Rate coefficients for the reaction of OH radicals with chlorinated ethenes at 298 K in one atmosphere of air vs. the number of chlorine atoms in the molecule.

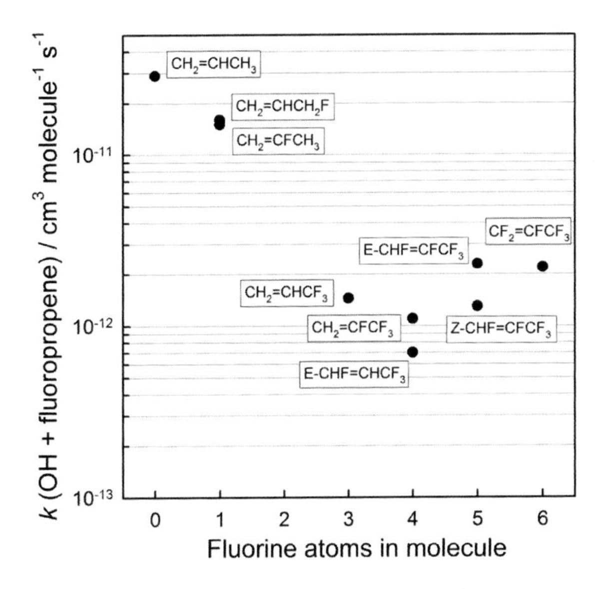

Fig. 5.20 Rate coefficients for the reaction of OH radicals with fluorinated propenes at 298 K in one atmosphere of air vs. the number of fluorine atoms in the molecule.

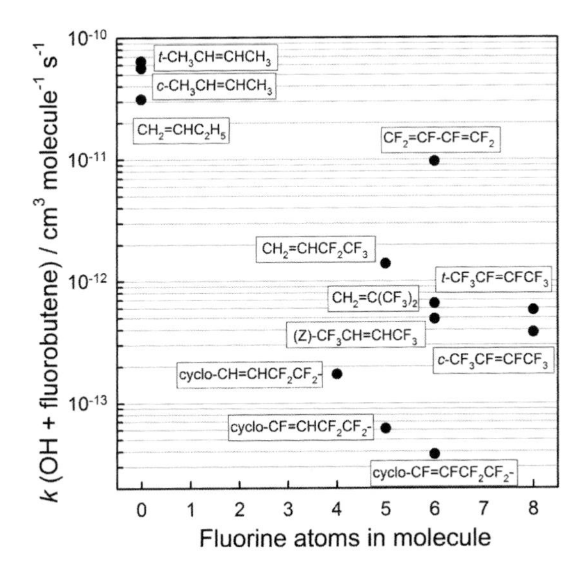

Fig. 5.21 Rate coefficients for the reaction of OH radicals with fluorinated butenes at 298 K in one atmosphere of air vs. the number of fluorine atoms in the molecule.

$CH_2{=}CH_2$. Clearly there are important factors at play in addition to the inductive effect of the halogen substituents.

At pressures less than ~5 Torr of helium diluent the kinetics of the reaction of OH with $CH_2{=}CF_2$, $CH_2{=}CHCl$, $CHCl{=}CCl_2$, $CFCl{=}CF_2$, and $CCl_2{=}CCl_2$ display a pressure dependence with the rate of reaction increasing with pressure. This pressure dependence is consistent with a mechanism where the OH radical adds to the double bond to form an excited adduct which can either decompose or be stabilized by collisions with the diluent gas (M). At pressures greater than 5 Torr of helium the rate of collisional stabilization of the excited adduct is sufficiently rapid that decomposition to reactants becomes unimportant and there is no further discernable effect of pressure. For haloalkenes with Cl, Br, or I substituents on the carbon atom at the site of the OH radical attack the elimination of a Cl, Br, or I atom (from the excited adduct, stabilized adduct, or both) is thermodynamically feasible leading to the formation of an enol. Indirect evidence for enol formation has been presented in studies of haloalkene oxidation[92] but the enol has not been directly observed and this aspect of haloolfin chemistry remains poorly understood. The atmospheric fate of stabilized carbon centered radicals is generally addition of molecular oxygen to give a peroxy radical. The rate constant for O_2 addition is typically of the order of 10^{-12}–10^{-11} cm^3 $molecule^{-1}$ s^{-1} and in the presence of an atmosphere of air the lifetime of carbon centered radicals with respect to reaction with O_2 is of the order of $(0.2–2.0) \times 10^{-9}$ s. The resulting peroxy radical undergoes further reactions with NO, NO_2, HO_2 or other peroxy radicals (RO_2) leading to a variety of products. Taking $CCl_2{=}CCl_2$ as an example, the mechanism of OH radical initiated oxidation is represented in Fig. 5.22.

As seen in Fig. 5.17–5.21, haloalkenes react with OH radicals with rate coefficients in the range of 10^{-13}–10^{-11} cm^3 mol^{-1} s^{-1}. Using a diurnal average $[OH] = 1.0 \times 10^6$ cm^{-3} [49] in the troposphere gives a range of atmospheric lifetimes for haloalkenes with respect to reaction with OH of ~1–100 days. In addition to reaction with OH radicals haloalkenes also react with NO_3 radicals, Cl atoms,

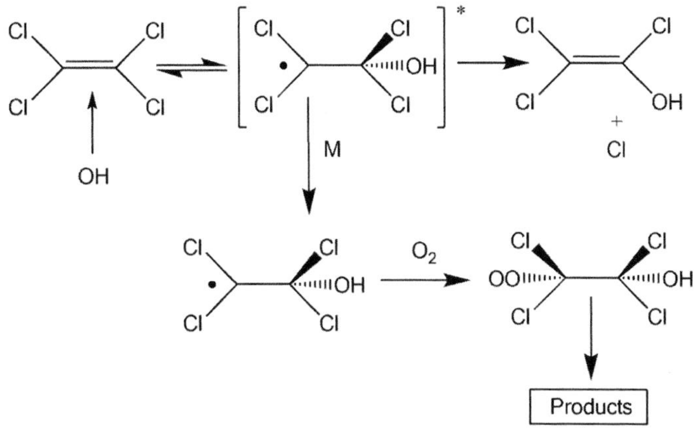

Fig. 5.22 Mechanism of OH radical initiated oxidation of $CCl_2{=}CCl_2$.

and O_3, however the rates of these reactions are too slow to play a significant role in atmospheric chemistry. For example $CF_3CF{=}CH_2$ (HFO-1234yf) reacts with O_3 with a rate coefficient of 2.8×10^{-21} cm^3 $molecule^{-1}$ s^{-1} [93] which combined with an average tropospheric ozone concentration of 35 ppb gives a lifetime of 13 years with respect to reaction with ozone. This can be compared to the ~11 day lifetime of $CF_3CF{=}CH_2$ with respect to reaction with OH.

Uniform mixing of a gas in the atmosphere requires a time scale of years. The lifetimes of haloalkenes are of the order of weeks and hence following their release they will not be well mixed in the atmosphere. Thus, for short lived compounds such as the haloalkenes it is not possible to assign a unique atmospheric lifetime. The lifetime will depend on the location of emissions and the chemical and physical conditions of the atmosphere. The atmospheric lifetimes of haloalkenes emitted in the winter at high latitudes where there is little photochemical activity will be substantially longer than when emitted in the summer in the tropics. The atmospheric lifetimes of selected haloalkenes as estimated using $[OH] = 1.0 \times 10^6$ cm^{-3} are given in Table 5.1. As seen from Table 5.1, haloalkenes have longer atmospheric lifetimes than the corresponding alkenes but shorter lifetimes than the corresponding haloalkanes.

5.6.1. *Halogenated Alkene GWPs*

Halogenated alkenes contain carbon–halogen bonds which absorb in the atmospheric window region and have the potential to contribute to radiative forcing of climate change. The radiative efficiency (units of $W\,m^{-2}\,ppb^{-1}$) of a molecule reflects its heat trapping ability in the atmosphere and can be calculated using a radiative transfer model. As seen from Table 5.1, the estimated REs for the haloalkenes are considerably smaller than the CFCs and HFCs they replace. The radiative efficiencies of haloalkenes are low because of their non-uniform horizontal and vertical mixing in the atmosphere.[44] Accounting for non-uniform horizontal mixing is important because radiative forcing is most efficient at lower latitudes where temperatures are higher. Accounting for non-uniform vertical mixing is important because radiative forcing is defined as the change in radiative flux at the tropopause which is located at ∼12–15 km altitude. Mixing of haloalkenes to lower latitudes and higher altitudes is inefficient and hence as seen in Table 5.1 the radiative efficiencies of haloalkenes are small.

The GWP is the integrated radiative forcing over a given time horizon following the pulsed emission of a kg of gas compared to the integrated radiative forcing for release of a kg of CO_2. As discussed above the radiative efficiencies of haloalkenes are small and the atmospheric lifetimes of haloalkenes are short. These two factors combine to give haloalkenes very small GWPs as seen in Table 5.1. The atmospheric degradation products of haloalkenes could in principle contribute to radiative forcing of climate change. However, it is well established that the oxidation products of haloalkenes are removed from the atmosphere by wet and dry deposition on a time scale of days to weeks.[58] Neither haloalkenes, nor their oxidation products, contribute significantly to radiative forcing of climate change.

5.6.2. *Halogenated Alkene POCPs*

The POCP is an index measuring the ability of a compound to contribute to ozone formation on an urban or regional scale. POCP is defined as the additional ozone formed in a multi-day model simulation when adding a given amount of volatile organic compound relative to adding the *same mass* of ethene (Derwent *et al.*).[94] The POCP scale is relative with the POCP for ethene defined as 100. POCP values are determined over a simulation period of ~5 days for an air parcel transported along an idealized straight line trajectory using a photochemical trajectory model.

POCPs for a range of HFOs have been evaluated by Wallington *et al.*[46] using the estimation method outlined by Jenkin.[95] POCPs for the commercially relevant haloalkenes are listed in Table 5.1. Haloalkenes have POCPs which are larger than those for analogous HFCs but much smaller than those for the parent alkenes. Many of the haloalkenes have POCPs which lie between those for methane (0.6) and ethane (8.8). Methane and ethane are oxidized sufficiently slowly that they do not contribute to any appreciable degree to local air quality issues and are generally exempt from air quality regulations (Dimitriades).[96] Carter[97] conducted Maximum Incremental Reactivity (MIR) calculations for $CF_3CF{=}CHF$ and found that ozone production from $CF_3CF{=}CHF$ is indistinguishable from ethane (C_2H_6). Luecken *et al.*[98] conducted an atmospheric modeling study and reported that replacing HFC-134a in vehicle air conditioning units with HFO-1234yf across the U.S. had essentially no impact ($<0.01\%$) on tropospheric ozone formation. The commercially relevant haloalkenes do not make a significant contribution to tropospheric ozone formation.

5.6.3. *Halogenated Alkene ODPs*

In contrast to chlorine and bromine, fluorine atoms rapidly abstract hydrogen from CH_4 and H_2O in the stratosphere to form as effective sink species (HF) and fluorine atoms do not participate in catalytic ozone destruction cycles.[54,55] Hence, hydrofluoroalkenes do not deplete stratospheric ozone and have an ozone depletion potential of zero. Chlorine and bromine containing haloalkenes will contribute to stratospheric ozone depletion although the magnitudes of their contributions are small because their short atmospheric lifetimes limit the fraction of the emissions which survive transport through the troposphere to the stratosphere.

Patten and Wuebbles[47] conducted a modeling study and derived an ODP for E-$CF_3CH{=}CHCl$ of 0.00034. The Z-isomer of $CF_3CH{=}CHCl$ has an atmospheric lifetime which is approximately half that of the E-isomer (Andersen *et al.* [99]) and hence the ODP for $Z{-}CF_3CH{=}CHCl$ will be even lower than that for E-$CF_3CH{=}CHCl$. Patten and Wuebbles[47] concluded that E-$CF_3CH{=}CHCl$ is unlikely to affect stratospheric ozone. The ODPs for tri- and tetrachloroethene have been calculated by Wuebbles *et al.*[41] to be 0.00037 and 0.0050, respectively. Haloalkenes have ODPs which are zero, or near zero, and will not impact stratospheric ozone.

5.6.4. *Formation of Noxious or Toxic Degradation Products*

Oxidation of haloalkenes is initiated by addition of OH to the double bond giving a β-hydroxy alkyl radical which in one atmosphere of air will rapidly (within $\sim 10^{-9}$ s) add O_2 to give a β-hydroxy alkylperoxy radical. The peroxy radicals are converted by reaction with NO into alkoxy radicals. The dominant fate of alkoxy radicals formed during the OH initiated oxidation of haloalkenes is unimolecular decomposition via C–C bond scission. This process as illustrated in Fig. 5.23 leads to the formation of halogenated carbonyl compounds. A list of such products in the atmospheric oxidation of the commercially important haloalkenes is given in Table 5.6. Halogenated carbonyl compounds are removed from the atmosphere via wet and dry deposition on a time scale of days to weeks. Hydrolysis of halogenated carbonyl compounds gives acid and CO_2 products. Trifluoroacetic acid (CF_3COOH) is formed either though hydrolysis of carbonyl oxidation products or via secondary photochemistry. For example, the atmospheric fate of $CF_3C(O)F$ (TFA) formed in the oxidation of $CF_3CF{=}CH_2$, is hydrolysis which occurs on a time scale of $\sim$10 days to give CF_3COOH.[58] CF_3CHO, which is formed in the oxidation of $CF_3CH{=}CH_2$, undergoes photolysis (lifetime of

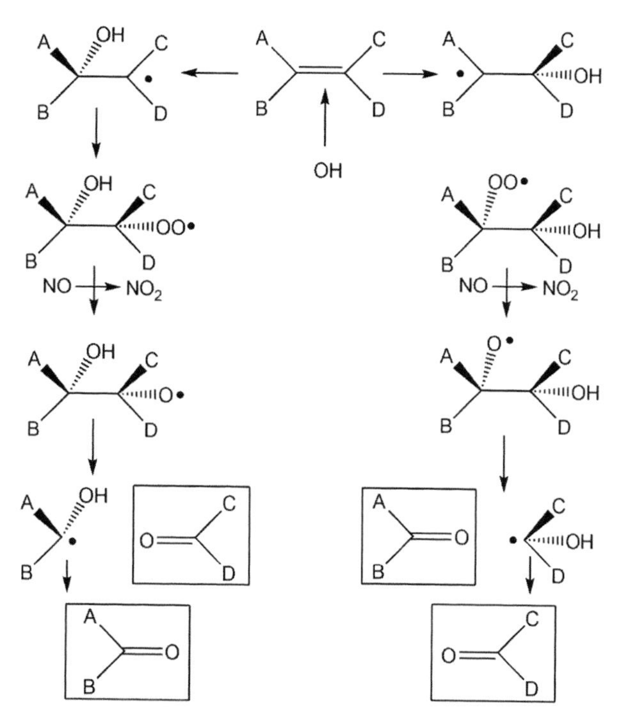

Fig. 5.23 Generic oxidation mechanism for the OH radical initiated oxidation of a haloalkene illustrating the scission of the >C=C< double bond. A, B, C, and D are H, F, CF_3, or alkyl groups.

≤2 days) giving CF_3 and HCO radicals while reaction with OH, which is of lesser importance, but also represents a sink for CF_3CHO, gives CF_3CO radicals.[100] Atmospheric degradation routes by which CF_3CO radicals can be transformed into CF_3COOH (TFA) as a minor product have been documented.[101]

In the case of HCFOs, chlorine substituted oxidation products are formed including HCOCl. The atmospheric fate of the oxidation product HCOCl is incorporation into rain, cloud and fog water followed by hydrolysis and removal by wet deposition, within 5–15 days. Hydrolysis of HCOCl gives formic acid which is a ubiquitous component of the environment and is of no concern. For the HCFOs there is also evidence of a primary oxidation channel leading to chlorine atom elimination and formation of an enol,

Table 5.6 Atmospheric oxidation products of commercial haloalkenes.

Halogenated alkene	Intermediate products	Final products
$CF_2{=}CF_2$	COF_2	CO_2, HF
$CH_2{=}CHCl$	HCHO, HC(O)Cl, $CH_2{=}CHOH$	CO_2, HC(O)OH, HCl
$CHCl{=}CCl_2$	HC(O)Cl, $COCl_2$, CHCl=CHOH	CO_2, HC(O)OH, HCl
CCl_2 =CCl_2	$COCl_2$, CCl_2 =CClOH	CO_2, HCl
CF_3-CF=CH_2 (HFO-1234yf)	$CF_3C(O)F$, HCHO	$CF_3C(O)OH$, CO_2, HF
trans-CF_3-CH=CHF (HFO-1234ze(E))	$CF_3C(O)H$, HC(O)F	CO_2, HC(O)OH, HF
CF_3-CF=CF_2 (HFO-1216)	$CF_3C(O)F$, COF_2	$CF_3C(O)OH$, CO_2, HF
E-CF_3-CH=CHCl (1233zd(E))	$CF_3C(O)H$, HC(O)Cl, HCl, $CF_3CH{=}CHOH$	CO_2, HF, HCl
Z-CF_3-CH=CHCl (1233zd(Z))	$CF_3C(O)H$, HC(O)Cl, HCl, $CF_3CH{=}CHOH$	CO_2, HF, HCl

$CF_3CH{=}CHOH$ (Sulbaek Andersen *et al.*[102]). The atmospheric fate of the enol is likely reaction with OH (directly, or indirectly via reaction with the tautomer form) to yield CF_3CHO, HCHO and HCOOH. The intermediate and final atmospheric degradation products of the commercially significant halogenated alkenes are listed in Table 5.6.

The oxidation products of the halogenated alkenes in Table 5.6 are similar to those from degradation of the commercially important HFCs and HFCs listed in Table 5.2. As discussed above there has been some discussion regarding the environmental impacts of trifluoroacetic acid formed in the degradation of HFCs and HCFCs. Modeling studies by Luecken *et al.*[98] and Russell *et al.*[103] concluded that levels of trifluoroacetic acid resulting from the projected use of HFO-1234yf over the next 50 years would not represent a risk to ecosystems in the U.S. The 2007 WMO Ozone Assessment Report concluded that "trifluoroacetic acid from the degradation of HCFCs and HFCs will not result in environmental concentrations capable of significant ecosystem damage". The same conclusion is applicable for short-chain halogenated alkenes. It should be noted that there is concern over the environmental impacts of the bioaccumulative long-chain

(<C6) perfluorocarboxylic acids which could be formed following the atmospheric oxidation of long-chain halogenated alkenes.[104] This is discussed in the halogenated carboxylic acid section below.

5.7. Halogenated Oxygenates

Partially oxidized organic compounds are often referred to as "oxygenates", hence halogenated oxygenates include those compounds which contain oxygen atoms as well as a carbon skeleton with halogen and possibly hydrogen atoms, occasionally combined with other optional hetero atoms (sulfur-, nitrogen-, phosphorus substituents). Halogenated oxygenates have a variety of industrial uses, including as solvents, cleaning agents, and in refrigeration and surfactant applications as listed in Table 5.7.

The common oxygenates discussed in this section include alcohols (ROH), ethers (ROR), esters (RC(O)OR), aldehydes and ketones (RC(O)H, RC(O)R), carboxylic acids (RC(O)OH,) and hetero-atom oxygenates. Halogenated oxygenates enter the atmosphere as fugitive emissions during industrial and commercial use and are formed in the atmosphere as oxidation products of halogenated organic compound emissions. The atmospheric oxidation of HCFCs, HFCs, HFOs and HFEs (hydrofluoroethers) gives halogenated alcohols, aldehydes, ketones, and acids. Halogenated carboxylic acids are formed in

Table 5.7 Commercial applications of halogenated oxygenates.

Halogenated oxygenates	Applications
Alcohols	Industrial solvents, synthesis reagents. Key ingredients for synthesis of surface coatings.
Ethers	Refrigerants and blowing agents, inhalation anesthetics, flame retardants (PBDEs), combustion by-products (PCDDs, PCDFs)
Aldehydes and ketones	Industrial synthesis reagents, fire extinguishing agents.
Esters	No commercial/industrial use.
Carboxylic acids	Processing aids in the polymerization of fluoropolymers, fire-fighting foams
Hetero-atom containing	Surfactants

the atmospheric oxidation of halogenated carbonyl compounds and in the hydrolysis of acyl halides. With the possible exception of trifluoroacetic acid in marine aerosols there are no natural emission sources of halogenated oxygenates into the atmosphere. As for the aliphatic oxygenates, the halogenated oxygenates can participate in atmospheric reaction cycles that lead to the formation of tropospheric ozone and other air pollutants. Reaction with OH radicals and direct photodecomposition of halogenated aldehydes lead to the formation of HO_2 radicals, which are a key transient species in the atmospheric conversion of NO to NO_2 and hence in the formation of tropospheric ozone:

$$HO_2 + NO \rightarrow HO + NO_2,$$
$$NO_2 + h\nu \rightarrow O(^3P) + NO,$$
$$O(^3P) + O_2 \rightarrow O_3.$$

Halogenated peroxyradicals and acyl peroxyradicals are also formed in the oxidation of fluorinated oxygenates, and participate in reactions that lead to O_3:

$$RO_2 + NO \rightarrow RO + NO_2,$$
$$RC(O)O_2 + NO \rightarrow RC(O)O + NO_2.$$

The presence of oxygen typically increases the atmospheric reactivity of organic molecules[48] and the lifetimes of oxygenated halogenated organic compounds are typically shorter than those of the corresponding non-oxygenated halogenated compounds. Research interest in halogenated oxygenates has been driven by concern over the environmental effects of long chain perfluorocarboxylic acids (PFCAs) such as perfluorooctanoic acid (PFOA; $C_7F_{15}COOH$). PFOA is detectable in the blood of humans and animals worldwide,[105–107] is only slowly eliminated in mammals,[108] is potentially toxic,[109,110] and has no known metabolic or environmental degradation pathway. PFCAs with carbon chain-lengths of between 9 and 14 are more bioaccumulative than PFOA and are present in wildlife at higher concentrations than PFOA. PFCA with carbon chains of six or less are more water soluble and hence less bioaccumulative. The chemical industry is in the process of reformulating products to use shorter chain molecules to avoid the bioaccumulative properties of the longer chain compounds.[111] In the following sections, we discuss the atmospheric chemistry and environmental fate of each of the six major groups of halogenated oxygenates; alcohols, ethers, aldehydes, ketones, esters, and acids.

5.7.1. *Halogenated Alcohols*

In the atmosphere α-halogen substituted alcohols (e.g., CF_3OH, $CH_3CHClOH$, $CF_3(CF)_4OH$)) undergo decomposition via elimination of HF/HCl to give corresponding carbonyl compounds.[58,112,113] Homogeneous gas-phase decomposition of α-fluoro and α-chloro alcohols such as CF_3OH,[114,115] CH_2ClOH,[116] $CHCl_2OH$,[116] CCl_3OH,[116] and $CH_3CHClOH$[117] is slow (lifetime on the order of 100 years) but heterogeneous decomposition on atmospheric cloud and aerosol surfaces results in an atmospheric lifetime of ~5–15 days. Theoretical work has shown the dominance of heterogeneous decomposition as the atmospheric fate of CF_3OH.[118] Short chain poly-halogenated alcohols (containing <5 carbon atoms) are used in industrial processes as solvents and reagents. Long chain fluorinated alcohols (containing >4 carbon atoms), including fluorotelomer alcohols, $C_xF_{2x+1}CH_2CH_2OH$ ($x = 6, 8, 10, 12$),[119] have attracted a significant amount of scientific interest over the past decade, driven by interest in the formation of perfluorocarboxylic acids (PFCAs) during their atmospheric degradation. The atmospheric lifetime of poly-halogenated alcohols is determined by reaction with OH radicals. A substantial literature database exists for the OH rate constants for halogenated alcohols (see Table 5.8). Figure 5.24 shows the short chain halogenated alcohol OH rate coefficients versus those for the analogous short chain un-substituted aliphatic alcohols. It is clear from Fig. 5.24 that halogen substitution leads to a decrease in the reactivity of alcohols by as much as 3 orders of magnitude depending on the position and the extent of substitution. This is consistent with the general trend observed for halogenation, that the C–H bond strength increases with the degree of halogenation with the more electronegative F atoms having a larger effect than Cl atoms, followed by Br and I.

Table 5.8 Rate coefficients (units of cm^3 $molecule^{-1}$ s^{-1}) for reactions of OH radicals with selected halogenated alcohols and carbonyls in one atmosphere of air at 298 K.[49]

Halogenated alcohol	k_{OH} (298 K)	Halogenated carbonyl	k_{OH} (298 K)
CH_2FCH_2OH	1.49×10^{-12}		
CHF_2CH_2OH	4.87×10^{-13}	CHF_2CHO	1.5×10^{-12}
CF_3CH_2OH	1.05×10^{-13}	CF_3CHO	5.77×10^{-13}
$C_xF_{2x+1}CH_2OH$	1.02×10^{-13}	$C_xF_{2x+1}CHO$	5.77×10^{-13}
$C_xF_{2x+1}CH_2CH2OH$	9.35×10^{-13}	$C_xF_{2x+1}CH_2CHO$	2.8×10^{-12}
CH_2ClCH_2OH	1.28×10^{-12}	CH_2ClCHO	3.15×10^{-12}
		$CHCl_2CHO$	2.52×10^{-12}
CCl_3CH_2OH	2.45×10^{-13}	CCl_3CHO	8.41×10^{-13}
		$CHClFCHO$	2.05×10^{-12}
		CF_2ClCHO	1.15×10^{-12}
		$CFCl_2CHO$	8.2×10^{-13}
$CF_3CH(OH)CF_3$	2.51×10^{-14}		
CF_3CHFCF_2CHOH	1.26×10^{-13}		
		$CH_2FC(O)CF_3$	2.15×10^{-13}
		$CF_3C(O)CH_3$	1.11×10^{-14}
		$CH_2FC(O)CH_3$	2.15×10^{-13}
		$CH_2ClCOCH_3$	4.38×10^{-13}
		$CHCl_2COCH_3$	4.02×10^{-13}
		CCl_3COCH_3	1.54×10^{-14}

In general, atmospheric degradation of halogenated alcohols is initiated by OH mediated hydrogen abstraction giving α-hydroxyalkyl radicals which react with O_2 to give the corresponding aldehyde. As an example, for CF_3CH_2OH:

$$CF_3CH_2OH + OH \rightarrow CF_3CHOH + H_2O,$$
$$CF_3CHOH + O_2 \rightarrow CF_3CHO + HO_2.$$

The alcohol group activates α C–H bonds towards OH radical attack and for small alcohols, e.g., CH_2ClCH_2OH, the majority of the reaction of OH radicals typically proceeds via attack at the α-hydroxy carbon.

$$CH_2ClCH_2OH + OH \rightarrow CH_2ClCHOH + H_2O,$$
$$CH_2ClCHOH + O_2 \rightarrow CH_2ClCHO + HO_2.$$

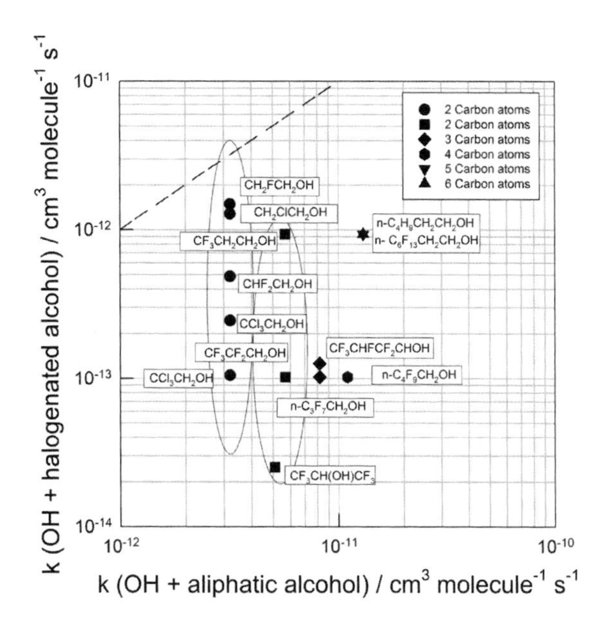

Fig. 5.24 Rate coefficients for reactions of OH radicals with partially halogenated alcohols (e.g., CH_2FCH_2OH) at 298 K in one atmosphere of air vs. rate coefficients for reaction of the corresponding aliphatic alcohols (e.g., CH_3CH_2OH). The dashed line indicates a one-to-one relationship. The ellipses group the 2- and 3-carbon molecules for ease of visual inspection.

Halogenated aldehydes are the primary oxidation products of the halogenated alcohols. The atmospheric fate of halogenated aldehyde is discussed in Sec. 5.6.4.

5.7.2. *Halogenated Ethers*

Halogenated ethers are used as refrigerants, inhalation anesthetics, and flame retardants (e.g., polybrominated diphenyl ethers, PBDEs). As with all ethers, there are no routes for *in situ* atmospheric formation of halogenated ethers. There are no known natural sources of halogenated ethers and these compounds are only present in the atmosphere as a result of anthropogenic emissions. In general the atmospheric oxidation of the halogenated ethers follows the same path as for halogenated alkanes in that reaction with OH radicals is the dominant sink, and the reaction occurs via H-abstraction

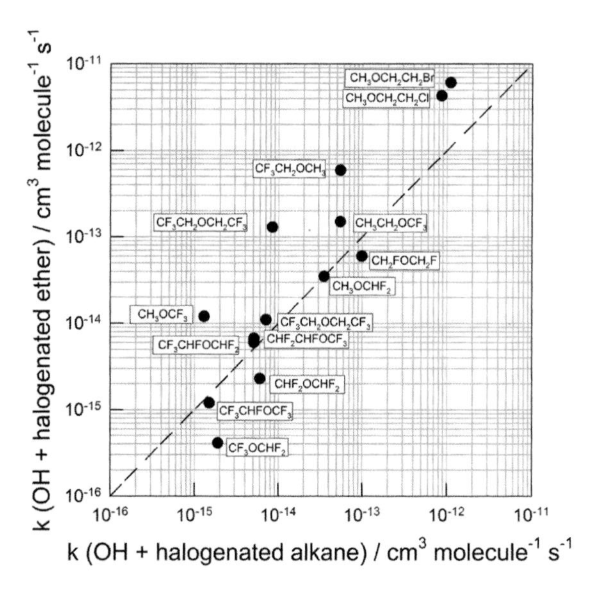

Fig. 5.25 Rate coefficients for reactions of OH radicals with halogenated ethers (e.g., CF_3OCHF_2) at 298 K in one atmosphere of air vs. those for the corresponding halogenated alkanes (e.g., CF_3CHF_2). The data are given in Table 5.9. The dashed line indicates the one-to-one relationship.

leading to the formation of peroxy and subsequently alkoxy radicals.

Figure 5.25 shows the halogenated ether OH rate coefficients versus those for the analogous halogenated alkanes. In aliphatic alkanes, the insertion of an ether oxygen imparts a weakening of the neighboring C–H bonds ("activation effect") which increases the reactivity towards H-abstraction. However, as seen in Fig. 5.25, for the halogenated ethers, the conflicting effects of the activating ether and the deactivating halogen substitution, makes the reactivity pattern difficult to predict. Where a halogen-free alkyl group is positioned next to the ether oxygen, the ether is more reactive than the halogenated alkane counterpart. However for halogenated ethers with fluorine substituents on both sides of the oxygen linkage, the rate coefficients for reaction with OH radicals can be larger or smaller than their halogenated alkane counterparts. This aspect makes the development of structure-activity relationships difficult for the halogenated ethers[120] and more research is needed to

Table 5.9 Rate coefficients (units of cm^3 $molecule^{-1}$ s^{-1}) for reactions of OH radicals with selected halogenated ethers and the corresponding halogenated alkanes in one atmosphere of air at 298 K.[1,49]

Halogenated ether	k_{OH} (298 K)	Halogenated alkane	k_{OH} (298 K)
CH_3OCHF_2	3.5×10^{-14}	CH_3CHF_2	3.51×10^{-14}
CH_2FOCH_2F	6×10^{-14}	CH_2FCH_2F	9.95×10^{-14}
CH_3OCF_3	1.2×10^{-14}	CH_3CF_3	1.3×10^{-14}
CF_3OCHF_2	4.1×10^{-16}	CF_3CHF_2	1.9×10^{-15}
CHF_2OCHF_2	2.3×10^{-15}	CHF_2CHF_2	6.0×10^{-15}
$CH_3CH_2OCF_3$	1.5×10^{-13}	$CH_3CH_2CF_3$	5.5×10^{-14}
$CF_3CH_2OCH_3$	5.9×10^{-13}	$CF_3CH_2CH_3$	5.5×10^{-14}
$CF_3OCH_2CHF_2$	1.1×10^{-14}	$CF_3CH_2CHF_2$	7.1×10^{-15}
$CHF_2CHFOCF_3$	6.7×10^{-15}	CHF_2CHFCF_3	5.1×10^{-15}
$CF_3CHFOCHF_2$	6.0×10^{-15}	$CF_3CHFCHF_2$	5.1×10^{-15}
$CF_3CHFOCF_3$	1.2×10^{-15}	CF_3CHFCF_3	1.5×10^{-15}
$CF_3CH_2OCH_2CF_3$	1.3×10^{-13}	$CF_3CH_2CH_2CF_3$	8.4×10^{-15}
$CH_3OCH_2CH_2Cl$	4.3×10^{-12}	$CH_3CH_2CH_2Cl$	8.6×10^{-13}
$CH_3OCH_2CH_2Br$	6.1×10^{-12}	$CH_3CH_2CH_2Br$	1.1×10^{-12}
$(CF_3)_2CFOCH_3$	1.55×10^{-14}		

further our understanding of patterns of reactivity of this class of compounds.

The major end product of the atmospheric oxidation of halogenated ethers is usually an ester. As illustrated in Fig. 5.26, the presence of the in-chain oxygen atom gives rise to unimolecular C–C bond and C–O bond decomposition pathways for the generated alkoxy radicals. In addition to esters, other carbonyl containing compounds, such as acyl halides and carbonyl halides, are also possible degradation products of ethers (see Fig. 5.26). Halogenated ethers, and in particular the long-chain length ethers and polyethers, possess strong infrared absorption bands which overlap with the atmospheric window region and hence act as greenhouse gases. Table 5.1 lists radiative efficiency and GWP estimates for several commercially important HFEs a more comprehensive listing is available in the review by Hodnebrog *et al.*[44]

5.7.3. *PCDDs, PCDFs, and PBDEs*

Polychlorinated dibenzodioxins (PCDDs), polychlorinated dibenzofurans (PCDFs) and polybrominated diphenyl ethers (PBDEs) are semi-volatile high-molecular mass halogenated oxygenates. The structures of PCDDs, PCDFs, and PBDEs are shown in Fig. 5.27. PCDDs and PCDFs are formed during combustion of chlorine-containing organic compounds, while PBDEs have seen extensive use as flame retardents. Hundreds of congeners of PCDDs and PCDFs exist reflecting different degrees and positions of Cl substitutions on dibenzodioxin or dibenzofuran carbon backbones. As the vapor pressures of these compounds are exceedingly low, the gas phase chemistry of these fluorinated ethers is very limited. Deposition to soils and oceans dominates their atmospheric fate, at least for the more extensively substituted congeners. Several hundred congeners of PBDEs are possible reflecting different 1–8 Br-atom substitutions onto the diphenyl ether carbon backbone. PBDEs are persistent organic pollutants and, like PCDDs and PCDFs, have very low vapor pressures. The atmospheric lifetime of PBDEs is

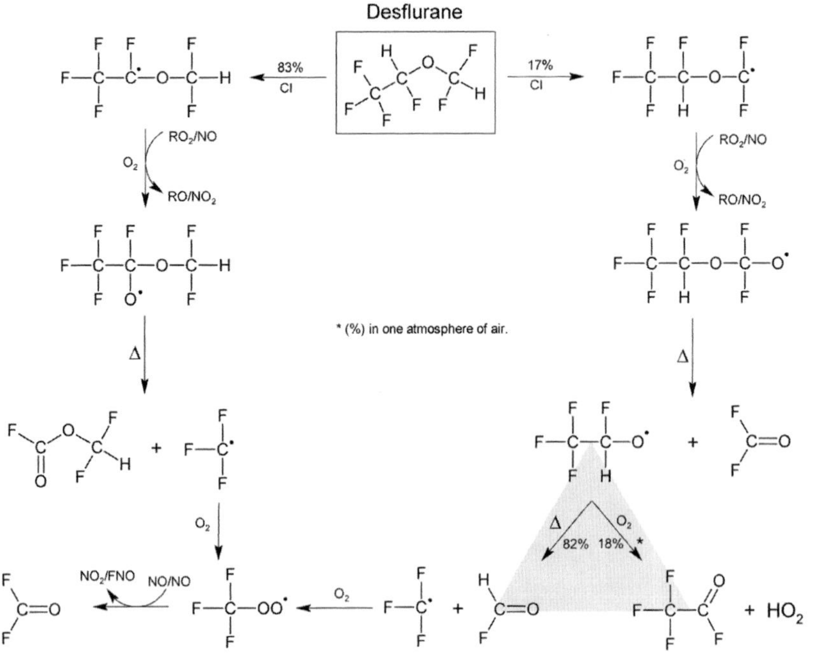

Fig. 5.26 OH radical initiated oxidation of the anesthetic desflurane illustrating the possibilities for C–C and C–O scission in the oxidation of halogenated ethers.

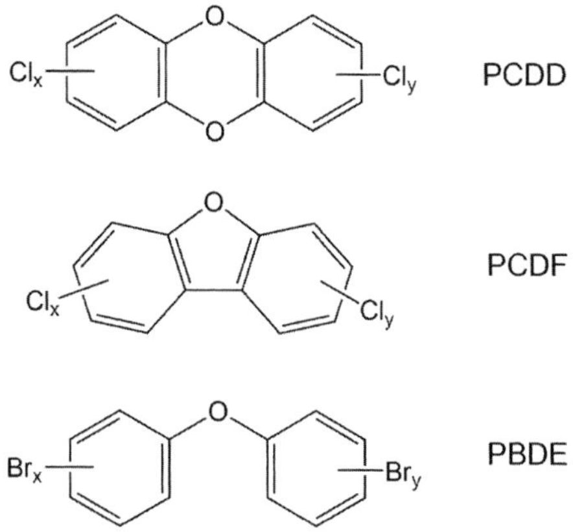

Fig. 5.27 Structures of PCDDs, PCDFs, and PBDEs

controlled by a combination of OH reaction, photolysis, and physical deposition processes, with gas-phase processes important for the lighter congeners and physical deposition important for the heavier congeners.

5.7.4. *Halogenated Carbonyls*

The halogenated carbonyls include aldehydes, ketones, acyl halides, and acyl carbonyls. While aliphatic aldehydes and ketones have multiple natural biogenic sources, their halogenated analogues do not have any significant natural sources. The main atmospheric source of fluorinated carbonyls is photochemical processing of halogenated precursors such as halogenated alkanes, ethers, and alcohols. Formaldehyde, HC(O)H, is the simplest carbonyl compound and it plays a very important role in atmospheric chemistry. Formaldehyde is formed in the oxidation of most organic compounds and it is very reactive. During the day the loss of formaldehyde is controlled by photolysis with a lifetime with respect to photolysis of ~3 h for an overhead sun.[49] Formaldehyde also reacts rapidly with OH radicals and has a lifetime of ~13 h with respect to reaction with OH.[49] The halogenated derivatives of formaldehyde HC(O)F, HC(O)Cl, HC(O)Br, $C(O)Cl_2$, $C(O)F_2$, and $C(O)Br_2$ are products of the oxidation of halogenated alkanes, alkenes, and ethers.

Halogenation of carbonyls typically introduces a shift in peak UV absorption (located around 300 nm) towards shorter wavelengths. Of the halogenated formaldehydes only HC(O)Br and $C(O)Br_2$ absorb significantly in the actinic region and undergo photolysis in the troposphere with lifetimes of 16 h and 4 days, respectively.[49] HC(O)Cl and HC(O)F have relatively slow OH reaction rate constants (3.2×10^{-13} [121] and $<5 \times 10^{-15}$ [122] cm^{-3} $molecule^{-1}$ s^{-1}) and are removed from the atmosphere by heterogeneous decomposition and hydrolysis on contact with cloud–rain–sea water on a time scale of ~10–15 days. HC(O)F hydrolyzes to give HC(O)OH and HF[58] while HC(O)Cl undergoes heterogeneous decomposition to give HCl and CO.[121]

While the acyl and carbonyl halides are well studied with respect to their short wavelength photodissociation quantum yields, there have been no measurements of the quantum yields at longer wavelengths relevant to photolysis in the troposphere. Calvert *et al.*[49] estimated atmospheric lifetimes from the published absorption spectra assuming that the high efficiencies (unity) observed at the shorter wavelength extend into the longer wavelength absorption regimes. Tropospheric photolysis lifetimes were estimated to be: HC(O)Br, 16h; $COBr_2$, 4 days; HC(O)Cl, 2.4 months; $COCl_2$, 16 years; $CCl_3C(O)Cl$, 4.4 days; $CHCl_2C(O)Cl$, 8 days; $CF_3C(O)Cl$, 23 days; $CH_2ClC(O)Cl$, 32 days; and $CH_3C(O)Cl$, 8.8 years.[49] COF_2, $COCl_2$, $CF_3C(O)F$ and CF_3COCl are unreactive with respect to gas phase atmospheric chemistry[58] and are lost from the atmosphere via uptake and hydrolysis in cloud water, rain water or ocean water[58,123] on a timescale of 5–30 days.

Halogenated aldehydes and ketones are important end-products in the oxidation of halogenated alkanes, alcohols, HFOs, and HFEs. As seen from Fig. 5.28, the OH radical reactivities of the halogenated aldehydes are up to a factor of 4 greater than their alcohol precursors. However, the aldehydes absorb sunlight in the troposphere and photolysis is a major atmospheric loss mechanism for these compounds with lifetimes ranging from a few hours to a few days.[124–126] Of the halogenated aldehydes, the perfluoroaldehydes, $CF_3(CF_2)_xCHO$, have been the focus of particular research interest

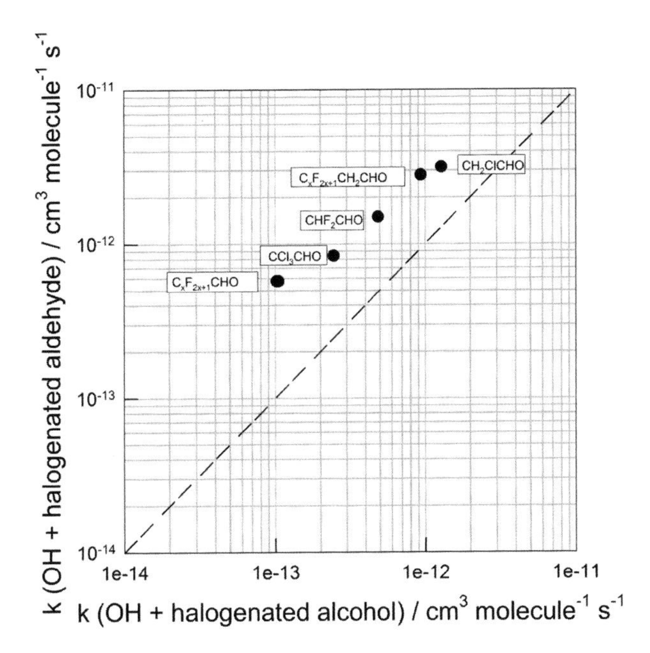

Fig. 5.28 Rate coefficients for reactions of OH radicals with halogenated aldehydes (e.g., CHF_2CHO) at 298 K in one atmosphere of air vs. rate coefficients for reaction of the corresponding halogenated alcohols (e.g., CHF_2CH_2OH). The dashed line indicates the one-to-one relationship.

due to the potential of $CF_3(CF_2)_xCHO$ photooxidation to give toxic bioaccumulative perfluorocarboxylic acids (PFCAs).

Figure 5.29 illustrates a simplified mechanism for the oxidation of perfluoroaldehydes. A range of possible $CF_3(CF_2)_xCHO$ precursor compounds exist and are shown in the top of Fig. 5.29. There are three atmospheric fates for $CF_3(CF_2)_xCHO$ and each can lead to the formation of PFCAs typically in a minor (<10%) yield.[104,127,128] First, OH radical initiated oxidation proceeds by abstraction of the aldehydic hydrogen. Hydrogen abstraction leads to formation of acyl peroxy radicals which can react with HO_2, NO or NO_2. Reaction with HO_2 radicals leads to the formation of PFCAs. Second, contact with liquid water gives hydrates which can react with OH radicals leading to the formation of PFCAs.[128] Third, photolysis generates perfluoroalkyl radicals which undergo a series of "unzipping" reactions in which successive COF_2 units are shed. $CF_3(CF_2)_xO_2$ radicals

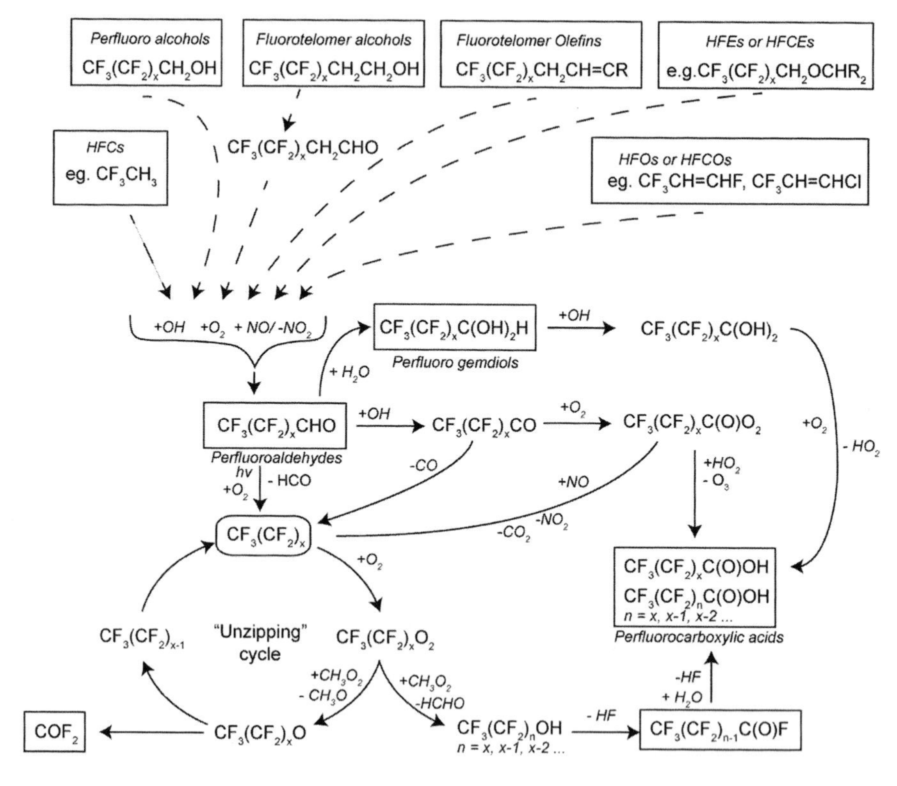

Fig. 5.29 Simplified oxidation mechanism for perfluoroaldehydes, illustrating the possible routes for PFCA formation. Atmospheric precursors to $CF_3(CF_2)_xCHO$ are indicated at the top of the figure.

are key species in the unzipping mechanism. It was established 60 years ago by Russell[129] that the disproportionation channel is important in the reaction between peroxy radicals in which at least one of the peroxy radicals contains an α-hydrogen atom. The perfluoroalkyl peroxy radicals, $CF_3(CF_2)_xO_2$, react with α-hydrogen containing peroxyradicals such as CH_3O_2 to give perfluoro alcohols.

$$CF_3(CF_2)_xO_2 + CH_3O_2 \rightarrow CF_3(CF_2)_xOH + HCHO + O_2,$$
$$CF_3(CF_2)_xOH \rightarrow CF_3(CF_2)_{x-1}C(O)F + HF,$$
$$CF_3(CF_2)_{x-1}C(O)F + H_2O \rightarrow CF_3(CF_2)_{x-1}C(O)OH + HF.$$

As highlighted earlier perfluoroalcohols decompose to yield acylfluorides which then are hydrolyzed to PFCAs, $CF_3(CF_2)_{x-1}$ $C(O)OH$. The fate of CF_3O radicals is reaction with NO or CH_4 to give COF_2:

$$CF_3O + NO \rightarrow COF_2 + FNO,$$
$$CF_3O + CH_4 \rightarrow CF_3OH + CH_3,$$
$$CF_3OH \rightarrow COF_2 + HF.$$

As indicated in Fig. 5.29, fluorotelomer alcohols ($CF_3(CF_2)_x$ CH_2CH_2OH) are precursors to halogenated aldehydes of the type $CF_3(CF_2)_xCH_2CHO$. These aldehydes undergo similar reactions as described above leading to the formation of perfluoroaldehydes:

$$CF_3(CF_2)_xCH_2CHO + hv \rightarrow CF_3(CF_2)_xCH_2 + CHO,$$
$$CF_3(CF_2)_xCH_2 + O_2 \rightarrow CF_3(CF_2)_xCH_2O_2,$$
$$CF_3(CF_2)_xCH_2O_2 + NO \rightarrow CF_3(CF_2)_xCH_2O + NO_2,$$
$$CF_3(CF_2)_xCH_2O + O_2 \rightarrow CF_3(CF_2)_xCHO + HO_2.$$

As for perfluoroaldehydes, in low NO_x environments reaction of $CF_3(CF_2)_xCH_2CHO$ with OH radicals can lead to the formation of $CF_3(CF_2)_xCH_2COOH$, and in high NO_x environments reaction of $CF_3(CF_2)_xCH_2CHO$ with OH radicals can lead to the formation of $CF_3(CF_2)_xCH_2C(O)OONO_2$. The main atmospheric sink for $CF_3(CF_2)_xCH_2CHO$ is photolysis followed by reaction with O_2 to yield $CF_3(CF_2)_xCHO$.

5.7.5. *Halogenated Esters*

Halogenated esters have not found significant commercial applications and are not emitted directly into the atmosphere in appreciable amounts from either natural or anthropogenic sources. Halogenated esters are formed in the atmosphere during the atmospheric oxidation of halogenated ethers. For example the oxidation of an ether with a methyl group adjacent to the oxygen bond, e.g., $C_xF_{2x+1}OCH_3$, gives the corresponding formate, $C_xF_{2x+1}OC(O)H$, while the oxidation of an ether with a ethyl group adjacent to the oxygen bond, e.g., $C_xF_{2x+1}OC_2H_5$, gives the corresponding acetate $C_xF_{2x+1}OC(O)CH_3$. Halogenated esters do not absorb in the actinic region and hence do not photolyze in the troposphere. The ester group deactivates the adjacent alkyl groups with regards to OH radical attack and esters are typically less reactive (k_{OH} $\sim 10^{-14} - 10^{-13}\,cm^{-3}$ molecule^{-1} s^{-1}) than the ethers from which they are derived. The atmospheric lifetimes of halogenated esters with respect to reaction with OH radicals are typically of the order of a few months to a few years. In addition to reaction with OH, loss via uptake to rain–cloud–sea water followed by hydrolysis is a competing atmospheric sink[130,131] for halogenated esters.

5.7.6. *Halogenated Carboxylic Acids*

Trifluoroacetic acid (TFA) is a product of the atmospheric degradation of many commercially important halogenated organic compounds including $CF_3CHClBr$ (halothane), $CF_3CHClOCHF_2$ (isoflurane), CF_3CHCl_2 (HCFC-123), CF_3CHFCl (HCFC-124), CF_3CHF_2 (HFC-134a), $CF_3CF{=}CH_2$ (HFO-1234yf), and CF_3 $CHFCF_3$ (HFC-227ea).[63] The oxidation of CF_3CHF_2 (HFC-134a) makes the largest contribution from known precursors to trifluoroacetic acid formation. Interestingly, the concentrations of TFA observed in the atmosphere and in rainwater are substantially greater than can be explained by the known atmospheric precursors. As an example, oxidation of HFC-134a explains ~2–13% of the TFA observed in atmosphere in Beijing.[132,133] There appear to be significant unknown sources of TFA in the atmosphere. Furthermore, as discussed above in Sec. 5.5.2 it is now well established that the TFA is a ubiquitous component of ocean water even at great depths in remote locations. The large amount of TFA (~300 million tonnes) and lack of a strong depth profile indicate that there are significant natural sources, but these sources are unknown. Further work is needed to establish the sources of the TFA observed in the atmosphere and hydrosphere. In contrast to TFA, there is no evidence for natural sources of long-chain PFCAs. Long-chain PFCAs have been directly emitted to the environment primarily via industrial processes. Such processes include use of PFCAs and their salts as processing aids in the polymerization of fluoropolymers, their largest use, and historically as an ingredient in fire-fighting foams.[134–137]

Trifluoroacetic acid (TFA) is removed from the atmosphere via reaction with OH radicals and via deposition. Reaction with OH proceeds with a rate constant of $k(OH{+}CF_3C(O)OH) = 1.24 \times 10^{-13}$ cm^3 molecule^{-1} s^{-1} leading to an atmospheric lifetime with respect to reaction with OH radicals of ~3 months.[49] The reactivity of OH radicals towards $C_2F_5C(O)OH$, $C_3F_7C(O)OH$, and $C_4F_9C(O)OH$ has been studied and no discernible difference in the reactivity of these acids was observed.[49] An average of the combined data gives $k(OH{+}C_xF_{2x+1}C(O)OH) = 1.55 \times 10^{-13}$ cm^3 molecule^{-1} s^{-1} $(x = 2, 3, 4)$[49] with an uncertainty of ±30%. As for TFA, the longer chain PFCAs have an atmospheric lifetime with respect to reaction with OH radicals of ~3 months. Reaction with OH leads to the conversion of $CF_3C(O)OH$ into CO_2 and HF:

$$OH + CF_3C(O)OH \rightarrow CF_3C(O)O + H_2O,$$
$$CF_3C(O)O + M \rightarrow CF_3 + CO_2 + M,$$
$$CF_3 + O_2 + M \rightarrow CF_3O_2 + M,$$
$$CF_3O_2 \rightarrow\rightarrow COF_2 \rightarrow \text{hydrolysis} \rightarrow HF + CO_2.$$

CF_3COOH is highly water soluble. The key physical property controlling the partitioning of TFA to water is the Henry's Law Constant $(K_H) = [CF_3COOH_{(aq)}]/[CF_3COOH_{(g)}]$ which at 298.15 K is $\sim 9 \times 10^3$ M atm^{-1}.[138,139] Dissociation in solution greatly decreases the concentration of the undissociated acid and hence substantially increases the solubility of CF_3COOH.

$$CF_3COOH_{(g)} \longleftrightarrow CF_3COOH_{(aq)},$$
$$CF_3COOH_{(aq)} \longleftrightarrow CF_3COO^-_{(aq)} + H^+_{(aq)}.$$

The solubility of carboxylic acids are calculated using an effective Henry's Law coefficient (K_H^{eff}) which is defined as $K_H^{\text{eff}} = K_H \times (1 + K_a/[H^+])$, where K_a is the acid dissociation constant. Assuming pKa = 0.47 for CF_3COOH[138] gives an effective Henry's Law coefficient (K_H^{eff}) at a pH of 5 relevant for rainwater of $\sim 3 \times 10^8$ M atm^{-1}. As the length of the perfluorinated chain increases, the molecule becomes more hydrophobic[140,141] and its water solubility decreases; whereas TFA is miscible with water, the water solubility of perfluorooctanoic acid is 3.4 g L^{-1}.[142] The Henry's Law Constant (K_H) for perfluorooctanoic acid (PFOA) at 298.15 K is $\sim$5 M atm^{-1} [139,143]; three orders of magnitude lower than that of TFA. Using pKa = 1.3 for PFOA[143] gives an effective Henry's Law coefficient (K_H^{eff}) at a pH of 5 relevant for rainwater of $\sim 2.5 \times 10^4$ M atm^{-1}.

Seinfeld and Pandis[144] have suggested that as a rough guide, species with Henry's Law constants below $\sim 10^3$ M atm^{-1} will partition strongly into the atmospheric gas phase, and wet/dry deposition will probably not be of major importance. In contrast, compounds with Henry's Law constants above $\sim 10^6$ M atm^{-1} will partition strongly into atmospheric aerosols and wet/dry deposition may be an important atmospheric loss process. TFA with $K_H^{eff} = 3 \times 10^8$ M atm^{-1} will partition strongly into atmospheric aerosols. For highly water soluble gases such as TFA the global average time scale for removal by wet deposition is limited by rain events and is $\sim$1 week.[113] In contrast, the longer chain PFCAs such as PFOA with $K_H^{eff} = 2.5 \times 10^4$ M atm^{-1} will be present in both the gas phase and in atmospheric aerosols. Transport in the gas-phase may play a role in explaining the observed levels of long-chain PFCAs in remote areas.[135,145] Our understanding of the atmospheric chemistry of PFCAs is far from complete and more work is needed in this area.

5.8. Hetero-Atom Containing Halogenated Oxygenates

The family of industrially relevant halogenated organic compounds include compounds, which in addition to halogens, carbon, oxygen, and perhaps hydrogen, include other elements, such as sulphur, nitrogen, silicon or phosphorous. This diverse group of compounds includes substances used in pesticides, fire-fighting foams, metal plating and the surfactant industry that produces polymers for treating textiles and carpets or for surfactants used in food contact paper applications. Examples include sulphonamides,[146] sulfonamido alcohols,[147] sulfonaic acids/sulfonates,[148] phosphoric acid diesters, phosphonates and phosphinates,[149] and siloxanes.[150] Halogen containing pesticides are discussed in the following section. The atmospheric chemistry of the hetero-atom containing halogenated oxygenates is complex. The unifying characteristic is that the perfluorinated parts of the compounds are recalcitrant and form persistent terminal transformation products, including PFCAs and perfluorosulfonic acids (PFSAs). PFSAs, including perfluorooctane sulfonate (PFOS), are unreactive towards the main atmospheric oxidants (OH, NO_3, Cl, O_3), and their main atmospheric sink is wet and dry deposition followed by hydrolysis/ionization (pKa < 0.3[151]) to yield the conjugate sulfonate ion. Long-chain PFCAs and PFSAs ($C_xF_{2x+1}SO_3H$, $x \geq 6$) are problematic because they are persistent, bio-accumulative, ubiquitous environmental contaminants.[152] This and the similar ubiquitous detection of potential precursors to long chain PFSAs,[153] substances related to the perfluorooctane sulfonyl fluoride (POSF) industry, led to restriction of production and use of these compounds under the Stockholm Convention of 2009.[154]

Figure 5.30 shows PFOS and its precursor compounds used in commercial products or present as contaminants in commercial products. The environmental occurrence of PFOS can arise from both direct emissions and abiotic and biological degradation of precursors. Martin *et al.*[146] studied the atmospheric oxidation of perfluoroalkyl sulfonamides (e.g., N-ethyl perfluorooctane sulphonamide, $F(CF_2)_8SO_2N(CH_2CH_3)]$) and found for the perfluorooctane analogue, N-ethyl perfluorobutanesulfonamide, that the OH initiated oxidation proceeds mainly through reaction at the secondary and primary carbons on the ethyl group. The subsequent reaction pathways were found to be a source of PFCAs; however no PFSAs were detected experimentally. In contrast, studies of the reaction of OH radicals with perfluoroalkyl sulfonamido alcohols, again using a perfluorobutane analogue, NEtFBSE N-ethyl-perfluorobutanesulfamido ethanol, $CF_3(CF_2)_3SO_2N(C_2H_5)CH_2CH_2OH)$),[147] was found to

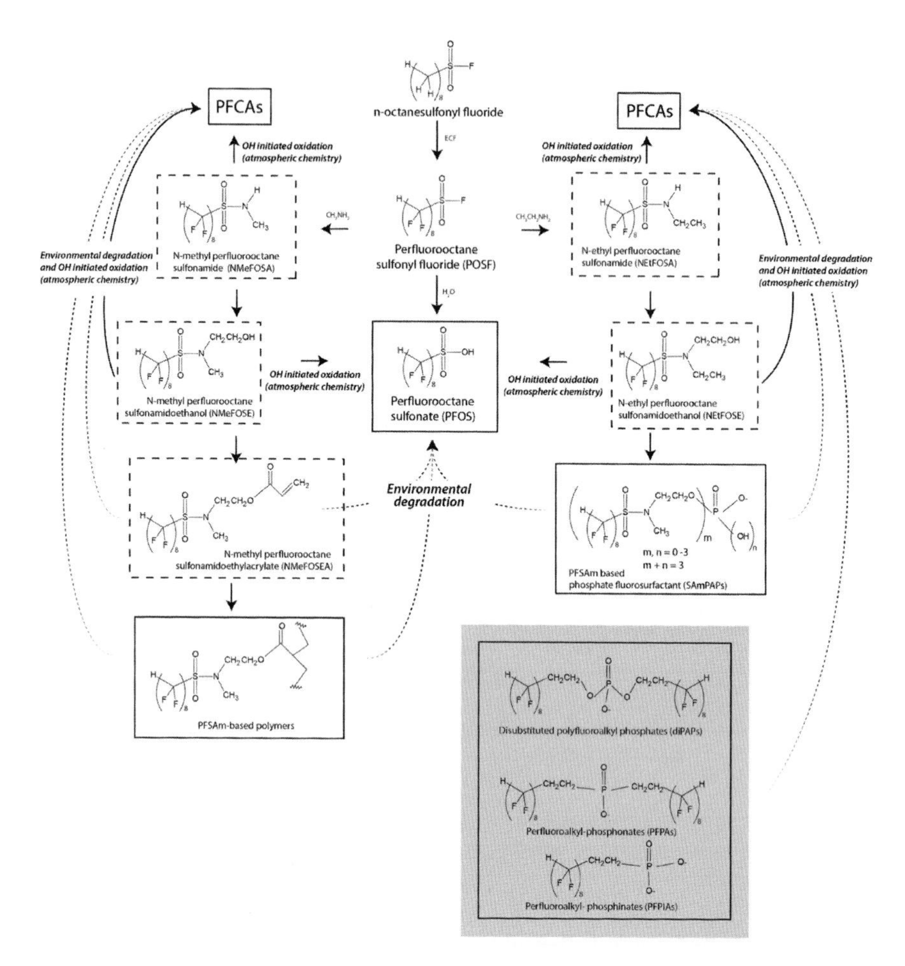

Fig. 5.30 Manufacturing scheme via electrochemical fluorination (ECF) and environmental transformation pathways of selected perfluorinated sulfonamide compounds. Commercial products are identified by solid boxes and known residual materials are identified by dashed boxes. Dotted arrow lines indicate plausible environmental degradation pathways, while solid arrow lines are atmospheric oxidation routes. The grey square insert contains additional hetero-atom containing halogenated oxygenates (fluorotelomer-based surfactants), which can also degrade to give PFCAs. Adapted from Refs. [148,149,155].

yield the PFOS analogue, $C_4F_9SO_3H$. Although the reaction of OH radicals with the sulfamido ethanol takes place primarily on the $-CH_2CH_2OH$ moiety, it can also proceed though OH addition to the sulfone double bond. The resulting sulfonyl radical will undergo either S–N bond scission to give the sulfonic acid and a nitrogen-centered radical, or C–S bond scission to give sulfamic acid and a perfluorinated radical, the latter being the most prevalent reaction pathway of the two. The gas-phase atmospheric lifetime for both the sulfonamides and the sulfamidoethanols is expected to be dictated by reaction with hydroxyl radicals. Lifetime estimates, which are expected to be independent of perfluorinated chain length, are shown in Table 5.10. Perfluorooctane sulfonamidoethanol-based phosphate (SAmPAP) esters are used in the treatment of food contact paper and packaging. These commercial polymers and surfactants (see Fig. 5.30 which may be among the largest potential historical reservoirs of PFOS,[156] have been shown to be sources of PFOS though biotransformation. Similarly, polyfluoroalkyl phosphates (PAPs), which are also primarily used in food-contact paper have been shown to undergo microbially mediated biodegradation to give PFCAs.[157] Partly due to their low volatility no studies of the atmospheric oxidation pathways of these phosphorous containing species are available in the literature.

Table 5.10 OH rate constants (units of cm^3 $molecule^{-1}s^{-1}$) and atmospheric lifetimes for selected hetero-atom containing oxygenates.

Compound	OH radical rate constant	Atmospheric lifetime
N-ethyl perfluoroalkyl sulfonamide $CF_3(CF_2)_x SO_2N(H)CH_2CH_3$	3.74×10^{-13}	20–50 days
N-methyl perfluoroalkyl sulfamidoethanol $CF_3(CF_2)_x SO_2N(CH_3)CH_2CH_2OH$	5.8×10^{-12}	2 days
Sulfuryl fluoride SO_2F_2	$<1.7 \times 10^{-14}$	36 years (limited by atmosphere-ocean exchange)

Finally, in the context of this section, it is germane to mention sulfuryl fluoride, SO_2F_2. Although it does not contain any carbon atoms and is not a halogenated organic compound, it is a halogenated oxygenated and purely covalently bonded compound, of industrial importance and a strong greenhouse gas. It is a widely used insecticide/rodenticide for whole-structure fumigation with annual emissions to the atmosphere on the order of 10^6 kg year^{-1}. SO_2F_2 is unreactive, with atmospheric lifetimes with respect to loss by OH, Cl, and O(^{1}D) reaction and UV photodissociation of >300, >10000, 700, and >4700 years, respectively.[158,159] Its main sink is ocean uptake, followed by hydrolysis which results in an effective atmospheric lifetime of ~36 years.[160] Sulfuryl fluoride possesses strong absorbance bands in the atmospheric window and the resulting

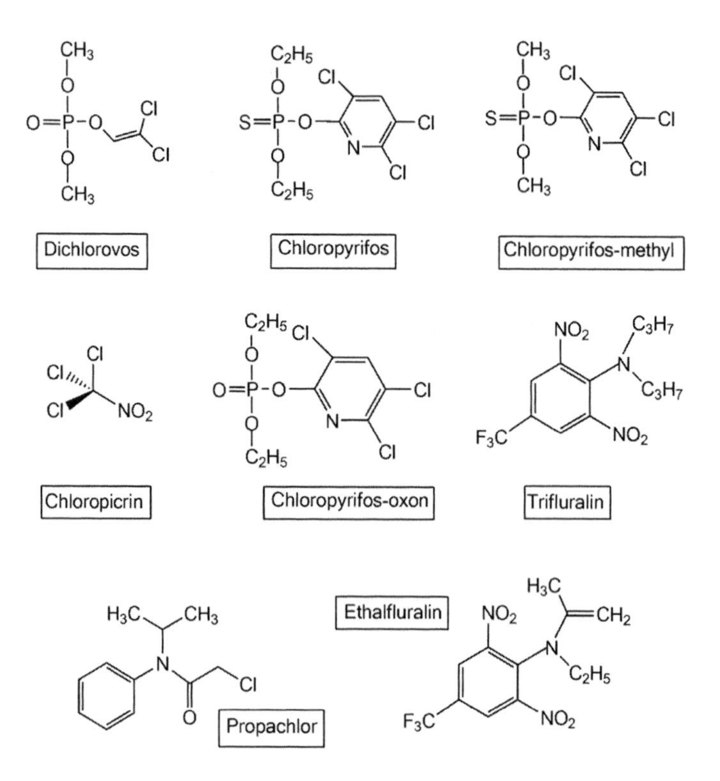

Fig. 5.31 Examples of halogenated organics that have found use as pesticides.

Table 5.11 Atmospheric photolysis rates, J, (units of s^{-1}), OH rate constants at 298 K in one atmosphere (units of cm^3 $molecule^{-1}$ s^{-1}), and atmospheric lifetimes for selected pesticides estimated by Muñoz and co-workers.[161]

Pesticide	J (s^{-1})	k_{OH} (cm^3 $molecule^{-1}$ s^{-1})	Lifetime/Main degradation pathway
Chlorpyrifos[162]	1.4×10^{-5}	$(9.1 \pm 1.8) \times 10^{-11}$	2 h/reaction with OH
Chlorpyrifos-oxon[162]	$<4.8 \times 10^{-5}$	$(1.6 \pm 0.8) \times 10^{-11}$	11 h/reaction with OH
Chlorpyrifos-methyl[163]	$<2 \times 10^{-5}$	$(4.1 \pm 0.4) \times 10^{-11}$	2.5 h/reaction with OH
Dichlorvos[164]	negligible	$(2.6 \pm 0.3) \times 10^{-11}$	5 h/reaction with OH
Ethalfluralin[165]	$(1.3 \pm 0.2) \times 10^{-3}$	$(9.6 \pm 1.8) \times 10^{-11}$	12 min/photolysis
Propachlo[166]	$<(2 \pm 0.5) \times 10^{-5}$	$(1.5 \pm 0.3) \times 10^{-11}$	20 h/reaction with OH
Trifluralin[167]	$(1.2 \pm 0.5) \times 10^{-3}$	$(1.7 \pm 0.4) \times 10^{-11}$	15 min/photolysis
Lindane[168]	$<3.5 \times 10^{-5}$	6.4×10^{-13}	20 days/reaction with OH
Chloropicrin[169]	3.6×10^{-5}		5.6 h/photolysis

radiative efficiency is substantial (0.196–0.222 $W\,m^{-2}$ ppb^{-1}). The 100-year GWP for SO_2F_2 is estimated at 4780.

5.9. Halogen Containing Pesticides

Halogenated organic compounds have found widespread use as pesticides; examples are shown in Fig. 5.31. The atmospheric chemistry of pesticides has been studied in detail by Muňoz and co-workers using the EUPHORE chamber. Estimates of the rates of atmospheric photolysis and reaction with OH radicals are given in Table 5.11. As seen from Table 5.11, the halogen containing pesticides generally are reactive in the troposphere and have lifetimes typically of 1 day or less, however some pesticides such as lindane are more persistent.

5.10. Conclusions

It is now more than 40 years since Molina and Rowland recognized the importance of chlorine-catalyzed ozone destruction initiated by photolysis of CFCs in the stratosphere.[3] This recognition led to a large international research effort to understand the chemistry and environmental impacts of halogenated organic compounds. Mario Molina and F. Sherwood ("Sherry") Rowland together with Paul Crutzen were awarded the Nobel Prize in 1995 for "their work in atmospheric chemistry, particularly in concerning the formation and decomposition of ozone". Interest in the environmental impacts of halogenated organic compounds together with the recognition afforded by the Nobel Prize helped establish atmospheric chemistry as a chemical discipline of its own. Halogenated organic compounds continue to have an important role in the chemistry of the atmosphere and have had an important role in establishing the field of atmospheric chemistry.

Ozone depleting substances including CFCs have been controlled under the Montreal Protocol and its Amendments. It is hard to overstate the success of Montreal Protocol. Absent the Protocol, it has been estimated that the Antarctic ozone hole would have grown 40% by 2013, a deep Arctic ozone hole would have developed in 2011, the ozone layer worldwide would have thinned by ~15%, and the climate effect of net annual ODS emissions by 2010 would have been equivalent to ~10 Gt CO_2/year. The Protocol has led to stabilization of stratosphere ozone levels and there is evidence of a small recovery with the 60°S–60°N average column ozone increasing by ~1% between 2000 and 2013.[15] The atmospheric concentration of ozone depleting substances is declining and stratospheric ozone levels are recovering, however because of the long atmospheric lifetimes of the CFCs the recovery will be gradual. Interestingly, because the cooling trend in the stratosphere (associated with increased levels of CO_2 and greater radiative cooling) slows down the ozone destruction chemistry, it is projected that there will be a global mean total column ozone "super-recovery" of ~5 DU in the second half of the century.[170]

Replacements for CFCs have been developed and a large international research effort has been conducted to understand their atmospheric chemistry and environmental impacts. In general the atmospheric chemistry of CFC replacements is well understood. Indeed it can be argued that of all the classes of organic compounds, the atmospheric chemistry of halogenated organic compounds is the best understood. That said, significant gaps still exist in our understanding. The source of the observed burden of trifluoroacetic acid in the atmosphere and hydrosphere is not understood. More research is needed to better understand the sources and sinks of perfluorocarboxylic acids and other halogenated persistent pollutants in the environment. As more CFC replacements are proposed (e.g., the recent interest in halogenated alkenes) their atmospheric chemistry and environmental impacts need to be evaluated by experimental and computational research before being adopted for large scale industrial use.

Acknowledgments

We thank Steve Montzka (NCAR) and Amalia Muñoz (CEAM) for helpful discussions.

References

1. J. G. Calvert, R. G. Derwent, J. J. Orlando, G. S. Tyndall and T. J. Wallington. *Mechanisms of Atmospheric Oxidation of the Alkanes.* (Oxford University Press, New York, 2008).
2. WMO (2011). Scientific assessment of stratospheric ozone: 2010. Report No. 52. *World Meteorological Organization Global Ozone Research and Monitoring Project*, Geneva, Switzerland.
3. M. J. Molina and F. S. Rowland. Stratospheric sink for chlorofluoromethanes: Chlorine atom catalyzed destruction of ozone. *Nature* **249**, 810–814 (1974).
4. J. C. Farman, B. G. Gardiner and J. D. Shanklin. Large losses of ozone in Antarctica reveal seasonal ClOx/NOx interaction. *Nature* **315**, 207–210 (1985).
5. IPCC, *Climate Change: The IPCC Scientific Assessment (1990).* (Cambridge University Press, Cambridge, 1990).
6. World Meteorological Organization, *Atmospheric Ozone 1985, World Meteorological Organization Research and Monitoring Project No 16.* (Geneva, Switzerland, 1985).
7. J. G. Calvert. Glossary of atmospheric chemistry terms. *Pure. Appl. Chem.* **62**, 2167–2219 (1990).
8. H. A. Favre and W. H. Powell. *Nomenclature of Organic Chemistry.* IUPAC Recommendations and Preferred Name 2013. Cambridge, UK: The Royal Society of Chemistry, 2013. ISBN 978-0-85404-182-4.
9. ANSI/ASHRAE Standard 34–1992.
10. O. J. Nielsen, M. S. Javadi, M. P. Sulbaek Andersen, M. D. Hurley, T.J. Wallington and R. Singh, Atmospheric chemistry of $CF_3CF{=}CH_2$: Kinetics and mechanisms of gas-phase reactions with Cl atoms, OH radicals and O_3. *Chem. Phys. Lett.* 439, 18–22 (2007).
11. IPCC. *Climate Change 2013: The physical science basis.* Contribution of Working Group I to the fifth assessment report of the Intergovernmental Panel on Climate Change. Intergovernmental Panel on Climate Change. (Cambridge University Press, Cambridge, UK, 2013).
12. BSR/ASHRAE Addendum z to ANSI/ASHRAE Standard 34–2007.
13. J. B. Burkholder, R. A. Cox, and A. R. Ravishankara, Atmospheric degradation of ozone depleting substances, their substitutes, and related species. *Chem. Rev.* **2015**, 3704–3759.
14. J. G. Calvert, J. J. Orlando, W. R. Stockwell and T. J. Wallington. *The Mechanisms of Reactions Influencing Atmospheric Ozone.* (Oxford University Press, New York, 2015).
15. World Meteorological Organization, *Scientific Assessment of Ozone Depletion: 2014*, World Meteorological Organization, Global Ozone Research and Monitoring Project-Report No. 55, pp. 416, (Geneva, Switzerland, 2014).
16. F. Swarts. Étude sur le fluorchloroforme. *Bull. Acad. Roy. Belg.* **24**, 474–484 (1892).

17. T. Midgley Jr. and A. L. Henne, Organic fluorides as refrigerants. *Ind. Eng. Chem.* **22**, 542–545 (1930).
18. Alternative Fluorocarbons Environmental Acceptability Study (AFEAS). http://www.afeas.org, downloaded June 2015.
19. A. McCulloch and P. M. Midgley. The of methyl chloroform emissions: 1951–2000. *Atmos. Environ.* **35**, 5311–5319 (2001).
20. S. A. Montzka, M. McFarland, S. O. Andersen, B. R. Miller, D. W. Fahey, B. D. Hall, L. Hu, C. Siso and J. W. Elkins. Recent trends in global emissions of hydrochlorofluorocarbons and hydrofluorocarbons: Reflecting on the 2007 adjustments to the Montreal Protocol. *J. Phys. Chem. A* **119**, 4439–4449 (2015).
21. J. L. Bullister. Atmospheric Histories (1765–2015) for CFC-11, CFC-12, CFC-113, CCl_4, SF_6 and N_2O. NDP-095 (2015). Carbon Dioxide Information Analysis Center, Oak Ridge National Laboratory, US Department of Energy, Oak Ridge, Tennessee. doi: 10.3334/CDIAC/otg.CFC_ATM_Hist_2015.
22. M. Meinshausen, S. J. Smith, K. Calvin, J. S. Daniel, M. L. T. Kainuma, J-F. Lamarque, K. Matsumoto, S. A. Montzka, S. C. B. Raper, K. Riahi, A. Thomson, G. J. M. Velders and D. P. P. van Vuuren. The RCP GHG concentrations and their extension from 1765 to 2300. *Climatic Change* **109**, 213–241 (2011).
23. S. Montzka, J. W. Elkins, G. Dutton, NOAA ESRL http://www.esrl.noaa.gov/gmd/hats, downloaded June 2015.
24. M. de Blas, Trichloroethylene, tetrachloroethylene and carbon tetrachloride in an urban atmosphere: Mixing ratios and temporal patterns, *Int. J. Environ. Anal. Chem.* **93**, 228–244 (2013).
25. P. J. Fraser, B. L. Dunse, A. J. Manning, S. Walsh, R. H. J. Wang, P. B. Krummel, L. P. Steele, L. W. Porter, C. Allison, S. O'Doherty, P. G. Simmonds, J. Mühle, R. F. Weiss and R. G. Prinn. Australian carbon tetrachloride emissions in a global context. *Environ. Chem.* **11**, 77–88 (2014).
26. Chapman, S. On ozone and atomic oxygen in the upper atmosphere. *Phil. Mag.* **10**, 369–383 (1930).
27. D. R. Bates and M. Nicolet. The photochemistry of atmospheric water vapor. *J. Geophys. Res.* **55**, 301–327 (1950).
28. P. J. Crutzen. The influence of nitrogen oxides on the atmospheric ozone content. *Quart. J. Roy. Met. Soc.* **96**, 320–325 (1970).
29. H. S. Johnston. Reduction of stratospheric ozone by nitrogen oxide catalysts from supersonic transport exhaust. *Science* **173**, 517–522 (1971).
30. R. S. Stolarski and R. J. Cicerone. Stratospheric chlorine: A possible sink for ozone, *Can. J. Chem.* **52**, 1610–1615 (1974).
31. NASA (2015). Ozone hole watch. Available at http://ozonewatch.gsfc.nasa.gov/.
32. M. J. Molina, L. T. Molina and F. C.-Y. Wang, Antarctic stratospheric chemistry of chlorine nitrate, hydrogen chloride, and ice: Release of active chlorine. *Science* **238**, 1253–1257 (1987).

33. L. T. Molina and M. J. Molina. Production of chlorine oxide (Cl_2O_2) from the self-reaction of the chlorine oxide (ClO) radical. *J. Phys. Chem.* **91**, 433–436 (1987).
34. M. P. Chipperfield, S. S. Dhomse, W. Feng, R. L. McKenzie, G. J. M. Velders and J. A. Pyle. Quantifying the ozone and ultraviolet benefits already achieved by the Montreal Protocol. *Nature Communications* **6**, doi:10.1038/ncomms8233 (2015).
35. J. D. Shanklin. *British Antarctic Survey.* Madingley Road, Cambridge, CB3 0ET, England, (2015). http://www.antarctica.ac.uk/met/jds/ozone/data/ZMIN5611.DAT.
36. V. Ramanathan. Greenhouse effect due to chlorofluorocarbons — climatic implications. *Science* **190**, 50–51 (1975).
37. W. C. Wang, Y. L. Yung, A. A. Lacis, T. Mo and J. E. Hansen. Greenhouse effects due to man-made perturbations of trace gases. *Science* **194**, 685–690 (1976).
38. S. Pinnock, M. D. Hurley, K. P. Shine, T. J. Wallington and T. J. Smyth. Radiative forcing of climate by hydrochlorofluorocarbons and hydrofluorocarbons. *J. Geophys. Res. Atmos.* **100**, 23227–23238 (1995).
39. F. Joos, R. Roth, J. S. Fuglestvedt, G. P. Peters, I. G. Enting, W. von Bloh, V. Brovkin, E. J. Burke, M. Eby, N. R. Edwards, T. Friedrich, T. L. Frölicher, P. R. Halloran, P. B. Holden, C. Jones, T. Kleinen, F. Mackenzie, K. Matsumoto, M. Meinshausen, G. K. Plattner, A. Reisinger, J. Segschneider, G. Shaffer, M. Steinacher, K. Strassmann, K. Tanaka, A. Timmermann and A. J. Weaver. Carbon dioxide and climate impulse response functions for the computation of greenhouse gas metrics: A multi-model analysis. *Atmos. Chem. Phys.* **13**, 2793–2825 (2013).
40. E. J. Highwood, K. P. Shine, M. D. Hurley and T. J. Wallington. Estimation of direct radiative forcing due to non-methane hydrocarbons. *Atmos. Environ.* **33**, 759 (1999).
41. D. J. Wuebbles , K. O. Patten , D. Wang, D. Youn, M. Martìnez-Avilès and J. S. Francisco. *Atmos. Chem. Phys.* **11**, 2371–2380 (2011).
42. R. G. Derwent, M. E. Jenkin, S. M. Saunders and M. J. Pilling. Photochemical ozone creation potentials for organic compounds. In northwest Europe calculated with a master chemical mechanism. *Atmos. Environ.* **32**, 2429–2441 (1998).
43. S. M. Saunders, M. E. Jenkin, R. G. Derwent and M. J. Pilling. Protocol for the development of the Master Chemical Mechanism, MCM v3 (Part A): Tropospheric degradation of non-aromatic volatile organic compounds. *Atmos. Chem. Phys.* **3**, 161–180 (2003).
44. Ø. Hodnebrog, M. Etminan, J. S. Fulglesvedt, G. Marston, G. Myhre, C. J. Nielsen, K. P. Shine and T. J. Wallington. Global warming potentials and radiative efficiencies of halocarbons and related compounds: A comprehensive review. *Rev. Geophys.* **51**, 300–378 (2013).
45. G. D. Hayman and R. G. Derwent. Atmospheric chemical reactivity and ozone-forming potentials of potential CFC replacements. *Environ. Sci. Technol.* **31**, 327–336 (1997).

46. T. J. Wallington, M. P. Sulbaek Andersen and O. J. Nielsen. Atmospheric chemistry of short-chain haloolefins: Photochemical ozone creation potentials (POCPs). global warming potentials (GWPs), and ozone depletion potentials (ODPs). *Chemosphere* **129**, 135–141 (2015).
47. K. O. Patten and D. J. Wuebbles. Atmospheric lifetimes and ozone depletion potentials of trans-1-chloro-3,3,3-trifluoropropylene and trans-1,2-dichloroethylene in a three-dimensional model. *Atmos. Chem. Phys.* **10**, 10867–10874 (2010).
48. A. Mellouki, T. J. Wallington and J. Chen. Atmospheric chemistry of oxygenated volatile organic compounds (OVOCs): Impacts on air quality and climate. *Chem. Rev.* **115**, 3984–4014 (2015).
49. J. G. Calvert, A. Mellouki, J. J. Orlando, M. J. Pilling and T. J. Wallington. *Mechanisms Of Atmospheric Oxidation of the Oxygenates.* (Oxford University Press, Oxford, UK, 2011).
50. G. J. M. Velders, S. O. Andersen, J. S. Daniel, D. W. Fahey and M. McFarland. The importance of the Montreal Protocol in protecting climate. *Proc. Natl. Acad. Sci.*, **104**, 4814–4819 (2007).
51. G. J. M., Velders, D. W. Fahey, J. S. Daniel, M. McFarland and S. O. Andersen. The large contribution of projected HFC emissions to future climate forcing. *Proc. Natl. Acad. Sci.* **106**, 10949–10954 (2009).
52. G. J. M. Velders, A. R. Ravishankara, M. K. Miller, M. J. Molina, J. Alcamo, J. S. Daniel, D. W. Fahey, S. A. Montzka and S. Reimann. Preserving Montreal Protocol climate benefits by limiting HFCs. *Science* **335**, 922–923 (2012).
53. T. J. Wallington, J. E. Anderson, S. A. Mueller, S. L. Winkler, O. J. Nielsen. Online Comment in *Science* (2012). http://comments.sciencemag.org/content/10.1126/science.1216414.
54. A. R. Ravishankara, A. A. Turnipseed, N. R. Jensen, S. Barone, M. Mills, C. J. Howard and S. Solomon. Do hydrofluorocarbons destroy stratospheric ozone? *Science* **263**, 71–75 (1994).
55. T. J. Wallington, W. F. Schneider, J. Sehested and O. J. Nielsen. Hydrofluorocarbons and stratospheric ozone. *Faraday Discuss* **100**, 55–64 (1995).
56. A. R. Ravishankara, S. Solomon, A. A. Turnipseed and R. F. Warren. Atmospheric lifetimes of long-lived halogenated species. *Science* **259**, 194–199 (1993).
57. M. M. Hurwitz, E. L. Fleming, P. A. Newman, F. Li, E. Mlawer, K. Cady-Pereira and R. Bailey. Ozone depletion by hydrofluorocarbons. *Geophys. Res. Lett.* **42**, (2015). doi:10.1002/2015GL065856.
58. T. J. Wallington, W. F. Schneider, D. R. Worsnop, O. J. Nielsen, J. Sehested, W. DeBruyn and J. A. Shorter. Atmospheric chemistry and environmental impact of CFC replacements: HFCs and HCFCs. *Environ. Sci. Technol.* **28**, 320A–326A (1994).
59. R. Atkinson, D. L. Baulch, R. A. Cox, J. N. Crowley, R. F. Hampson, R. G. Hynes, M. E. Jenkin, M. J. Rossi, J. Troe and T. J. Wallington. Evaluated kinetic and photochemical data for atmospheric chemistry: Volume IV —

Gas phase reactions of organic halogen species. *Atmos. Chem. Phys.* **8**, 4141–4496 (2008).

60. S. P. Sander, R. R. Friedl, J. R. Barker, D. M.Golden, M. J. Kurylo, P. H. Wine, J. P. D. Abbatt, J. B. Burkholder, C. E. Kolb, G. K. Moortgat, R. E. Huie and V. L. Orkin. *Chemical kinetics and photochemical data for use in atmospheric studies, Evaluation Number 17*, Jet Propulsion Laboratory, Pasadena, California, JPL Publication 10-6, http://jpldataeval.jpl.nasa.gov/pdf/JPL 10-6 Final 15June2011.pdf. (2011).
61. J. Sehested, T. Møgelberg, K. Fagerstrom, G. Mahmoud and T. J. Wallington. Absolute rate constants for the self reactions of HO_2, CF_3CFHO_2 and CF_3O_2 radicals and the cross reactions of HO_2 with FO_2, HO_2 with CF_3CFHO_2, and HO_2 with CF_3O_2 at 295 K. *Int. J. Chem. Kinet.* **29**, 673–682 (1997).
62. T. J. Wallington, M. D. Hurley, J. M. Fracheboud, J. J. Orlando, G. S. Tyndall, J. Sehested, T. E. Møgelberg and O. J. Nielsen. Role of excited CF_3CFHO radicals in the atmospheric chemistry of HFC-134a. *J. Phys. Chem.* **100**, 18116–18122 (1996).
63. X. Tang, S. Madronich, T. Wallington and D. Calamari. Changes in tropospheric composition and air quality. *J. Photochem. Photobiol. B* **46**, 83–95 (1998).
64. K. R. Solomon, X. Tang, S. R. Wilson, P. Zanis and A. F. Bais. Changes in tropospheric composition and air quality due to stratospheric ozone depletion. *Photochem. Photobiol. Sci.* **2**, 62–67 (2003).
65. WMO (World Meteorological Organization), 2007. Scientific assessment of ozone depletion: 2006, Global ozone, research and monitoring project — Report 50, Geneva, Switzerland.
66. J. C. Boutonnet, P. Bingham, D. Calamari, C. de Rooij, J. Franklin, T. Kawano, J-M. Libre, A. McCulloch, G. Malinverno, J. M. Odom, G. M. Rusch, K. Smythe, I. Sobolev, R. Thompson and J. M. Tiedje. Environmental risk assessment of trifluoroacetic acid. *Hum. Ecol. Risk Assess.* **5**, 59–124 (1999).
67. M. Berg, S. R. Müller, J. Mühlemann, A. Wiedmer and R. P. Scharzenbach. Concentrations and fluxes of chloroacetic acids and trifluoroacetic acid in rain and natural waters in Switzerland. *Environ. Sci. Technol.* **34**, 2675–2683 (2000).
68. H. Frank, E. H. Christoph, O. Holm-Hansen and J. L. Bullister. Trifluoroacetate in ocean waters. *Environ. Sci. Technol.* **36**, 12–15 (2002).
69. B. F. Scott, R.W. Macdonald, K. Kannan, A. Fisk, A. Witter, N. Yamashita, L. Durham, C. Spencer and D. C. G. Muir. Trifluoroacetate profiles in the Arctic, Atlantic and Pacific Oceans. *Environ. Sci. Technol.* **39**, 6555–6560 (2005).
70. B. F. Scott, C. Spencer, S. A. Mabury and D. C. G. Muir. Poly and perfluorinated carboxylates in North American precipitation. *Environ. Sci. Technol.* **40**, 7167–7174 (2006).

71. L. M. Von Sydow, A. B. Grimvall, H. B. Borén, K. Laniewski and A. T. Nielsen. Natural background levels of trifluoroacetate in rain and snow. *Environ. Sci. Technol.* **34**, 3115–3118 (2000).
72. T. K. Tromp, M. K. W. Ko, J. M. Rodríguez and N. D. Sze. Potential accumulation of a CFC-replacement degradation product in seasonal wetlands. *Nature* **376**, 327–330 (1995).
73. J. A. Benesch, M. S. Gustin, G. R. Cramer and T. M. Cahill. Investigation of effects of trifluoroacetate on vernal pool ecosystems. *Environ. Toxicol. Chem.* **21**, 640–647 (2002).
74. S. A. Montzka, M. Krol, E. Dlugokencky, B. Hall, P. Jöckel and J. Lelieveld. Small interannual variability of global atmospheric hydroxyl. *Science.* **331**, 67–69 (2011).
75. C. M. Spivakovsky, J. A. Logan, S. A. Montzka, Y. J. Balkanski, M. Foreman-Fowler, D. B. A. Jones, L. W. Horowitz, A. C. Fusco, C. A. M. Brenninkmeijer, M. J. Prather, S. C. Wofsy and M. B. McElroy. Three-dimensional climatological distribution of tropospheric OH: Update and evaluation. *J. Geophys. Res.* **105**, 8931–8980 (2000).
76. K. A. Read, A. S. Mahajan, L. J. Carpenter, M. J. Evans, B. V. E. Faria, D. E. Heard, J. R. Hopkins, J. D. Lee, S. J. Moller, A. C. Lewis, L. Mendes, J. B. McQuaid, H. Oetjen, A. Saiz-Lopez, M. J. Pilling and J. M. C. Plane. Extensive halogen-mediated ozone destruction over the tropical Atlantic Ocean. *Nature* **453**, 1232–1235 (2008).
77. A. Saiz-Lopez, J-F. Lamarque, D. E. Kinnison, S. Tilmes, C. Ordóñez, J. J. Orlando, A. J. Conley, J. M. C. Plane, A. S. Mahajan, G. Sousa Santos, E. L. Atlas, D. R. Blake, S. P. Sander, S. Schauffler, A. M. Thompson and G. Brasseur. Estimating the climate significance of halogen-driven ozone loss in the tropical marine troposphere. *Atmos. Chem. Phys.* **12**, 3939–3949 (2012).
78. J. S. Brown. HFOs new, low global warming potential refrigerants. *ASHRAE Journal.* **51**, 22–29 (2009).
79. E. C. Tuazon, R. Atkinson and S. M. Aschmann. Kinetics and products of the gas-phase reactions of the OH radical and O_3 with allyl chloride and benzyl chloride at room temperature. *Int. J. Chem. Kinet.* **22**, 981–998 (1990).
80. J. Albaladejo, B. Ballesteros, E. Jimenez, Y. Diaz de Mera and E. Martinez. Gas-phase OH radical-initiated oxidation of the 3-halopropenes studied by PLP-LIF in the temperature range 228–388 K. *Atmos. Environ.* **37**, 2919–2926 (2003).
81. E. C. Tuazon, R. Atkinson, S. M. Aschmann, M. A. Goodman and A. M. Winer. Atmospheric reactions of chloroethenes with the OH radical. *Int. J. Chem. Kinet.* **20**, 241–265 (1988).
82. M. P. Sulbaek Andersen, E. J. K. Nilsson, O. J. Nielsen, M. S. Johnson, M. D. Hurley, T. J. Wallington and R. Singh. Atmospheric chemistry of trans-$CF_3CH{=}CHCl$: Kinetics of the gas-phase reactions with Cl atoms, OH radicals, and O_3. *J. Photochem. Photobiol. A: Chem.* **199**, 92–97 (2008).

83. R. A. Perry, R. Atkinson and J. N. Pitts, Jr. Rate constants for the reaction of OH radicals with $CH_2=CHF$, $CH_2=CHCl$, and $CH_2=CHBr$ over the temperature range 299–426 K. *J. Chem. Phys.* **67**, 458–462 (1977).
84. T. J. Wallington and M. D. Hurley. Atmospheric chemistry of hexafluorocyclobutene, octafluorocyclopentene, and hexafluoro-1,3-butadiene. *Chem. Phys. Lett.* **507**, 19–23 (2011).
85. M. Baasandorj, A. R. Ravishankara and J. B. Burkholder. Atmospheric chemistry of (Z)-$CF_3CH=CHCF_3$: OH radical reaction rate coefficient and global warming potential. *J. Phys. Chem. A* **115**, 10539–10549 (2011).
86. J. P. D. Abbatt and J. G. Anderson. High-pressure discharge flow kinetics and frontier orbital mechanistic analysis for OH + CH_2CCl_2, *cis*-CHClCHCl, *trans*-CHClCHCl, CFClCF2, and CF_2CCl_2 → products. *J. Phys. Chem.* **95**, 2382–2390 (1991).
87. V. L. Orkin, F. Louis, R. E. Huie and M. J. Kurylo. Photochemistry of bromine-containing fluorinated alkenes: Reactivity toward OH and UV spectra. *J. Phys. Chem. A* **106**, 10195–10199 (2002).
88. V. L. Orkin, G. A. Poskrebyshev and M. J. Kurylo. Rate constants for the reactions between OH and perfluorinated alkenes. *J. Phys. Chem. A* **115**, 6568–6574 (2011).
89. X. Jia, L. Chen, J. Mizukado , S. Kutsuna and K. Tokuhashi. Rate constants for the gas-phase reactions of cyclo-$CX=CXCF_2CF_2$–(X = H, F) with OH radicals at a temperature range of 253–328 K. *Chem. Phys. Lett.* **572**, 21–25 (2013).
90. C. M. Tovar, M. B. Blanco, I. Barnes, P. Wiesen and M. A. Teruel. Gas-phase reactivity study of a series of hydrofluoroolefins (HFOs) toward OH radicals and Cl atoms at atmospheric pressure and 298 K, *Atmos. Environ.* **88**, 107–114 (2014).
91. J. P. A. Abrate, I. Pisso, S. A. Peirone, P. M. Cometto and S. I. Lane. Relative rate coefficients of OH radical reactions with $CF_3CF=CClCF_3$ and $CF_3CH=CHCH_2OH$. Ozone depletion potential estimate for $CF_3CF=CClCF_3$. *Atmos. Environ.* **67**, 85–92 (2013).
92. M. S. Javadi, R. Søndergaard, O. J. Nielsen, M. D. Hurley and T. J. Wallington. Atmospheric chemistry of trans-$CF_3CH=CHF$: Products and mechanisms of hydroxyl radical and chlorine atom initiated oxidation. *Atmos. Chem. Phys.* **8**, 3141–3147 (2008).
93. O. J. Nielsen, M. S. Javadi, M. P. Sulbaek Andersen, M. D. Hurley, T. J. Wallington and R. Singh. Atmospheric chemistry of $CF_3CF=CH_2$: Kinetics and mechanisms of gas-phase reactions with Cl atoms, OH radicals and O_3. *Chem. Phys. Lett.* **439**, 18–22 (2007).
94. R. G. Derwent, M. E. Jenkin and S. M. Saunders. Photochemical ozone creation potentials for a large number of reactive hydrocarbons under European conditions. *Atmos. Environ.* **30**, 181–199 (1996).
95. M. E. Jenkin. Photochemical ozone and PAN creation potentials: Rationalisation and methods of estimation, AEA Technology plc, Report AEAT-4182/20150/003, AEA Technology plc, National Environmental Technology Centre, Culham, Oxfordshire OX14 3DB, UK (1998).

96. B. Dimitriades, Scientific basis of an improved EPA policy on control of organic emissions for ambient ozone reduction. *J. Air & Waste Man. Assoc.* **49**, 831–838 (1999).
97. W. P. L. Carter. Investigation of atmospheric ozone impacts of trans 2,3,3,3-tetrafluoropropene, Final Report for contract UCR-09010016, http://www.cert.ucr.edu/∼carter/pubs/YFrept.pdf (2009).
98. D. L. Luecken, R. L. Waterland, S. Papasavva, K. Taddonio, W. T. Hutzell, J. P. Rugh and S. O. Andersen. Ozone and TFA impacts in North America from degradation of 2,3,3,3-tetrafluoropropene (HFO-1234yf), a potential greenhouse gas replacement. *Environ. Sci. Technol.* **44**, 343–348 (2010).
99. L. L. Andersen, F. F. Østerstrøm, M. P. Sulbaek Andersen, O. J. Nielsen and T. J. Wallington. Atmospheric chemistry of cis-$CF_3CH{=}CHCl$ (HCFO-1233zd(Z)): Kinetics of the gas-phase reaction with Cl atoms, OH radicals, and O_3. *Chem. Phys. Lett.* **639**, 289–293 (2015).
100. M. P. Sulbaek Andersen, O. J. Nielsen, M. D. Hurley, J. C. Ball, T. J. Wallington, J. E. Stevens, J. W. Martin, D. A. Ellis and S. A. Mabury. Atmospheric chemistry of n-$C_2F_{2x+1}CHO$ ($x = 1, 3, 4$): reaction with Cl atoms, OH radicals and IR spectra of $C_xF_{2x+1}C(O)O_2NO_2$. *J. Phys. Chem. A* **108**, 5189–5196 (2004).
101. M. D. Hurley, J. C. Ball, T. J. Wallington, M. P. Sulbaek Andersen, O. J. Nielsen, D. A. Ellis, J. W. Martin and S. A. Mabury. Atmospheric chemistry of n-$C_xF_{2x+1}CHO$ ($x = 1, 2, 3, 4$): Fate of n-$C_xF_{2x+1}C(O)$ radicals. *J. Phys. Chem. A* **110**, 12443–12447 (2006).
102. M. P. Sulbaek Andersen, O. J. Nielsen, M. J. Hurley and T. J. Wallington. Atmospheric chemistry of t-$CF_3CH{=}CHCl$: Products and mechanisms of the gas-phase reactions with chlorine atoms and hydroxyl radicals. *Phys. Chem. Chem. Phys.* **14**, 1735–1748 (2012).
103. M. H. Russell, G. Hoogeweg, E. M. Webster, D. A. Ellis, R. L. Waterland and R. A. Hoke. TFA from HFO-1234yf: Accumulation and aquatic risk in terminal water bodies. *Environ. Toxicol. Chem.* **31**, 1957–1965 (2012).
104. T. J. Wallington, M. D. Hurley, J. Xia, D. J. Wuebbles, S. Sillman, A. Ito, J. Penner, D. A. Ellis, J. Martin, S. A. Mabury, O. J. Nielsen and M. P. Sulbaek Andersen. Formation of $C_7F_{15}COOH$ (PFOA) and other perfluorocarboxylic acids during the atmospheric oxidation of 8:2 fluorotelomer alcohol. *Environ. Sci. Technol.* 40, 924–930 (2006).
105. J. W. Martin, M. M. Smithwick, B. M. Braune, P. F. Hoekstra, D. C. G. Muir and S. A. Mabury. Identification of long-chain perfluorinated acids in biota from the Canadian Arctic. *Environ. Sci. Tech.* **38**, 373–380 (2004).
106. K. Kannan, J. W. Choib, N. Isekic, K. Senthilkumarc, D. H. Kima, S. Masunagac and J. P. Giesy. Concentrations of perfluorinated acids in livers of birds from Japan and Korea. *Chemosphere* **49**, 225–231 (2002).
107. K. J. Hansen, L. A. Clemen, M. E. Ellefson and H.O. Johnson. Compound-specific, quantitative characterization of organic: Fluorochemicals in biological matrices. *Environ. Sci. Tech.* **35**, 766–770 (2001).
108. N. Kudo and Y. Kawashima. Toxicity and toxicokinetics of perfluorooctanoic acid in humans and animals. *J. Toxicol Sci.* **28**, 49–57 (2003).

109. J. Berthiaume and K. B. Wallace. Perfluorooctanoate, perflourooctanesulfonate, and N-ethyl perfluorooctanesulfonamido ethanol; peroxisome proliferation and mitochondrial biogenesis. *Tox. Lett.* **129**, 23–32 (2002).
110. B. L. Upham, N. D. Deocampo, B. Wurl and J. E. Trosko. Inhibition of gap junctional intercellular communication by perfluorinated fatty acids is dependent on the chain length of the fluorinated tail. *Int. J. Cancer* **78**, 491–495 (1998).
111. S. K. Ritter. The shrinking case for fluorochemicals. *Chem. Eng. News* **93**, 27–29 (2015).
112. J. Lelieveld and P. J. Crutzen. Influences of cloud photochemical processes on tropospheric ozone. *Nature* **343**, 227–233 (1990).
113. G. P. Brasseur, J. J. Orlando and G. S. Tyndall. *Atmospheric Chemistry and Global Change*, Oxford University Press, Oxford (1999).
114. J. Sehested and T. J. Wallington. Atmospheric chemistry of hydrofluorocarbon 134a; fate of the alkoxy radical CF_3O. *Environ. Sci. Technol.* **27**, 146–152 (1993).
115. L. G. Huey, D. R. Hanson and E. R. Lovejoy. Atmospheric fate of CF_3 OH 1: Gas phase thermal decomposition. *J. Geophys. Res.* **100**, 18771–18774 (1995).
116. T. J. Wallington, W. F. Schneider, I. Barnes, K. H. Becker, J. Sehested and O. J. Nielsen. Stability and IR spectra of mono-, di-, and trichloromethanol. *Chem. Phys. Lett.* **322**, 97–102 (2000).
117. C. A. Taatjes, L. K. Christensen, M. D. Hurley and T. J. Wallington. Absolute rate coefficients and site-specific abstraction rates for reactions of Cl with CH_3CH_2OH, CH_3CD_2OH, and CD_3CH_2OH between 295 and 600 K. *J. Phys. Chem. A* **103**, 9805–9814 (1999).
118. W. F. Schneider, T. J. Wallington and R. E. Huie. Energetics and mechanism of decomposition of CF_3OH: Evidence for water-catalyzed homogeneous decomposition. *J. Phys. Chem.* **100**, 6097–6103 (1996).
119. D. A. Ellis, J. W. Martin, A. O. De Silva, S. A. Mabury, M. D. Hurley, M. P. Sulbaek Andersen and T. J. Wallington. Degradation of fluorotelomer alcohols: A likely atmospheric source of perfluorinated carboxylic acids. *Environ. Sci. Technol.* **38**, 3316–3321 (2004).
120. E. S. C. Kwok and R. Atkinson. Estimation of hydroxyl radical reaction rate constants for gas-phase organic compounds using a structure-reactivity relationship: An update. *Atmos. Environ.* **29**, 1685–1695 (1995).
121. H. G. Libuda, F. Zabel, E. H. Fink and K. H. Becker. Formyl chloride: UV absorption cross sections and rate constants for the reactions with Cl and OH. *J. Phys. Chem.* **94**, 5860–5865 (1990).
122. T. J. Wallington and M. D. Hurley. Atmospheric chemistry of HC(O)F: Reaction with OH radicals. *Envrion. Sci. Technol.* **27**, 1448–1452 (1993).
123. W. J. Debruyn, S. X. Duan, X. Q. Shi, P. Davidovits, D. R. Worsnop, M. S. Zahniser and C. E. Kolb. Tropospheric heterogeneous chemistry of haloacetyl and carbonyl halides. *Geophys. Res. Lett.* **19**, 1939–1942 (1992).
124. S. R Sellevag, V. Stenstrom, T. Helgaker and C. J. Nielsen. Atmospheric chemistry of CHF_2CHO: Study of the IR and UV-vis absorption cross

sections, photolysis and OH, Cl- and NO_3-initiated oxidation. *J. Phys. Chem. A* **109**, 3652–3662 (2005).

125. M. S. Chiappero, F. E. Malanca, G. A. Arguello, S. T. Woodridge, M. D. Hurley, J. C. Ball, T. J. Wallington, R. L. Waterland and R. C. Buck. Atmospheric chemistry of perfluoraldehyde ($C_xF_{2x+1}CHO$) and fluorotelomer aldehydes ($C_xF_{2x+1}CH_2CHO$): Quantification of the importance of photolysis. *J. Phys. Chem. A* **110**, 11944–11953 (2006).
126. R. K. Talukdar, A. Mellouki, J. B. Burkholder, M. K. Gilles, G. Le Bras, A. R. Ravishankara. Quantification of the tropospheric removal of chloral (CCl_3CHO): Rate coefficient for the reaction of OH, UV absorption cross sections, and quantum yields. *J. Phys. Chem A* **105**, 5188–5196 (2001).
127. M. P. Sulbaek Andersen, C. Stenby, O. J. Nielsen, M. D. Hurley, J. C. Ball, T. J. Wallington, J. W. Martin, D. A. Ellis and S. A. Mabury. Atmospheric chemistry of n-$C_xF_{2x+1}CHO$ ($x = 1, 3, 4$): Mechanism of the $C_xF_{2x+1}C(O)O_2 + HO_2$ reaction. *J. Phys. Chem. A* **108**, 6325–6330 (2004).
128. M. P. Sulbaek Andersen, A. Toft, O. J. Nielsen, M. D. Hurley, T. J. Wallington, K. Tonokura, S. A. Mabury, J. W. Martin and D. A. Ellis. Atmospheric chemistry of perfluorinated aldehyde hydrates (n-$C_xF_{2x+1}CH(OH)_2$, $x = 1, 3, 4$): Hydration, dehydration, and kinetics and mechanism of the Cl atom and OH radical initiated oxidation. *J. Phys. Chem. A* **110**, 9854–9860 (2006).
129. G. A. Russell. Deuterium-isotope effects in the autoxidation of aralkyl hydrocarbons — mechanism of the interaction of peroxy radicals. *J. Am. Chem. Soc.* **79**, 3871–3877 (1957).
130. S. Kutsuna, L. Chen, T. Abe, J. Mizukado, T. Uchimaru, K. Tokuhashi and A. Sekiya. Henry's law constants of 2,2,2-trifluoroethyl formate, ethyl trifluoroacetate, and non-fluorinated analogous esters. *Atmos. Environ.* **39**, 5884–5892 (2005).
131. S. Kutsuna, L. Chen, K. Ohno, K. Tokuhashi and A. Sekiya. Henry's law constants and hydrolysis rate constants of 2,2,2-trifluoroethyl acetate and methyl trifluoroacetate. *Atmos. Environ.* **38**, 725–732 (2004).
132. J. Wu, J. W. Martin, Z. Zhai, K. Lu, X. Fang and H. Jin. Airborne trifluoroacetic acid and its fraction from the degradation of HFC-134a in Beijing, China, 2014. *Environ. Sci. Technol.* **48**, 3675–3681 (2014).
133. T. J. Wallington, J. J. Orlando, G. S. Tyndall and O. J. Nielsen. Comment on airborne trifluoroacetic acid and its fraction from the degradation of HFC-134a in Beijing, China, *Environ. Sci. Technol.* **48**, 9948–9948 (2014).
134. C. A. Moody, J. W. Martin, W. C. Kwan, D. C. G. Muir and S. A. Mabury. Monitoring perfluorinated surfactants in biota and surface water samples following an accidental release of fire-fighting foam into Etobicoke Creek. *Environ Sci. Technol.* **36**, 545–551 (2002).
135. K. Prevedouros, I. T. Cousins, R. C. Buck and S. H. Korzeniowski. Sources, fate and transport of perfluorocarboxylates. *Environ. Sci. Technol.* **40**, 32–44 (2006).

136. C. A. Moody, G. N. Hebert, S. H. Strauss and J. A. Field. Occurrence and persistence of perfluorooctanesulfonate and other perfluorinated surfactants in groundwater at a fire-training area at Wurtsmith Air Force Base, Michigan, USA. *J. Environ. Mon.* **5**, 341–345 (2003).
137. C. A. Moody and J. A. Field. Perfluorinated surfactants and the environmental implications of their use in fire-fighting foams. *Environ Sci. Technol.* **34**, 3864–3870 (2000).
138. D. J. Bowden, S. L. Clegg and P. Brimblecombe. The Henry's law constant of trifluoroacetic acid and its partitioning into liquid water in the atmosphere. *Chemosphere.* **32**, 405–420 (1996).
139. R. Sander. Compilation of Henry's law constants (version 4.0) for water as solvent. *Atmos. Chem. Phys.* **15**, 4399–4981 (2015).
140. E. Kissa. *Fluorinated Surfactants: Synthesis, Properties, Applications.* (Marcel Dekker Inc., New York, 1994).
141. D. A. Ellis, T. M. Cahill, S. A. Mabury, I. T. Cousins and D. Mackay. Partitioning of organofluorine compounds in the environment. In *The Handbook of Environmental Chemistry Vol. 3, Part N Organofluorines*, Neilson, A. H., Ed. (Springer-Verlag: Berlin, 2002).
142. U. S. Environmental Protection Agency, *Revised draft hazard assessment of perfluorooctanoic acid and its salts*, Office of Pollution Prevention and Toxics; Risk Assessment Division (2003).
143. S. Kutsuna and H. Hori. Experimental determination of Henry's law constant of perfluorooctanoic acid (PFOA) at 298 K by means of an inert-gas stripping method with a helical plate. *Atmos. Environ.* **42**, 8883–8892 (2008).
144. J. H. Seinfeld and S. N. Pandis. *Atmospheric Chemistry and Physics: From Air Pollution to Climate Change*, 2nd Ed. (Wiley, New York, 2006).
145. C. J. McMurdo, D. A. Ellis, E. Webster, J. Butler, R. Christensen and L. K. Reid. Aerosol enrichment of the surfactant PFO and mediation of the water-air transport of gaseous PFOA. *Environ. Sci. Technol.* **42**, 3969–3974 (2008).
146. J. W. Martin, D. A. Ellis, S. A. Mabury, M. D. Hurley and T. J. Wallington. Atmospheric chemistry of perfluoroalkanesulfonamides: Kinetic and product studies of the OH and Cl atom initiated oxidation of N-ethyl perfluorobutanesulfonamide. *Environ. Sci. Technol.* **40**, 864–872 (2006).
147. J. C. D'eon, M. D. Hurley, T. J. Wallington and S. A. Mabury. Atmospheric chemistry of N-methyl perfluorobutane sulfonamidoethanol, $C_4F_9SO_2N(CH_3)CH_2CH_2OH$: Kinetics and mechanism of reaction with OH. *Environ. Sci. Technol.* **40**, 1862–1868 (2006).
148. A. O. De Silva, C. N. Allard, C. Spencer, G. M. Webster and M. Shoeib. Phosphorus-containing fluorinated organics: Polyfluoroalkyl phosphoric acid diesters (diPAPs), perfluorophosphonates (PFPAs), and perfluorophosphinates (PFPIAs) in residential indoor dust. *Environ. Sci. Technol.* **46**, 12575–12582 (2012).

149. J. C. D'eon and S. M. Mabury. Is indirect exposure a significant contributor to the burden of perfluorinated acids observed in humans? *Environ. Sci. Technol.* **45**, 7974–7984 (2011).
150. Z. Wang, I. T. Cousins, M. Scheringer and K. Hungerbühler. Fluorinated alternatives to long-chain perfluoroalkyl carboxylic acids (PFCAs), perfluoroalkane sulfonic acids (PFSAs) and their potential precursors. *Environ. Int.* **60**, 242–248 (2013).
151. L. Vierke, U. Berger and I. T. Cousins. Estimation of the acid dissociation constant of perfluoroalkyl carboxylic acids through an experimental investigation of their water-to-air transport. *Environ. Sci. Technol.* **47**, 11032–11039 (2013).
152. J. P. Giesy and K. Kannan. Global distribution of perfluorooctane sulfonate in wildlife. *Environ. Sci. Technol.* **35**, 1339–1342 (2001).
153. J. W. Martin, D. C. G. Muir, C. A. Moody, D. A. Ellis, W. Kwan, K. R. Solomon and S. A. Mabury. Collection of airborne fluorinated organics and analysis by gas chromatography/chemical ionization mass spectrometry. *Anal. Chem.* **4**, 584–590 (2002).
154. R. C. Buck, J. Franklin, U. Berger, J. M. Conder, I. T. Cousins, P. de Voogt, A. A. Jensen, K. Kannan, S. A. Mabury and S. P. J. van Leeuwen. Perfluoroalkyl and polyfluoroalkyl substances in the environment: Terminology, classification, and origins. *Integ. Environ. Assess Management* **7**, 513–541 (2011).
155. P. De Voogt, (ed.). *Reviews of Environmental Contamination and Volume 208: Perfluorinated alkylated substances.* (Springer, New York, 2010).
156. J. P. Benskin, M. G. Ikonomou, F. A. P. C. Gobas, M. B. Woudneh and J. R. Cosgrove. Observation of a novel PFOS-precursor, the perfluorooctane sulfonamido ethanol-based phosphate (SAmPAP) diester, in marine sediments. *Environ. Sci. Technol.* **46**, 6505–6514 (2012).
157. H. Lee, J. C. D'eon and S. A. Mabury. Biodegradation of polyfluoroalkyl phosphates as a source of perfluorinated acids to the environment. *Environ. Sci. Technol.* **44**, 3305–3310 (2010).
158. M. P. Sulbaek Andersen, D. R. Blake, F. S. Rowland, M. D. Hurley and T. J. Wallington. Atmospheric chemistry of sulfuryl fluoride: Reaction with OH radicals, Cl atoms and O_3, Atmospheric lifetime, IR spectrum, and global warming potential. *Environ. Sci. Technol.* **43**, 1067–1070 (2009).
159. V. C. Papadimitriou, R. W. Portmann, D. W. Fahey, J. Muhle, R. F. Weiss and J. B. Burkholder. Experimental and theoretical study of the atmospheric chemistry and global warming potential of SO_2F_2. *J. Phys. Chem. A* **112**, 12657–12666 (2008).
160. J. Muhle, J. Huang, R. F. Weiss, R. G. Prinn, B. R. Miller, P. K. Salameh, C. M. Harth, P. J. Fraser, L. W. Porter, B. R. Greally, S. O'Doherty and P. G. Simonds. Sulfuryl fluoride in the global atmosphere. *J. Geophys. Res.* **114**, D05306, (2009). doi:10.1029/2008JD011162.
161. A. Muñoz. Personal communication (2015).

162. A. Muñoz, M. Ródenas, E. Borrás, M. Vázquez and T. Vera. The gas-phase degradation of chlorpyrifos and chlorpyrifos-oxon towards OH radical under atmospheric conditions. *Chemosphere* **111**, 522–528 (2014).
163. A. Muñoz, T. Vera, H. Sidebottom, A. Mellouki, E. Borrás, M. Ródenas, E. Clemente and M. Vázquez. Studies on the atmospheric degradation of chlorpyrifos-methyl. *Environ. Sci. Technol.* **45**, 1880–1886 (2011).
164. V. Feigenbrugel, A. Le Person, S. Le Calvé, A. Mellouki, A. Muñoz and K. Wirtz. Atmospheric fate of dichlorvos: Photolysis and OH-initiated oxidation studies. *Environ. Sci. Technol.* **40**, 850–857 (2006).
165. A. Muñoz, T. Vera, M. Ródenas, E. Borrás, A. Mellouki, J. Treacy and H. Sidebottom. Gas-phase degradation of the herbicide ethalfluralin under atmospheric conditions. *Chemosphere* **95**, 395–401 (2014).
166. A. Muñoz, T. Vera, H. Sidebottom, M. Ródenas, E. Borrás, M. Vázquez, M. Raro and A. Mellouki. Studies on the atmospheric fate of propachlor (2-chloro-N-isopropylacetanilide) in the gas-phase. *Atmos. Environ.* **49**, 33–40 (2012).
167. A. Le Person, A. Mellouki, A. Muñoz, E. Borrás, M, Martin-Reviejo and K. Wirtz. Trifluralin: Photolysis under sunlight conditions and reaction with HO radicals. *Chemosphere* **67**, 376–383 (2007).
168. T. Vera, E. Borrás, J. Chen, C. Coscollá, V. Daële, A. Mellouki, M. Ródenas, H. Sidebottom, X. Sun, V. Vicent Yusá, X. Zhang and A. Muñoz. Atmospheric degradation of lindane and 1,3-dichloroacetone in the gas phase. Studies at the EUPHORE simulation chamber. *Chemosphere* **138**, 112–119 (2015).
169. T. Vera, A. Muñoz, M. Ródenas, E. Borrás, M. Vázquez, A. Mellouki, J. Treacy, I. Al Mulla and H. Sidebottom. Photolysis of trichloronitromethane (chloropicrin) under atmospheric conditions. *Z. Phys. Chem.* **224**, 1039–1057 (2010).
170. F. Li, R. S. Stolarski and P. A. Newman. Stratospheric ozone in the post-CFC era. *Atmos. Chem. Phys.* **9**, 2207–2213 (2009).

Chapter 6

Atmospheric Reaction Rate Constants and Kinetic Isotope Effects Computed Using the HEAT Protocol and Semi-Classical Transition State Theory

Thanh Lam Nguyen,[*,‡] John R. Barker,[†,§] and John F. Stanton[*,¶]

[*] *Department of Chemistry, The University of Texas, Austin, Texas 78712-0165, USA*

[†] *Climate and Space Sciences and Engineering, University of Michigan, Ann Arbor, MI 48109-2143, USA*

[‡] *lnguyen@cm.utexas.edu*

[§] *jrbarker@umich.edu*

[¶] *jfstanton@mail.utexas.edu*

Thermal rate constants and kinetic isotope effects (KIEs) are required in chemical kinetics modeling to understand chemical processes in the atmosphere. These quantities can often be obtained experimentally, but sometimes this is not achievable in the laboratory under atmospheric conditions. High-level theoretical calculations are needed in such cases to support and complement experimental results. In this work, we have reviewed the use of Miller's semiclassical transition state theory (SCTST), in conjunction with second-order vibrational perturbation theory (VPT2) and a very high level treatment of electronic structure (the "HEAT" protocol). Some important reactions chosen as examples are presented and show that the combination of the SCTST/VPT2/HEAT approaches can provide results that are comparable in quality to available experiments. As compared to other theoretical techniques, the method appears to be a practical tool that can offer a good compromise between accuracy and computational cost. Limitations of the method are also discussed.

6.1. Introduction

Quantitative modeling of the reaction networks important in the atmosphere is based mostly on solving the coupled ordinary differential equations that are derived from explicit elementary reaction mechanisms and the associated elementary rate constants. The reaction models can be used to interpret observational data, to assess environmental impacts of pollutants, to predict chemical changes due to radiative forcing by greenhouse gases, and for many other purposes.

The accuracy of models depends critically on the quality of the rate constants, all of which depend on temperature and some of which also depend on pressure. The International Standard Atmosphere [International Organization for Standardization, *Standard Atmosphere*, ISO 2533:1975, 1975] and U. S. Standard Atmosphere[1] summarize average atmospheric temperature and pressure as functions of altitude, month, and latitude. Global average temperatures at sea level, the tropopause, stratopause, and mesopause are 288, 217, 271, and 187 K, respectively. The observed extremes range to much higher and lower temperatures. For example, extreme surface temperatures range up to ~330 K (Death Valley, CA in 1913) and down to ~184 K (Vostok, Antarctica, in 1983) [http://wmo.asu.edu/#global]. Temperatures at the mesopause range down to perhaps 135 K. Atmospheric pressures range downward from about 1.1 bar, becoming much smaller at higher altitudes.

Many rate constants have been measured as functions of temperature and pressure in laboratory experiments. Critical evaluations of laboratory data relevant to atmospheric chemistry are published periodically. The NASA/JPL evaluation[2] emphasizes reactions important in the stratosphere and upper troposphere, while the IUPAC evaluations[3–5] place more emphasis on the lower troposphere; there is considerable overlap between the two. Many rate constant measurements have been reported at ~300 K, some have been reported down to ~250 K, but, due to experimental limitations, only a few extend to lower temperatures. Thus either the existing data must be extrapolated to low atmospheric temperatures, or additional rate constant data must be obtained. New data can be obtained by performing additional measurements, but theoretical methods have advanced to the point that new data can be obtained in some cases by theoretical calculations. The latter is the subject of this review.

In this review of work from our groups, we describe high-level theoretical calculations that were carried out with two aims. First, we aimed to assess whether the methods are sufficiently accurate for the purpose of modeling atmospheric reactions. Thus, some reactions that we have studied have little relevance to the atmosphere, but they serve as important test cases. We show that in favorable cases, the theoretical absolute rate constants are as accurate as experimental measurements, but in other cases they fall somewhat short and improvements are still needed. Our second aim was to address specific atmospheric reactions in order to provide better quantitative descriptions of how their rate constants depend on environmental conditions.

This review is organized as follows. Theoretical methods are presented in Section 2. These include discussions of the SCTST/VPT2 approach, the HEAT protocol, Atmospheric Isotope Effects, and the master-equation technique. In Section 3, results for several reactions are presented. These include reactions selected for testing theoretical methods ($H + H_2$ and $O + H_2$) and reactions of atmospheric relevance ($HO + H_2$, $Cl + CH_4$, and $HO + CO$). In the final section, the general performance of the theoretical methods is discussed, along with suggestions for future work.

6.2. Theory

6.2.1. *Theoretical Background*

The field of theoretical chemistry has grown in significance in the past few decades.[6–13] In addition to deriving benefits from improved theoretical methodology, the burgeoning field has also benefitted from rapid developments in computational hardware (high-performance computing) and advanced numerical algorithms.[8,10,14] One area that has witnessed significant and exciting developments is the field of chemical kinetics. In principle, all information pertaining to a chemical reaction from a state characterized by the vibrational and rotational state (ν_1, J_1) to another state (ν_2, J_2),[7,9–11,15] including quantum mechanical effects (such as zero-point vibration energy, non-classical reflection, tunneling, and resonances) can be obtained from calculations. Unfortunately, such sophisticated work can only be carried out for very small molecular systems (up to about five atoms)[14] because the complexity of all aspects of the computation grow extremely rapidly with the numbers of electrons and vibrational degrees of freedom (dimensionality of the potential energy surface, or PES). Some approaches have been suggested to treat such problems in a reduced-dimension framework,[16–20] although such a framework that does not compromise the exquisite accuracy of full-dimensional quantum calculations has yet to be found.

As an alternative to the formally exact approaches alluded to above, one can also compute rate constants using either transition state theory (TST)[21–23] or Rice–Ramsperger–Kassel–Marcus (RRKM) theory.[24–26] In such calculations, one only needs rovibrational parameters and barrier heights of stationary points on the PES; knowledge of the *global* PES is not needed. The information associated with the stationary points (including their geometries) can now be computed very accurately with sophisticated quantum-chemical methodology.[27–30] The restriction to information about the PES to those regions "near" the stationary points (minima and first-order saddle points) means that the quantum-chemical methods tend to be used in regions where they usually perform quite well. In contrast, computing properties of the PES in "bond-breaking" regions and other "remote" parts of the PES can be considerably more complicated and challenging.

With the PES information in hand, conventional TST or RRKM calculations are very simple to do. However, conventional TST does not include some important quantum mechanical effects such as tunneling and non-classical reflection (e.g., barrier re-crossing).[31–37] The former effect is often seen when the motion of light atom(s) is important in a chemical reaction at low temperatures (or, from a microcanonical perspective, at low energies that are below the energy barrier),[35,38] whereas the re-crossing effect often occurs at very high temperatures (or at high energies that are well above the energy barrier).[31–33] The tunneling effect can be estimated by empirically multiplying the calculated rate constant by a correction factor, a practice that comprises a so-called tunneling correction.[35] The non-classical reflection effect, on the other hand, can be recovered using variational transition state theory (VTST).[35,37] A large number of techniques, most of which are largely empirical in nature, have been developed to compute these two effects, and they are implemented in the well-known POLYRATE package.[39] However, to compute these two effects using POLYRATE, in addition to two stationary points on the PES, more detailed information — including the reaction energy profile and the dependence of the Hessian (matrix of second derivatives) with respect to the reaction path coordinate — is needed.[39]

An alternative way to account for these quantum-mechanical effects was proposed by Miller in 1977.[40] By recognizing that good action-angle variables exist at a transition state as well as at minima on the PES, Miller and coworkers developed a SCTST. Given an expression for the energy in terms of a set of vibrational quantum numbers, the semiclassical correspondence between the reaction coordinate "quantum number" and the barrier transmission coefficient can be used to compute state-specific reaction probabilities that include tunneling and reaction path curvature (anharmonic terms that couple the reaction coordinate and the bath coordinates) in a framework that is effectively non-empirical.

Later, together with Handy, Miller and coworkers[41–43] developed a practical SCTST based on the energy formula taken from VPT2,[44] and reported preliminary calculations using this approach. This variant of SCTST, which is based on VPT2 (hereafter, designated as SCTST/VPT2), is a simple approach (but still gives a high accuracy) because it requires neither a global PES nor detailed information along the reaction path; instead, the local regions at (and nearby) the stationary points on the PES are sufficient for calculations of reaction rate constants. According to VPT2,[41,44–46] these local regions on the PES are characterized using fully quadratic, fully cubic, and semi-quartic force fields that can be calculated relatively routinely[47] from high-level quantum chemistry methods.[27,30] This is certainly more information than traditional TST or RRKM calculations (where only the Hessian matrix is needed), but the extra information is not excessive and is certainly warranted by the improvement of the theoretical results.

Recently, we developed a new method[45] based on the Wang–Landau algorithm[48–50] to compute quantum vibrational states of a molecule efficiently. This technique allows us to study medium-sized molecular reaction systems using the SCTST/VPT2 approach.[45,46] In a number of recent works,[46,51–59] it has been amply demonstrated that SCTST/VPT2 can provide *ab initio* thermal rate constants that are in very good agreement with reliable experimental values, *provided that SCTST/VPT2 is used in conjunction with very high-level quantum chemical calculations to characterize the potential surface and its derivatives.* Therefore, the SCTST/VPT2 approach may be used to accurately predict thermal rate constants under a variety of conditions, including those that are not accessible to experiments. In addition, SCTST/VPT2 can provide microcanonical rate constants that are dependent upon both energy and angular momentum, $k(E, J)$. The resulting $k(E, J)$ can be used in a two-dimensional (2D; depending on both E and J), or a one-dimensional (1D; depending on E alone, after averaging over J) master equation treatment[58–61] that is applied to various gas-phase reactions (e.g., in atmosphere, combustion and planetary) to compute thermally averaged rate constants and product branching ratios as functions of both pressure and temperature.

It is also important to estimate possible errors in the calculated results using SCTST/VPT2. For this purpose, a sensitivity analysis needs to be carried out. Apart from limitations of SCTST/VPT2 itself, the calculated rate constants are quite sensitive to the input data (including barrier height, the barrier frequency, rovibrational parameters, and anharmonic constants) that can all be obtained directly from high-level quantum chemistry. Our experience with spectroscopic calculations for local minima[62] shows that coupled-cluster methods in combination with VPT2 and appropriate basis sets can be used to reproduce measured fundamental frequencies within $10\,\text{cm}^{-1}$. This suggests that the rovibrational parameters and anharmonic constants calculated with coupled-cluster methods are quite accurate.

For the barrier height, the form of the Arrhenius equation[63] shows that the rate constant depends exponentially on this quantity, and it is essential to use an accurate value in meaningful kinetics calculations. For example, with a change of 0.25 kcal/mol (*ca.* 1 kJ/mol, which is termed "subchemical accuracy") in the energy barrier, the calculated thermal rate constant is altered by a factor of 1.5 at room temperature. Thus, the accurate calculation of the barrier height plays an extremely important role in computing accurate reaction rate constants, especially at low temperatures. Improving the accuracy of the barrier energy from subchemical to sub-kJ/mol accuracy (i.e., reducing the error from $\sim 80\,\mathrm{cm}^{-1}$ to $\sim 20\,\mathrm{cm}^{-1}$) is very challenging and costly because calculations of molecular energies at this level of accuracy are dependent upon many factors. We have chosen to employ the HEAT method[64–66] of computational thermochemistry for these purposes.

High-accuracy extrapolated *ab initio* thermochemistry (HEAT) was originally developed to compute bond energies and heats of formation, but we have found that it serves well as a computational framework for transition state properties, as well. HEAT is a systematic approach: it endeavors to estimate the exact electronic energy at the full CI limit, computed with an infinite basis set. Of course, such a calculation is based on a number of approximations, including basis set extrapolation and the assumption that all correlation effects are recovered in a coupled-cluster framework that includes some treatment of quadruple excitations. In addition to this, HEAT includes corrections for special relativity, spin-orbit (SO) effects, and the diagonal Born–Oppenheimer correction (DBOC).[64–66] Finally, the zero-point energy is determined beyond the harmonic approximation (with VPT2).[44]

6.2.2. *Thermochemistry: The HEAT Protocol*

To a very good approximation, the ground state energy of a molecular system can be obtained by summing the following quantities: (1) the electronic energy, obtained in the usual non-relativistic clamped-nucleus (Born–Oppenheimer) approximation[67]; (2) the zero-point energy of molecular vibrations that take place on the PES defined by (1); (3) the corrections to the energy due the effects of special relativity; and (4) the effects of nuclear motion on the PES (the adiabatic correction).[68] The HEAT[64–66] was developed about a decade ago to approximate each of these contributions as accurately as one can practically achieve, given the constraints provided by the cost of doing the requisite calculation on molecular systems. How each of these contributions is approximated by HEAT is described in the following paragraphs.

The most difficult of the contributions above to compute accurately is the electronic energy. In practical terms, the best estimate of this quantity is that which most closely approximates the so-called full configuration interaction (FCI) calculation using wave functions constructed from a complete set of orthonormal basis functions (i.e., the infinite basis set). To the reader unfamiliar with the detailed arcana of quantum chemistry, it suffices to say that this is a variational calculation with the most general and flexible trial wave function possible. However, such a calculation is not practical, and must be approximated. The way that this is done in the HEAT protocol is to use coupled-cluster theory[69,70] (a many-body approach that is not variational, but which provides a rapidly convergent sequence of approximations to a full matrix diagonalization). Any calculation that uses coupled-cluster theory is reliant upon a basis set, which usually comprises atom-centered Gaussian functions[71,72] As this size of this basis increases, and the coupled-cluster expansion is extended, the electronic energy will converge to "exactness" (the Born–Oppenheimer result alluded to above), and the onus is on quantum chemistry to develop accurate, yet economical, approximations. In the HEAT method, the complete basis-set limit (the second consideration above) is approached through calculations with a well-defined sequence of basis sets and then extrapolating the result. This process starts with the CCSD(T) coupled-cluster treatment[73,74] of electron correlation, and then further corrections are made that involve additional triple and quadruple "excitation" effects.

In HEAT, the zero-point energy is obtained with VPT2, using potential functions obtained with levels of theory well below those that are used to obtain the electronic energy (contribution 1, described above), because it is known that the higher levels of theory are not needed for calculating accurate vibrational parameters. In the initial description of the method, the zero-point energy associated with HEAT is obtained at the CCSD(T) level of theory, using the cc-pVQZ basis set of Dunning.[75] Such calculations can be quite expensive for polyatomic molecules with a large number of vibrational modes because quadratic, cubic and a subset of quartic force constants are needed. We have recently also performed some calculations at the CCSD(T) level, where the ANO1 basis set[76,77] (which is smaller than cc-pVQZ, but approximately the same in accuracy) for the harmonic contribution, and ANO0 (which is smaller still) for the anharmonic contribution.

The simplest relativistic contributions to the ground-state energy of molecules are of two types, scalar and SO. The former include the mass-velocity correction (which accounts for the difference in the relativistic and non-relativistic kinetic energy operators),[78] and the one-and two-electron Darwin terms, which account — in a general way — for *zitterbewegung* ("trembling motion") which is an effect from quantum electrodynamics that accounts for electron spin and the electron magnetic moment. Both of these are estimated in HEAT by means of first-order perturbation theory, with numerical values obtained by contracting appropriate integrals with the electronic density matrix. This calculation is at the CCSD(T) level, using the aug-cc-pCVTZ basis set. In lowest order, all that is considered in HEAT, SO contributions split degenerate states into multiplets, and the physical ground state energy of the molecule is that of the lowest component of the multiplet. In HEAT, this correction is needed only for degenerate states (usually atoms and linear molecules only), and is taken from experiment when available. Otherwise, it can be calculated.[79]

Finally, and certainly least important in magnitude, is the adiabatic correction,[68] which is also approximated by first-order perturbation theory by computing the expectation value of the nuclear kinetic energy operator over the electronic wave function. Such calculations can be performed at the SCF or CCSD level, using the aug-cc-pVTZ basis set.

By summing the contributions above, the HEAT energy is obtained. A considerable amount of benchmarking has shown that bond energies calculated by the HEAT approach are within 1 kJ/mol of the exact values in most cases.[64–66] However, when HEAT is extended to transition states — as we have done in recent years for our SCTST/VPT2 calculations[46,51–59] — one often needs to be somewhat more conservative in the error estimate. The reason for this is that the electronic structures of stable minima are usually less complicated than for transition states, and the calculation of the correlation energy (the first contribution discussed above) is very difficult. In an extreme case, the use of coupled-cluster theory (the foundation of the HEAT method) may even be inappropriate, thus requiring very expensive multi-reference methods.[28,80] However, by extending the calculations that are done in HEAT to those including so-called quadruple excitation effects,[28,81] the range of minima and transition states that are adequately treated is vastly extended.

6.2.3. *SCTST/VPT2 Rate Constants*

6.2.3.1. *Statistical Rate Constants*

According to statistical chemical kinetics theory,[82–86] microcanonical (energy-dependent) and canonical (temperature-dependent) rate constants are expressed as:

$$k(E, J) = \frac{\sigma}{h} \times \frac{G^{\neq}(E, J)}{\rho(E, J)}, \tag{6.1}$$

$$k(T) = \frac{1}{h} \times \frac{Q_t^{\neq} Q_e^{\neq}}{Q_t^{re} Q_e^{re}} \times \frac{\sum_{J=0}^{\infty} \int_0^{+\infty} G^{\neq}(E,J) \exp\left(-\frac{E}{RT}\right) dE}{\sum_{J=0}^{\infty} \int_0^{+\infty} \rho(E,J) \exp\left(-\frac{E}{RT}\right) dE}, \tag{6.2}$$

where h is Planck's constant, σ is the reaction path degeneracy, Q_t is the translational partition function, Q_e is the electronic partition function (the superscripts "re" and "$\neq$" designate reactants and transition state, respectively), T is temperature, and E is the internal energy measured from the reactant(s). $G^{\neq}(E, J)$ and $\rho(E, J)$ are the sums and densities of rovibrational states of transition state and reactant(s), respectively. The former quantity is also frequently termed the cumulative reaction probability (CRP).[41] Assuming that three external rotors are active at the stationary points, vibrations and rotations can exchange energies freely. As a result, $G^{\neq}(E, J)$ and $\rho(E, J)$ can be convoluted[84] using Eqs. (6.3) and (6.4).

$$G^{\neq}(E, J) = \int_0^E G_v^{\neq}(E_v)\rho_r^{\neq}(E - E_v)dE_v, \tag{6.3}$$

$$\rho(E, J) = \int_0^E \rho_v(E_v)\rho_r(E - E_v)dE_v, \tag{6.4}$$

where v and r stand for vibrational and rotational degrees of freedom, respectively.

For a nonlinear molecule, A, B, and C are the three rotational constants. By assuming $B \approx C$ it is possible to describe the stationary points as symmetric tops with rotational energy levels given by

$$E_r = \bar{B}J(J + 1) + (A - \bar{B})K^2, \tag{6.5}$$

where $\bar{B} = \sqrt{B \cdot C}$ and $|K| \leq J$. This is a conventional (and expedient) approximation in unimolecular rate theory.[86]

The problem of computing reaction rate constants with Eqs. (6.1) and (6.2) reduces to computing sums and densities of vibrational states for the transition state and reactants, respectively. Within the harmonic oscillator approximation where all vibrations are separable, these quantities can be computed efficiently using the standard combination of the Beyer–Swinehart[87] and Stein–Rabinovitch[88] (BSSR) algorithms. Unfortunately, the BSSR method cannot be used for coupled vibrations, and other methods are required. For small molecules, the brute-force direct-count method can be used to obtain sums and densities of vibrational states, but this technique rapidly becomes expensive and ultimately impossible as the molecule becomes larger.

Recently, we have developed a new technique[45] that is based on the Wang–Landau algorithm[48–50] to compute sums and densities of states for non-separable vibrations. This technique can be applied to medium-sized molecules that have a few dozen vibrational degrees of freedom.[89] A "parallelized" version of this algorithm (intended for parallel execution on large multiple-processor computer systems) has been reported, which is suitable for coupled anharmonic vibrational models with >100 degrees of freedom.[90] Very recently, the Barker research group has developed an efficient hybrid code ("bdens") that combines the direct-count method, which produces exact counts in the low energy regime, with the Wang–Landau method, which produces quite accurate results at higher energies. This new code, which is available in the MultiWell Program Suite,[91] offers another good compromise between accuracy and computational cost.

6.2.3.2. *SCTST/VPT2*

According to reaction rate theory,[40,41] the sum of vibrational states of the transition state (or cumulative reaction probability, CRP) is given by:

$$G_v^{\neq}(E) = \sum_{n_1} \sum_{n_2} \cdots \sum_{n_{F-2}} \sum_{n_{F-1}} P_n(E), \tag{6.6}$$

where n_i is the ith vibrational quantum number. $F - 1$ is the number of orthogonal vibrational degrees of freedom exclusive of the imaginary frequency vibration along the reaction coordinate. P_n is the tunneling probability, which according to the uniform semiclassical approximation is given by:

$$P_n(E) = \frac{1}{1 + \exp[2\theta(n, E)]}, \tag{6.7}$$

where the barrier penetration integral $\theta(n, E)$ and associated quantities are obtained by the semiclassical correspondence between the quantum number of the reaction coordinate and $\theta(n, E)$, viz.

$$\frac{i\theta(n, E)}{\pi} \Leftrightarrow \left(n_F + \frac{1}{2}\right). \tag{6.8}$$

Within VPT2, the energy expression assumes the form

$$E = \Delta V_o + G_o + \sum_{k\in\perp} \omega_k \left(n_k + \frac{1}{2}\right) + \omega_F \left(n_F + \frac{1}{2}\right)$$

$$+ \sum_{(k,l)\in\perp} x_{kl} \left(n_k + \frac{1}{2}\right)\left(n_l + \frac{1}{2}\right)$$

$$+ \sum_{k\in\perp} x_{kF} \left(n_k + \frac{1}{2}\right)\left(n_F + \frac{1}{2}\right) + x_{FF}\left(n_F + \frac{1}{2}\right)^2, \quad (6.9)$$

where the sums extend over the modes perpendicular to the reaction coordinate, the barrier penetration integral can be written as

$$\theta(n, E) = \frac{-\pi \Delta E}{\Omega_F} \cdot \frac{2}{1+\sqrt{1-4x_{FF}\Delta E/\Omega_F^2}}, \quad (6.10)$$

$$\Delta E = E - \Delta V_o - G_o - \sum_{k=1}^{F-1} \omega_k \left(n_k + \frac{1}{2}\right)$$

$$-\sum_{k=1}^{F-1}\sum_{l=k}^{F-1} x_{kl}\left(n_k + \frac{1}{2}\right)\left(n_l + \frac{1}{2}\right), \quad (6.11)$$

$$\Omega_F = -i\omega_F - \sum_{k=1}^{F-1} i x_{kF}\left(n_k + \frac{1}{2}\right). \quad (6.12)$$

In these expressions, ω_k is the kth harmonic vibrational frequency, ω_F is the imaginary frequency associated with the reaction coordinate, x_{kl} are the anharmonic constants that involve coupling among the orthogonal vibration degrees of freedom, x_{kF} are the (pure imaginary) coupling terms between the reaction coordinate and the orthogonal vibration degrees of freedom, x_{FF} is the (real-valued) anharmonic constant for the reaction coordinate, and ΔV_o is the classical barrier height. The term G_o is a constant, which cannot be determined from spectroscopic measurements (where only energy differences are measured) but that must be included for thermochemistry (and, of course, kinetics).

ΔE in Eq. (6.11) represents the vibrational energy (that is measured from the top of the classical barrier) available along the reaction coordinate. According to classical mechanics, the tunnel effect does not exist; therefore, P_n classically is equal to zero and unity when ΔE is negative and positive, respectively. In contrast, in quantum mechanics, the tunneling probability P_n does not vanish below the top of the barrier; P_n from Eq. (6.7) is easily verified to be in the range [0, 1]. On the basis of these remarks, thermally averaged rate constants with $(k(T)_w)$ and without $(k(T)_{wo})$ tunneling corrections can be computed; and then quantum mechanical tunneling corrections as a function of temperature can be defined as the ratio of $k(T)_w/k(T)_{wo}$.

Ω_F in Eq. (6.12) is the effective (positive real-valued) reaction coordinate frequency that accounts for the effects of coupled anharmonic constants x_{kF} by modifying the magnitude of imaginary frequency ω_F. Thus, Ω_F may be associated with the effective width of a barrier that sits atop the multi-dimensional, curved reaction path.

6.2.3.3. *Limitations of SCTST/VPT2*

In this section, we discuss three scenarios where SCTST/VPT2 is not sufficient and describe pragmatic ways to still obtain realistic results. Two of these scenarios start to become important only at extremely high and extremely low energies and temperatures. The third scenario becomes significant only at very high energies. None of these scenarios significantly affects rate constants at the temperatures that are important for most atmospheric and combustion applications.

Because the VPT2 approach is used, the SCTST/VPT2 model is subject to failures of this quite simple, but successful, treatment of anharmonicity. Consequently, it is also important to know when the SCTST/VPT2 model breaks down. Generally speaking, one may speculate that the SCTST/VPT2 model can fail when the energy available to the reaction coordinate (ΔE) is either far above the barrier or far below it (the deep tunneling regime). As can be seen in Eq. (6.10), SCTST/VPT2 will fail catastrophically when the term $1-4x_{FF}\Delta E/\Omega_F^2$ becomes negative, or $\Omega_F^2 < 4x_{FF}\Delta E$. In such a case, the quadratic equation that governs the barrier penetration integral

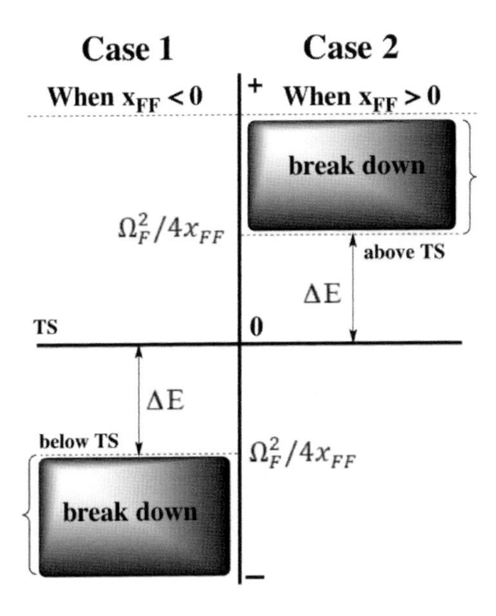

Scheme 6.1 Two cases where the SCTST/VPT2 approach fails.

has no physically meaningful solutions. Two possible scenarios lead to this failure of SCTST/VPT2 (see Scheme 6.1): (i) First, when $x_{FF} < 0$ (as is usually the case in calculations on real systems), it can happen when ΔE is less than $\Omega_F^2/4x_{FF}$ and becomes negative ($\Delta E < \Omega_F^2/4x_{FF} < 0$). This case, which corresponds to ΔE values being "*too negative*", occurs in the deep-tunneling regime below the TS, and Wagner[92] has given an extensive analysis of the problem, along with a suggested solution. When this occurs in our implementation, however, P_n is set to zero; (ii) Second, when $x_{FF} > 0$ (which is very uncommon), leading to $\Delta E > \Omega_F^2/4x_{FF} > 0$. This case, which corresponds to ΔE values being "*too positive*", can be seen in the high-energy regime above the TS. Our practical workaround to this case is to set P_n to unity.

The SCTST/VPT2 approach can also experience a third type of failure. When the energy is high enough, the quantum numbers of the orthogonal degrees of freedom may be quite large. If, in addition, the off-diagonal anharmonicities (i.e., the x_{kF} terms) that couple the reaction coordinate to the orthogonal modes are negative, Ω_F may be reduced to zero, or even to negative values (see Eq. (6.12)). Since Ω_F is the effective magnitude of the imaginary frequency, a reduced value of Ω_F suggests that the barrier thickness has increased and the probability of tunneling has therefore decreased. When $\Omega_F = 0$, the barrier is essentially infinite and there is no tunneling. The "negative omega" scenario (for $\Omega_F \leq 0$) is considered to be unphysical: SCTST/VPT2 breaks down because there is no effective maximum. Possibly a higher order of VPT can remedy this failure in future, but for present purposes a practical workaround is needed for the "negative omega" scenario. Our practical approach is to regard the barrier is infinitely thick if $\Omega_F \leq 0$ and we thus assume in that case that the system behaves classically: the tunneling probability $P_n = 0$ when $\Delta E \leq 0$, and $P_n = 1$ when $\Delta E > 0$.

6.2.4. *Rate Theory for Pressure-Dependent Reactions*

Energy transfer collisions play an important role controlling the rates of recombination, unimolecular dissociation, and isomerization reactions. The theory describing these processes is quite mature; it has been described in several monographs[82,85,86,93] and has been implemented in several computer codes.[94–99] (Note that the list of codes referenced here is not comprehensive and it is beyond the scope of the present work to analyze their relative strengths and weaknesses.) It suffices to say that most of the kinetics calculations reported here were performed using the MultiWell Program Suite.[91]

In a unimolecular dissociation reaction, the rate constant increases rapidly as the internal energy of an energized reactant molecule increases above the reaction threshold (i.e., the critical energy E_0). The rate constant for a microcanonical ensemble of excited reactant molecules with internal energy E and angular momentum J is described accurately (under most conditions of interest) by Eq (6.1) (Section 6.2.3), which predicts that the rate constant $k(E, J)$ increases very rapidly with E. Because $k(E, J)$ is largest at high energies, the population distribution tends to become depleted at high energy. This effect tends to reduce the average rate of reaction. The depleted population can be restored by collisional activation, if the rate of collisional energy transfer is fast enough. Because the latter depends on bath gas pressure, the average rate constant is pressure dependent.

Similar collisional effects play a role in recombination reactions (the reverse of unimolecular dissociations) and isomerization reactions. All of these processes can be described mathematically with a "master equation",[86] which is a set of coupled rate equations that account for the sources and sinks (input and output) of population with a specified energy E. The master equation can be written as follows[55,91]:

$$\begin{aligned}\frac{dN(E',J',t)}{dt} &= F'(E',J',t) + \sum_{J}\int_{0}^{\infty} R(E',J';E,J)N(E,J;t)dE \\ &\quad - \sum_{J}\int_{0}^{\infty} R(E,J;E',J')N(E',J';t)dE \\ &\quad - \sum_{i=1}^{\text{channels}} k_i(E',J')N(E',J';t), \end{aligned} \tag{6.13}$$

where the primed E' and J' refer to the current state and the unprimed quantities refer to a different state of the excited species that is undergoing reaction. In Eq. (6.13), $N(E',J',t)dE'$ is the concentration of a chemical species with energy in the range E' to $E' + dE'$; $R(E,J;E',J')$ is the (pseudo-first-order) rate coefficient for collisional energy transfer from initial energy E' to energy E and quantum number J' to J; $F'(E',J',t)dE'$ is a source term (e.g., thermal, chemical, or photo activation, or isomerization); and $k_i(E',J')$ is the unimolecular reaction rate constant for molecules at energy E' and rotational quantum number J' reacting via the ith reaction channel.

Collisional energy transfer involving rovibrationally excited species is usually treated as the product of the Lennard–Jones collision frequency multiplied by the probability density function for the conventional exponential–down step size distribution[97]:

$$P(E, E') = N^{-1} \exp[(E - E')/\alpha] \quad \text{for } E \leq E', \qquad (6.14)$$

where E and E' are the final and initial energies, respectively, N is a normalization factor, and α (not be confused with an isotopic fraction factor!) is an adjustable parameter that is approximately equal to the magnitude of $\langle \Delta E \rangle_{\text{down}}$, the average amount of energy transferred in deactivating collisions. Although it had little effect on these calculations, the collision frequency may be attenuated at low energies, where the density of states is sparse.[100]

Numerical solutions of the master equation usually proceed by reducing the "2D" equation (depending on E, J) to an approximate "1D" version (depending only on E), followed by an appropriate numerical solution of the 1D equation. However, if collisional energy transfer is infinitely fast (i.e., at pressure $P = \infty$), then the population distributions do not evolve with time and a thermal energy distribution is maintained. Under these conditions, an exact solution of the 2D equation exists. Consider reaction systems like HO + CO (Sec. 6.4.3), where an intermediate (HOCO) is connected to reactants by one transition state and to products by a second transition state:

$$\text{HO} + \text{CO} \rightleftharpoons \text{HOCO} \rightarrow \text{H} + \text{CO}_2.$$

For reaction systems like this one, a second non-trivial exact solution exists when there are no energy transfer collisions (i.e., at $P = 0$). For further details, see Sec. 6.4.3.

The rate constants were computed using the CRP obtained from SCTST and densities of states computed for fully coupled anharmonic vibrational models, as described above and implemented[45,48,50,89] in the MultiWell Program Suite.[60,97] The most recent version of this suite of programs includes improved versions of the codes.[91]

6.2.5. *Atmospheric Isotope Effects*

The observed isotopic composition of an atmospheric species gives clues that help to constrain its atmospheric budget, identify contributions to chemical cycles in which it is involved, and identify the relative importance of various sources and sinks. The basic approach for analyzing such isotope effects is to use sets of equations describing production and loss of a chemical species and its isotopologues. Losses (L_i for the ith process) often involve reactions of the target species with hydroxyl (HO) free radicals, chlorine atoms (Cl), and excited oxygen atoms ($O^1(D)$). Production processes (P_j for the jth process) include emission sources and reactions that take place *in situ*. Each source has a unique isotopic signature, as does each sink. Furthermore, the reaction rate constants are different for each reaction isotopologue and isotopomer. The ambient isotopic abundances of a target species reflect all of these influences. In cases when the reaction rate constant for an isotopologue has not been measured, theoretical rate constants are a useful alternative.

As an example, consider molecular hydrogen (H_2) in the atmosphere. The H_2 isotopologue, which occurs with a mixing ratio of ~0.5 parts per million by volume (ppmv),[101] is by far the most abundant; second most abundant is the HD isotopologue. The global distribution of H_2 is not uniform, since its atmospheric concentration in the southern hemisphere is about 3% greater than in the northern hemisphere. Neither the sources nor the sinks of H_2 are distributed uniformly around the globe.

To quantitatively determine the relative global importance of the various sources and sinks of H_2, it is necessary to use a global transport/reaction model. Here, for the purpose of discussion, we consider only the chemical transformations. In each volume element of the atmospheric model, a mass balance equation can be written[102] for the total mass of molecular hydrogen (all isotopologues):

$$\frac{dM_{\text{hyd}}}{dx} = P_{\text{ff}} + P_{\text{bb}} + P_{\text{h}\nu} - L_{\text{surf}} - L_{\text{rxn}}, \qquad (6.15)$$

where the source terms are for fossil fuel combustion (P_{ff}), biomass burning (P_{bb}), and photolysis of formaldehyde and other precursors ($P_{\text{h}\nu}$). The sink terms are for uptake in soils and on other surfaces (L_{surf}) and gas phase reaction with HO (and possibly other) free radicals ($L_{\rm rxn}$). If additional reactions are important, the $L_{\rm rxn}$ term would be an appropriate sum of terms. Following Gerst and Quay,[102] a similar equation can be written for the deuterium component:

$$\frac{d[(D/H)_{\rm atm}M_{\rm hyd}]}{dx} = (D/H)_{\rm ff}P_{\rm ff} + (D/H)_{\rm bb}P_{\rm bb} + (D/H)_{\rm h\nu}P_{\rm h\nu} - (D/H)_{\rm atm}\alpha_{\rm surf}L_{\rm surf} - (D/H)_{\rm atm}\alpha_{\rm rxn}L_{\rm rxn}, \quad (6.16)$$

where $(D/H)_{atm}$ is the deuterium to hydrogen ratio (i.e., the mole ratio n_D/n_H) in an atmospheric volume element and the (D/H) ratios with source term subscripts are characteristic of each source. The fractionation factors α_i ($i = \text{surf}, rxn$) are characteristics of the individual sinks. In particular, the fractionation factor α_{rxn} is equal to the ratio of the reaction rate constants for the reaction of interest. For the reaction of HO radicals with HD and H_2: ${}^{\rm D}\alpha_{OH} = k_{HD}/k_{HH}$. The kinetic isotope effect (KIE) for these reactions is conventionally defined as the ratio of the rate constant for the light isotopologue divided by that for the heavy isotopologue. Thus the kinetic isotope effect is given by $\text{KIE} = k_{HH}/k_{HD} = 1/{}^{\rm D}\alpha_{OH}$ for the reactions of HO radicals with H_2 and HD.

When most of the KIEs and (D/H) ratios are known with some certainty, it is possible to use this modeling approach to determine parameters that may not be known as accurately, as mentioned later in this review.

6.3. Benchmark Tests

6.3.1. $H + H_2 \rightarrow H_2 + H$

The reaction of a hydrogen atom with a H_2 molecule is a benchmark system for both experimental and theoretical studies.[103–118] This reaction, which is an "automerization", is well characterized and understood.[103] The reaction proceeds by passing over (and tunneling through) a collinear transition structure (TS) of a hydrogen abstraction leading to the same species as the initial reactants, $H_2 + H$. Therefore, to observe products in the laboratory, isotopic substitution is necessary,[103,104,112–114,116–118] including $H + D_2$, $D + H_2$, $H + HD$, etc.

With only three electrons, the global PES for this reaction can now be obtained with exquisite accuracy using a FCI method in conjunction with very large basis sets. Such a global PES has been reported and is designated as the complete configuration–interaction (CCI) PES.[105,108] On the basis of the CCI PES, accurate quantum mechanical scattering calculations[104] were performed and the results obtained (state-to-state reaction cross-sections, thermally averaged rate constants, and others) agree extremely well with measurable quantities where they are available.

As a part of this review, SCTST/VPT2 will be used to compute thermally averaged rate constants for four isotopic reactions involving $H + H_2$, $H + D_2$, $D + H_2$, and $D + D_2$. The calculated results will then be compared to those of accurate quantum mechanical calculations[104] and experiments.[103–118] Given that a very high-accuracy PES[104,105,108,110,119] will be used, deviations between various theoretical models or between theory and experiment are essentially due to inherent limits in an approximate model that is applied. Therefore, such a comparison is useful to understand advantages and possible limits of SCTST/VPT2.

Geometries of various species were optimized using an FCI method (equivalently, the CCSD method[119,120] was used for H_2 and CCSDT[121–123] for the collinear TS of H-abstraction) in combination with cc-pV5Z basis set. A series of single point energy calculations with the DBOC[124] and scalar relativistic corrections[125] were then performed to obtain a total energy according to the HEAT-456 model. Note that in this case, the CCSD(T) method in the HEAT protocol is replaced by the FCI method. Hereafter we denote this method as FCI/CBS(456). The classical barrier height calculated with the FCI/CBS(456) method is 9.60 kcal/mol, in excellent agreement with 9.602 ± 0.01 kcal/mol of the CCI PES[104,105,108] and the best estimate[108,110] of 9.608 ± 0.01 kcal/mol that was suggested earlier. It should be mentioned that such a small variance –0.01 kcal/mol (*ca.* 3.5 cm^{-1}) — in the calculated barrier height would change the thermally averaged rate constant calculated at room temperature by less than 2%.

Zero-point vibrational energy (ZPE) corrections were obtained with second-order VPT2, which requires harmonic and anharmonic oscillator force fields as input. The FCI/cc-pV5Z level of theory was used for these calculations.

Tables 6.1 and 6.2 show that our results are in excellent agreement both with the exact quantum calculations obtained from the CCI PES and with experiments,[126] suggesting that the FCI/CBS(456) method used to construct the PES in this work produces results that are very close to the CCI surface reported previously. To compare $k(T)$ from the SCTST/VPT2 approach with those of accurate quantum mechanical calculations, the same CCI PES should be used. Therefore, we also computed SCTST rate constants using the CCI PES and we then compared the obtained CCI results with those computed with the FCI/CBS(456) method (see Table 6.3). The agreement is excellent, within 5%. It should be mentioned that anharmonic vibration constants that are required in SCTST/VPT2 depend sensitively on the quality of the fitted PES. So, here we choose to use a set of anharmonic vibration constants that were computed with coupled-cluster methods using the CFOUR program.

Table 6.4 shows the individual contributions of various terms in vibrationally adiabatic barrier heights. The SCF method gives a much too high barrier (17.61 kcal/mol) because electron correlation is neglected. Inclusion of a quantitative and exact treatment of electron correlation afforded by the FCI method reduces the barrier by 8.01 kcal/mol. Anharmonic ZPE corrections also lower the barrier by 1.23 kcal/mol, whereas the DBOC that is computed using CCSD/aug-cc-pVQZ raises the barrier by 0.16 kcal/mol. It should be mentioned that the DBOC term is more important for lighter species (see Table 6.5). Scalar relativistic corrections are negligibly small in this system ($< 0.2\,\text{cm}^{-1}$). Taking all these contributions into account, the vibrationally adiabatic barrier heights (i.e., the barrier height, including ZPE corrections) calculated with FCI/CBS(456) method are 8.52, 8.06, 9.50, and 8.89 kcal/mol for $H + H_2$, $D + H_2$, $H + D_2$, and $D + D_2$, respectively. The differences in the barriers are mainly due to two factors, the ZPE and the DBOC (see Table 6.5).

Table 6.1 Calculated (FCI/cc-pV5Z) TS parameters for the $H + H_2$ isotopologues.

	H—H—H		H—D—D		D—H—H		D—D—D	
Parameter	This work	CCI PES[a]	This work	CCI PES[a]	This work	CCI PES[a]	This work	@CCI PES[a]
ω_F (cm^{-1})	1498i	1501i	1125i	1127i	1421i	1424i	1059i	1062i
ω_{bend} (cm^{-1})	885	878	676	671	847	840	626	621
ω_{sym} (cm^{-1})	2053	2052	1765	1765	1767	1766	1452	1451
B (2D-rotor, cm^{-1})	9.6687	n/a	6.9116	n/a	7.0368	n/a	4.8400	n/a
R (Bohr)	1.7572	1.7572	1.7572	1.7572	1.7572	1.7572	1.7572	1.7572
ΔDBOC (cm^{-1})[b]	54.32	53.57	31.97	31.57	49.49	48.80	27.14	26.80

[a]Taken from Refs. [105,108].
[b]Calculated at FC-CCSD/aug-cc-pVQZ level of theory.

Table 6.2 Harmonic spectroscopic parameters of H_2 and its isotopes from FCI/cc-pV5Z.

	H_2		HD		D_2	
Parameter	This work	Exptl.[a]	This work	Exptl.[a]	This work	Exptl.[a]
ω_e (cm^{-1})	4404.7021	4401.21	3815.0726	3813.1	3115.7916	3115.50
$\omega_e x_e$ (cm^{-1})	−122.8740	−121.33	−92.1655	−91.65	−61.4644	−61.82
B_e (cm^{-1})	60.8383	60.853	45.6404	45.655	30.4425	30.443
R_e (Å)	0.74153	0.74144	0.74153	0.74142	0.74153	0.74152

[a]Taken from Ref. [126].

Table 6.3 Comparison of FCI/CBS(456) k(T) with CCI k(T) for $H + H_2$.

T (K)	FCI/CBS(456)	CCI	Dev. (%)
200	1.77×10^{-18}	1.69×10^{-18}	4.88
250	3.83×10^{-17}	3.78×10^{-17}	1.51
300	3.39×10^{-16}	3.38×10^{-16}	0.26
350	1.72×10^{-15}	1.73×10^{-15}	−0.20
400	6.08×10^{-15}	6.11×10^{-15}	−0.35
500	3.81×10^{-14}	3.83×10^{-14}	−0.31
600	1.37×10^{-13}	1.37×10^{-13}	−0.11
800	7.54×10^{-13}	7.50×10^{-13}	0.49
1000	2.29×10^{-12}	2.27×10^{-12}	1.25
1200	5.14×10^{-12}	5.04×10^{-12}	2.05
1400	9.57×10^{-12}	9.32×10^{-12}	2.77
1600	1.58×10^{-11}	1.53×10^{-11}	3.32
2000	3.37×10^{-11}	3.25×10^{-11}	3.82
2500	6.57×10^{-11}	6.34×10^{-11}	3.56

Table 6.4 Individual contributions of various terms to barrier height calculated using FCI/CBS(4,5,6) level of theory.

Species	δ_{HF}	δ_{FCI}	δ_{ZPE}	δ_{DBOC}	δ_{REL}	δ_{Total}
$H + H_2$	0.0000	0.0000	0.0000	0.0000	0.0000	0.0000
TS	17.6130	−8.0130	−1.2330	0.1553	0.0004	8.5227

Note: All are in kcal/mol.

Table 6.5 Calculated (FCI/CBS(4,5,6)) classical barrier height (V_o), DBOC, scalar relativity, zero-point vibration energies, vibrationally adiabatic barrier heights (V_a), and reaction enthalpies for the H + H_2 isotopologues (kcal/mol).

Reaction	V_o	DBOC	REL	ZPE	V_a
H + H_2	0.0000	0.0000	0.0000	0.0000	0.0000
TS (H–H–H)	9.6000	0.1553	0.0004	−1.2330	8.5227
H + D_2	0.0000	0.0000	0.0000	0.0000	0.0000
HD + D	0.0000	−0.0036	0.0000	0.9837	0.9801
TS (H–D–D)	9.6000	0.0914	0.0004	−0.1899	9.5019
D + H_2	0.0000	0.0000	0.0000	0.0000	0.0000
DH + H	0.0000	0.0036	0.0000	−0.8269	−0.8233
TS (D–H–H)	9.6000	0.1415	0.0004	−1.6850	8.0569
D + D_2	0.0000	0.0000	0.0000	0.0000	0.0000
TS (D–D–D)	9.6000	0.0777	0.0004	−0.7900	8.8881

SCTST/VPT2 rate constants are plotted in Figs. 6.1 and 6.2 for H/D + D_2 and H/D + H_2 reactions, respectively. Experimental[104,112,114,116–118,127] and accurate quantum mechanical results,[104] where available, are also shown for comparison. Figure 6.1 shows excellent agreement between our results and both experimental data and accurate theoretical results for a wide temperature range (200–2500 K), in which the thermal rate constants vary by about nine orders of magnitude. Compared to experiments, the SCTST rate constants are in line with the laboratory data when $T < 700$ K; at higher temperatures, they fall between two recent measurements, which exhibit appreciable scatter. Compared to the most accurate quantum mechanical calculations,[104] our SCTST values are systematically ~15% too high (see Table 6.6). For H + D_2 the overall agreement is remarkable. Considering the simplicity and economy of SCTST/VPT2, as compared to rigorous quantum mechanical reactive scattering calculations,[104] one cannot expect better performance.

For D + H_2, perhaps because of the existence of Fermi-resonance effects (i.e., $\omega_{\text{sym}} \approx 2\omega_{\text{bend}}$, see Table 6.1), the SCTST/VPT2 results are not as good as those for H + D_2. Figure 6.2 shows that the SCTST rate constants are significantly too high at low temperatures,

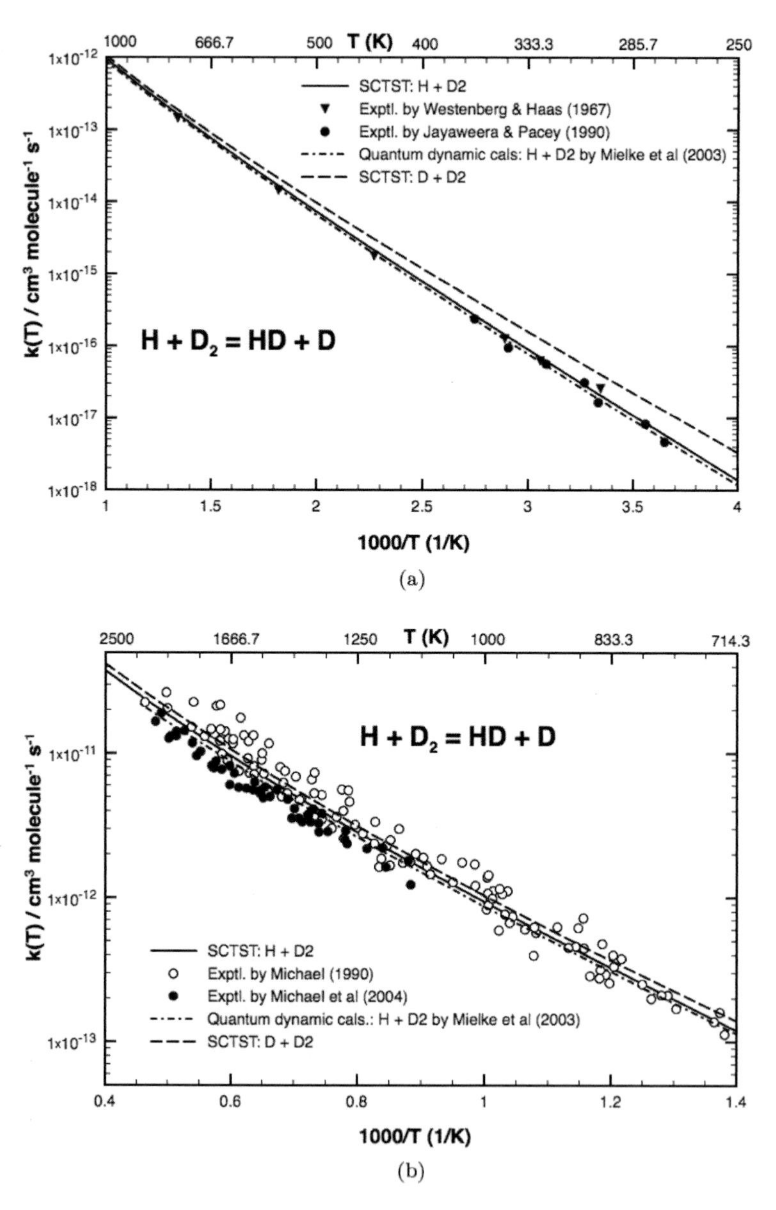

Fig. 6.1 Thermal rate constants for the $H + D_2$ reaction. (a) Low temperature range, (b) High temperature range. Experimental measurements noted in the figure legend: Westenberg and Haas (1967),[113] Jayaweera and Pacey (1990),[116] Mielke (2003),[104] Michael (1990),[117] Michael (2004).[118]

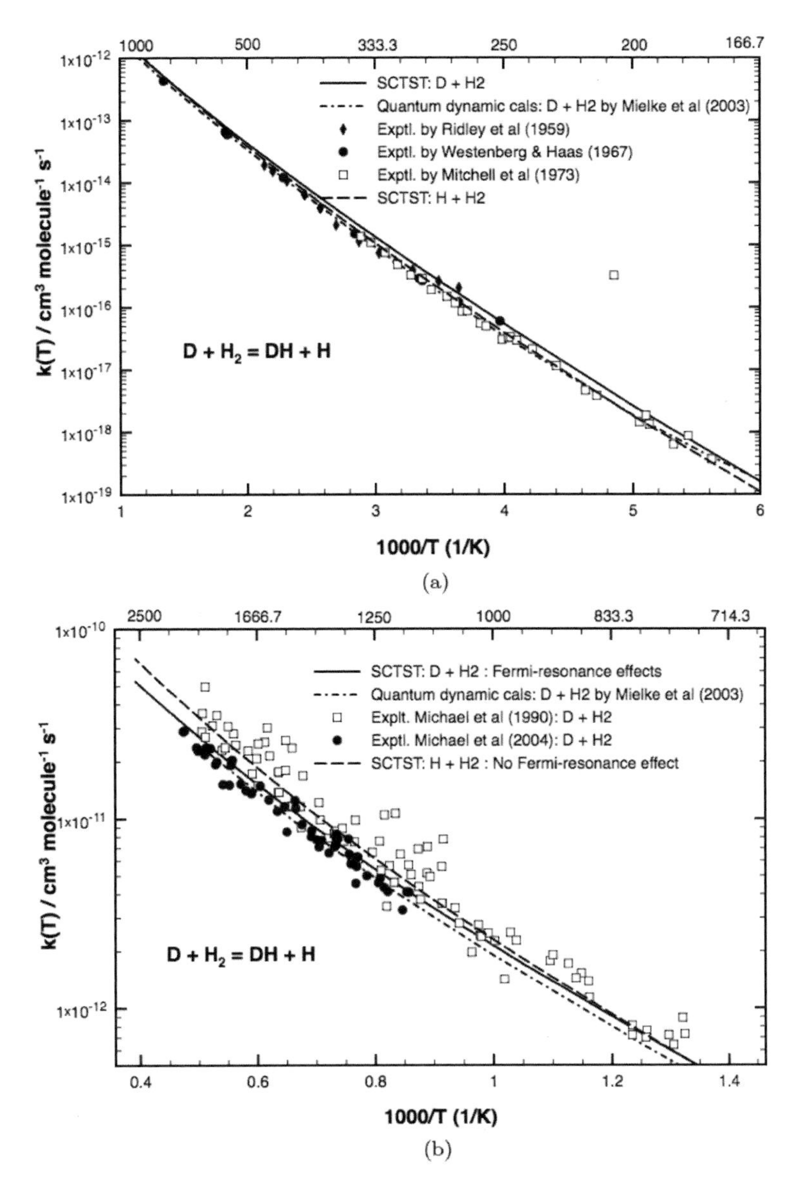

Fig. 6.2 Thermal rate constants for the $D + H_2$ reaction. (a) Low temperatures; (b) High temperatures. Experimental measurements noted in the figure legend: Ridley (1959),[112] Westenberg and Haas (1967),[113] Mitchell (1973),[114] Mielke (2003),[104] Michael (1990),[117] Michael (2004).[118]

Table 6.6 SCTST/VPT2 rate constants compared with those of accurate quantum mechanical dynamics calculations.

Temp. (K)	$H + D_2$ This work	$H + D_2$ QD[a]	$H + D_2$ %Dev.[a]	$D + H_2$ This work	$D + H_2$ QD[a]	$D + H_2$ %Dev.[a]
200	2.58×10^{-20}	2.51×10^{-20}	2.8	2.57×10^{-18}	1.86×10^{-18}	38.2
250	1.37×10^{-18}	1.19×10^{-18}	15.1	5.32×10^{-17}	3.40×10^{-17}	56.5
300	2.17×10^{-17}	1.86×10^{-17}	16.7	4.46×10^{-16}	2.96×10^{-16}	50.7
350	1.65×10^{-16}	1.43×10^{-16}	15.4	2.16×10^{-15}	1.51×10^{-15}	43.0
400	7.78×10^{-16}	6.85×10^{-16}	13.6	7.31×10^{-15}	5.38×10^{-15}	35.9
500	7.28×10^{-15}	6.54×10^{-15}	11.3	4.29×10^{-14}	3.38×10^{-14}	26.9
600	3.41×10^{-14}	3.10×10^{-14}	10.0	1.47×10^{-13}	1.21×10^{-13}	21.5
800	2.59×10^{-13}	2.37×10^{-13}	9.3	7.55×10^{-13}	6.46×10^{-13}	16.9
1000	9.49×10^{-13}	8.65×10^{-13}	9.7	2.19×10^{-12}	1.90×10^{-12}	15.3
1200	2.37×10^{-12}	2.16×10^{-12}	9.7	4.71×10^{-12}	4.12×10^{-12}	14.3
1500	6.33×10^{-12}	5.75×10^{-12}	10.1	1.08×10^{-11}	9.53×10^{-12}	13.3
2000	1.85×10^{-11}	1.64×10^{-11}	12.8	2.74×10^{-11}	2.45×10^{-11}	11.8
2200	2.52×10^{-11}	2.21×10^{-11}	14.0	3.61×10^{-11}	n/a	n/a

[a]Taken from the accurate quantum mechanical dynamics (QD) calculations[104] %Dev = $(k_{SCTST} - k_{QD}) \times 100\%/k_{QD}$.

but the agreement becomes better at higher temperatures (see Table 6.6).

Thermal rate constants for $D + D_2$ and $H + H_2$ were also computed and are shown in Figs. 6.1 and 6.2. This information may be useful for studying secondary KIEs. Generally speaking, rate constants for heavier reactants are slightly larger than for the lighter counterparts (e.g., for a given T, $k_{D+D2} > k_{H+D2}$ and $k_{D+H2} > k_{H+H2}$). The same behavior was also observed previously.

6.3.2. $O(^3P) + H_2 \rightleftharpoons OH + H$

The title reaction is an important chain-propagation step in combustion of H_2 and hydrogen-containing fuels.[128–133] However, because of the high barrier for reaction (about 13.0 kcal/mol)[134–137] and the rather minute concentration of atmospheric atomic oxygen, it is not very important in Earth's atmosphere. The title reaction has been both extensively reviewed and studied.[128,131–135] Experimentally, thermal rate constants have been measured by numerous techniques at temperatures that range from 298 K to 3500 K.[113,115,131,138–151] On the theoretical side,[134–137] global PES with an accuracy of better than 0.5 kcal/mol have recently become available and thermally averaged rate constants, quasi-classical trajectory (QCT) and quantum mechanical reactive scattering calculations[134,136,137] have been reported using these surfaces. In general, the theoretical results are consistent with the experiments.

6.3.2.1. *Thermal Rate Constants*

In this section, we calculate thermally averaged rate constants for the title reactions, computed using SCTST. The calculated results will then be used to compare with experimental and theoretical data where available.

Energies of various species in the reaction of ground-state oxygen atom with molecular hydrogen were calculated using HEAT-345(Q), HEAT-456(Q), and HEAT-456QP methods. These three sets of results are consistent to within 0.1 kcal/mol (see Table 6.7). Zero-point vibrational energies (ZPE) were calculated using the CCSD(T) method (in the frozen-core approximation) in combination with the atomic natural orbital (ANO) basis set of Taylor and Almlöf.[62,76,77] Harmonic force fields were calculated using the ANO2 set of contractions, while ANO1 was used for the anharmonic force fields. Second-order VPT2 was then used to compute anharmonic constants and the associated anharmonic ZPE. Experimental SO corrections were used for the ground-state oxygen atom ($SO_O = 78\,cm^{-1}$) and hydroxyl radical ($SO_{OH} = 69.6\,cm^{-1}$).[126] A SO correction for the linear electronically degenerate TS is neglected, however, as our previous calculations[56] showed that the predicted kinetics are very insensitive to the magnitude of the SO correction at the TS. Hence, the barrier we compute corresponds to the average of two components of the transition state, that is, we ignore the degeneracy — and with it, the Renner-Teller effect, in our SCTST calculations of $P_n(E)$.

Table 6.7 displays contributions of various terms to the total HEAT energies. As can be seen, Hartree-Fock energies are (typically)

Table 6.7 Calculated relative energies (kcal/mol) of various terms in HEAT protocol.

Species	δ_{HF}	$\delta_{CCSD(T)}$	δ_{CCSDT}	δ_{HLC}	δ_{REL}	δ_{ZPE}	δ_{DBOC^a}	δ_{SO}	δ_{HEAT}
$O(^3P) + H_2$	0.0000	0.0000	0.0000	0.0000	0.0000	0.0000	0.0000	0.0000	0.0000
									1.4463[b]
$OH(^2\Pi) + H$	15.3913	–13.0763		–0.1239					(1.3351)[c]
	(15.3679)	(–13.1640)	0.0308	[–0.1150]	0.1255	–0.9420	0.0168	0.0241	[1.3441][d]
									11.2306
TS	32.1324	–18.7682		–0.1064					(11.1799)
O—H—H ($^2\Pi$)	(32.1097)	(–18.7961)	–0.1756	[–0.1074]	0.0405	–2.2853	0.1704	0.2228	[11.1789]

[a]Obtained with CCSD/aug-cc-pVTZ level of theory
[b]Calculated with HEAT-345(Q) method
[c]Calculated with HEAT-456(Q) method
[d]Calculated with HEAT-456QP method

far from adequate; the CCSD(T) corrections, as always, play the most important role. The second most important correction is the ZPE, which contributes –0.94 and –2.28 kcal/mol to the reaction enthalpy and the barrier height, respectively. Although the other corrections are much smaller (<0.2 kcal/mol), they represent crucial fine-tunings that result in excellent agreement with experiment. HEAT calculations predict endothermic by 1.45 kcal/mol, which agrees well with the Active Thermochemical Tables (ATcT) value of 1.54 ± 0.01 kcal/mol,[152–155] which differ by only ~0.1 kcal/mol. HEAT also provides a vibrationally adiabatic energy barrier of 11.2 kcal/mol, with an estimated maximum error of *ca.* 0.25 kcal/mol.

Vibrationally adiabatic energy barriers for the associated isotopologues were also calculated and are presented in Table 6.8. The differences between the title reaction and its isotopic variants are due to two mass-dependent contributions, specifically the DBOC and the ZPE.

$O(^3P) + H_2 \rightarrow OH + H$: For this reaction, experimental rate constants are available for a wide range of temperature,[113,115,131,138–151]

Table 6.8 Classical barrier heights (V_o), vibrationally adiabatic barrier heights (V_a), DBOC and zero-point vibration energies (ZPE) for $O(^3P) + H_2$ and its variant isotopes.

Reactions	V_o	DBOC	ZPE	V_a
$O(^3P) + H_2$	0.0000	0.0000	0.0000	0.000
OH + H	2.3714	0.0168	−0.9420	1.446
TS (O—H—H)	13.3455	0.1704	−2.2853	11.231
$O(^3P) + D_2$	0.0000	0.0000	0.0000	0.000
OD + D	2.3714	−0.0199	−0.5600	1.792
TS (O—D—D)	13.3455	0.0865	−1.5847	11.847
$O(^3P)$ + HD	0.0000	0.0000	0.0000	0.000
OH + D	2.3714	0.0133	−0.1148	2.270
TS (O—H—D)	13.3455	0.1583	−2.2276	11.276
$O(^3P)$ + DH	0.0000	0.0000	0.0000	0.000
OD + H	2.3714	−0.0163	−1.5440	0.811
TS (O—D—H)	13.3455	0.0986	−1.7425	11.702

Note: All are in kcal/mol.

from room temperature up to about 3500 K. The thermal rate constants span eight orders of magnitude from *ca.* 10^{-17} to *ca.* 10^{-10} cm^3 mol^{-1} s^{-1}. For the purpose of a careful comparison of our calculated k(T) values with experimental data, we display the results in Figs. 6.3–6.5.

For T = 298 − 1000 K (Fig. 6.3), our calculated $k(T)$ values are in line with most experimental data, except two sets of rate constants reported by Dubinsky and McKenney[140] and Marshall and Fontijn.[143] The difference between calculation and experiment is generally less than 10%, which is comparable to the typical uncertainty of the rate measurements. From 298 K to 1500 K, there are two distinct trends seen in the experimental data (see Figs. 6.3 and 6.4): one is seen in that of Dubinsky and McKenney and Marshall & Fontijn; the other is that seen in all remaining works. The current SCTST calculations support the latter set of experiments. In

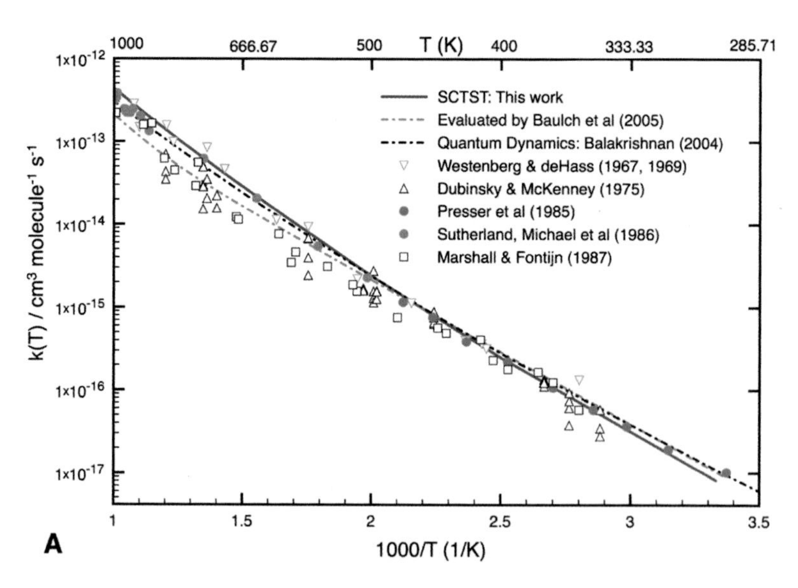

Fig. 6.3 Thermally averaged rate constants for $O(^3P) + H_2$: 298–1000. Experimental measurements and theoretical results noted in the figure legend: Westenberg and deHaas (1967, 1969),[138,139] Dubinsky and McKenney (1975),[140] Presser (1985),[141] Sutherland (1986),[142] Marshall (1987),[143] Balakrishnan (2003),[156] Baulch (2005).[131]

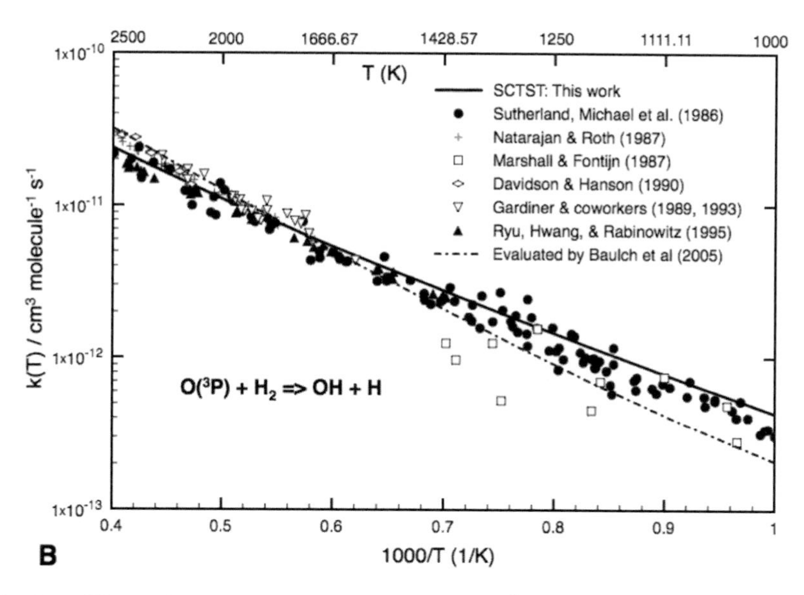

Fig. 6.4 Thermally averaged rate constants for $O(^3P)+H_2$: 1000–2500 K. Experimental measurements noted in the figure legend: Sutherland (1986),[142] Natarajan (1987),[144] Marshall (1987),[143] Davidson (1990),[147] Gardiner (1989, 1993),[145,149] Ryu (1995),[150] Baulch (2005).[131]

contrast, Baulch's recommendation[131] appears to fit the Marshall *et al.* and Dubinsky *et al.* data.

From 1000 K to 2500 K (see Fig. 6.4), the experimental data are scattered; generally speaking, agreement between theory and experiment is excellent, within 10–15%. For T = 2500–4000 K (see Fig. 6.5), the calculated thermal rate constants underestimate experimental values moderately, by about 15–35%. The disagreement becomes worse with increasing temperature. The underestimation at very high temperatures may be due to the fact that there are a number of other possible effects that have not been included in our calculations. These may include (i) reactions of non-statistical distributions of vibrationally excited H_2 molecules with oxygen atoms; (ii) the Renner–Teller effect that we have ignored here; (iii) the intrinsic limits of VPT2 in treating anharmonicity (one might suspect that higher order VPT might be needed; (iv) inadequacy of the rovibrational parameters; and finally, (v) coupling of vibrations and rotations in the TS. It is however expected that the last of these

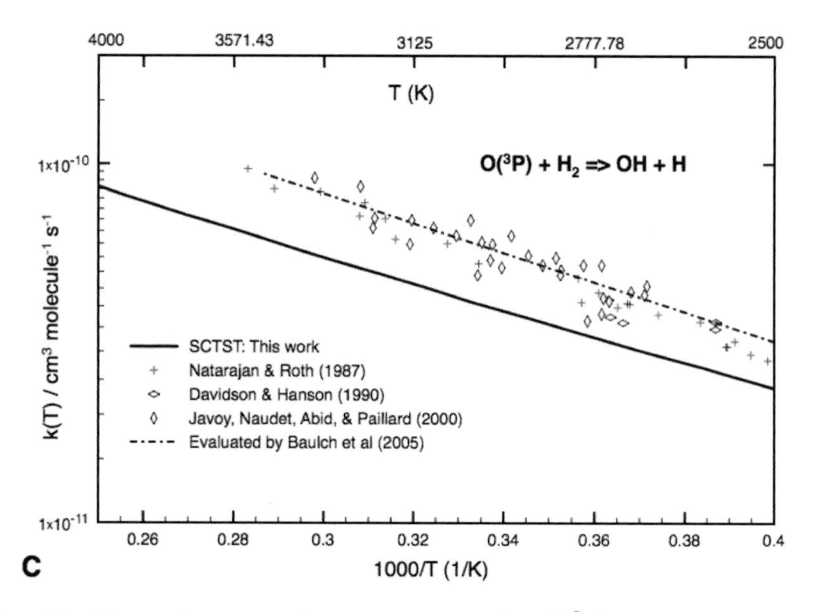

Fig. 6.5 Thermally averaged rate constants for $O(^3P) + H_2$: 2500–4000 K. Experimental measurements noted in the figure legend: Natarajan (1987),[144] Davidson (1990),[147] Javoy (2000),[151] Baulch (2005).[131]

should have only a small effect because it should be largely cancelled out by the corresponding terms for the H_2 reactant. Additionally, the effects of coupling the vibrations with rotations in H_2 were found to be small (less than 4% of the rovibrational partition function at 3500 K).

It is of interest to compare the present SCTST rate constants with previous quantum mechanical reactive scattering calculations using GLDP surface reported by Balakrishnan.[156] Apart from differences in the surfaces used, Fig. 6.3 suggests that SCTST rates may slightly underestimate results from quantum dynamics at $T < 400$ K, but are too high at $T > 500$ K. Overall, however, the SCTST rates agree reasonably well with Balakrishnan's results. A small difference (likely $\lesssim$20%) between the two sets of rate constants may be due to differences in PESs and methodologies that are applied. In addition, we also compared our SCTST results with ICVT/LAG,[134] where excellent agreement is also observed.

$O(^3P) + D_2 \rightarrow OD + D$: For convenience, we divide the temperature range into two regions for this isotopic variant. Figure 6.6(a)

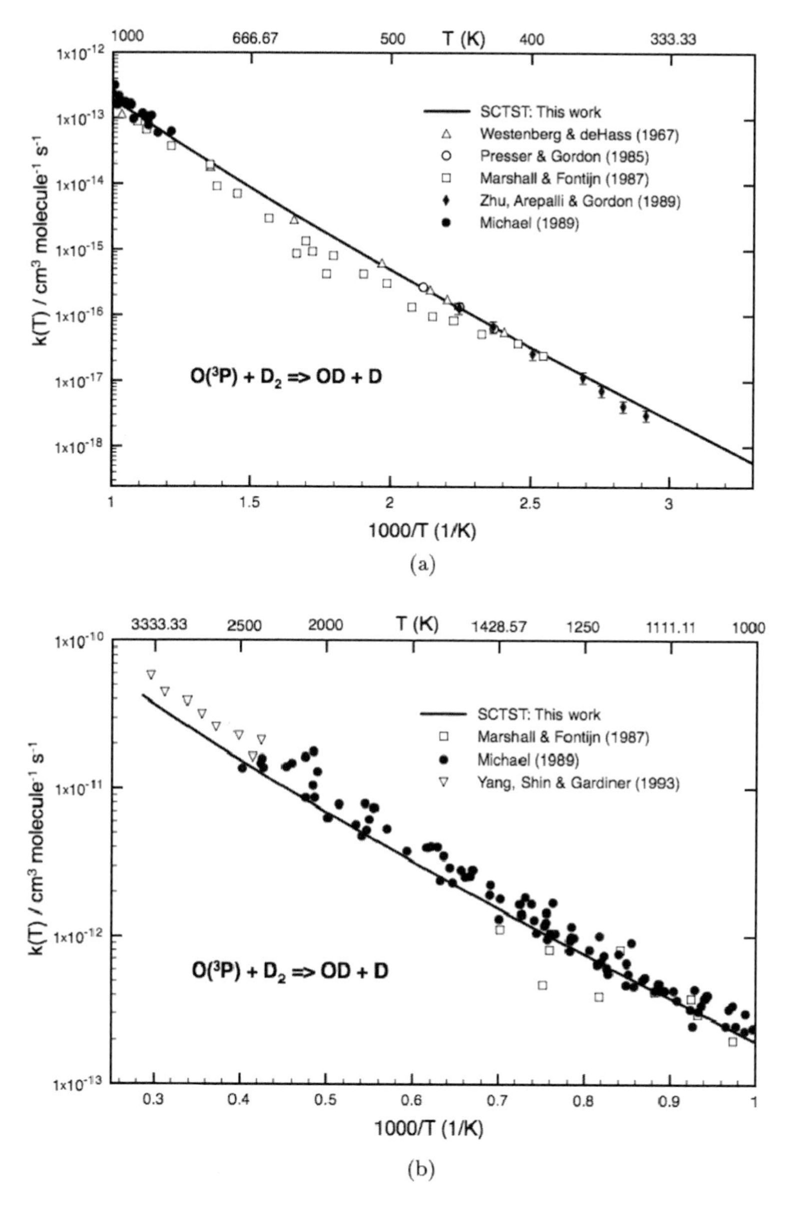

Fig. 6.6 Thermally averaged rate constants for $O(^3P) + D_2$: 298–1000 K (a) and 1000–3500 K (b). Experimental measurements noted in the figure legend: Westenberg and deHaas (1967),[138] Presser (1985),[141] Marshall (1987),[143] Zhu (1989),[146] Michael (1989),[115] Yang (1993).[149]

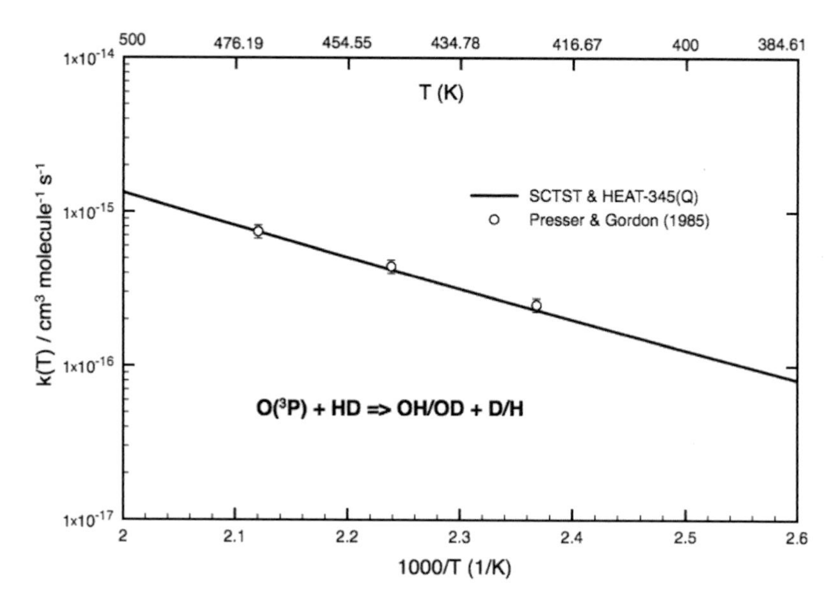

Fig. 6.7 Thermally averaged rate constants for $O(^3P)$ + HD. Experimental measurements noted in the figure legend: Presser (1985).[141]

displays thermal reaction rate constants for temperatures from 298 K to 1000 K, while $k(T)$ values with $T \geq 1000$ K are illustrated in Fig. 6.6(b). As can be seen, excellent agreement (within 10%) between theory and experiment is found for the whole range of temperatures. As for the normal isotopic variant of the reaction, the calculated $k(T)$ appears to systematically underestimate experiments at very high temperatures ($T \geq 2500$ K).

$O(^3P) + HD/DH \rightarrow OH/OD + D/H$: Thermal rate constants from these two reactions were computed and the sum (total rate constant) is shown in Fig. 6.7. Again, excellent agreement is seen between theory and experiment (within 5%); although the temperature range of the measurements is only from 420 K to 475 K.[141,148]

6.3.2.2. *KIEs*

In Fig. 6.8, we compare our calculated KIEs with those derived from experimental measurements performed by Gordon and co-workers[141,146,148] over a small range of temperature, 340–475 K.

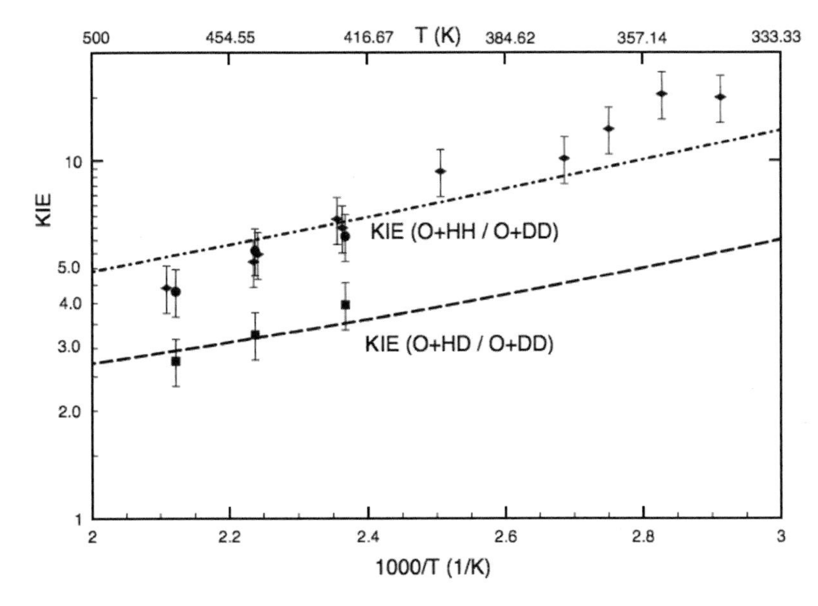

Fig. 6.8 A comparison of theoretical with experimental KIE. Dash and dot-dash lines are theoretical results calculated with SCTST approach. The symbols are experimental data from Gordon's group: the red symbols are taken from Ref. [141] and the green symbols are from Ref. [146].

Figure 6.8 shows that above 420 K, our predicted KIE values are in excellent agreement (within the experimental error bar) with values derived from experiment. However, the KIE (i.e., k_{O+H2}/k_{O+D2}) is slightly underestimated at lower temperatures ($T < 350$ K). This seems to be due to the fact that our results for $O + D_2$ are marginally too high in this temperature range (see Fig. 6.6(a)). Overall, our KIE values agree reasonably well with the experiments — perhaps as well as one can expect.

6.4. Atmospheric Reactions

6.4.1. $HO + H_2 \rightarrow H_2O + H$

Molecular hydrogen (H_2) is emitted to the atmosphere from nitrogen-fixing bacteria in soils, from fossil fuel combustion, and from biomass burning.[157] It is produced *in situ* by the photolysis of atmospheric formaldehyde (H_2CO),[158] which is produced by the photo-oxidation of CH_4 and other volatile organic compounds (VOCs). It is removed from the atmosphere by deposition to soils and by reaction with $O(^1D)$, but mostly by reaction with HO free radicals[101,132,159]:

$$HO + H_2 \rightarrow H_2O + H. \tag{R1}$$

The atmospheric H_2 budget has been studied in detail by using the modeling approach outlined earlier in Sec. 6.2.5. Most recently, Pieterse and coworkers[101,160] used data for production and loss along with experimental values for M_{hyd}, $(D/H)_{atm}$, and ${}^{D}\alpha_{OH}$, couched in the "two-way nested" TM5 global atmospheric model,[161] to analyze the atmospheric H_2 cycle and budget (for technical details and citations of previous studies, see Pieterse *et al.*).[101] They concluded that the most important source term is $P_{h\nu}$, which is a little more than twice as large as P_{ff}, which is closely followed by P_{bb}. The most important sink term is L_{surf}, which is about twice as large as L_{rxn}. They also found that the global average atmospheric residence time of H_2 is 2–3 years.

There is considerable interest[162] in using H_2 to partly replace fossil fuels, because H_2 combustion does not produce CO_2, which is the principal source of anthropogenic radiative forcing. The ambient atmospheric mixing ratio of H_2 is currently ~550 ppb.[101,163] If H_2 is widely used as a replacement fuel, it is expected that significant amounts of unburned H_2 will be released into the atmosphere in addition to the atmospheric sources already present. As a result, the average concentration of HO free radicals will be reduced due to reaction (R1). Since HO radicals are key atmospheric reactants, their depletion would significantly affect tropospheric O_3 and other atmospheric constituents.[164] This analysis shows how processes in the atmosphere are closely coupled, and how a seemingly beneficial change in a fuel source (H_2) can have wide-ranging and not necessarily beneficial consequences.

The rate constant for reaction (R1) has been studied experimentally over a very wide temperature range (~200 – ~2500 K).[165–170] and rate constants of the isotopologues have been reviewed.[171,172] The various experimental determinations are in mutual agreement over the entire temperature range, thus providing an excellent dataset for testing theoretical methods. Because the experimental data are so extensive and precise, the $HO + H_2$ reaction is a benchmark for theoretical kinetics calculations (also for quantum reactive scattering).[173] Prior to our work, numerous theoretical kinetics calculations had been reported.[174–188] Because this reaction involves significant motions of light atoms, quantum effects are exceptionally important: ZPVE are large and quantum mechanical tunneling is dominant over much of the temperature range. Thus this reaction provides a severe test for theoretical methods.

In this review, we describe our recent theoretical calculations on reaction (R1) and its isotopologues.[45,46] Generally, there is good agreement among the theoretical models, but our calculations differ in several important respects from the previous ones. The most important difference is that our calculations are fully from first principles and contain no adjustments or empirical factors, except for the measured SO splitting constant of the HO free radical.

We used a very high level *ab initio* method and large basis set to compute optimized geometries (from which come the corresponding rotational constants) and the harmonic vibrational frequencies of the reactants and transition states. We used the same high level of theory and VPT2 to compute all of the $x_{i,j}$ vibrational anharmonicity coefficients for each species, and employed the HEAT-345Q theory for computing highly accurate reaction energetics. And we used all of these theoretical quantities in the framework of SCTST to compute *ab initio* rate constants. As will be shown below, the results are in exceptionally good agreement with the experimental data over the entire temperature range. Because the theory accurately reproduces the experimental rate constants over the entire range, it would in principle allow one to estimate rate constants with reasonable confidence at atmospheric temperatures lower than the lowest experimental measurement ($\sim$200 K).

6.4.1.1. *Technical Details*

All technical details can be found in the original publications.[45,46] Here, we restrict our attention to the main points. Geometries and harmonic vibrational frequencies of all stationary points on the PES were obtained using coupled-cluster theory involving single, double, and a perturbative treatment of triple excitations (CCSD(T)),[73] in combination with atomic natural orbital basis sets,[76] truncated to 4s3p2s1f for oxygen and 4s2p1d for hydrogen; this basis is hereafter designated as ANO1. The core electrons were kept frozen in all calculations. Anharmonic force fields were computed at the CCSD(T)/ANO1 level of theory for all species (H_2, OH, H_2O, the TSs, and their isotopic variants) using VPT2. In all cases where harmonic frequencies are within $100\,cm^{-1}$ of zeroth-order level positions corresponding to two quantum transitions, standard deperturbation techniques were applied in order to account for potential Fermi resonances.

Energies of all species considered were refined using high accuracy extrapolated *ab initio* thermochemistry (HEAT-345Q, hereafter denoted as HEAT) theory, which was described elsewhere in this review. From the HEAT protocol, the reaction enthalpy at 0 K for the $H_2 + OH \rightarrow H + H_2O$ reaction is computed to be -14.35 kcal mol^{-1}, within the uncertainty of the experimental value of -14.33 ± 0.04 kcal mol^{-1} (see Ref. [45]). It is expected that an accuracy within $\sim$0.2 kcal mol^{-1} is achieved for the computed reaction barrier heights. Note that a barrier height difference of only 0.1 kcal mol^{-1} alters the computed thermal rate by only $\sim$2% at 2500 K, but by $\sim$30% at 200 K! Thus accurate thermochemistry is an extremely important requisite for accurate predictions.

6.4.1.2. *Results and Discussion*

For the $HO + H_2$ reaction, there are six reaction isotopologues containing H and D, but we will discuss only four of them here; further discussion can be found in the original publications.[45,46] Results for the $HO + H_2$ and $HO + D_2$ reaction isotopologues are shown (Figs. 6.9(a) and 6.9(b)) because they have been the most studied in experiments. From the perspective of the atmosphere, the most important isotopologues in addition to $HO + H_2$ are the reactions $HO + HD$ and $DO + H_2$; results for the latter two reactions are shown in Figs. 6.9(c) and 6.9(d). The agreement between the present *ab initio* SCTST rate constants for these three reactions and the experimental data is very good to excellent over the entire temperature range for which data exist.

Of course, in the atmosphere, all possible combinations of the H,D reaction isotopologues occur and SCTST rate constants for all of them have been reported.[45,46] In the atmosphere, the reaction isotopologues that contain two or more D-atoms are negligible because the atmospheric D/H ratio is small ($\sim 1.8 \times 10^{-4}$). The theoretical rate constants for the less important isotopologues are in excellent agreement with the laboratory data and further discussion of them can be found in the original publications. Here we confine our discussion to only the three atmospherically important isotopologues. Because recent reviews of the reactions have appeared,[171,172] we will focus on only the most recent results.

Dynamics calculations by Chakraborty *et al.*[188] using the WSLFH PES of Wu *et al.*[181] produced thermal rate constants that agree within about a factor of two with the experiments in the $T = 250\text{–}1000\,\text{K}$ range. Troya *et al.*[186] reported VTST (ICVT/μOMT) calculations using PLOYRATE and the WSLFH PES; their calculations agree with experiments within about 35%. Part of the difference between the performance of our *ab initio* SCTST method and the other methods is because of small energy differences between using different PESs and the aformentioned sensitivity of the results with respect to barrier height. The HO+H_2PES computed at the UCCSD(T)/aug-cc-pVQZ level of theory by Yang *et al.*[182] has a classical barrier height in closer agreement with our HEAT-345Q value, and gives thermal rate constants in closer agreement with the experimental data: within 16% at $T = 238\,\text{K}$.[182] Generally speaking, the theoretical kinetics methods, as a group, perform well on this reaction, but all are exquisitely dependent on accurate thermochemistry.

Thermal rate constants for the reaction HO + HD were measured by Talukdar *et al.* between 248 K and 418 K using pulsed photolysis and laser-induced fluorescence to monitor HO radicals.[168] Their results for the total rate constant are highly precise, as shown in Fig. 6.9(c), and the present *ab initio* SCTST results are in excellent

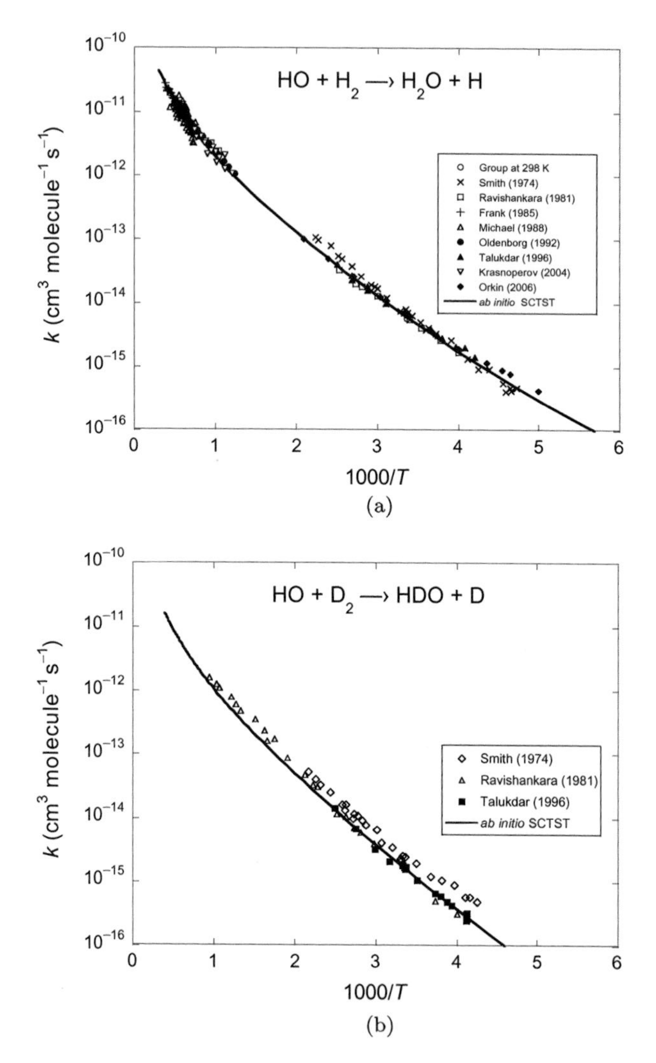

Fig. 6.9 (a) *Ab initio* SCTST and Experimental Rate Constants for $HO + H_2 \rightarrow H_2O + H$. Experimental measurements noted in the figure legend: Group at 98 K,[189–192] Smith (1974),[193] Ravishankara (1981),[167] Frank (1985),[194] Michael (1988),[195] Oldenborg (1992),[196] Talukdar (1996),[168] Krasnoperov (2004),[169] and Orkin (2006).[170] (b) *Ab initio* SCTST and experimental rate constants for $HO + D_2 \rightarrow HOD + D$. Experimental measurements noted in the figure legend: Smith (1974),[193] Ravishankara (1981),[167] and Talukdar (1996).[168] (c) *Ab initio* SCTST and experimental rate constants for $HO + HD \rightarrow$ Products. Experimental measurements noted in the figure legend: Talukdar (1996).[168] (d) *Ab initio* SCTST and experimental rate constants for $DO + H_2 \rightarrow DOH + H$. Experimental measurement noted in the figure legend: Talukdar (1996).[168]

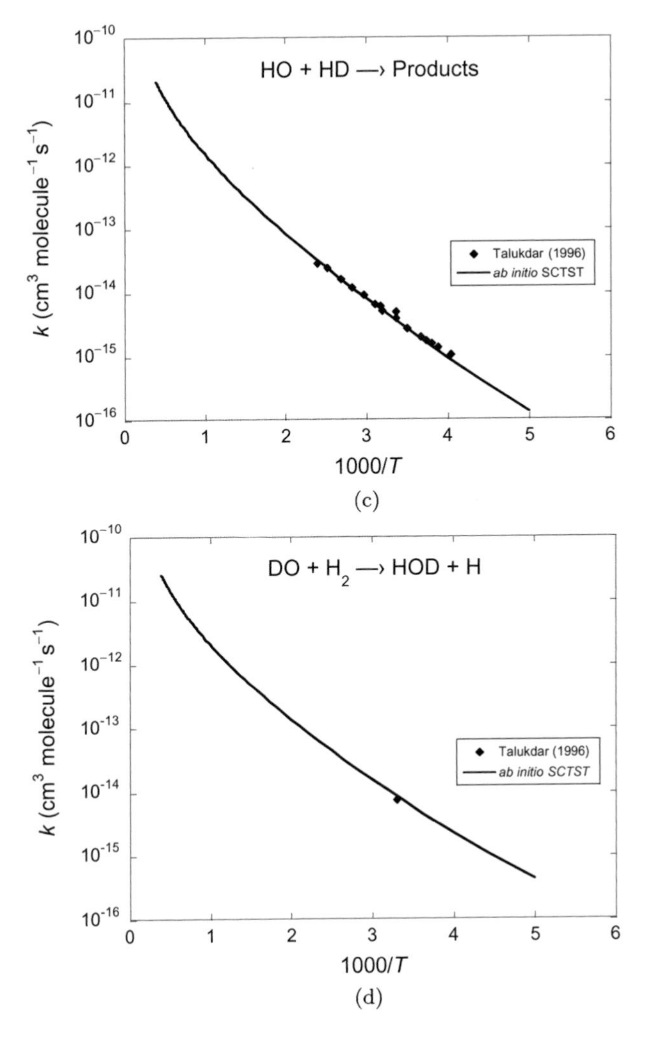

Fig. 6.9 (*Contiuned*)

agreement. Two sets of products are produced in the reaction:

$$\mathrm{HO} + \mathrm{HD} \rightarrow \mathrm{HOD} + \mathrm{H}, \quad \text{(R2a)}$$

$$\rightarrow \mathrm{HOH} + \mathrm{D}. \quad \text{(R2b)}$$

For a complete description of the isotopic chemistry, the branching ratio for the reaction products is also needed and is shown in

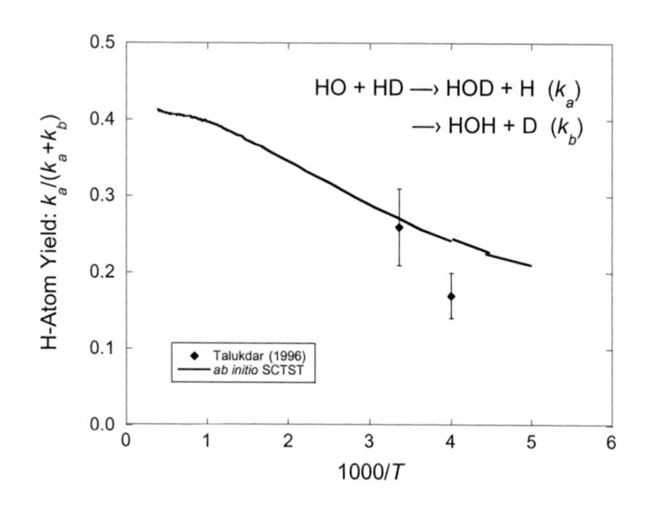

Fig. 6.10 H-Atom yield from *ab initio* SCTST rate constants for HO + HD → Products. Experiments (2σ error bars): Talukdar (1996).[168]

Fig. 6.10 as a function of temperature. The results show that at atmospheric temperatures, the reaction mostly produces $H_2O + D$. These results are also in very good quantitative agreement with experiments carried out by Talukdar *et al.*, who used Lyman-α resonance fluorescence to measure H-atom yields (for each HO radical reacted) of 0.17 ± 0.03 and 0.26 ± 0.05 at 250 K and 298 K, respectively.[168] The present *ab initio* SCTST results (0.24 and 0.27 at 250 K and 298 K, respectively) are in good agreement with their experiments.

The reaction HO + HD has also been investigated theoretically by other groups, who have used various theoretical techniques and PESs and have also achieved excellent agreement with the experimental data. These include investigations by Troya *et al.*,[186] who used VTST and the WSLFH PES surface,[181] and Zhang *et al.*,[183] who used a full-dimensional quantum dynamic treatment on the YZCL2 PES.[182] In their calculations, Zhang *et al.* obtained H-atom yields of 0.17 and 0.23 at 250 K and 298 K, respectively, which are in excellent agreement with the measurements.[183] Garcia *et al.*[185] performed QCT calculations also using the YZCL2 PES and obtained 0.29 for the relative yield of H-atoms at 298 K, also in excellent agreement with experiment. Thus the *ab initio* SCTST approach produces results with accuracy comparable to other high-level theoretical approaches.

In the $HO + H_2$ reaction, the H-atom in the hydroxyl radical reactant does not participate directly, although it participates indirectly through its influence on the vibrational energy levels, rotational constants, and mass of the reactant. When the H-atom is replaced by a Deuterium, the reaction rate constant is affected by this "secondary isotope effect". The thermal rate constant for $DO + H_2$ computed using *ab initio* SCTST agree very well at the single temperature at which the reaction has been measured.[168] The reaction rate constant involving DO is slightly faster than that involving HO, especially at low temperatures. This difference in rate constants is due to differences in zero-point vibration energies ($0.26\,\mathrm{kcal\,mol^{-1}}$), rovibrational parameters, anharmonicity constants, and quantum mechanical tunneling.

At temperatures below $\sim$500 K, where tunneling dominates, the rate constants for DO reacting with H_2, HD, and D_2 are almost indistinguishable from the corresponding reactions involving HO. Thus we conclude the "spectator" hydrogen isotope present in the hydroxyl radical does not influence tunneling significantly. At high temperatures, the rate constants are less sensitive to tunneling and zero-point energy differences, but they still differ slightly from each other due to differences in rotational constants.

Overall, *ab initio* SCTST and other theoretical approaches are quite successful in reproducing the experimental measurements for this reaction system. This level of success suggests that theoretical rate constants can be used with confidence for the reaction isotopologues for which experimental data are scarce. In particular, the rate constant for $DO + H_2$ has been measured only near 298 K, but the *ab initio* SCTST results are available and are expected to be of comparable accuracy over the entire temperature range from 200 K to 2500 K.

For convenience, algebraic expressions for the *ab initio* SCTST thermal rate constants for all of the reaction isotopologues reported previously[46] have been fitted by least squares and are listed in

Table 6.9 Thermal rate constants[a] fitted to the expression $k(T) = A \cdot T^c \cdot \exp(-B/T)$ cm^3 molecule^{-1} s^{-1}.[46]

Reaction	A	c	B
$HO + H_2$	4.2×10^{-17}	1.78	1453
$HO + HD$	3.2×10^{-17}	1.80	1627
$HO + D_2$	3.3×10^{-17}	1.77	1830
$DO + H_2$	2.8×10^{-17}	1.83	1419
$DO + HD$	2.0×10^{-17}	1.86	1525
$DO + D_2$	1.6×10^{-17}	1.87	1724

[a]Accurate within ±~10% from 200 K to 2500 K

Table 6.9. These expressions are accurate to within ±~10% over the temperature range from 200 K to 2500 K.

6.4.2. $Cl + CH_4 \rightarrow HCl + CH_3$

Atmospheric methane, which is the second-most important long-lived greenhouse gas (after CO_2), has both biogenic and anthropogenic sources. The natural sources include wetlands, ocean biosphere, geological sources, biomass burning, termites, and ruminating wild animals. Anthropogenic sources include leakage from fossil fuel extraction, livestock, landfills, waste treatment, fossil fuel combustion, biomass burning, and emissions due to agricultural practices.[197–199] Its atmospheric lifetime of ~12 years is controlled mostly by its reaction with hydroxyl radicals and secondarily by its reactions with chlorine atoms and $O(^1D)$.[200] In the stratosphere, reaction with methane is the principal fate of Cl-atoms. The reaction and its reverse have been the subject of many experimental[2,4] and theoretical studies. A more direct source of knowledge about the atmospheric sinks in comparison to each other and to other physical processes, like transport and mixing, can be obtained from the isotopic composition of atmospheric methane.[198,199,201–205]

The primary sources of atmospheric methane have distinct isotopic signatures, which are affected by fractionation in the atmosphere. Isotopic fractionation occurs because different isotopologues react at different rates. The isotopologue that reacts most quickly is depleted and the one that reacts most slowly is enriched in the atmosphere. The reaction of an isotopologue is characterized by the KIE, or fractionation factor α (see Sec. 6.3).

Quantum mechanical tunneling is important in governing the rate of this reaction, as evidenced by the curved Arrhenius plot. Quantum mechanical tunneling is one of the causes of "mass-independent" isotope effects, which are of considerable atmospheric interest.[206] Because of its atmospheric importance and because of the extensive experimental rate constant data available for comparisons, this reaction provides a useful test for SCTST. Here, we describe our recent calculations using *ab initio* SCTST.[207]

The $Cl + CH_4$ reaction provides a demanding test for SCTST and its implementation. The reaction transition state contains of 12 vibrational degrees of freedom and includes 27 electrons, compared with 6 and 11, respectively for the $OH + H_2$ reaction. The thermochemistry of the latter reaction was successfully treated using HEAT-345Q theory.[64–66] For $Cl + CH_4$, HEAT-345Q theory is prohibitively expensive. However, Eskola *et al.*[208] used the CCSD(T) level of theory, extrapolated to the complete basis set limit (CBS), to compute the reaction thermochemistry, with estimated errors of $\lessapprox 2$ kJ/mol. However, small errors in classical barrier heights are usually not important when calculating KIEs, because the errors tend to cancel when considering ratios of rate constants. Thus, KIEs are not very sensitive to errors in the classical barrier height and *ab initio* rate constants can be used without adjustments to obtain *ab initio* KIEs. In any event, the original work[51] shows that the thermochemistry computed by Eskola *et al.* is highly accurate, with apparent errors of only $\sim$0.13 kJ/mol.

In the original work,[51] *ab initio* SCTST rate constants were computed for 25 isotopologues of the title reaction for various combinations of ^{35}Cl, ^{37}Cl, ^{12}C, ^{13}C, ^{14}C, H, and D. From the *ab initio* rate constants, many different KIEs can be calculated.

6.4.2.1. *Technical Details*

The *ab initio* SCTST rate constants are based on harmonic vibrational frequencies calculated at the CCSD(T)/aug-cc-pVTZ level of theory and X_{ij} vibrational anharmonicity coefficients calculated at the CCSD(T)/aug-cc-pVDZ level of theory. For some reactions, anharmonicity coefficients were also computed at the CCSD(T)/aug-cc-pVTZ level of theory. Anharmonicities were computed using the CFOUR electronic structure code.[27] The classical reaction barrier was taken from Eskola *et al.*,[209] who extrapolated CCSD(T) calculations to the complete basis set limit. They found classical reaction barriers in the forward and reverse directions for the title reaction of 32.11 kJ/mol and 7.80 kJ/mol, respectively, including the electronic SO correction[210] of 3.51 kJ/mol for the Cl atom.

To save expense, most anharmonicities were computed using the smaller aug-cc-pVDZ basis set (designated as aVDZ), but some were obtained with the aVTZ basis. When combined with harmonic vibrational frequencies obtained with the aVTZ basis, this leads to two combinations of vibrational data generated by CFOUR, designated TcTc and TcDc (where T and D denote the aVTZ and aVDZ basis sets, respectively, and the "c" designates CFOUR). The latter set, a hybrid, is analogous to the hybrid used to obtain accurate results by Lee, Martin and Taylor (LMT) in a study of anharmonic force fields of CH_4 isotopologues.[211] (In addition, the GAUSSIAN program[30] was used for some of the calculations by the Barker group at Michigan, leading to a third set of vibrational parameters, which is designated with a "g".[51])

The CCSD(T) method is based on the Born–Oppenheimer approximation, which generates mass-independent PESs. Mass-dependent PESs, which are more accurate, are obtained by applying the DBOC obtained using perturbation theory. The DBOC for all isotopic species was obtained using the HF/aVTZ level of theory. In addition, the CCSD/aVTZ level of theory was used to obtain the DBOC for all of the species involved in two reactions: $^{35}Cl+^{12}CH_4$ and $^{35}Cl+^{13}CH_4$. All of the DBOC calculations were carried out for optimized geometries obtained at the CCSD(T)/aVTZ level of theory. For reaction barriers, the differences in DBOCs obtained by the HF method are less than 35 cm^{-1} and those obtained using the CCSD level of theory are less than 25 cm^{-1}. These DBOC energy differences are small, compared with possible errors of $\sim 2\,kJ\,mol^{-1}$ in the computed barrier heights, but they are significant for the H/D KIEs at low temperatures. For $^{12}C/^{13}C$ KIEs, the DBOC is essentially insignificant ($<0.3\,cm^{-1}$).

As discussed by Eskola *et al.*,[208] a hydrogen-bonded adduct (a post-reaction complex, or PRC) exists between CH_3 and HCl. Its energy (including zero-point energy, ZPE) is lower than the energy of the initial reactants Cl + CH_4. If the PRC is ignored, the reaction to form separated $CH_3 + HCl$ is endothermic. Since formation of the PRC is exothermic, it is possible for the system to tunnel through the energy barrier in the forward direction from $Cl + CH_4$ to produce the PRC. It is also possible for $CH_3 + HCl$ to form the thermalized PRC, when the reaction proceeds in the reverse direction, if collisional stabilization is significant. However, the probability of collisional stabilization is very small, because the lifetime of the weakly bound PRC is expected to be much shorter than the average time between collisions, even at pressures $\gg 1$ bar.

6.4.2.2. *Results and Discussion*

The *ab initio* SCTST rate constant for $^{35}Cl + {}^{12}CH_4$ is compared to experiment in Fig. 6.11. (This comparison is appropriate, because ^{35}Cl is more abundant than ^{37}Cl and the computed ratio of the rate constants for the two isotopologues is very nearly equal to unity.[51]) The figure shows that the *ab initio* SCTST rate constant (with no adjustments at all) is in excellent agreement with the experimental data. The *ab initio* SCTST rate constants shown in Fig. 6.9 are also in very good agreement with previous high quality theoretical calculations that employed other methods.[212–214] The *ab initio* SCTST rate constants are quite acceptable, but fall very slightly below the centroid of the data at low temperatures. Lowering the barrier height by only $\sim 0.13\,kJ/mol$ would bring the computed rate constants into line (not shown), suggesting that the thermochemistry computed by Eskola *et al.*[209] is accurate to within this small energy difference. Since the *ab initio* rate constants are already acceptable without corrections and because test calculations

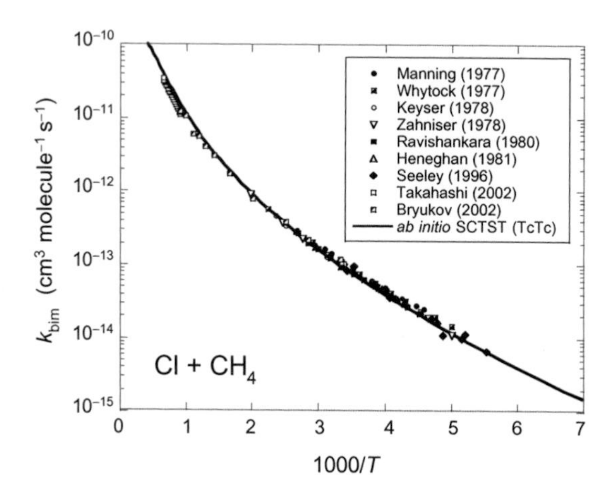

Fig. 6.11 *Ab initio* SCTST rate constant for $^{35}Cl+^{12}CH_4$ (the TcTc data set, solid line) compared to experimental measurements (symbols). The experimental data are: Manning (1977),[215] Whytock (1977),[216] Keyser (1978),[217] Zahniser (1978),[218] Ravishankara (1980),[219] Heneghan (1981),[220] Seeley (1966),[221] Takahashi (2002),[222] and Bryukov (2002).[223]

showed no sensitivity to the classical barrier height, we did not apply any corrections in the present work.

All of the H/D reaction isotopologues were investigated,[51] but only the reactions of CH_3D are important in the atmosphere, since the atmospheric H/D ratio is quite low, as mentioned in the previous section. The *ab initio* KIE for this isotopologue is compared to experimental data in Fig. 6.12. It is apparent from the figure that the computed KIEs fall right in the midst of the experimental measurements. The temperature dependence is in good overall agreement with the data of Saueressig *et al.*[224] and Sauer *et al.*[225] The KIEs calculated from the *ab initio* SCTST rate constants appear to be at least as accurate as the experimental measurements, which are somewhat scattered, and are more accurate than those from previous calculations,[212,213] as discussed in the original paper.[51]

The KIEs for ^{13}C and ^{14}C have also been investigated. In Fig. 6.11, the existing experimental data for ^{13}C are shown as points with 2σ error bars. The lines display theoretical results. The short-dashed line shows the KIE predicted for ^{14}C using *ab initio* SCTST (no measurements have been reported). The long-dashed and solid

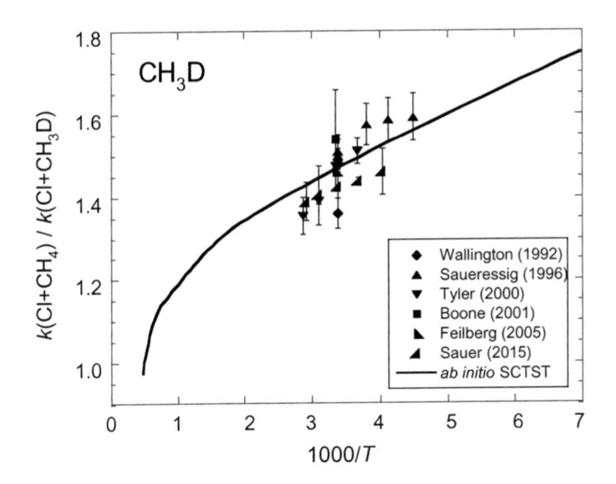

Fig. 6.12 KIEs for $Cl + CH_3D$. Experimental data[13,224–228] are shown as points (2σ error bars) and the line is the *ab initio* KIE.

lines are both for ^{13}C. Both dashed-lines are based on vibrational anharmonicities computed using the smaller basis set. The solid-line is the KIE for ^{13}C based on vibrational anharmonicities computed using the larger basis set. Thus the difference between the two lines for ^{13}C is due to just the basis sets, suggesting that use of a still larger basis set may further improve the agreement with experiments.

We conclude by observing the *ab initio* H/D KIEs for all of the reaction isotopologues agree well with the experiments, which are scattered by 5–10%; we surmise that the theoretical KIEs could be used with no loss of accuracy in place of experimental data for analyzing atmospheric budgets and cycling. The theoretical $^{12}C/^{13}C$ KIEs are in fairly good agreement with the experimental data, but in this case the magnitude of the effect is much smaller and an error of 1% is not tolerable. Thus the accuracy of theoretical KIEs must be improved still more. This can be achieved by using larger basis sets. An improved, higher-order VPT might also be needed.

6.4.3. $HO + CO \rightleftharpoons$ *trans*-$HOCO \rightleftharpoons$ *cis*-$HOCO \rightarrow H + CO_2$

The oxidation of carbon monoxide to carbon dioxide by the hydroxyl radical is important in the terrestrial and planetary atmospheres,

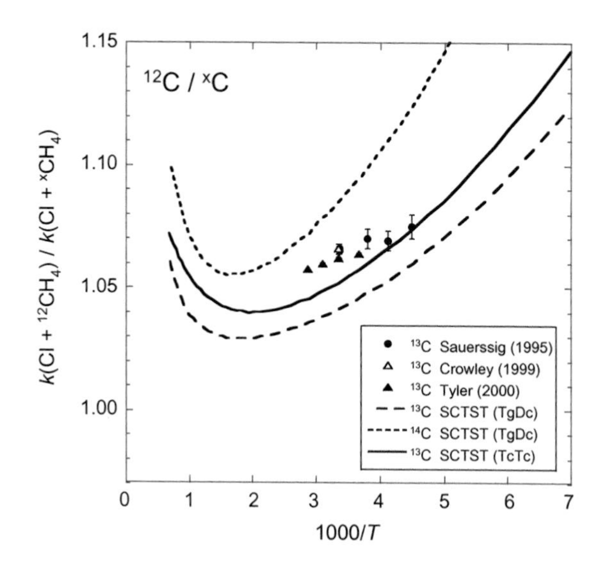

Fig. 6.13 ^{13}C and ^{14}C KIEs for $Cl + CH_4$. Experimental data[229–231] are shown as points (2σ error bars); the lines are the present SCTST results. The short-dashed line is a prediction for ^{14}C, since no measurements have been reported.

and at high temperatures in combustion chemistry.[232] As a sink for HO free radicals in the terrestrial atmosphere, the reaction is of comparable importance to that of $HO + CH_4$.[233] The surface of Venus is very hot ($\sim$730 K), because of radiative forcing (i.e., the natural greenhouse effect) acting via the >90 bar of CO_2 in the atmosphere. The surface of Venus is dry, because of the "runaway greenhouse effect": the positive feedback of the greenhouse effect resulted in evaporation of all liquid water on the surface. In the Venusian atmosphere, the HO + CO reaction plays a key role in the cycling of carbon dioxide.[234]

Because the HO + CO reaction is a key step in so many important chemical systems, it is regularly reviewed[2,4] and it has been the subject of many experiments, in which rate constants for the forward reaction have been obtained over a wide range of temperatures and pressures.[166,191,235–285] When the experimental investigations were being performed, it soon became apparent that the reaction rate was dependent on total pressure and that the temperature dependence is non-Arrhenius. Smith and Zellner[166,279] were the first to suggest that the reaction involves formation of an HOCO intermediate that can be stabilized by collisions in a competition with dissociating back to reactants or forming products.

Experiments have determined the effect of deuteration,[248,252,256,264,286] and of substituting ^{17}O, ^{18}O, or ^{13}C in carbon monoxide.[251,280–283] The dynamics of the reaction have been investigated by various methods. Early attempts[242] to find vibrational excitation in the carbon dioxide product were unsuccessful. Later, tunable diode laser spectroscopy enabled Smith and co-workers[258–260] to determine that not more than 6% of the available energy was found in the CO_2 product. They also compared the reaction rates of HO and DO in ground and excited vibrational states.[256,286] Molecular beam experiments have provided information about the dynamics of the reaction.[284] In coincidence photodetachment experiments by Continetti and coworkers on the $HOCO^-$ anion, the $HO + CO$, HOCO, and $H + CO_2$ products sets were observed.[287–291]

The reaction system has become a benchmark for kinetics and dynamics calculations.[292] Largely, this is because the number of electrons is small enough for accurate quantum mechanical methods to be applied, all of the reactions in the system have intrinsic energy barriers, quantum mechanical tunneling is important, the system consists of multiple wells and multiple reaction channels (but is not too complex), and pre-reactive complexes are present. In short, it is an elegant microcosm of the complicated chemical activation reaction systems that are prominent in atmospheric chemistry and combustion.

The stationary points the PES for this reaction are shown in Fig. 6.14. Energies of important points on this surface have been calculated with *ab initio* methods at various levels of theory by many research groups.[57,293–302] Statistical or RRKM calculations of rate constants have been based on the results of *ab initio* quantum chemical calculations and on more or less *ad hoc* values for the parameters needed in the calculations.[263,303–308] Rate constants have also been obtained using quantum reactive scattering methods[292,309–312] and quasi-classical trajectory calculations have been performed (usually) on analytical PESs derived from *ab initio* calculations.[311,313–317]

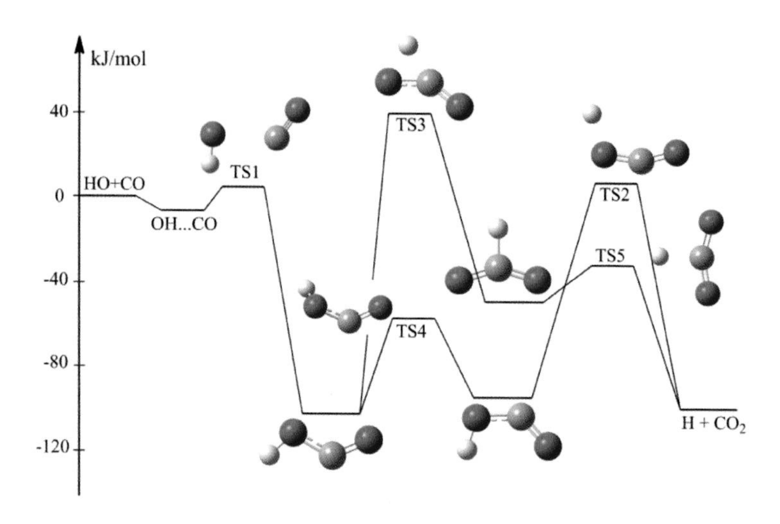

Fig. 6.14 PES of the HOCO reaction system.[57]

Studies of "deep tunneling" have also been investigated in this reaction system.[318]

The global reaction expresses the loss of HO radicals:

$$\mathrm{HO} + \mathrm{CO} \rightarrow \text{Products.} \tag{R3}$$

Neglecting for the moment the pre-reactive complex (OH$\cdots$CO), the lowest energy pathway on the PES involves both the cis- and the trans-HOCO intermediates, which react differently as is apparent from the PES shown in Fig. 6.14. The full mechanism must also include collisional activation and deactivation of rovibrationally excited species diluted in a chemically inert bath gas.

As mentioned in Section 6.2.4, when the *cis*- and *trans*-HOCO isomers are in microcanonical equilibrium, analytical solutions[263,319,320] of the 2D master equation can be obtained for the $P = 0$ and $P = \infty$ limits[55,57]:

$$k_{P=0}(T) = \frac{1}{h}\frac{Q^{\neq}_{\text{trans,elec}}}{Q_{\text{HO}}Q_{\text{CO}}}\sum_{J=0}^{+\infty}(2J+1) \times \int_0^{+\infty} \frac{G_1^{\neq}(E,J)G_{23}^{\neq}(E,J)}{G_1^{\neq}(E,J)+G_{23}^{\neq}(E,J)}\exp\left(-\frac{E}{k_BT}\right)dE, \tag{6.17a}$$

$$G^{\neq}_{23}(E,J) = G^{\neq}_{2}(E,J) + G^{\neq}_{3}(E,J), \tag{6.17b}$$

$$k_{P=\infty}(T) = \frac{1}{h}\frac{Q^{\neq}_{\text{trans,elec}}}{Q_{\text{HO}}Q_{\text{CO}}}\sum_{J=0}^{+\infty}(2J+1) \times \int_0^{+\infty} G^{\neq}_{1}(E,J)\exp\left(-\frac{E}{k_BT}\right)dE, \tag{6.18a}$$

$$k_{P=\infty}(T) = \frac{k_BT}{h}\frac{Q^{\neq}_{TS1}}{Q_{\text{HO}}Q_{\text{CO}}}\exp\left(-\frac{\Delta E^{\neq}_{TS1}}{k_BT}\right). \tag{6.18b}$$

For the HO + CO reaction, $G^{\neq}_{1}$, $G^{\neq}_{2}$, and $G^{\neq}_{3}$ are the cumulative reaction probabilities (CRPs) for transition states $TS1$, $TS2$, and $TS3$, respectively (see Fig. 6.14). In these equations, k_B is Boltzmann's constant and Q_{HO}, Q_{CO}, and Q_{TS1} are total partition functions for HO, CO, and transition state $TS1$, respectively. The other factors were introduced in Secs. 6.2.3 and 6.2.4.

At intermediate pressures, the reduction of the 2D master equation to 1D is only an approximation, as mentioned in Sec. 6.2.4. The specific approximation used in the MultiWell Program Suite is described in the User Manual[91] and in the literature.[55] When that approximation is used at P = 0, the solutions are in error by ~15%,[55] which is noticeable, but tolerable for most applications.

The aim of the calculations reported here was to use *ab initio* SCTST microcanonical rate constants in master equation simulations of the HO + CO chemical reaction system to obtain total rate constants for Reaction R3, which can be compared with experimental data. One of the advantages of SCTST is that it can easily be used to determine the microcanonical rate constants required for master equation calculations. The work described here was the first application of SCTST to pressure-dependent master equation simulations.

Master equation simulations are needed in order to model the complex interplay of chemical reactions and energy transfer, especially over the wide ranges of temperature and pressure important for atmospheric chemistry and combustion. The 1D master equation simulations (see Sec. 6.2.4), which utilize *ab initio* SCTST rate constants and empirical energy transfer parameters, are shown to provide a very good description of the experimental data over the range of temperatures and pressures found in the terrestrial atmosphere and in combustion.

A second motivation was to determine whether the *ab initio* SCTST rate constants can predict the H/D KIEs accurately for the global HO + CO reaction. In the application of SCTST to the study of KIEs in the $Cl + CH_4$ reaction (see Section 6.4.2), it was found that the predicted H/D KIEs are as accurate as the experimental data, but the $^{12}C/^{13}C$ KIEs are not.[51] In previous work, the H/D KIEs for the global HO + CO reaction have resisted theoretical interpretation. Chen and Marcus reported that standard RRKM theory (with Eckart tunneling corrections) could not account successfully for the H/D KIEs for reaction (R1), and a novel and non-statistical model was required.[301,321,322] In contrast, it is found in the present work that the H/D KIEs calculated by using the *ab initio* SCTST rate constants are in reasonable agreement with the experimental data, showing that difficulties encountered in earlier studies were possibly due to less accurate treatments of tunneling.

6.4.3.1. *Technical Details*

Rate constants in the intermediate pressure regime (i.e., in the fall-off) are computed using the MultiWell Software Suite of programs.[61,97,100] In the MultiWell master equation code, angular momentum conservation is included approximately, via centrifugal corrections,[323] implemented as described elsewhere.[55] However, the 1D MultiWell master equation code (depending explicitly on active energy) is in exact agreement with Eq. (6.2) at $P = \infty$ limit and is in very good, but not exact, agreement with the solution at the $P = 0$ limit, as discussed in Sec. 6.2.4.

For the 1D master equation simulations, the multi-well, multichannel model consisted of cis- and trans-HOCO and reactions via $TS1$, $TS2$, $TS3$, and $TS4$; HCO_2 and the weakly bound prereactive complexes were neglected. At low energies (i.e., during and after vibrational deactivation), it cannot be assumed that *cis*- and *trans*-HOCO are in equilibrium and thus the isomerization reaction (*TS*4) was considered explicitly and treated as reversible.

The densities of states for the fully-coupled anharmonic vibrations were computed using computer code ADENSUM (part of the MultiWell Software Suite), which uses an algorithm[89] that is based on the work of Basire *et al.*[50] and Wang and Landau.[48] The CRPs were computed using computer code SCTST (part of the MultiWell Software Suite), which implements Semi Classical Transition State Theory, as documented elsewhere.[45,46] The stationary point energies computed using the HEAT protocol were used without adjustment.

When computing the CRPs and densities of states, one external rotational degree of freedom (the K-rotor) was assumed to be completely active (i.e., its energy can randomize, along with the energy in the vibrational modes) and completely independent of the total angular momentum. When centrifugal corrections are used, this "separable-rotors" approximation[324] is commonly employed.[85,86,325] In the present work, all chemical species were approximated as symmetric tops with $A \neq B = C$ (rotational constants A, B, and C). Current and Rabinovitch showed that the separable-rotors approximation is most accurate when the rotational constants for the K-rotor and the 2D adiabatic rotor are given by $B_K = A$ and $B_{2D} = (BC)^{1/2}$, respectively.[324]

The value of the energy transfer parameter α was adjusted in order to obtain agreement with the experimental data for the global reaction near 300 K, as discussed in the next section. The same value of α was used for both HOCO isomers at all temperatures. The Lennard–Jones (LJ) parameters used for the bath gases are standard values found in the literature.[55]

The MultiWell stochastic simulations were initiated by assuming that trans-HOCO is initially formed by reaction (R3) with a nascent chemical activation energy distribution.[97,326] An energy grain of $5\,\text{cm}^{-1}$ was used when computing the CRPs and densities of states; the energy ceiling for the master equation calculations was $49500\,\text{cm}^{-1}$. Gillespie's stochastic simulation algorithm[327,328] was used to produce the simulations. No special constraints or assumptions were imposed on the simulations, which produced records of the time-dependent relative concentrations of the intermediates (i.e., cis- and trans-HOCO) and products (i.e., HO + CO and H + CO_2). Rate constants were obtained by analyzing the time-dependent concentration profiles, or by direct modeling of the chemical activation process, as described elsewhere.[55,97,326]

6.4.3.2. *Results and Discussion*

Master equation simulations were performed for very wide ranges of temperature and pressure. For comparisons with experimental data, it is convenient to consider the rate constants measured and calculated near the temperatures used by Fulle *et al.*[263] In Figs. 6.15–6.17, the computed rate constants are compared with experimental measurements from the literature. In these semi-log plots, the rate constant has a sigmoidal shape as it increases from the $P = 0$ intercept, passes through intermediate pressures, and then approaches the high-pressure limit. In each plot, the 1D master equation results are shown as a heavy solid line corresponding to Helium collider gas $\alpha(\mathrm{He}) = 200\,\mathrm{cm}^{-1}$ (slightly better agreement

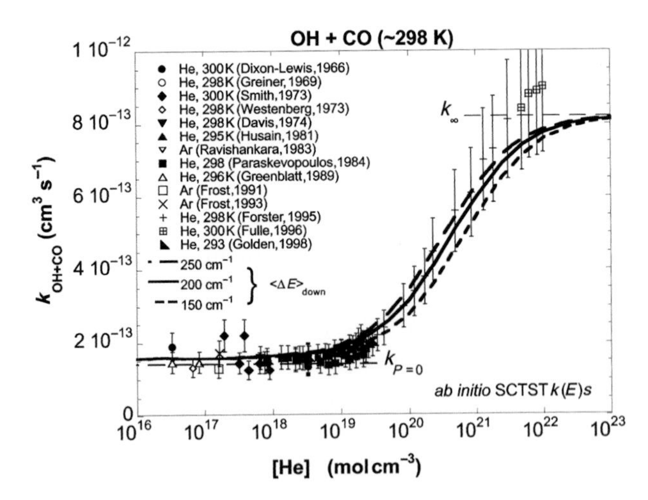

Fig. 6.15 Rate constants for HO + CO in He bath at 298 K. Experimental data (errors of ~20%) are shown as points. The heavy solid line is the rate constant predicted by the 1D master equation with $\alpha(\mathrm{He}) = 200\,\mathrm{cm}^{-1}$ (the upper and lower dashed lines correspond to $250\,\mathrm{cm}^{-1}$ and $150\,\mathrm{cm}^{-1}$, respectively).[55] Also shown are the exact $P = 0$ and $P = \infty$ rate constants.[57]

with experiments can be obtained at 298 K and $P \lesssim 1$ bar, with $\alpha(\text{He}) = 150\,\text{cm}^{-1}$ as shown in Fig. 6.15). The exact results for the low and high-pressure limits (see Sec. 6.2.4) are shown as horizontal thin dashed lines in each figure. The energy transfer parameter α is the only empirical parameter employed in the model.

At the lowest pressures, the experimental rate constants are in good agreement with the calculated $P=0$ limit. Even at ~98 K, where the experimental scatter is at least ±25%, the computed rate constants are accurate to within a factor of two (Fig. 6.16). If the factor of two at 98 K is attributed to errors in the calculated energies, it corresponds to $|\Delta E| = 0.14\,\text{kcal mol}^{-1}$, which is smaller than the energy errors conservatively estimated for the HEAT 345Q protocol.

Each of the plots shows a small discrepancy between the 1D master equation and the exact result at $P=0$. As explained in Sec. 6.2.4, this small discrepancy arises at least in part from how the centrifugal corrections are implemented in the 1D master equation. Although noticeable, the discrepancy is relatively small.

At 298 K and very high pressures (Fig. 6.15), the data of Forster *et al.*[261] and Fulle *et al.*[263] have a tendency to roll over as if they are

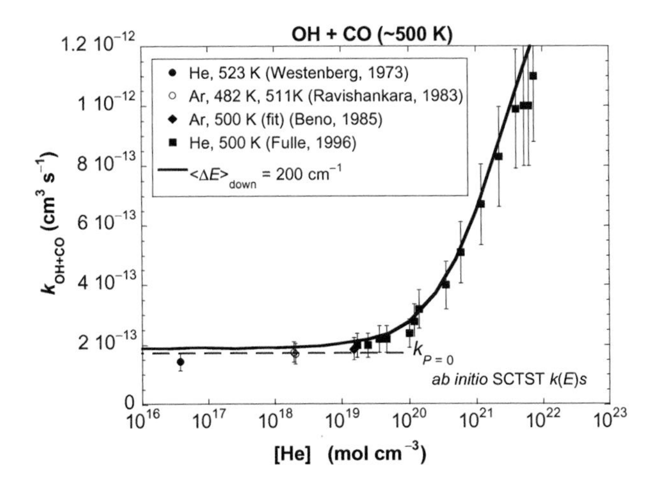

Fig. 6.16 Rate constants for HO + CO in He bath at 500 K. Experimental data are shown as points (errors of ~20%). The heavy solid line is the rate constant predicted by the 1D master equation.[55] Also shown are the exact $P = 0$ and $P = \infty$ rate constants.[57]

approaching the high-pressure limit, but it is not entirely clear that this is a real effect, since the stated experimental errors are ~20%. At temperatures >300 K (Fig. 6.16), the experimental He pressures are not high enough to reach the predicted high-pressure limits. Furthermore, the agreement with experimental data from 300 K to >800 K is excellent when using $\alpha(\text{He}) = 200\,\text{cm}^{-1}$, suggesting that $\alpha(\text{He})$ is independent of temperature over this range.

At low temperatures (Fig. 6.17), the agreement with experiments is not as good. The experimental rate constants do not show any clear indications of reaching the high-pressure limit, but their magnitudes at high pressures greatly exceed the calculated rate constants. At low pressure, however (which is relevant to the terrestrial atmosphere), the agreement is still quite good. The discrepancy between theory and experiment at high pressures gradually becomes perceptible at temperatures below 300 K, and is significant around ~250 K and below.

In the original work, the reason for the large discrepancy between theory and the low temperature experiments conducted at high pressures was not identified.[55] Recently, however, a possible

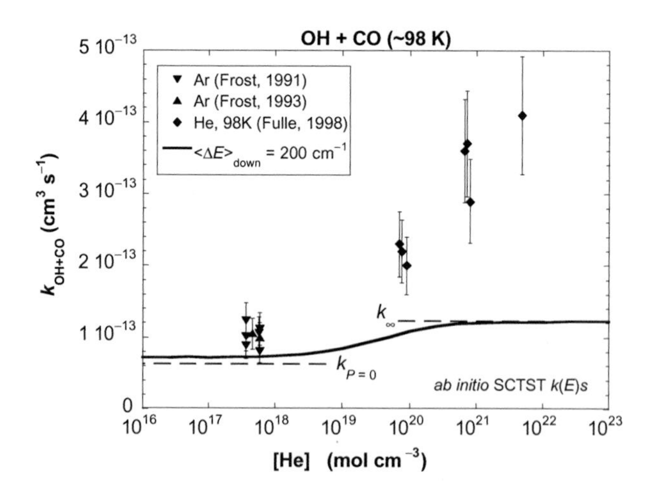

Fig. 6.17 Rate constants for HO + CO in He bath at 98 K. Experimental data are shown as points (errors of ~20%). The heavy solid line is the rate constant predicted by the 1D master equation.[55] Also shown are the exact $P = 0$ and $P = \infty$ rate constants.[57]

explanation has emerged. The large increase in the measured high pressure HO + CO rate constants at very low temperatures resembles the behavior observed for bimolecular reactions measured at extremely low temperatures: as the temperature is decreased far below room temperature, the bimolecular rate constant increases dramatically.[329–332] Heard and coworkers[333,334] have attributed this effect to the presence of a pre-reactive complex (PRC) in the reaction entrance channel, prior to encountering the energy barrier to reaction. In essence, their explanation is that although the PRC is weakly bound and is not important when k_BT is greater than the well depth, at low temperatures k_BT is small and the binding energy is sufficient to trap the reactants in the PRC local minimum on the PES. At higher temperatures, the thermalized PRC simply re-dissociates to the separated reactants, but at low temperatures the trapped reactants can react further by tunneling through the base of the energy barrier to produce final products, thus enhancing the over-all bimolecular rate constant.

In the context of the PES for HO + CO (Fig. 6.14), the PRC, which was ignored in the master equation calculations, has a well-depth of ~2.6 kcal/mol. At 100 K, k_BT is only 0.2 kcal/mol, suggesting that the lifetime of the thermalized PRC may be long enough to allow significant tunneling and a consequent enhancement of the bimolecular rate constant. This scenario must be investigated further in the context of HO + CO, but accurate quantitative results may be difficult to obtain, since it is known that SCTST has shortcomings in the deep tunneling region.[92,318]

The linear scales for rate constants in the figures help to reveal subtle features in the HO + CO data that are not as visible in the semi-log plots. First, the agreement between models and most experiments would be improved if the computed rate constant at $P = 0$ was lowered by ~15% to the value obtained[57] using the exact solution of the 2D master equation. This suggests that the *ab initio* SCTST rate constants are highly accurate. The experimental data and calculations agree reasonably well near helium pressures of 1 atm and below for $\alpha(\text{He}) = 125$ or $150\,\text{cm}^{-1}$, but if the ~15% offset was eliminated, the optimal value of $\alpha(\text{He})$ would be close to $175\,\text{cm}^{-1}$.

However, the data show considerable scatter between and even within data sets, except for the set of Golden *et al.*,[264] which shows the modest curvature that is also seen in the theoretical results.

Other bath gases have been investigated at 298 K and pressures of $\lesssim$1 bar, and we have modeled experimental data for Ar, N_2, CF_4, and SF_6 for those conditions. It was found that $\alpha(\mathrm{Ar}) \approx \alpha(\mathrm{N_2})$, in general agreement with the findings of Liu and Sander,[285] when the collision frequencies for the two bath gases are taken into account. Furthermore, both $\alpha(\mathrm{CF_4})$ and $\alpha(\mathrm{SF_6})$ are large, suggesting that collisional energy transfer involving CF_4 and SF_6 is very efficient, as expected because of their low-frequency vibrational modes.[335]

H/D KIEs were also considered.[55] In Fig. 6.18, the experimental data at 298 K and theoretical calculations are compared as functions of pressure. The rate constant for DO + CO has been determined in a number of experiments, at pressures up to about one atmosphere, and mostly at temperatures around ~298 K. In some cases, the same group has studied both the HO and the DO reaction, but in separate experiments, not in rate measurements with both species present simultaneously at the same bath gas pressure and temperature.[248,252,259,264,286] For this reason, we have simply plotted the two rate constants separately. This approach also makes the experimental scatter and inconsistencies readily apparent.

The theoretical rate constants are based on harmonic frequencies, vibrational anharmonicities, and reaction enthalpies for the DO + CO PES computed using the same levels of theory and basis sets that were used for the normal isotope. Since the vibrational frequencies and rotational constants for the DOCO isomers differ from those of the HOCO species, it was *not* assumed that the energy transfer parameters are the same. Instead, rate constants for both isotopes were computed for three assumed values of $\alpha(\mathrm{He})$ noted in the figure. Like the HO + CO rate data, the rate constants for DO + CO exhibit modest curvature that is reproduced by the 1D master equation simulations. In all cases, the slopes of the theoretical rate constants with buffer gas concentration are greatest at the lowest pressures. This behavior combined with experimental scatter makes it difficult to identify the zero pressure intercept of the experimental

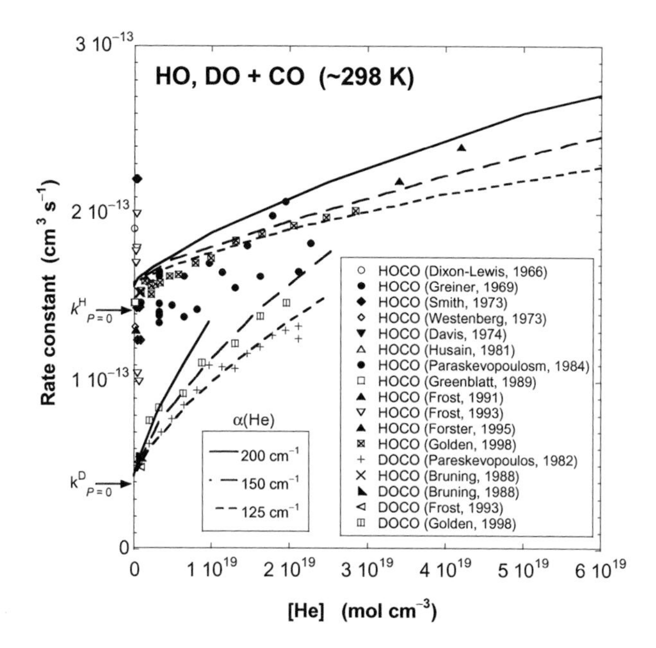

Fig. 6.18 Pressure-dependent rate constants (helium bath gas) for HO + CO and for DO + CO. In each case, the solid and dashed lines correspond to simulations obtained using three values of α(He), as shown. The exact $P = 0$ intercepts for the HO and DO isotopologues predicted using the exact limiting values are shown by the arrows on the Y-axis.

data. In addition, there is a small displacement of the data of Golden *et al.*[264] from those of Paraskevopolous *et al.*[248] However, most of the data are consistent with $\alpha(\mathrm{He}) = 125$–$150\,\mathrm{cm}^{-1}$ for the DOCO isotopologues.

For the DO + CO isotopologue, there is again a small discrepancy between the $P = 0$ intercepts predicted by the exact expression (Sec. 6.2.4) and the 1D master equation at 300 K: 3.87×10^{-14} vs. $4.25 \times 10^{-14}\,\mathrm{cm}^3\,\mathrm{s}^{-1}$, respectively, as shown in the figure (the lower value is more accurate). The H/D KIEs at 300 K predicted using the two methods are 3.66 and 3.80, respectively (the lower value is more accurate).

Considering the scatter displayed in Fig. 6.18, the uncertainty in the experimental H/D KIE near $P = 0$ is significant. The pressure-dependent data of Golden *et al.*[264] have relatively high precision and appear to have an intercept near $1.5 \times 10^{-13}\,\mathrm{cm^3\,s^{-1}}$, but if the plot is curved like the theoretical results, the intercept may be somewhat lower. The data for DO + CO are more limited: the pressure-dependent data of Golden *et al.*[264] and Paraskevopolous *et al.*[248] require risky extrapolations to $P = 0$, but we estimate that the intercepts are roughly 0.5×10^{-13} and $0.43 \times 10^{-13}\,\mathrm{cm^3\,s^{-1}}$, respectively. Taking all of these results into consideration, we think it is likely that the H/D KIE at zero pressure is ~3.2 with an uncertainty of perhaps 10%. This value for the KIE is ~13% lower than the 3.66 computed from theory in the present work.

Chen and Marcus have argued that the H/D KIE at low pressure can only be explained by abandoning RRKM theory.[301,321] However, the H/D KIE at $P = 0$ computed in the present work, which employed significantly higher levels of theory, is in much better agreement with the experimental data. Thus, we conclude that there is no compelling reason to postulate non-statistical behavior in the HO + CO reaction. At the same time, however, it is important to bear in mind that non-statistical effects may still be present, since intramolecular vibrational energy redistribution may be incomplete due to the brief lifetimes of the HOCO isomers.[284,336–338] We expect that future improvements in both experiments and theory will further reduce the discrepancies that remain.

6.5. Conclusions

In this work, we have reviewed SCTST, as formulated within the context of VPT2 and based on HEAT. SCTST is a non-separable transition state theory that inherently includes fully coupled anharmonic quantum vibrations and multi-dimensional tunneling transmission coefficients. As compared to other techniques (such as quantum mechanical methods, conventional TST, or classical trajectory methods), the combination of SCTST with VPT2 is an efficient theoretical method that offers a good compromise between accuracy and computational cost. The SCTST/VPT2 model requires only local (not global) PES regions that are in the vicinity of stationary points. A number of examples given in this work show that the SCTST/VPT2 approach can be used to compute thermally averaged rate constants with a high accuracy, as well as accurate isotope effects. It is also easily used to obtain the energy-dependent microcanonical rate constants that are needed for master-equation simulations.

Because of the attributes mentioned in the previous paragraph, SCTST/VPT2 is a very powerful approach. It can be used to predict bimolecular rate constants and KIEs. With only one empirical parameter to account for collisional energy transfer, it can be used to describe or predict pressure dependent reactions rate constants, which are so important in atmospheric chemistry. Thus the theory, when used judiciously, can supplement and extend experimental measurements. When the computed thermochemistry is sufficiently accurate, the *ab initio* theory is as accurate as experiments, even at very low atmospheric temperatures.

Every theory should be examined in light of its limitations. The reliance on highly accurate thermochemistry is an important limitation, since it is a challenging task to compute energies with sufficient accuracy when a molecular species is larger than a handful of atoms. This is a critical area for more research. High-level methods like the HEAT protocol are practical for small molecules, free radicals, and transition states, but they are not feasible for most of the larger species that are important in the atmosphere. This is especially true for chemical species in which multi-reference effects are important.

The computation of vibrational anharmonicity coefficients poses another challenge. It is probably true that at least 90% of the computer time expended on the systems outlined in this chapter was spent on computing energies and vibrational parameters. Recent work by Mackie et al. shows that careful use of density functional methods may provide a way to compute accurate vibrational parameters more cheaply.[339] More work is needed in this area.

Because VPT2 is only a second-order perturbation theory, it is expected to be accurate only in the vicinity of the stationary points and is limited to regions where the perturbation corrections are rather small. These limitations are must be considered carefully when using the SCTST/VPT2 model. A higher order perturbation theory approach (VPT4) that is under development is expected to extend the range of SCTST calculations to higher and lower energies, where VPT2 may fail.

It is less straightforward to address the limitations of perturbation theory when applied to large amplitude motions. Except in the limit of small quantum numbers and high torsion barriers (i.e., when the amplitudes are small), VPT2 is not suitable for treating hindered internal rotors. Thus we have resorted to treating internal rotations as separable from the set of coupled vibrations. This pragmatic approach is clearly an approximation, but it has produced reasonably accurate results for the few cases in which it has been applied so far. This is an area where further work is needed.

"Deep tunneling", which takes place at energies far below the top of an energy barrier, is another phenomenon that calls for further development. Al Wagner at Argonne National Laboratory has made some notable efforts to address this issue,[92] but more work is needed in this area.

The methods reviewed in this chapter (especially when developed further) can be applied to many important and interesting atmospheric problems. These may include the reactions of Criegee intermediates, weakly bound complexes, poly-functional compounds, and exotic species like HO_3. In principle, SCTST can enable predictions at the very low temperatures and pressures found at the mesopause and in the upper troposphere/lower stratosphere — conditions that are very difficult to study in the laboratory.

Acknowledgments

We wish to acknowledge our colleague Ralph E. Weston, Jr., who was a co-author of our papers on the HO + CO reaction, and Albert F. Wagner for very useful conversations regarding the limitations of SCTST/VPT2. JFS and TLN are supported by the Robert A. Welch Foundation (Grant F-1283). This material is based on the work supported by the U.S. Department of Energy, Office of Basic Energy Sciences under Award Number

DE-FG02-07ER15884. JRB acknowledges financial support from the National Science Foundation (Atmospheric and Geospace Sciences, Contract No. 1231842) and NASA (Upper Atmospheric Research Program, Contract No. NNX12AE03G).

References

1. *U.S. Standard Atmosphere, 1976.* (U.S. Government Printing Office, Washington DC, 1976).
2. J. B. Burkholder, S. P. Sander, J. Abbatt, J. R. Barker, R. E. Huie, C. E. Kolb, M. J. Kurylo, V. L. Orkin, D. M. Wilmouth and P. H. Wine. Chemical Kinetics and Photochemical Data for Use in Atmospheric Studies, Evaluation No. 18, in JPL Publication 15–10. (Jet Propulsion Laboratory, Pasadena, CA, 2015).
3. R. Atkinson, D. L. Baulch, R. A. Cox, J. N. Crowley, R. F. Hampson, R. G. Hynes, M. E. Jenkin, M. J. Rossi and J. Troe. Evaluated kinetic and photochemical data for atmospheric chemistry: Volume I — gas phase reactions of O_x, HO_x, NO_x and SO_x species. *Atmos. Chem. Phys.* **4**, 1461–1738 (2004).
4. R. Atkinson, D. L. Baulch, R. A. Cox, J. N. Crowley, R. F. Hampson, R. G. Hynes, M. E. Jenkin, M. J. Rossi and J. Troe. Evaluated kinetic and photochemical data for atmospheric chemistry: Volume II — gas phase reactions of organic species. *Atmos. Chem. Phys.* **6**, 3625–4055 (2006).
5. R. Atkinson, D. L. Baulch, R. A. Cox, J. N. Crowley, R. F. Hampson, R. G. Hynes, M. E. Jenkin, M. J. Rossi and J. Troe. Evaluated kinetic and photochemical data for atmospheric chemistry: Volume III — gas phase reactions of inorganic halogens. *Atmos. Chem. Phys.* **7**, 981–1191 (2007).
6. U. Manthe. The multi-configurational time-dependent Hartree approach revisited. *J. Chem. Phys.* **142**, (2015).
7. M. T. Cvitas and S. C. Althorpe. Quantum wave packet method for state-to-state reactive scattering calculations on AB + CD → ABC + D reactions. *J. Phys. Chem. A* **113**, 4557–4569 (2009).
8. J. Mayneris, M. Gonzalez and S. K. Gray. Real wavepacket code for ABC plus D → AB plus CD reactive scattering. *Comput. Phys. Commun.* **179**, 741–747 (2008).
9. S. C. Althorpe and D. C. Clary. Quantum scattering calculations on chemical reactions. *Ann. Rev. Phys. Chem.* **54**, 493–529 (2003).
10. D. Skouteris, J. F. Castillo and D. E. Manolopoulos. ABC: A quantum reactive scattering program. *Comput. Phys. Commun.* **133**, 128–135 (2000).
11. A. Kuppermann and G. C. Schatz. Quantum-mechanical reactive scattering — accurate 3-dimensional calculation. *J. Chem. Phys.* **62**, 2502–2504 (1975).
12. D. G. Truhlar and A. Kupperman. Quantum mechanics of the $H+H_2$ reaction — exact scattering probabilities for collinear collisions. *J. Chem. Phys.* **52**, 3841–3843 (1970).

13. L. Vereecken, D. R. Glowacki and M. J. Pilling. Theoretical chemical kinetics in tropospheric chemistry: Methodologies and applications. *Chem. Rev.* **115**, 4063–4114 (2015).
14. T. Wu, H. J. Werner and U. Manthe. First-principles theory for the $H+CH_4 \rightarrow H_2+CH_3$ reaction. *Science* **306**, 2227–2229 (2004).
15. R. Welsch, F. Huarte-Larranaga and U. Manthe. State-to-state reaction probabilities within the quantum transition state framework. *J. Chem. Phys.* **136**, 034103 (2012).
16. J. M. Bowman. Reduced dimensionality theory of quantum reactive scattering. *J. Phys. Chem.* **95**, 4960–4968 (1991).
17. J. Palma, J. Echave and D. C. Clary. Rate constants for the $CH_4 + H \rightarrow CH_3 + H_2$ reaction calculated with a generalized reduced-dimensionality method. *J. Phys. Chem. A* **106**, 8256–8260 (2002).
18. B. Kerkeni and D. C. Clary. A simplified reduced-dimensionality study to treat reactions of the type $X + CZ_3Y$->$XY + CZ_3$, *J. Phys. Chem. A* **107**, 10851–10856 (2003).
19. H. F. von Horsten, S. T. Banks and D. C. Clary. An efficient route to thermal rate constants in reduced dimensional quantum scattering simulations: Applications to the abstraction of hydrogen from alkanes. *J. Chem. Phys.* **135**, 094311 (2011).
20. R. Welsch and U. Manthe. Full-dimensional and reduced-dimensional calculations of initial state-selected reaction probabilities studying the $H + CH_4 \rightarrow H_2 + CH_3$ reaction on a neural network PES. *J. Chem. Phys.* **142**, 064309 (2015).
21. H. Eyring, The activated complex in chemical reactions. *J. Chem. Phys.* **3**, 107–115 (1935).
22. M. G. Evans and M. Polanyi. Some applications of the transition state method to the calculation of reaction velocities, especially in solution. *Trans. Faraday Soc.* **31**, 875–893 (1935).
23. E. Wigne. The transition state method. *Trans. Faraday Soc.* **34**, 29–40 (1938).
24. R. A. Marcus and O. K. Rice. The kinetics of the recombination of methyl radicals and iodine atoms. *J. Phys. Colloid Chem.* **55**, 894–908 (1951).
25. R. A. Marcus. Unimolecular dissociations and free radical recombination reactions. *J. Chem. Phys.* **20**, 359–364 (1952).
26. H. M. Rosenstock, M. B. Wallenstein, A. L. Wahrhaftig and H. Eyring. Absolute rate theory for isolated systems and the mass spectra of polyatomic molecules. *Proc. Natl. Acad. Sci. U. S. A.* **38**, 667–678 (1952).
27. J. F. Stanton, J. Gauss, M. E. Harding, P. G. Szalay, A. A. Auer, R. J. Bartlett, U. Benedikt, C. Berger, D. E. Bernholdt, Y. J. Bomble, O. Christiansen, M. Heckert, O. Heun, C. Huber, T.-C. Jagan, D. Jonsson, J. Jusélius, K. Klein, W. J. Lauderdale, D. A. Matthews, T. Metzroth, D. P. O'Neill, D. R. Price, E. Prochnow, K. Ruud, F. Schiffmann, S. Stopkowicz, J. Vázquez, F. Wang and J. D. Watts; *and the integral packages MOLECULE (J. Almlöf and P.R. Taylor); PROPS (P.R. Taylor); ABACUS (T. Helgaker, H.J. Aa. Jensen, P. Jørgensen, and J. Olsen); and*

ECP routines by A. V. Mitin and C. van Wüllen. CFOUR, a quantum chemical program package (2009). http://www.cfour.de/.
28. Z. Rolik, L. Szegedy, I. Ladjánszki, B. Ladóczki and M. Kállay. An efficient linear-scaling CCSD(T) method based on local natural orbitals. *J. Chem. Phys.* **139**, 094105 (2013).
29. H.-J. Werner, P. J. Knowles, R. Lindh, F. R. Manby, M. Schütz, P. Celani, T. Korona, G. Rauhut, R. D. Amos, A. Bernhardsson, A. Berning, D. L. Cooper, M. J. O. Deegan, A. J. Dobbyn, F. Eckert, C. Hampel, G. Hetzer, A. W. Lloyd, S. J. McNicholas, W. W. Meyer, M. E. Mura, A. Nicklass, P. Palmieri, R. Pitzer, U. Schumann, H. Stoll, A. J. Stone, R. Tarroni and T. Thorsteinsson; *MOLPRO version 2006.1 (a package of ab initio programs)*; see http://www.molpro.net. (2006). http://www.molpro.net.
30. M. J. Frisch, G. W. Trucks, H. B. Schlegel, G. E. Scuseria, M. A. Robb, J. R. Cheeseman, G. Scalmani, V. Barone, B. Mennucci, G. A. Petersson, H. Nakatsuji, M. Caricato, X. Li, H. P. Hratchian, A. F. Izmaylov, J. Bloino, G. Zheng, J. L. Sonnenberg, M. Hada, M. Ehara, K. Toyota, R. Fukuda, J. Hasegawa, M. Ishida, T. Nakajima, Y. Honda, O. Kitao, H. Nakai, T. Vreven, J. J. A. Montgomery, J. E. Peralta, F. Ogliaro, M. Bearpark, J. J. Heyd, E. Brothers, K. N. Kudin, V. N. Staroverov, R. Kobayashi, J. Normand, K. Raghavachari, A. Rendell, J. C. Burant, S. S. Iyengar, J. Tomasi, M. Cossi, N. Rega, J. M. Millam, M. Klene, J. E. Knox, J. B. Cross, V. Bakken, C. Adamo, J. Jaramillo, R. Gomperts, R. E. Stratmann, O. Yazyev, A. J. Austin, R. Cammi, C. Pomelli, J. W. Ochterski, R. L. Martin, K. Morokuma, V. G. Zakrzewski, G. A. Voth, P. Salvador, J. J. Dannenberg, S. Dapprich, A. D. Daniels, O. Farkas, J. B. Foresman, J. V. Ortiz, J. Cioslowski and D. J. Fox. *Gaussian 09, Revision A.1.* (Gaussian, Inc., Wallingford CT (2009).
31. D. G. Truhlar and B. C. Garrett. Variational transition-state theory. *Ann. Rev. Phys. Chem.* **35**, 159–189 (1984).
32. D. G. Truhlar and B. C. Garrett. Variational transition-state theory. *Acc. Chem. Res.* **13**, 440–448 (1980).
33. W. L. Hase. Variational unimolecular rate theory. *Acc. Chem. Res.* **16**, 258–264 (1983).
34. D. G. Truhlar, W. L. Hase and J. T. Hynes. Current status of transition-state theory. *J. Phys. Chem.* **87**, 2664–2682 (1983).
35. D. G. Truhlar, B. C. Garrett and S. J. Klippenstein. Current status of transition-state theory. *J. Phys. Chem.* **100**, 12771–12800 (1996).
36. A. Fernandez-Ramos, J. A. Miller, S. J. Klippenstein and D. G. Truhlar. Modeling the kinetics of bimolecular reactions. *Chem. Rev.* **106**, 4518–4584 (2006).
37. S. J. Klippenstein, V. S. Pande and D. G. Truhlar. Chemical kinetics and mechanisms of complex systems: A perspective on recent theoretical advances. *J. Am. Chem. Soc.* **136**, 528–546 (2014).
38. W. H. Miller. Tunneling corrections to unimolecular rate constants, with application to formaldehyde. *J. Am. Chem. Soc.* **101**, 6810–6814 (1979).

39. J. Zheng, S. Zhang, B. J. Lynch, J. C. Corchado, Y. Y. Chuang, P. L. Fast, W. P. Hu, Y. P. Liu, G. C. Lynch, K. A. Nguyen, C. F. Jackels, A. F. Ramos, B. A. Ellingson, V. S. Melissas, J. Villa, I. Rossi, E. L. Coition, J. Pu, T. V. Albu, R. Steckler, B. C. Garrett, A. D. Isaacson and D. G. Truhlar. *POLYRATE-v2008* (University of Minnesota, Minneapolis, Minnesota, 2008).
40. W. H. Miller. Semiclassical theory for non-separable systems — construction of good action-angle variables for reaction-rate constants. *Faraday Discuss.* **62**, 40–46 (1977).
41. W. H. Miller, R. Hernandez, N. C. Handy, D. Jayatilaka and A. Willetts. Abinitio calculation of anharmonic constants for a transition-state, with application to semiclassical transition-state tunneling probabilities. *Chem. Phys. Lett.* **172**, 62–68 (1990).
42. M. J. Cohen, N. C. Handy, R. Hernandez and W. H. Miller. Cumulative reaction probabilities for $H + H_2 \rightleftharpoons H_2 + H$ from a knowledge of the anharmonic force field. *Chem. Phys. Lett.* **192**, 407–416 (1992).
43. R. Hernandez and W. H. Miller. Semiclassical transition-state theory — a new perspective. *Chem. Phys. Lett.* **214**, 129–136 (1993).
44. I. M. Mills. Vibration-rotation structure in asymmetric- and symmetric-top molecules. In *Molecular Spectroscopy: Modern Research*; Rao, K. N., Mathews, C. W., eds., Vol. 1, p. 115, (Academic Press, New York, 1972).
45. T. L. Nguyen, J. F. Stanton and J. R. Barker. A practical implementation of semi-classical transition state theory for polyatomics. *Chem. Phys. Lett.* **499**, 9–15 (2010).
46. T. L. Nguyen, J. F. Stanton and J. R. Barker. *Ab initio* reaction rate constants computed using semiclassical transition-state theory: $HO + H_2 \rightarrow H_2O + H$ and Isotopologues. *J. Phys. Chem. A* **115**, 5118–5126 (2011).
47. W. H. Miller, R. Hernandez, N. C. Handy, D. Jayatilaka and A. Willets. *Ab initio* calculation of anharmonic constants for a transition state, with application to semiclassical transition state tunneling probabilities. *Chem. Phys. Lett.* **172**, 62–68 (1990).
48. F. G. Wang and D. P. Landau. Efficient, multiple-range random walk algorithm to calculate the density of states. *Phys. Rev. Lett.* **86**, 2050–2053 (2001).
49. F. Wang and D. P. Landau. Efficient, multiple-range random walk algorithm to calculate the density of states. *Phys. Rev. Letters* **86**, 2050–2053 (2001).
50. M. Basire, P. Parneix and F. Calvo. Quantum anharmonic densities of states using the Wang–Landau method. *J. Chem. Phys.* **129**, 081101 (2008).
51. J. R. Barker, T. L. Nguyen and J. F. Stanton. Kinetic Isotope Effects for $Cl+CH_4 \rightleftharpoons HCl + CH_3$ Calculated Using *ab Initio* semiclassical transition state theory. *J. Phys. Chem. A* **116**, 6408–6419 (2012).
52. T. L. Nguyen, J. Li, R. Dawes, J. F. Stanton and H. Guo. Accurate determination of barrier height and kinetics for the $F + H_2O \rightarrow HF + OH$ Reaction. *J. Phys. Chem. A* **117**, 8864–8872 (2013).

53. T. L. Nguyen and J. F. Stanton. *Ab initio* thermal rate calculations of HO+ HO $\rightleftharpoons$ O(^{3}P) + H_2O reaction and isotopologues. *J. Phys. Chem. A* **117**, 2678–2686 (2013).
54. T. L. Nguyen, B. C. Xue, G. B. Ellison and J. F. Stanton. Theoretical study of reaction of ketene with water in the gas phase: Formation of acetic acid?, *J. Phys. Chem. A* **117**, 10997–11005 (2013).
55. R. E. Weston, Jr., T. L. Nguyen, J. F. Stanton and J. R. Barker. HO + CO reaction rates and H/D kinetic isotope effects: Master equation models with *ab initio* SCTST rate constants. *J. Phys. Chem. A* **117**, 821–835 (2013).
56. T. L. Nguyen and J. F. Stanton. Accurate *ab initio* thermal rate constants for reaction of O(^{3}P) with H_2 and isotopic analogues. *J. Phys. Chem. A* **118**, 4918–4928 (2014).
57. T. L. Nguyen, B. Xue, R. E. Weston, Jr., J. R. Barker and J. F. Stanton. Reaction of HO with CO: Tunneling is indeed important *J. Phys. Chem. Lett.* **3**, 1549–1553 (2012).
58. T. L. Nguyen, H. Lee, D. A. Matthews, M. C. McCarthy and J. F. Stanton. Stabilization of the simplest criegee intermediate from the reaction between ozone and ethylene: A high-level quantum chemical and kinetic analysis of ozonolysis. *J. Phys. Chem. A* **119**, 5524–5533 (2015).
59. T. L. Nguyen and J. F. Stanton. A steady-state approximation to the two-dimensional master equation for chemical kinetics calculations. *J. Phys. Chem. A* **119**, 7627–7636 (2015).
60. J. R. Barker, N. F. Ortiz, J. M. Preses, L. L. Lohr, A. Maranzana, P. J. Stimac, T. L. Nguyen and T. J. D. Kumar. *MultiWell-2010.1 Software*. (Ann Arbor, Michigan, USA, 2010). http://aoss.engin.umich.edu/multiwell/.
61. J. R. Barker, N. F. Ortiz, J. M. Preses, L. L. Lohr, A. Maranzana, P. J. Stimac, T. L. Nguyen and T. J. D. Kumar. *MultiWell-2012.2 Software*; J. R. Barker, (University of Michigan, Ann Arbor, Michigan, USA) (http://aoss.engin.umich.edu/multiwell/), 2012. http://aoss.engin.umich.edu/ multiwell/.
62. L. McCaslin and J. Stanton. Calculation of fundamental frequencies for small polyatomic molecules: A comparison between correlation consistent and atomic natural orbital basis sets. *Mol. Phys.* **111**, 1492–1496 (2013).
63. S. Arrhenius. Über die Reaktionsgeschwindigkeit bei der Inversion von Rohrzucker durch Säuren. *Z. Physik. Chem.* **4**, 226–248 (1889).
64. A. Tajti, P. G. Szalay, A. G. Csaszar, M. Kallay, J. Gauss, E. F. Valeev, B. A. Flowers, J. Vazquez and J. F. Stanton. HEAT: High accuracy extrapolated ab initio thermochemistry. *J. Chem. Phys.* **121**, 11599–11613 (2004).
65. Y. J. Bomble, J. Vazquez, M. Kallay, C. Michauk, P. G. Szalay, A. G. Csaszar, J. Gauss and J. F. Stanton. High-accuracy extrapolated *ab initio* thermochemistry. II. Minor improvements to the protocol and a vital simplification. *J. Chem. Phys.* **125**, 064108 (2006).
66. M. E. Harding, J. Vazquez, B. Ruscic, A. K. Wilson, J. Gauss and J. F. Stanton, High-accuracy extrapolated ab initio thermochemistry. III. Additional improvements and overview. *J. Chem. Phys.* **128**, 114111 (2008).

67. M. Born and R. Oppenheimer. Quantum theory of molecules. *Ann. Phys.-Berlin* **84**, 0457–0484 (1927).
68. M. Born and K. Huang. *Dynamical Theory of Crystal Lattices.* (Clarendon Press & Oxford University Press, Oxford and New York, 1988).
69. R. J. Bartlett. Coupled-cluster theory in quantum chemistry: The emergence of a new paradigm. *Abstr. Pap. Am. Chem. Soc.* **233**, (2007).
70. R. J. Bartlett and M. Musial. Coupled-cluster theory in quantum chemistry. *Rev. Mod. Phys.* **79**, 291–352 (2007).
71. S. F. Boys. Electronic wave functions. 1. A general method of calculation for the stationary states of any molecular system. *Proc. Roy. Soc. London Ser. A* **200**, 542–554 (1950).
72. S. F. Boys. Electronic wave functions. 2. A calculation for the ground state of the beryllium atom. *Proc. Roy. Soc. London Ser. A* **201**, 125–137 (1950).
73. K. Raghavachari, G. W. Trucks, J. A. Pople and M. Headgordon. A 5th-order perturbation comparison of electron correlation theories. *Chem. Phys. Lett.* **157**, 479–483 (1989).
74. R. J. Bartlett, J. D. Watts, S. A. Kucharski and J. Noga. Non-iterative 5th-order triple and quadruple excitation-energy corrections in correlated methods. *Chem. Phys. Lett.* **165**, 513–522 (1990).
75. T. H. Dunning. Gaussian-basis sets for use in correlated molecular calculations. 1. The atoms boron through neon and hydrogen. *J. Chem. Phys.* **90**, 1007–1023 (1989).
76. J. Almlof and P. R. Taylor. General contraction of Gaussian basis sets. I. Atomic natural orbitals for first- and second-row atoms. *J. Chem. Phys.* **86**, 4070–4077 (1987).
77. J. Almlof and P. R. Taylor. General contraction of Gaussian basis sets. II. Atomic natural orbitals and the calculation of atomic and molecular properties. *J. Chem. Phys.* **92**, 551–560 (1990).
78. L. Cheng, J. Gauss and J. F. Stanton. Treatment of scalar-relativistic effects on nuclear magnetic shieldings using a spin-free exact-two-component approach. *J. Chem. Phys.* **139**, 054105 (2013).
79. J. L. Tilson, W. C. Ermler and R. M. Pitzer. Parallel spin-orbit coupled configuration interaction. *CoPhC* **128**, 128–138 (2000).
80. T. Shiozaki and H. J. Werner. Explicitly correlated multireference configuration interaction with multiple reference functions: Avoided crossings and conical intersections. *J. Chem. Phys.* **134**, 184104 (2011).
81. D. A. Matthews and J. F. Stanton. Non-orthogonal spin-adaptation of coupled cluster methods: A new implementation of methods including quadruple excitations. *J. Chem. Phys.* **142**, 064108 (2015).
82. P. J. Robinson and K. A. Holbrook. *Unimolecular Reactions.* (Wiley-Interscience, London; New York, 1972).
83. W. Forst. *Theory of Unimolecular Reactions.* (Academic Press, New York, 1973).
84. T. Baer and W. L. Hase. *Unimolecular Reaction Dynamics. Theory and Experiments.* (Oxford University Press, New York, 1996).

85. K. A. Holbrook, M. J. Pilling and S. H. Robertson. *Unimolecular Reactions.* (Wiley, Chichester, 1996).
86. W. Forst. *Unimolecular Reactions: A Concise Introduction.* (Cambridge University Press, Cambridge, 2003).
87. T. Beyer and D. F. Swinehart. Number of multiply-restricted partitions, *Comm. Assoc. Comput. Machines* **16**, 379 (1973).
88. S. E. Stein and B. S. Rabinovitch. Accurate evaluation of internal energy level sums and densities including anharmonic oscillators and hindered rotors. *J. Chem. Phys.* **58**, 2438–2445 (1973).
89. T. L. Nguyen and J. R. Barker. Sums and densities of fully-coupled anharmonic vibrational states: A comparison of three practical methods. *J. Phys. Chem. A* **114**, 3718–3730 (2010).
90. C. Aieta, F. Gabas and M. Ceotto. An efficient computational approach for the calculation of the vibrational density of states. *J. Phys. Chem. A.* **submitted**, (2016).
91. J. R. Barker, T. L. Nguyen, J. F. Stanton, C. Aieta, M. Ceotto, F. Gabas, T. J. D. Kumar, C. G. L. Li, L. L. Lohr, A. Maranzana, N. F. Ortiz, J. M. Preses and P. J. Stimac; *MultiWell-2016 Software Suite*, (University of Michigan, Ann Arbor, Michigan, USA, 2016). http://clasp-research.engin.umich.edu/multiwell/.
92. A. F. Wagner. Improved multidimensional semiclassical tunneling theory. *J. Phys. Chem. A* **117**, 13089–13100 (2013).
93. W. Forst. Unimolecular rate theory test in thermal reactions. *J. Phys. Chem.* **76**, 342–348 (1972).
94. R. G. Gilbert, M. J. T. Jordan and S. C. Smith. *UNIMOL Program Suite*, (Sydney, Australia, 1993).
95. L. Vereecken, G. Huyberechts and J. Peeters. Stochastic simulation of chemically activated unimolecular reactions. *J. Chem. Phys.* **106**, 6564–6573 (1997).
96. S. J. Klippenstein, A. F. Wagner, S. H. Robertson, R. Dunbar and D. M. Wardlaw. VariFlex Software, 1.0 ed.; Argonne National Laboratory, 1999.
97. J. R. Barker. Multiple-well, multiple-reaction-path unimolecular reaction systems. I. MultiWell computer program suite. *Int. J. Chem. Kinet.* **33**, 232–245 (2001).
98. D. R. Glowacki, C. H. Liang, C. Morley, M. J. Pilling and S. H. Robertson. MESMER: An open-source master equation solver for multi-energy well reactions. *J. Phys. Chem. A* **116**, 9545–9560 (2012).
99. M. V. Duong, H. T. Nguyen, N. Truong, T. N. M. Le and L. K. Huynh. Multi-species multi-channel (MSMC): An ab initio-based parallel thermodynamic and kinetic code for complex chemical systems. *Int. J. Chem. Kinet.* **47**, 564–575 (2015).
100. J. R. Barker. Energy transfer in master equation Simulations: A new approach. *Int. J. Chem. Kinet.* **41**, 748–763 (2009).
101. G. Pieterse, M. C. Krol, A. M. Batenburg, C. A. M. Brenninkmeijer, M. E. Popa, S. O'Doherty, A. Grant, L. P. Steele, P. B. Krummel, R. L. Langenfelds, H. J. Wang, A. T. Vermeulen, M. Schmidt, C. Yver, A. Jordan,

A. Engel, R. E. Fisher, D. Lowry, E. G. Nisbet, S. Reimann, M. K. Vollmer, M. Steinbacher, S. Hammer, G. Forster, W. T. Sturges and T. Röckmann. Reassessing the variability in atmospheric H_2 using the two-way nested TM5 model. *J. Geophys. Res. (Atmos.)* **118**, 3764–3780 (2013).
102. S. Gerst and P. Quay. Deuterium component of the global mulecular hydrogen cycle. *J. Geophys. Res. (Atmos.)* **106**, 5021–5031 (2001).
103. J. Jankunas, M. Sneha, R. N. Zare, F. Bouakline, S. C. Althorpe, D. Herraez-Aguilar and F. J. Aoiz. Is the simplest chemical reaction really so simple? *Proc. Natl. Acad. Sci. U. S. A.* **111**, 15–20 (2014).
104. S. L. Mielke, K. A. Peterson, D. W. Schwenke, B. C. Garrett, D. G. Truhlar, J. V. Michael, M.-C. Su and J. W. Sutherland. H + H_2 thermal reaction: A convergence of theory and experiment. *Phys. Rev. Lett.* **91**, 063201 (2003).
105. S. L. Mielke, B. C. Garrett and K. A. Peterson. A hierarchical family of global analytic Born-Oppenheimer potential energy surfaces for the $H + H_2$ reaction ranging in quality from double-zeta to the complete basis set limit. *J. Chem. Phys.* **116**, 4142–4161 (2002).
106. D. G. Fleming, D. J. Arseneau, O. Sukhorukov, J. H. Brewer, S. L. Mielke, G. C. Schatz, B. C. Garrett, K. A. Peterson and D. G. Truhlar. Kinetic isotope effects for the reactions of muonic helium and muonium with H_2. *Science* **331**, 448–450 (2011).
107. M. H. Alexander, Chemical kinetics under test. *Science.* **331**, 411–412 (2011).
108. S. L. Mielke, D. W. Schwenke, G. C. Schatz, B. C. Garrett and K. A. Peterson. Functional representation for the Born–Oppenheimer diagonal correction and Born–Huang adiabatic potential energy surfaces for isotopomers of H_3. *J. Phys. Chem. A* **113**, 4479–4488 (2009).
109. K. A. Peterson, D. E. Woon and T. H. Dunning. Benchmark calculations with correlated molecular wave-functions. 4. The Classical Barrier Height of the $H+H_2 \rightleftharpoons H_2+H$ Reaction. *J. Chem. Phys.* **100**, 7410–7415 (1994).
110. K. E. Riley and J. B. Anderson. Higher accuracy quantum Monte Carlo calculations of the barrier for the $H+H_2$ reaction. *J. Chem. Phys.* **118**, 3437–3438 (2003).
111. D. G. Truhlar and R. E. Wyatt. History of H_3 Kinetics. *Ann. Rev. Phys. Chem.* **27**, 1–43 (1976).
112. B. A. Ridley, W. R. Schulz and D. J. Le Roy. Kinetics of reaction $D + H_2 \rightleftharpoons HD + H$. *J. Chem. Phys.* **44**, 3344 (1966).
113. A. A. Westenberg and N. De Haas. Atom-molecule kinetics using ESR detection 3. Results for $O + D_2 \rightleftharpoons OD + D$ and theoretical comparison with $O + H_2 \rightleftharpoons OH + H$. *J. Chem. Phys.* **47**, 4241- (1967).
114. D. N. Mitchell and D. J. Le Roy. Rate constants for reaction $D + H_2 \rightleftharpoons DH + H$ at low-temperatures using ESR detection. *J. Chem. Phys.* **58**, 3449–3453 (1973).
115. J. V. Michael. Rate constants for the reaction $O + D_2 \rightleftharpoons OD+D$ by the Flash Photolysis-Shock Tube Technique over the Temperature-Range 825-2487-K — the H_2 to D_2 Isotope Effect. *J. Chem. Phys.* **90**, 189–198 (1989).

116. I. S. Jayaweera and P. D. Pacey. Electron-spin-resonance study of the reaction of H with D2 at 274-K-364-K. *J. Phys. Chem.* **94**, 3614–3620 (1990).
117. J. V. Michael and J. R. Fisher. Rate constants for the reaction $D + H_2 \rightleftharpoons HD + H$, over the temperature-range 655–1979-K, by the flash photolysis-shock tube technique. *J. Phys. Chem.* **94**, 3318–3323 (1990).
118. J. V. Michael, M. C. Su and J. W. Sutherland. New rate constants for $D + H_2$ and $H + D_2$ between ∼1150 and 2100 K. *J. Phys. Chem. A* **108**, 432–437 (2004).
119. G. D. Purvis and R. J. Bartlett. A full coupled-cluster singles and doubles model — the inclusion of disconnected triples. *J. Chem. Phys.* **76**, 1910–1918 (1982).
120. G. E. Scuseria, C. L. Janssen and H. F. Schaefer. An efficient reformulation of the closed-shell coupled cluster single and double excitation (CCSD) equations. *J. Chem. Phys.* **89**, 7382–7387 (1988).
121. R. J. Bartlett and J. Noga. The expectation value coupled-cluster method and analytical energy derivatives. *Chem. Phys. Lett.* **150**, 29–36 (1988).
122. G. E. Scuseria and H. F. Schaefer. A new implementation of the full CCSDT model for molecular electronic-structure. *Chem. Phys. Lett.* **152**, 382–386 (1988).
123. J. D. Watts and R. J. Bartlett. The coupled-cluster single, double, and triple excitation model for open-shell single reference functions. *J. Chem. Phys.* **93**, 6104–6105 (1990).
124. J. Gauss, A. Tajti, M. Kallay, J. F. Stanton and P. G. Szalay. Analytic calculation of the diagonal Born–Oppenheimer correction within configuration-interaction and coupled-cluster theory. *J. Chem. Phys.* **125**, (2006).
125. L. Cheng, S. Stopkowicz and J. Gauss. Analytic energy derivatives in relativistic quantum chemistry. *Int. J. Quant. Chem.* **114**, 1108–1127 (2014).
126. NIST Chemistry WebBook, NIST Standard Reference Database Number 69, Eds. P.J. Linstrom and W.G. Mallard, National Institute of Standards and Technology, Gaithersburg MD, 20899. http://webbook.nist.gov (retrieved November 13, 2015).
127. A. A. Westenberg and N. De Haas. Atom-molecule kinetics using ESR detection 2. Results for $D + H_2 \rightleftharpoons HD + H$ and $H + D_2 \rightleftharpoons HD{+}D$. *J. Chem. Phys.* **47**, 1393 (1967).
128. W. C. Gardiner. *Gas-Phase Combustion Chemistry.* (Springer, New York, 2000).
129. N. G. Dautov and A. M. Starik. Kinetics of combustion of $H_2 + O_2$ mixture with participation of vibrationally excited molecules. *Combust. Explos. Shock Waves* **30**, 571–581 (1994).
130. N. G. Dautov and A. M. Starik. Investigation of the effect of vibrational-excitation of molecules on the kinetics of combustion of an $H_2 + O_2$ mixture. *High Temp.* **32**, 210–217 (1994).
131. D. L. Baulch, C. T. Bowman, C. J. Cobos, R. A. Cox, T. Just, J. A. Kerr, M. J. Pilling, D. Stocker, J. Troe, W. Tsang, R. W. Walker and J. Warnatz.

Evaluated kinetic data for combustion modeling: Supplement II. *J. Phys. Chem. Ref. Data* **34**, 757–1397 (2005).
132. J. A. Miller, M. J. Pilling and E. Troe. Unravelling combustion mechanisms through a quantitative understanding of elementary reactions, *Proc. Combust. Inst.* **30**, 43–88 (2005).
133. A. A. Konnov. Remaining uncertainties in the kinetic mechanism of hydrogen combustion. *Combust. Flame* **152**, 507–528 (2008).
134. T. Joseph, D. G. Truhlar and B. C. Garrett. Improved potential-energy surfaces for the reaction $O(^3P) + H2 \rightleftharpoons OH + H$. *J. Chem. Phys.* **88**, 6982–6990 (1988).
135. S. Rogers, D. S. Wang, A. Kuppermann and S. Walch. Chemically accurate *ab initio* potential energy surfaces for the lowest $^3A'$ and $^3A''$ electronically adiabatic states of $O(^3P) + H_2$. *J. Phys. Chem. A* **104**, 2308–2325 (2000).
136. J. Brandao, C. Mogo and B. C. Silva. Potential energy surface for $H_2O(^3A'')$ from accurate *ab initio* data with inclusion of long-range interactions. *J. Chem. Phys.* **121**, 8861–8868 (2004).
137. H. S. Zhai, P. Y. Zhang and P. W. Zhou. Quantum wave packet calculation of the $O(^3P)$ + H_2 reaction on the new potential energy surfaces for the two lowest states. *Comput. Theor. Chem.* **986**, 25–29 (2012).
138. A. A. Westenberg and N. De Haas, Atom-Molecule Kinetics at High Temperature Using ESR Detection. Technique and Results for $O + H_2$, $O + CH_4$, and $O + C_2H_6$, *J. Chem. Phys.* **46**, 490- (1967).
139. A. A. Westenberg and N. De Haas, Reinvestigation of Rate Coefficients for $O + H_2$ and $O + CH_4$, *J. Chem. Phys.* **50**, 2512 (1969).
140. R. N. Dubinsky and D. J. Mckenney. Determination of rate constant of $O + H_2 \rightleftharpoons OH + H$ reaction using atomic oxygen resonance fluorescence and air afterglow techniques. *Can. J. of Chem.* **53**, 3531–3541 (1975).
141. N. Presser and R. J. Gordon. The kinetic isotope effect in the reaction of $O(^3P)$ with H_2, D_2 and HD. *J. Chem. Phys.* **82**, 1291–1297 (1985).
142. J. W. Sutherland, J. V. Michael, A. N. Pirraglia, F. L. Nesbitt and R. B. Klemm, Rate constants for the reaction of $O(^3P)$ with H_2 by the flash photolysis-shock tube and flash photolysis-resonance fluorescence technique; $504K < T < 2495K$. *Proc. Combust. Inst.* **21**, 929–941 (1986).
143. P. Marshall and A. Fontijn. Htp kinetics studies of the reactions of $O(^3P_j)$ atoms with H_2 and D_2 over wide temperature ranges. *J. Chem. Phys.* **87**, 6988–6994 (1987).
144. K. Natarajan and P. Roth. High-temperature rate coefficient for the reaction of $O(^3P)$ with H_2 Obtained by the resonance-absorption of O-Atoms and H-Atoms. *Combust. Flame* **70**, 267–279 (1987).
145. K. S. Shin, N. Fujii and W. C. Gardiner. Rate-constant for $O + H_2 \rightleftharpoons OH + H$ by laser-absorption spectroscopy of OH in shock-heated H_2-O_2-Ar Mixtures. *Chem. Phys. Lett.* **161**, 219–222 (1989).
146. Y. F. Zhu, S. Arepalli and R. J. Gordon. The rate-constant for the reaction $O(^3p) + D_2$ at low-temperatures. *J. Chem. Phys.* **90**, 183–188 (1989).

147. D. F. Davidson and R. K. Hanson. A direct comparison of shock-tube photolysis and pyrolysis methods in the determination of the rate coefficient for O + H2-]Oh + H. *Combust. Flame* **82**, 445–447 (1990).
148. D. C. Robie, S. Arepalli, N. Presser, T. Kitsopoulos and R. J. Gordon. The intramolecular kinetic isotope effect for the reaction $O(^3P) + HD$. *J. Chem. Phys.* **92**, 7382–7393 (1990).
149. H. X. Yang, K. S. Shin and W. Gardiner. Rate coefficients for $O + H_2 \rightleftharpoons OH + H$ and $O + D_2 \rightleftharpoons OD + D$ by kinetic laser-absorption spectroscopy in shock-waves. *Chem. Phys. Lett.* **207**, 69–74 (1993).
150. S. O. Ryu, S. M. Hwang and M. J. Rabinowitz. Rate coefficient of the $O + H_2 \rightleftharpoons OH + H$ reaction determined via shock-tube laser-absorption spectroscopy. *Chem. Phys. Lett.* **242**, 279–284 (1995).
151. S. Javoy, V. Naudet, S. Abid and C. E. Paillard. Rate constant for the reaction of O with H_2 at high temperature by resonance absorption measurements of O atoms. *Int. J. Chem. Kinet.* **32**, 686–695 (2000).
152. B. Ruscic, J. E. Boggs, A. Burcat, A. G. Csaszar, J. Demaison, R. Janoschek, J. M. L. Martin, M. L. Morton, M. J. Rossi, J. F. Stanton, P. G. Szalay, P. R. Westmoreland, F. Zabel and T. Berces. IUPAC critical evaluation of thermochemical properties of selected radicals. Part I. *J. Phys. Chem. Ref. Data.* **34**, 573–656 (2005).
153. B. Ruscic, R. E. Pinzon, G. von Laszewski, D. Kodeboyina, A. Burcat, D. Leahy, D. Montoya and A. F. Wagner. Active thermochemical tables: Thermochemistry for the 21st century. *J. Phys. Conf. Ser.* **16**, 561–570 (2005).
154. B. Ruscic. Uncertainty quantification in thermochemistry, benchmarking electronic structure computations, and active thermochemical tables. *Int. J. Quantum. Chem.* **114**, 1097–1101 (2014).
155. B. Ruscic. Active thermochemical tables: Sequential bond dissociation enthalpies of methane, ethane, and methanol and the related thermochemistry. *J. Phys. Chem. A* **119**, 7810–7837 (2015).
156. N. Balakrishnan. Quantum mechanical investigation of the $O + H_2 \text{->} OH + H$ reaction. *J. Chem. Phys.* **119**, 195–199 (2003).
157. F. A. Haumann, A. M. Batenburg, G. Pieterse, C. Gerbig, M. C. Krol and T. Röckmann. Emission ratio and isotopic signatures of molecular hydrogen emissions from tropical biomass burning. *Atmos. Chem. Phys.* **13**, 9401–9413 (2013).
158. T. Rahn, J. M. Eller, K. A. Boering, P. O. Wennberg, M. C. McCarthy, S. Tyler, S. Schauffer, S. Donnelly and E. Atlas. Extreme deuterium enrichment in stratospheric hydrogen and the global atmospheric budget of H2. *Nature* **424**, 918–921 (2003).
159. G. Pieterse, M. C. Krol and T. Röckmann. A consistent molecular hydrogen isotope chemistry scheme based on an independent bond approximation. *Atmos. Chem. Phys.* **9**, 8503–8529 (2009).
160. G. Pieterse, M. C. Krol, A. M. Batenburg, L. P. Steele, P. B. Krummel, R. L. Langenfelds and T. Röckmann. Global modelling of H_2 mixing ratios

and isotopic compositions with the TM5 model. *Atmos. Chem. Phys.* **11**, 7001–7026 (2011).

161. M. Krol, S. Houweling, B. Bregman, M. v. d. Broek, A. Segers, P. v. Velthoven, W. Peters, F. Dentener and P. Bergamaschi. The two-way nested global chemistry-transport zoom model TM5: Algorithm and applications. *Atmos. Chem. Phys.* **5**, 417–432 (2005).
162. M. Balat. Possible methods for hydrogen production. *Energy Sources: Part A-Recovery, Utilization, and Environmental Effects* **31**, 39–50 (2009).
163. P. C. Novelli, P. M. Lang, K. A. Masarie, D. F. Hurst, R. Myers and J. W. Elkins. Molecular hydrogen in the troposphere: Global distribution and budget. *J. Geophys. Res. (Atmos.)* **104**, 30427–30444 (1999).
164. M. E. Popa, A. J. Segers, H. A. C. Denier van der Gon, M. C. Krol, A. J. H. Visschedijk, M. Schaap and T. Röckmann. Impact of a future H_2 transportation on atmospheric pollution in Europe. *Atmos. Eviron.* **113**, 208–222 (2015).
165. B. R. Strazisar, C. Lin and H. F. Davis. Mode-specific energy disposal in the four-atom reaction $OH + D_2 \rightarrow HOD + D$. *Science* **290**, 958–961 (2000).
166. I. W. M. Smith and R. Zellner. Rate measurements of reactions of OH by resonance-absorption. 2. Reactions of OH with CO, C_2H_4 and C_2H_2. *J. Chem. Soc. Faraday Trans. II* **69**, 1617–1627 (1973).
167. R. Ravishankara, J. M. Nicovich, R. L. Thompson and F. P. Tully. Kinetic study of the reaction of OH with H_2 and D_2 from 250 to 1050 K. *J. Phys. Chem.* **85**, 2498–2503 (1981).
168. R. K. Talukdar, T. Gierczak, L. Goldfarb, Y. Rudich, B. S. M. Rao and R. Ravishankara. Kinetics of hydroxyl radical reactions with isotopically labeled hydrogen. *J. Phys. Chem.* **100**, 3037–3043 (1996).
169. L. N. Krasnoperov and J. V. Michael. Shock tube studies using a novel multipass absorption cell: Rate constant results for $OH + H_2$ and $OH + C_2H_6$. *J. Phys. Chem. A* **108**, 5643–5648 (2004).
170. V. L. Orkin, S. N. Kozlov, G. A. Poskrebyshev and M. J. Kurylo. Rate constant for the reaction of OH with H_2 between 200 and 480 K. *J. Phys. Chem. A.* **110**, 6978–6985 (2006).
171. I. W. M. Smith and F. F. Crim. The chemical kinetics and dynamics of the prototypical reaction: $OH + H_2 \rightarrow H_2O + H$. *Phys. Chem. Chem. Phys.* **4**, 3543–3551 (2002).
172. J. F. Castillo. The dynamics of the H + H2O reaction. *Chem. Phys. Chem.* **3**, 320–332 (2002).
173. D. C. Clary. Quantum dynamics of chemical reactions. *Science* **321**, 789–791 (2008).
174. T. N. Truong and T. J. Evans. Direct *ab initio* dynamics calculations of thermal rate constants and kinetic isotopic effects for the $H + H_2O \rightleftharpoons OH + H_2$ reaction. *J. Phys. Chem.* **98**, 9558–9564 (1994).
175. T. N. Truong. Direct *ab initio* dynamics studies of vibrational-state selected reaction rate of the $OH + H_2 \rightleftharpoons H + H_2O$ reaction. *J. Chem. Phys.* **102**, 5335–5341 (1995).

176. F. Matzkies and U. Manthe. Accurate quantum calculations of thermal rate constants employing MCTDH: $H_2 + OH \rightleftharpoons H + H_2O$ and $D_2 + OH \rightleftharpoons D + DOH$. *J. Chem. Phys.* **108**, 4828–4836 (1998).
177. E. Garcia, A. Saracibar, A. Rodriquez, A. Lagara and G. Lendvay. Calculated versus measured product distributions of the OH + D2 reaction. *Mol. Phys.* **104**, 839–846 (2006).
178. X. Tian, T. Gao, N. He and Z. Zhang. *Ab initio* molecular dynamics studies of the $OH + D_2 \rightleftharpoons HOD + D$ reaction: Direct classical trajectory calculations by MP2. *Chem. Phys.* **354**, 142–147 (2008).
179. J. D. Sierra, R. Martinez, J. Hernando and M. Gonzalez. The $OH + D_2 \rightleftharpoons HOD + D$ angle-velocity distribution: Quisi-classical trajectory calculations on the YZCL2 and WSLFH potential energy surfaces and comparision with experiments at $E_T = 0.28\,\text{eV}$. *Phys. Chem. Chem. Phys.* **11**, 11520–11527 (2009).
180. J. Espinosa-Garcia, L. Bonnet and J. C. Corchado. Classical description in a quantum spirit of the prototype four-atom reaction $OH + D_2$, *Phys. Chem. Chem. Phys.* **12**, 3873–3877 (2010).
181. G. S. Wu, G. C. Schatz, G. Lendway, D. C. Fang and L. B. Harding. A new potential surface and quasiclassical trajectory study of $H + H_2O \rightarrow OH + H_2$. *J. Chem. Phys.* **113**, 3150–3161 (2000).
182. M. Yang, D. H. Zhang, M. A. Collins and S. Y. Lee. Quantum dynamics on new potential energy surfaces for the $H_2 + OH \rightarrow H_2O + H$ reaction. *J. Chem. Phys.* **114**, 4759–4762 (2001).
183. D. H. Zhang, M. Yang and S.-Y. Lee. Branching ratio in the HD + OH reaction: A full-dimensional quantum dynamics study on a new *ab initio* potential energy surface. *J. Chem. Phys.* **114**, 8733–8736 (2001).
184. D. H. Zhang, M. Yang and S.-Y. Lee. Quantum dynamics of the $D_2 + OH$ reaction. *J. Chem. Phys.* **116**, 2388–2394 (2002).
185. E. Garcia, A. Saracibar, C. Sanchez and A. Lagana. A multiproperty analysis of the $OH + H_2$ (D_2,HD) potential energy surface. *Chem. Phys.* **308**, 201–210 (2005).
186. D. Troya, M. J. Lakin, G. C. Schatz and M. Gonzalez. Variational transition state theory and quasiclassical trajectory studies of the $H_2 + OH \rightarrow H + H_2O$ reaction and some isotopic variants. *J. Chem. Phys.* **115**, 1828–1842 (2001).
187. P. Defazio and S. K. Gray. A quantum dynamics study of $D_2 + OH \rightarrow DOH + D$ on the WSLFH potential energy function. *J. Phys. Chem. A* **107**, 7132–7137 (2003).
188. A. Chakraborty and D. G. Truhlar. Quantum mechanical reaction rate constants by vibrational configuration interaction: The $OH + H_2 \rightarrow H_2O + H$ reaction as a function of temperature. *Proc. Nat. Acad. Sci.* **102**, 6744–6749 (2005).
189. F. Kaufman and F. P. del Greco. Lifetime and reactions of OH radicals in discharge-flow systems. *Discuss. Faraday Soc.* **3**, 128–138 (1962).

190. G. Dixon-Lewis, W. E. Wilson and A. A. Westenberg. Studies of hydroxyl radical kinetics by quantitative ESR. *J. Chem. Phys.* **44**, 2877–2884 (1966).
191. F. Stuhl and H. Niki. Pulsed vacuum-uv photochemical study of reactions of OH with H_2, D_2, and CO using a resonance-fluorescent detection method. *J. Chem. Phys.* **57**, 3671–3677 (1972).
192. R. P. Overend, G. Paraskevopoulos and R. J. Cvetanović, Rates of OH radical reactions. I. Reactions with H_2, CH_4, C_2H_6, and C_3H_8 at 295 K. *Can. J. Chem.* **53**, 3374–3382 (1975).
193. I. W. M. Smith and R. Zellner. Rate measurements of reactions of OH by resonance absorption. Part 3 — reactions of OH with H_2, D_2 and hydrogen and deuterium halides. *J. Chem. Soc. Faraday Trans. 2* **70**, 1045–1056 (1974).
194. P. Frank and T. Just. High temperature reaction rate for $H + O_2 \rightleftharpoons OH + O$ and $OH + H_2 \rightleftharpoons H_2O + H$. *Ber. Bunsenger. Phys. Chem.* **89**, 181–187 (1985).
195. J. V. Michael and J. W. Sutherland. Rate constant for the reaction of H with H_2O and OH with H_2 by the flash photolysis-shock tube technique over the temperature range 1246–2297 K. *J. Phys. Chem. A. 2006*, 110 (16), **92**, 3853–3857 (1988).
196. R. C. Oldenborg, G. W. Loge, D. M. Harradine and K. R. Winn. Kinetic Study of the $OH + H_2$ Reaction from 800 to 1550 K. *J. Phys. Chem.* **96**, 8426–8430 (1992).
197. P. Ciais, C. Sabine, G. Bala, L. Bopp, V. Brovkin, J. Canadell, A. Chhabra, R. DeFries, J. Galloway, M. Heimann, C. Jones, C. L. Quéré, R. B. Myneni, S. Piao and P. Thornton. Carbon and Other Biogeochemical Cycles, in *Climate Change 2013: The Physical Science Basis. Contribution of Working Group I to the Fifth Assessment Report of the Intergovernmental Panel on Climate Change.* (Cambridge University Press, Cambridge, United Kingdom and New York, NY, USA, 2013).
198. S. Kirschke, P. Bousquet, P. Ciais, M. Saunois, J. G. Canadell, E. J. Dlugokencky, P. Bergamaschi, D. Bergmann, D. R. Blake, L. Bruhwiler, P. Cameron-Smith, S. Castaldi, F. Chevallier, L. Feng, A. Fraser, M. Heimann, E. L. Hodson, S. Houweling, B. Josse, P. J. Fraser, P. B. Krummel, J.-F. Lamarque, R. L. Langenfelds, C. L. Quéré, V. Naik, S. O'Doherty, P. I. Palmer, I. Pison, D. Plummer, B. Poulter, R. G. Prinn, M. Rigby, B. Ringeval, M. Santini, M. Schmidt, D. T. Shindell, I. J. Simpson, R. Spahni, L. P. Steele, S. A. Strode, K. Sudo, S. Szopa, G. R. v. d. Werf, A. Voulgarakis, M. v. Weele, R. F. Weiss, J. E. Williams and G. Zeng. Three decades of global methane sources and sinks. *Nature Geosci.* **6**, 813–823 (2013).
199. A. Ghosh, P. K. Patra, K. Ishijima, T. Umezawa, A. Ito, D. M. Etheridge, S. Sugawara, K. Kawamura, J. B. Miller, E. J. Dlugokencky, P. B. Krummel, P. J. Fraser, L. P. Steele, R. L. Langenfelds, C. M. Trudinger, J. W. C. White, B. Vaughn, T. Saeki, S. Aoki and T. Nakazawa. Variations in global methane sources and sinks during 1910–2010. *Atmos. Chem. Phys.* **15**, 2595–2612 (2015).

200. IPCC. *Climate Change 2007: The Physical Science Basis, Contribution of Working Group I to the Fourth Assessment Report of the Intergovernmental Panel on Climate Change.* (Cambridge University Press, Cambridge, U.K., and New York, NY, U.S.A., 2007).
201. C. M. Stevens and F. E. Rust. The carbon isotopic composition of atmospheric methane. *J. Geophys. Res.* **87**, 4879–4882 (1982).
202. M. Gupta, S. Tyler and R. Cicerone. Modeling atmospheric $\delta^{13}C\text{-}CH_4$ and the causes of recent changes in atmospheric CH_4 amounts. *J. Geophys. Res.* **101**, D22923–D22932 (1996).
203. A. L. Rice, S. C. Tyler, M. C. McCarthy, K. A. Boering and E. Atlas. Carbon and hydrogen isotopic compositions of stratospheric methane: 1. High-precision observations from the NASA ER-2 aircraft. *J. Geophys. Res.* **108**, 4460 (2003).
204. S. C. Tyler, A. L. Rice and H. O. Ajie. Stable isotope ratios in atmospheric CH_4: Implications for seasonal sources and sinks. *J. Geophys. Res.* **112**, D03303 (16 pages) (2007).
205. T. Röckmann, M. Brass, R. Borchers and A. Engel. The isotopic composition of methane in the stratosphere: High-altitude balloon sample measurements. *Atmos. Chem. Phys. Discuss.* **11**, 12039–12102 (2011).
206. R. E. Weston, Jr. Anomalous or mass-independent isotope effects. *Chem. Rev.* **99**, 2115–2136 (1999).
207. J. R. Barker, N. F. Ortiz, J. M. Preses and L. L. Lohr. *MultiWell-1.4.1 Software.* (Ann Arbor, Michigan, USA, 2004). http://aoss. engin.umich.edu/ multiwell/.
208. A. J. Eskola, R. S. Timonen, P. Marshall, E. N. Chesnokov and L. N. Krasnoperov, Rate constants and hydrogen isotope substitution effects in the CH_3+ HCl and $CH_3 + Cl_2$ reactions. *J. Phys. Chem. A* **112**, 7391–7401 (2008).
209. A. J. Eskola, J. A. Seetula and R. S. Timonen. Kinetics of the CH_3 + HCl/DCl $\rightarrow$ CH_4/CH_3D + Cl and CD_3+ HCl/DCl $\rightarrow$ $CD_3H/CD_4 + Cl$ reactions: An experimental H atom tunneling investigation. *Chem. Phys.* **331**, 26–34 (2006).
210. NIST Electronic Spin Splitting Corrections, Computational Chemistry Comparison and Benchmark DataBase, NIST Standard Reference Database 101, Release 15b, August 2011 ed.; National Institute of Standards and Technology, 2011, http://cccbdb.nist.gov/.
211. T. J. Lee, J. M. L. Martin and P. R. Taylor. An accurate ab initio quartic force field and vibrational frequencies for CH_4 and isotopomers. *J. Chem. Phys.* **102**, 254–261 (1995).
212. J. C. Corchado, D. G. Truhlar and J. Espinosa-Garcia, Potential energy surface, thermal, and state-selected rate coefficients, and kinetic isotope effects for $Cl + CH_4 \rightarrow HCl + CH_3$, *J. Chem. Phys.* **112**, 9375–9389 (2000).
213. C. Rangel, M. Navarrete, J. C. Corchado and J. Espinosa-García, Potential energy surface, kinetics, and dynamics study of the $Cl + CH_4 \rightarrow HCl + CH_3$ reaction. *J. Chem. Phys.* **124**, 124306 (1–19) (2006).

214. M.-Y. Yang, C.-L. Yang, J.-Z. Chen and Q.-G. Zhang, Modified potential energy surface and time-dependent wave packet dynamics study for Cl + $CH_4 \rightarrow HCl + CH_3$ reaction. *Chem. Phys.* **354**, 180–185 (2008).
215. R. Manning and M. J. Kurylo. Flash photolysis resonance fluorescence investigation of the temperature dependencies of the reactions of $Cl(^2P)$ atoms with methane, chloromethane, fluoromethane, excited fluoromethane, and ethane. *J. Phys. Chem.* **81**, 291–296 (1977).
216. D. A. Whytock, J. H. Lee, J. V. Michael, W. A. Payne and L. J. Stief. Absolute rate of the reaction of $Cl(^2P)$ with methane from 200–500 K, *J. Chem. Phys.* **66**, 2690–2695 (1977).
217. L. F. Keyser, Absolute rate and temperature dependence of the reaction between chlorine (^2P) atoms and methane. *J. Chem. Phys.* **69**, 214–218 (1978).
218. M. S. Zahniser, B. M. Berquist and F. Kaufman. Kinetics of the Reaction $Cl + CH_4 \rightarrow CH_3 + HCl$ from 200° to 500°K. *Int. J. Chem. Kinetics* **10**, 15–29 (1978).
219. A. R. Ravishankara and P. H. Wine. A laser flash photolysis-resonance fluorescence kinetics study of the reaction $Cl(^2P) + CH_4 \rightarrow CH_3 + HCI$. *J. Chem. Phys.* **72**, 25–30 (1980).
220. S. P. Heneghan, P. A. Knoot and S. W. Benson. The temperature coefficient of the rates in the system $Cl + CH_4 \rightarrow CH_3 + HCl$, thermochemistry of the methyl radical. *Int. J. Chem. Kinetics* **13**, 677–691 (1981).
221. J. V. Seeley, J. T. Jayne and M. J. Molina. Kinetic studies of chlorine atom reactions using the turbulent flow tube technique. *J. Phys. Chem. A.* **100**, 4019–4025 (1996).
222. K. Takahashi, O. Yamamoto and T. Inomata. Direct measurements of the rate coefficients for the reactions of some hydrocarbons with chlorine atoms at high temperatures. *Proc. Combust. Inst.* **29**, 2447–2453 (2002).
223. M. G. Bryukov, I. R. Slagle and V. D. Knyazev. Kinetics of reactions of Cl atoms with methane and chlorinated methanes. *J. Phys. Chem. A.* **106**, 10532–10542 (2002).
224. G. Saueressig, P. Bergamaschi, J. N. Crowley, H. Fischer and G. W. Harris, D/H kinetic isotope effect in the reaction $CH_4 + Cl$, *Geophys. Res. Lett.* **23**, 3619– 3622 (1996).
225. F. Sauer, R. W. Portmann, A. R. Ravishankara and J. B. Burkholder. Temperature dependence of the Cl atom reaction with deuterated methanes. *J. Phys. Chem. A* **119**, 4396–4407 (2015).
226. T. J. Wallington and M. D. Hurley. A kinetic study of the reaction of chlorine atoms with CF_3CHCl_2, CF_3CH_2F, $CFCl_2CH_3$, CF_2ClCH_3, CHF_2CH_3, CH3D, CH_2D_2, CHD_3, CD_4, and CD_3Cl at 295 ± 2 K. *Chem. Phys. Letters* **189**, 437–442 (1992).
227. G. D. Boone, F. Agyin, D. J. Robichaud, F.-M. Tao and S. A. Hewitt. Rate constants for the reactions of chlorine atoms with deuterated methanes: Experiment and theory. *J. Phys. Chem. A.* **105**, 1456–1464 (2001).

228. K. L. Feilberg, D. W. T. Griffith, M. S. Johnson and C. J. Nielsen. The ^{13}C and D Kinetic isotope effects in the reaction of CH_4 with Cl. *Int. J. Chem. Kinetics* **37**, 110–118 (2005).
229. G. Saueressig, P. Bergamaschi, J. N. Crowley, H. Fischer and G. W. Harris. Carbon kinetic isotope effect in the reaction of CH_4 with Cl atoms. *Geophys. Res. Lett.* **22**, 1225–1228 (1995).
230. J. N. Crowley, G. Saueressig, P. Bergamaschi, H. Fischer and G. W. Harris. Carbon kinetic isotope effect in the reaction $CH_4 + Cl$: a relative rate study using FTIR spectroscopy. *Chem. Phys. Letters* **303**, 268–274 (1999).
231. S. C. Tyler, H. O. Ajie, A. L. Rice and R. J. Cicerone. Experimentally determined kinetic isotope effects in the reaction of CH_4 with CI: Implications for atmospheric CH_4, *Geophys. Res. Letters* **27**, 1715–1718 (2000).
232. W. C. Gardiner, Jr. ed. Combustion Chemistry. (Springer-Verlag, New York, 2012), pp. Pages.
233. B. J. Finlayson-Pitts and J. N. Pitts. Jr., *Chemistry of the Upper and Lower Atmosphere.* (Academic Press, San Diego, 2000).
234. Y. L. Yung and W. B. DeMore. *Photochemistry of Planetary Atmospheres.* (Oxford University Press, Oxford, 1999).
235. N. R. Greiner. Hydroxyl radical kinetics by kinetic spectroscopy. V. Reactions with H2 and CO in the range 300–500°K. *J. Chem. Phys.* **51**, 5049–5051 (1969).
236. A. M. Dean and G. B. Kistiakowsky. Oxidation of carbon monoxide/methane mixtures in shock waves. *J. Chem. Phys* **54**, 1718–1725 (1971).
237. F. Dryer, D. Naegeli and I. Glassman. Temperature dependence of reaction $CO + OH \rightleftharpoons CO_2 + H$. *Combust. Flame* **17**, 270–272 (1971).
238. T. P. J. Izod, G. B. Kistiakowsky and S. Matsuda. Oxidation of carbon monoxide mixtures with added ethane or azomethane studied in incident shock waves. *J. Chem. Phys.* **55**, 4425–4432 (1971).
239. A. A. Westenberg and N. De Haas. Rates of CO + OH and H_2+ OH over an extended temperature range. *J. Chem. Phys.* **58**, 4061–4065 (1973).
240. C. J. Howard and K. M. Evenson. Laser magnetic resonance study of the gas phase reactions of OH with CO, NO, and NO2. *J. Chem. Phys.* **61**, 1943–1952 (1974).
241. D. D. Davis, S. Fischer and R. Schiff. Flash photolysis-resonance fluorescence kinetics study: Temperature-dependence of reactions $OH + CO \rightarrow CO_2 + H$ and $OH + CH_4 \rightarrow H_2O + CH_3$. *J. Chem. Phys.* **61**, 2213–2219 (1974).
242. D. W. Trainor and C. W. von Rosenberg, Jr. Energy partitioning in reaction $OH + CO \rightarrow CO_2 + H$. *Chem. Phys. Lett.* **29**, 35–38 (1974).
243. S. Gordon and W. A. Mulac. Reaction of OH(X $^2\Pi$) Radical produced by pulse-radiolysis of water-vapor. *Int. J. Chem. Kinet.* **7**, 289–299 (1975).
244. R. A. Perry, R. Atkinson and J. N. Pitts. Jr. Kinetics of reactions of OH radicals with C_2H_2 and CO. *J. Chem. Phys.* **67**, 5577–5584 (1977).
245. R. Overend and G. Paraskevopoulos. Question of a pressure effect in reaction OH + CO at room-temperature. *Chem. Phys. Lett.* **49**, 109–111 (1977).

246. H. W. Biermann, C. Zetzsch and F. Stuhl. *Ber. Bunsenges. Phys. Chem.* **82**, 633 (1978).
247. M. A. A. Clyne and P. M. Holt. Reaction kinetics involving ground $X^2\Pi$ and excited $A^2\Sigma^+$ hydroxyl radicals. Part 1. — Quenching kinetics of OH $A^2\Sigma^+$ and rate constants for reactions of OH $X^2\Pi$ with CH_3CCl_3 and CO. *J. Chem. Soc. Faraday Trans. 2* **75**, 569–581 (1979).
248. G. Paraskevopoulos and R. S. Irwin. The pressure-dependence of the rate-constant of the reaction of OD radicals with CO. *Chem. Phys. Lett.* **93**, 138–143 (1982).
249. A. R. Ravishankara and R. L. Thompson. Kinetic-study of the reaction of OH with CO from 250-K to 1040-K. *Chem. Phys. Lett.* **99**, 377–381 (1983).
250. A. Hofzumahaus and F. Stuhl. Rate-Constant of the Reaction HO + CO in the Presence of N_2 and O_2. *Ber. Bunsenges. Phys. Chem.* **88**, 557–561 (1984).
251. H. Niki, P. D. Maker, C. M. Savage and L. P. Breitenbach. Fourier transform infrared spectroscopic study of the kinetics for the hydroxyl radical reaction of $^{13}C^{16}O$-carbon monoxide and $^{12}C^{18}O$-carbon monoxide. *J. Phys. Chem.* **88**, 2116–2119 (1984).
252. G. Paraskevopoulos and R. S. Irwin. Rates of OH radical reactions. The pressure-dependence of the rate-constant of the reaction of OH radicals with CO. *J. Chem. Phys.* **80**, 259–266 (1984).
253. C. D. Jonah, W. A. Mulac and P. Zeglinski. Rate constants for the reaction of OH + CO; OD + CO; and OH + methane as a function of temperature. *J. Phys. Chem.* **88**, 4100–4104 (1984).
254. W. B. DeMore, Rate-constant for the Oh + Co reaction: pressure-dependence and the effect of oxygen. *Int. J. Chem. Kinet.* **16**, 1187–1200 (1984).
255. M. F. Beno, C. D. Jonah and W. A. Mulac. Rate constants for the reaction OH + CO as functions of temperature and water concentration, *Int. J. Chem. Kinet.* **17**, 1091–1101 (1985).
256. I. W. M. Smith and M. D. Williams. Kinetics of OH(v = 0 and 1) and OD(v = 0 and 1) studied by time-resolved laser-induced fluorescence. *Ber. Bunsenges Phys. Chem.* **89**, 319–320 (1985).
257. A. J. Hynes, P. H. Wine and A. R. Ravishankara. Kinetics of the OH + CO reaction under atmospheric conditions. *J. Geophys. Res. (Atmos.)* **91**, 1815–1820 (1986).
258. M. J. Frost, J. S. Salh and I. W. M. Smith. Vibrational-state distribution of CO_2 produced in the reaction between OH radicals and CO. *J. Chem. Soc. Faraday Trans.* **87**, 1037–1038 (1991).
259. M. J. Frost, P. Sharkey and I. W. M. Smith. Reaction between OH (OD) radicals and CO at temperatures down to 80 K: Experiment and theory. *J. Phys. Chem.* **97**, 12254–12259 (1993).
260. I. W. M. Smith. Rate of the Reaction: $OH + CO \rightarrow CO_2 + H$ at interstellar temperatures. *Mon. Not. R. Astron. Soc.* **234**, 1059–1063 (1988).
261. R. Forster, M. Frost, D. Fulle, H. F. Hamann, H. Hippler, A. Schlepegrell and J. Troe. High-pressure range of the addition of HO to HO; NO; NO_2;

and CO. 1. Saturated laser-induced fluorescence measurements at 298 K. *J. Chem. Phys.* **103**, 2949–2958 (1995).

262. V. Lissianski, H. Yang, Z. Qin, M. R. Mueller, K. S. Shin and W. C. Gardiner, Jr. High-temperature measurements of the rate coefficient of the $H + CO_2 \rightleftharpoons CO + OH$ reaction. *Chem. Phys. Lett.* **240**, 57–62 (1995).
263. D. Fulle, H. F. Hamann, H. Hippler and J. Troe. High pressure range of addition reactions of HO. 2. Temperature and pressure dependence of the reaction $HO + CO \rightleftharpoons HOCO \rightarrow H + CO_2$, *J. Chem. Phys.* **105**, 983–1000 (1996).
264. D. M. Golden, G. P. Smith, A. B. McEwen, C. L. Yu, B. Eiteneer, M. Frenklach, G. L. Vaghjiani, A. R. Ravishankara and F. P. Tully. OH(OD) + CO: Measurements and an optimized RRKM fit. *J. Phys. Chem. A* **102**, 8598–8606 (1998).
265. H. C. Hottel, G. C. Williams, N. M. Nerheim and G. R. Schneider. Kinetic studies in stirred reactors: Combustion of carbon monoxide and propane. *Proc. Combust. Inst.* **10**, 111–121 (1964).
266. A. A. Westenberg and R. M. Fristrom. H and O atom profiles measured by ESR in C_2 hydrocarbon-O_2 flames. *Proc. Combust. Inst.* **10**, 473–487 (1964).
267. G. Dixon-Lewis, M. M. Sutton and A. Williams. Some reactions of hydrogen atoms and simple radicals at high temperatures. *Proc. Combust. Inst.* **10**, 495–502 (1964).
268. G. A. Heath and G. S. Pearson. Perchloric acid flames: Part III. Chemical structure of methane flames. *Proc. Combust. Inst.* **11**, 967–977 (1966).
269. W. E. Wilson, Jr., J. T. O'Donovan and R. M. Fristrom. Flame inhibition by halogen compounds. *Proc. Combust. Inst.* **12**, 929–942 (1968).
270. R. R. Baldwin, R. W. Walker and S. J. Webster. The carbon monoxide-sensitized decomposition of hydrogen peroxide. *Combust. Flame* **15**, 167–172 (1970).
271. G. M. Atri, R. R. Baldwin, D. Jackson and R. W. Walker. The reaction of OH radicals and HO_2 radicals with carbon monoxide. *Combust. Flame* **30**, 1–12 (1977).
272. J. Peeters and G. Mahnen. Reaction mechanisms and rate constants ofelementary steps in methane-oxygen flames. *Proc. Combust. Inst.* **14**, 133–146 (1972).
273. K. H. Eberius, K. Hoyermann and H. G. Wagner. Structure of lean acetylene-oxygen flames. *Proc. Combust. Inst.* **14**, 147–156 (1972).
274. J. Vandooren, J. Peeters and P. J. Van Tiggelen, Rate constant of the elementary reaction of carbon monoxide with hydroxyl radical. *Proc. Combust. Inst.* **15**, 745 (1974).
275. M. S. Wooldridge, R. K. Hanson and C. T. Bowman. A shock tube study of the $CO + OH \rightarrow CO_2 + H$ reaction. *Proc. Combust. Inst.* **25**, 741–748 (1994).
276. R. A. Brabbs, F. E. Belles and R. S. Brokaw. Shock-tube measurements of specific reaction rates in the branched-chain H_2-CO-O_2 system, *Proc. Combust. Inst.* **13**, 129–136 (1970).

277. W. C. Gardiner, Jr., W. G. Mallard, M. McFarland, K. Morinaga, J. H. Owen, W. T. Rawlins, T. Takeyama and B. F. Walker. Elementary reaction rates from post-induction-period profiles in shock-initiated combustion. *Proc. Combust. Inst.* **14**, 61–75 (1972).
278. J. C. Biordi, C. Lazzara and J. F. Papp. Flame structure studies of CF_3Br-Inhibited methane flames: II. Kinetics and mechanisms. *Proc. Combust. Inst.* **15**, 917–932 (1974).
279. I. W. M. Smith. Mechanism of OH + CO reaction and stability of HOCO radical. *Chem. Phys. Lett.* **49**, 112–115 (1977).
280. C. M. Stevens, L. Kaplan, R. Gorse, S. Durkee, M. Compton, S. Cohen and K. Bielling. The kinetic isotope effect for carbon and oxygen in the reaction CO + OH. *Int. J. Chem. Kinetic.* **12**, 935–948 (1980).
281. T. Röckmann, C. A. M. Brenninkmeijer, G. Saueressig, P. Bergamaschi, J. N. Crowley, H. Fischer and P. J. Crutzen. Mass-independent oxygen isotope fractionation in atmospheric CO as a result of the reaction CO + OH. *Science* **281**, 544–546 (1998).
282. K. L. Feilberg, S. R. Sellevåg, C. J. Nielsen, D. W. T. Griffith and M. S. Johnson. $CO + OH \rightarrow CO_2 + H$: The relative reaction rate of five CO isotopologues. *Phys. Chem. Chem. Phys.* **4**, 4687–4693 (2002).
283. K. L. Feilberg, M. S. Johnson and C. J. Nielsen. Relative rates of reaction of C-13 O-16, C-12 O-18, C-12 O-17 and C-13 O-18 with OH and OD radicals. *Phys. Chem. Chem. Phys.* **7**, 2318–2323 (2005).
284. M. Alagia, N. Balucani, P. Casavecchia, D. Stranges and G. G. Volpi. Crossed beam studies of four-atom reactions: The dynamics of OH + CO. *J. Chem. Phys.* **98**, 8341–8344 (1993).
285. Y. Liu and S. P. Sander. Rate constant for the OH + CO reaction at low temperatures. *J. Phys. Chem. A* **119**, 10060–10066 (2015).
286. J. Brunning, D. W. Derbyshire, I. W. M. Smith and M. D. Williams. Kinetics of OH(v = 0; 1) and OD(v = 0; 1) with CO and the mechanism of the OH + CO reaction. *J. Chem. Soc. Faraday Trans. 2* **84**, 105–119 (1988).
287. T. G. Clements, R. E. Continetti and J. S. Francisco. Exploring the $OH + CO \rightarrow H + CO_2$ potential surface via dissociative photodetachment of $(HOCO)^-$, *J. Chem. Phys.* **117**, 6478–6488 (2002).
288. C. J. Johnson and R. E. Continetti. Dissociative photodetachment studies of cooled $HOCO^-$ anions revealing dissociation below the barrier to H + CO_2. *J. Phys. Chem. Lett.* **1**, 1895–1899 (2010).
289. C. J. Johnson, B. L. J. Poad, B. B. Shen and R. E. Continetti. Communication: New insight into the barrier governing CO2 formation from OH + CO. *J. Chem. Phys.* **134**, 171106 (2011).
290. J. Y. Ma and H. Guo. Full-dimensional quantum state resolved predissociation dynamics of HCO2 prepared by photodetaching HCO2. *Chem. Phys. Lett.* **511**, 193–195 (2011).
291. C. J. Johnson, R. Otto and R. E. Continetti. Spectroscopy and dynamics of the HOCO radical: Insights into the OH + CO − H + CO_2 reaction. *Phys. Chem. Chem. Phys.* **16**, 19091–19105 (2014).

292. H. Guo. Quantum dynamics of complex-forming bimolecular reactions. *Int. Rev. Phys. Chem.* **31**, 1–68 (2012).
293. A. D. McLean and Y. Ellinger. An *Ab Initio* Configuration-Interaction Study of Cis and Trans Ground-State HOCO Radical Using Localized Orbitals: Structural-Analysis of Correlation-Effects. *Chem. Phys.* **94**, 25–41 (1985).
294. K. Kudla, A. G. Koures, L. B. Harding and G. C. Schatz. A quasi-classical trajectory study of OH rotational-excitation in OH + CO collisions using *ab initio* potential surfaces. *J. Chem. Phys.* **96**, 7465–7473 (1992).
295. J. S. Francisco. Molecular structure; vibrational frequencies; and energetics of the HOCO + ion. *J. Chem. Phys.* **107**, 9039–9045 (1997).
296. Y. Li and J. S. Francisco. High level ab initio studies on the excited states of HOCO radical. *J. Chem. Phys.* **113**, 7963–7970 (2000).
297. T. V. Duncan and C. E. Miller. The HCO_2 potential energy surface: Stationary point energetics and the HOCO heat of formation. *J. Chem. Phys.* **113**, 5138–5140 (2000).
298. H.-G. Yu, J. T. Muckerman and T. J. Sears. A theoretical study of the potential energy surface for the reaction OH + CO -> H + CO2. *Chem. Phys. Lett.* **349**, 547–554 (2001).
299. M. Aoyagi and S. Kato. A Theoretical-study of the potential-energy surface for the reaction Oh + Co-]Co_2 + H. *J. Chem. Phys.* **88**, 6409–6418 (1988).
300. R. S. Zhu, E. G. W. Diau, M. C. Lin and A. M. Mebel. A computational study of the OH(OD) plus CO reactions: Effects of pressure; temperature; and quantum-mechanical tunneling on product formation. *J. Phys. Chem. A* **105**, 11249–11259 (2001).
301. W.-C. Chen and R. A. Marcus. On the theory of the CO + OH reaction, including H and C kinetic isotope effects. *J. Chem. Phys.* **123**, 094307 (2005).
302. X. Song, J. Li, H. Hou and B. Wang. Ab initio study of the potential energy surface for the $OH + CO \rightarrow H + CO_2$ reaction. *J. Chem. Phys.* **125**, 094301 (2006).
303. J. P. Senosiain, C. B. Musgrave and D. M. Golden. Temperature and pressure dependence of the reaction of OH and CO: Master equation modeling on a high-level potential energy surface. *Int. J. Chem. Kinet.* **35**, 464–474 (2003).
304. J. P. Senosiain, S. J. Klippenstein and J. A. Miller. A complete statistical analysis of the reaction between OH and CO. *Proc. Combus. Inst.* **30**, 945–953 (2005).
305. C. W. Larson, P. H. Stewart and D. M. Golden. Pressure and temperature-dependence of reactions proceeding via a bound complex: An approach for combustion and atmospheric chemistry modelers: Application to $HO + CO \rightleftharpoons HOCO \rightleftharpoons H + CO_2$. *Int. J. Chem. Kinet.* **20**, 27–40 (1988).
306. J. J. Lamb, M. Mozurkewich and S. W. Benson. Negative activation energies and curved Arrhenius plots. 2. OH + CO. *J. Phys. Chem.* **88**, 6435–6441 (1984).

307. R. Valero and G.-J. Kroes. Theoretical rate constants for the $OH + CO \rightarrow H + CO_2$ reaction using variational transition state theory on analytical potential energy surfaces. *J. Chem. Phys.* **117**, 8736 (2002).
308. A. V. Joshi and H. Wang. Master equation modeling of wide range temperature and pressure dependence of CO + OH -> products, *Int. J. Chem. Kinet.* **38**, 57–73 (2006).
309. G. C. Schatz and J. Dyck. A reduced dimension quantum reactive scattering study of $OH + CO \rightleftharpoons H + CO_2$. *Chem. Phys. Lett.* **188**, 11–15 (1992).
310. D. C. Clary and G. C. Schatz. Quantum and quasiclassical calculations on the $OH + CO \rightarrow CO_2 + H$ reaction. *J. Chem. Phys.* **99**, 4578–4589 (1993).
311. K. L. Feilberg, G. D. Billing and M. S. Johnson. Quantum dressed classical mechanics: Application to the $HO + CO \rightarrow H + CO_2$ reaction. *J. Phys. Chem. A* **105**, 11171–11176 (2001).
312. J. Ma, J. Li and H. Guo. Quantum dynamics of the $HO + CO \rightarrow H + CO_2$ reaction on an accurate potential energy surface. *J. Phys. Chem. Lett.* **3**, 2482–2486 (2012).
313. C. J. Xie, J. Li, D. Q. Xie and H. Guo. Quasi-classical trajectory study of the $H + CO_2 \rightarrow HO + CO$ reaction on a new ab initio based potential energy surface. *J. Chem. Phys.* **137**, (2012).
314. J. Li, Y. Wang, B. Jiang, J. Ma, R. Dawes, D. Xie, J. M. Bowman and H. Guo. Communication: A chemically accurate global potential energy surface for the $HO + CO \rightarrow H + CO_2$ reaction. *J. Chem. Phys.* **136**, 041103 (2012).
315. E. Garcia, A. Saracibar, L. Zuazo and A. Laganà. A detailed trajectory study of the $OH + CO \rightarrow H + CO_2$ reaction. *Chem. Phys.* **332**, 162–175 (2007).
316. M. J. Lakin, D. Troya, G. C. Schatz and L. B. Harding. A quasiclassical trajectory study of the reaction $OH + CO \rightarrow H + CO_2$. *J. Chem. Phys.* **119**, 5848–5859 (2003).
317. K. Kudla, G. C. Schatz and A. F. Wagner. A quasi-classical trajectory study of the $OH + CO$ reaction. *J. Chem. Phys.* **95**, 1635–1647 (1991).
318. A. F. Wagner, R. Dawes, R. E. Continetti and H. Guo. Theoretical/experimental comparison of deep tunneling decay of quasi-bound H(D)OCO to H(D) + CO2. *J. Chem. Phys.* **141**, 054304 (2014).
319. W. H. Miller. Importance of non-separability in quantum mechanical transition-state theory. *Acc. Chem. Res* **9**, 306–312 (1976).
320. J. Troe. The Polanyi Lecture — The colorful world of complex-forming bimolecular reactions. *J. Chem. Soc. Faraday Trans.* **90**, 2303–2317 (1994).
321. W. C. Chen and R. A. Marcus. On the theory of the reaction rate of vibrationally excited CO molecules with OH radicals. *J. Chem. Phys.* **124**, 024306 (2006).
322. R. A. Marcus. Interaction between experiments, analytical theories, and Computation. *J. Phys. Chem. C* **113**, 14598–14608 (2009).

323. R. A. Marcus. Dissociation and isomerization of vibrationally excited species. III. *J. Chem. Phys.* **43**, 2658–2661 (1965).
324. J. H. Current and B. S. Rabinovitch. Decomposition of chemically activated Ethyl-d_3 radicals. Primary intramolecular kinetic isotope effect in a non-equilibrium system. *J. Chem. Phys.* **38**, 783–795 (1963).
325. E. E. Aubanel, D. M. Wardlaw, L. Zhu and W. L. Hase. Role of angular momentum in statistical unimolecular rate theory. *Int. Rev. Phys. Chem.* **10**, 249–286 (1991).
326. J. R. Barker. Monte-Carlo calculations on unimolecular reactions, energy-transfer, and IR-multiphoton decomposition. *Chem. Phys.* **77**, 301–318 (1983).
327. D. T. Gillespie. A general method for numerically simulating the stochastic time evolution of coupled chemical reactions. *J. Comp. Phys.* **22**, 403–434 (1976).
328. D. T. Gillespie. Exact stochastic simulation of coupled chemical reactions. *J. Phys. Chem.* **81**, 2340–2361 (1977).
329. I. R. Sims and I. W. M. Smith. Gas-phase reactions and energy transfer at very low temperatures. *Annu. Rev. Phys. Chem.* **46**, 109–138 (1995).
330. I. W. M. Smith. Reactions at very low temperatures: Gas kinetics at a new frontier. *Angew. Chem. Int. Ed.* **45**, 2842–2861 (2006).
331. B. Hansmann and B. Abel. Kinetics in cold laval nozzle expansions: From atmospheric chemistry to oxidation of biomolecules in the gas phase. *Chem. Phys. Chem.* **8**, 343–356 (2007).
332. K. M. Hickson and A. Bergeat. Low temperature kinetics of unstable radical reactions. *Phys. Chem. Chem. Phys.* **14**, 12057–12069 (2012).
333. K. Acharyya, E. Herbst, R. L. Caravan, R. J. Shannon, M. A. Blitz and D. E. Heard. The importance of OH radical–neutral low temperature tunnelling reactions in interstellar clouds using a new model. *Mol. Phys.* **113**, 2243–2254 (2015).
334. R. L. Caravan, R. J. Shannon, T. Lewis, M. A. Blitz and D. E. Heard. Measurements of Rate Coefficients for Reactions of OH with ethanol and propan-2-ol at very low temperatures. *J. Phys. Chem. A* **119**, 7130–7137 (2015).
335. J. R. Barker, L. M. Yoder and K. D. King. Feature article: Vibrational energy transfer modeling of non-equilibrium polyatomic reaction systems. *J. Phys. Chem. A* **105**, 796–809 (2001).
336. N. F. Scherer, C. Sipes, R. B. Bernstein and A. H. Zewail. Real-time clocking of bimolecular reactions: Application to $H + CO_2$. *J. Chem. Phys.* **92**, 5239–5259 (1990).
337. S. I. Ionov, G. A. Brucker, C. Jaques, L. Valachovic and C. Wittig. Subpicosecond resolution studies of the $H + CO_2 \rightarrow CO + OH$ reaction photoinitiated in CO_2-HI complexes. *J. Chem. Phys.* **99**, 6553–6561 (1993).
338. M. Brouard, D. W. Hughes, K. S. Kalogerakis and J. P. Simons. The product rovibrational and spin–orbit state dependent dynamics of the complex

reaction H + CO_2 → OH($^2\Pi$;ν,N,Ω,f) + CO: Memories of a lifetime. *J. Chem. Phys.* **112**, 4557–4571 (2000).

339. C. J. Mackie, A. Candian, X. Huang, E. Maltseva, A. Petrignani, J. Oomens, W. J. Buma, T. J. Lee and A. G. G. M. Tielens. The anharmonic quartic force field infrared spectra of three polycyclic aromatic hydrocarbons: Naphthalene, anthracene, and tetracene. *J. Chem. Phys.* **143**, 224314 (2015).

Chapter 7

Recent Advances in the Chemistry of OH and HO_2 Radicals in the Atmosphere: Field and Laboratory Measurements

Sebastien Dusanter[*,‡] and Philip S. Stevens[†,§]

[*] *Sciences de l'Atmosphère et Génie de l'Environnement, Mines Douai, France*

[†] *School of Public and Environmental Affairs and Department of Chemistry, Indiana University, Bloomington, Indiana*

[‡] *sebastien.dusanter@mines-douai.fr*

[§] *pstevens@indiana.edu*

The hydroxyl radical (OH) and peroxy radicals, both the hydroperoxy (HO_2) and organic peroxy radicals (RO_2), play a central role in the chemistry of the atmosphere. In addition to controlling the lifetimes of many trace gases important to issues of global climate change, OH radical reactions initiate the oxidation of volatile organic compounds (VOCs) which in the presence of nitrogen oxides can lead to the production of ozone and secondary organic aerosols in the atmosphere. Recent ambient measurements of OH and HO_2 (HO_x) radicals have revealed serious discrepancies with model predictions, especially in forested environments with high emissions of biogenic VOCs. These results suggest that either our understanding of the sources and sinks of these radicals is incomplete or that there are unknown interferences associated with measurements of these radicals in the atmosphere. This review summarizes recent advances in our understanding of HO_x radical chemistry, including a summary of current detection techniques and recent efforts to characterize interferences. Chamber and laboratory measurements designed to explore new HO_x radical recycling mechanisms are discussed in addition to a review of newly developed OH reactivity instruments, including field and chamber measurements designed to test our understanding of OH radical sinks in the atmosphere.

7.1. Introduction

The hydroxyl radical (OH) and the hydroperoxy radical (HO_2) play an important role in the chemistry of the atmosphere. The OH radical reacts with volatile organic compounds (VOCs) usually leading to the formation of peroxy radicals, both HO_2 and organic peroxy radicals (RO_2) which in the presence of nitric oxide (NO) are primarily converted back to OH. This fast cycling of radicals controls many aspects of atmospheric chemistry such as the formation of ozone and secondary organic aerosols and the removal of methane and other greenhouse gases that affect the radiative balance of the atmosphere.

Because of this central role, measurements of these radicals together with measurements of their sources and sinks can provide a critical test of our understanding of the fast photochemistry of the atmosphere, as the concentration of OH in the atmosphere is determined by the chemistry of VOCs and NO_x ($NO + NO_2$) rather than the result of transport.[1] However, recent measurements of OH and HO_2 radicals in the atmosphere often show significant discrepancies with model predictions, especially in forested environments characterized by high mixing ratios of biogenic volatile organic compounds (BVOCs), such as isoprene, and low mixing ratios of NO_x.[2,3] These results have challenged our understanding of the complex chemistry of OH radicals in the atmosphere, bringing into question our ability to accurately project how changes in emissions as a result of climate change and human activity will impact the future composition of the atmosphere.

Advances in both field measurements and laboratory studies have led to improvements in our understanding of the sources, sinks, and chemistry of these radicals in the atmosphere as well as improvements in ambient measurement techniques, especially related to understanding OH radical chemistry in areas with high BVOC and low NO_x concentrations. Several previous critical reviews have examined OH and HO_2 measurements in urban, rural, and remote environments.[1,3,4] Heard and Pilling[1] provided a comprehensive review of field measurements of tropospheric OH and HO_2 including detailed descriptions of the different measurement techniques and calibration methods. More recently, Stone *et al.*[4] provided a detailed review of tropospheric OH and HO_2 measurements in the marine boundary layer, low NO_x forested environments, polluted environments, and polar regions, focusing on comparing the field measurements with the results of box models, while Whalley *et al.* provided a review of the tropospheric chemistry of isoprene and its impact on HO_x radical chemistry.[3] In addition, Yang *et al.* recently reviewed techniques for measuring total OH reactivity measurements and their use in the field.[5]

In this review, we focus on recent measurements designed to improve our understanding of the tropospheric chemistry of OH and HO_2 radicals in various environments, including advances in our understanding of potential instrument interferences, and a review of recent measurements of total OH reactivity. Comparisons of measurements to model predictions will be discussed, including model predictions of major radical sources and sinks. The results of laboratory measurements that address some of the discrepancies with model predictions will also be discussed.

7.2. OH and HO_2 Radical Chemistry in the Troposphere

Figure 7.1 provides a schematic of the chemistry of OH and HO_2 radicals in the troposphere. As illustrated in this figure, an important source of the OH radical is the photolysis of ozone to form $O(^1D)$ followed by reaction with H_2O:

$$O_3 + h\nu(< 340\,\text{nm}) \rightarrow O(^1D) + O_2 \tag{7.1}$$

$$O(^1D) + H_2O \rightarrow OH + OH \tag{7.2}$$

Once produced, OH radicals initiate a series of radical propagation reactions that lead to the production of peroxy radicals, both HO_2 and RO_2 through reactions with carbon monoxide, ozone, and

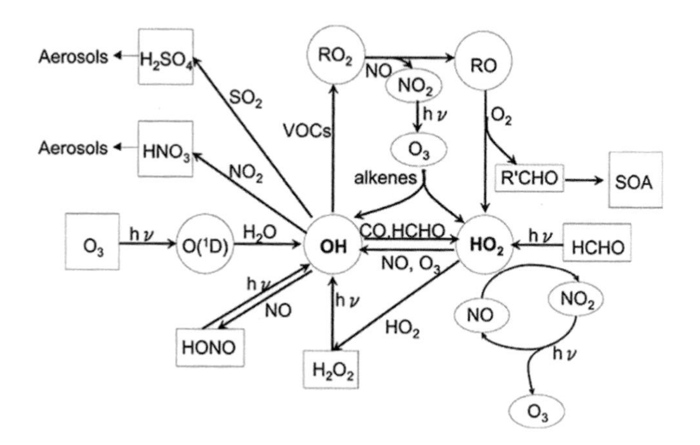

Fig. 7.1 Simplified schematic diagram illustrating OH and HO_2 radical chemistry in the troposphere (reproduced from Ref. [6]).

volatile organic compounds (RH where R = CH_3, C_2H_5, etc.):

$$OH + CO \xrightarrow{O_2} HO_2 + CO_2 \tag{7.3}$$

$$OH + O_3 \rightarrow HO_2 + O_2 \tag{7.4}$$

$$OH + RH \xrightarrow{O_2} RO_2 + H_2O \tag{7.5}$$

In environments with relatively high concentrations of NO, these peroxy radicals oxidize NO leading to the formation of NO_2 and oxygenated VOCs (OVOCs):

$$HO_2 + NO \rightarrow OH + NO_2 \tag{7.6}$$

$$RO_2 + NO \rightarrow RO + NO_2 \tag{7.7}$$

$$RO + O_2 \rightarrow R'CHO + HO_2 \tag{7.8}$$

$$NO_2 + h\nu(< 420\text{nm}) \xrightarrow{O_2} O_3 + NO \tag{7.9}$$

$$NO + O_3 \rightarrow NO_2 + O_2 \tag{7.10}$$

The production of NO_2 perturbs the ozone photostationary state (reactions (7.9) and (7.10)), leading to an increase in the steady-state concentration of ozone, with the rate of ozone production determined by the rate of NO_2 production by reactions (7.6) and (7.7). Under these conditions, termination of the radical chain is dominated by the formation of nitric acid from the reaction of OH with NO_2, as well as the formation of organic nitrates, both alkyl nitrates ($RONO_2$) and peroxy nitrates (RO_2NO_2) from the reaction of organic peroxy radicals with NO and NO_2:

$$OH + NO_2 \xrightarrow{M} HNO_3 \tag{7.11}$$

$$RO_2 + NO \xrightarrow{M} RONO_2 \tag{7.12}$$

$$RO_2 + NO_2 \xrightarrow{M} RO_2NO_2 \tag{7.13}$$

In environments with relatively low concentrations of NO_x, radical propagation and termination of the radical chain involves self and cross reactions of peroxy radicals, in addition to peroxy radical isomerization reactions:

$$HO_2 + HO_2 \rightarrow H_2O_2 + O_2 \tag{7.14}$$

$$HO_2 + O_3 \rightarrow OH + 2O_2 \tag{7.15}$$

$$RO_2 + HO_2 \rightarrow ROOH + O_2 \tag{7.16}$$

$$RO_2 + RO_2 \rightarrow RO + RO + O_2 + \text{other products} \tag{7.17}$$

In the remote atmosphere, reactions (7.1) and (7.2) can be the dominant OH radical initiation route. However, in areas impacted by anthropogenic and/or biogenic emissions, photolysis of nitrous acid, hydrogen peroxide, and OVOCs such as formaldehyde and dicarbonyls as well as the ozonolysis of alkenes can be significant sources of radicals during the daytime:[6–13]

$$HONO + h\nu \rightarrow OH + NO \tag{7.18}$$

$$H_2O_2 + h\nu \rightarrow 2OH \tag{7.19}$$

$$HCHO + h\nu \xrightarrow{O_2} 2HO_2 + CO \tag{7.20}$$

$$O_3 + \text{alkene} \rightarrow \alpha OH + \beta HO_2 + \gamma RO_2 + \text{products} \tag{7.21}$$

The reactions of HO_2 with halogen oxides XO (where X = Br, I) have been shown to impact the chemistry of OH and HO_2 in the marine boundary layer, with photolysis of HOX leading to radical propagation, or radical loss through aerosol uptake:[14]

$$HO_2 + XO \rightarrow HOX + O_2 \tag{7.22}$$

$$HOX + h\nu \rightarrow HO + X \tag{7.23}$$

Because of its high reactivity, concentrations of the OH radical in the troposphere reach steady-state on the order of seconds resulting in a balance between the rate of OH radical production and the rate of OH radical destruction:

$$P_{\mathrm{OH}} = L_{\mathrm{OH}} = \sum_i k_{\mathrm{OH}+R_i}[OH][R_i] \tag{7.24}$$

Here, P_{OH} is the sum of the rates of all OH production reactions, including radical initiation from the photolysis of ozone (reactions (7.1) and (7.2)), HONO, H_2O_2, OVOCs, and the ozonolysis of alkenes (7.18–7.21), as well as OH production from radical propagation reactions of HO_2 with NO and ozone (7.6 and 7.15). L_{OH} is the sum of the rates of all OH loss reactions with species R_i, including reactions with CO, ozone, NO_2, and various VOCs and OVOCs (7.3, 7.4, 7.5, and 7.11). Thus the concentration of OH in the troposphere can be determined if all the sources and sinks are known:

$$[\mathrm{OH}] = \frac{P_{\mathrm{OH}}}{\sum_i k_{\mathrm{OH}+R_i}[R_i]} = \frac{P_{\mathrm{OH}}}{k_{\mathrm{OH}}} \tag{7.25}$$

In this equation, k_{OH} is the total OH reactivity defined as the sum of the products of the second-order rate constant (or for termolecular reactions the bimolecular rate constant calculated at atmospheric pressure) for the OH reaction of each chemical species R_i with OH ($k_{\mathrm{OH}+Ri}$) and the concentration of R_i. With this definition, total OH reactivity is often referred to as the inverse of the OH chemical lifetime or the OH loss frequency.

Measurements of OH and k_{OH} can thus provide a rigorous test of our understanding of the sources, sinks, and propagation of radicals in the atmosphere. Comparison of measured OH concentrations and k_{OH} to that predicted by chemical models can help to identify missing sources and sinks of radicals that can have a significant impact on our understanding of ozone and secondary organic aerosol production, as well as our understanding of the oxidative capacity of the atmosphere. Predictions of OH concentrations by chemical models require accurate measurements of the individual radical precursors as well as the individual radical sinks. Although measurements of the individual contributions to P_{OH} and k_{OH} have improved, a complete characterization of all the individual species is a challenging task. An underestimation of both P_{OH} and k_{OH} could result in a fortuitous agreement between the model and measurements, while unknown interferences with measurements of OH radical concentrations can lead to an overestimation of the discrepancies with model predictions. However, recent techniques designed to identify potential interferences with ambient measurements of OH radical concentrations in the troposphere in addition to recently developed techniques to measure total OH reactivity have improved our understanding of the OH radical chemistry in the troposphere.

7.3. Measuring OH and HO_2 Radicals in the Troposphere

Direct measurements of the concentration of OH and HO_2 radicals have been and remain a considerable analytical challenge. Because of its high reactivity, ambient concentrations of OH and HO_2 radicals in the troposphere are very small, typically on the order of 10^6 molecules cm^{-3} (0.04 parts per trillion at sea level) for OH and on the order of 10^8 molecules cm^{-3} for HO_2. As a result, techniques to measure OH and HO_2 require both high sensitivity and selectivity. In the past, several instrumental techniques have attempted to make reliable measurements of OH radicals in the troposphere.[1,15] However, recent ambient measurements have either involved instruments employing the laser-induced fluorescence (LIF) at low pressure technique, or the chemical ionization mass spectrometry (CIMS) technique, while measurements using long path differential optical absorption spectroscopy (DOAS) are presently only done as part of studies inside the SAPHIR (Simulation of Atmospheric PHotochemistry In a large Reaction chamber) facility at Forschungszentrum Jülich.[4] These techniques and their corresponding calibration methods have been reviewed previously,[1] and only a brief description of each will be provided here, including recent measurements and characterizations of interferences with each technique.

7.3.1. *Laser-Induced Fluorescence at Low Pressure (LIF-FAGE)*

Laser-induced fluorescence was first identified as a promising technique for the detection of OH radicals in the atmosphere in 1972 and some of the first attempts at measuring ambient OH concentrations were reported in 1974.[1,4,16] These first attempts used excitation at atmospheric pressure of the $A^2\Sigma^+(v' = 1) \leftarrow X^2\Pi_i(v'' = 0)$ band of OH at 282 nm with detection of the red-shifted fluorescence from the $A^2\Sigma^+(v' = 0) \rightarrow X^2\Pi_i(v'' = 0)$ band near 308 nm. This technique allowed separation of the OH fluorescence from Rayleigh and Mie scattering from the laser pulse. However, collisional quenching of the OH excited state is very efficient at atmospheric pressure, requiring high excitation energies to observe fluorescence from ambient OH. Unfortunately, it was quickly discovered that these early measurements suffered from a significant interference from the photolysis of ozone by the laser pulse leading to the production of OH from reactions (7.1) and (7.2).[17,18] Although this method has been used successfully to measure OH concentrations in the stratosphere,[15,19,20] the higher concentration of water vapor and higher fluorescence quenching rates in the troposphere result in interferences from laser-generated OH that were often much higher than ambient concentrations.[15]

Laser-induced fluorescence at low pressure, commonly known as LIF-FAGE (Fluorescence Assay by Gas Expansion) was developed to reduce the interference from laser generated OH.[21–23] By sampling ambient air through an inlet, the concentration of ozone, water vapor, and other potentially interfering species are reduced, reducing the rate of production of OH by reactions (7.1) and (7.2). Although the concentration of OH is also reduced upon expansion, the total

OH fluorescence signal does not change significantly because the reduction in OH concentration is offset by the decreased collisional quenching rate at the lower pressure, which results in an increased quantum yield for fluorescence. In addition, the decrease in collisional quenching also results in an increased fluorescence lifetime beyond the length of the laser pulse, allowing the OH fluorescence to be discriminated from laser scatter.

Although the FAGE technique helped to reduce the laser-generated interference, excitation at 282 nm still resulted in high background signals due to laser-generated OH that limited the overall sensitivity of the technique.[23,24] This interference can be further reduced by excitation and detection of OH using the $A^2\Sigma^+(v' = 0) \rightarrow X^2\Pi_i(v'' = 0)$ band near 308 nm using a high repetition rate (3–10 kHz) laser system.[25] The low energy per pulse at the high repetition rate reduces the rate of ozone photolysis from reaction (7.1) while maintaining high average power that result in a greater number of OH excitations per second. Excitation of OH at 308 nm also reduces the rate of ozone photolysis by a factor of approximately 30 compared to excitation at 282 nm due to the lower absorption cross section of ozone and reduced quantum yield of $O(^1D)$ at 308 nm compared to 282 nm.[25,26] The disadvantage of this approach is the difficulty in isolating the OH fluorescence from laser scatter, which requires careful gating of the detector to reduce the memory of the high laser scatter in a photomultiplier tube that can lead to high background signals long after the laser pulse.[1,26]

Currently all LIF instruments involved in ambient measurements employ the FAGE technique with excitation and detection at 308 nm. Each instrument incorporates a vacuum system to expand ambient air through an inlet into a low pressure detection chamber, and each uses a high repetition rate laser system for excitation with a gated detector. However, the specific design and operating parameters (cell pressure, flow rates) are different for each instrument. Figure 7.2 illustrates some of the individual designs. Several groups (University of Leeds,[14] Forschungszentrum Jülich,[31] Research System Global Change, Yokohama[32]) employ a design that uses a single pass of the laser across the airstream, with detection of the fluorescence using

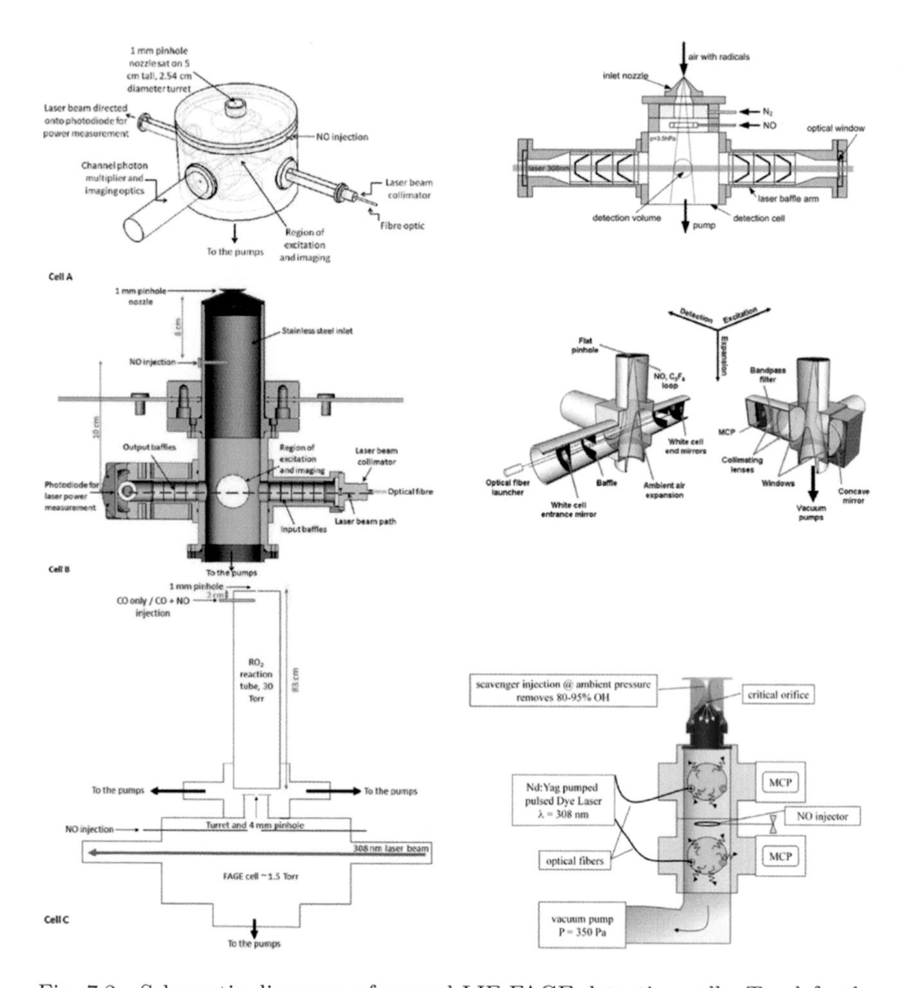

Fig. 7.2 Schematic diagrams of several LIF-FAGE detection cells. Top left, the Leeds single pass instrument used during the OP3 campaign. Middle left, the Leeds single pass instrument used during the HCCT campaign. Bottom left, the Leeds single pass instrument coupled to an external reactor for detection of RO_2 radicals.[27] Top right, the Forschungszentrum Jülich single pass instrument.[28] Middle right, the Indiana University multipass instrument.[29] Bottom right, the Max Planck Institute for Chemistry multipass instrument for simultaneous detection of OH and HO_2[30] (Reproduced from Refs. [27–30]).

sensitive photomultiplier tubes. Other groups (Penn State University,[33] Max Planck Institute for Chemistry,[34] Indiana University,[29] University of Lille[35]) use a design that incorporates multiple passes of the laser across the airstream, with detection of the fluorescence using less sensitive but more easily gated microchannel plate detectors (MCP). Limits of detection for OH are typically on the order of $1 - 2 \times 10^5$ molecules cm^{-3} for an averaging time of less than 5 min, depending on the instrument.[1]

These instruments can also measure HO_2 radicals indirectly through chemical conversion to OH using reaction (7.6) followed by detection of OH using LIF. Addition of NO is accomplished through an injector inside the detection cell and measurements are made either sequentially with measurements of OH using the same detection cell, or simultaneously with measurements of OH using two different detection cells operating either in parallel with separate inlets or sequentially with a single inlet (Fig. 7.2). Sequential measurements of OH and HO_2 using the same detection cell has the advantage of allowing for measurements of the HO_2/OH ratio (which can provide important information regarding our understanding of radical propagation) with higher accuracy given that these measurements of the ratio are independent of the calibration of the instrument.[30] However, simultaneous measurements of this ratio using two separate detection axes can provide better time resolution given the short chemical lifetimes of both OH and HO_2.[36] Limits of detection for HO_2 by this technique are similar to that for OH.[1,4]

This technique can also measure total peroxy radical concentrations (RO_2+HO_2) through chemical conversion to OH by reaction with NO in an external reactor (Fig. 7.2).[27,37] To detect all RO_2 radicals, the chemical conversion must be done at higher pressures and with longer reaction times to efficiently convert RO_2 radicals to HO_2 prior to entering the low pressure FAGE detection cell, where the resulting HO_2 radicals are then converted to OH.[27,37] In this technique, an external reactor is incorporated above the FAGE inlet and ambient air is sampled to pressures between 20 and 30 Torr. NO is added through an injector to convert RO_2 radicals to HO_2

inside the reactor and CO is added simultaneously to convert OH radicals produced in the reactor from reaction (7.6) back to HO_2 through reaction (7.3) in addition to converting ambient OH radicals entering the external reactor to HO_2. This technique can measure total ambient RO_x radical concentrations ($RO_2 + HO_2 + OH$) when both NO and CO are added to the external reactor, or total HO_x concentrations ($OH + HO_2$) when only CO is added.[37]

Although the LIF-FAGE technique provides a direct measurement of OH radicals in the atmosphere, the measurement requires an external calibration to determine the sensitivity of the fluorescence signal towards OH and to account for potential radical losses through the sampling process. This requires the generation of a known concentration of OH in air at atmospheric pressure under various conditions of temperature and relative humidity with an apparatus that is portable so that it can be used during field measurements. Several calibration methods have been used in the past, including monitoring the decay rate of hydrocarbons in a continuously stirred tank reactor and measuring the production of OH from the ozonolysis of alkenes.[38–40] However, because of their associated uncertainties and poor accuracies, as well as their more complex and cumbersome procedures, these methods have only been used occasionally.[1,40]

All current LIF-FAGE instruments utilize the photolysis of water vapor to produce a known concentration of OH at atmospheric pressure, and this calibration method has been described in detail elsewhere.[1–4] Briefly, the photolysis of water vapor at 185 nm produces equal amounts of OH and H radicals, and in the presence of O_2 the latter are quickly converted to HO_2:

$$H_2O + h\nu(185\text{nm}) \rightarrow OH + H \qquad (7.26)$$

$$H + O_2 \xrightarrow{M} HO_2 \qquad (7.27)$$

This method produces equal concentrations of both OH and HO_2 and their concentrations can be determined from the following equation:

$$[OH] = [HO_2] = [H_2O]\sigma_{H_2O}\phi_{OH+H}(F \times t) \qquad (7.28)$$

Here σ_{H_2O} is the absorption cross section of water at 185 nm and ϕ_{OH+H} is the quantum yield for photodissociation leading to the production of OH and H, F is the photon flux at 185 nm, and t is the exposure time. Potential loss of excited H atoms produced from reaction (7.26) through reaction with water vapor has been shown to be negligible, confirming that both OH and H (and therefore HO_2) are produced with a quantum yield of one.[28] The product $F \times t$ is determined either directly through absolute measurements of the photon flux through a calibrated detector and calculation of exposure time using the flow rate through the calibrator,[33] or indirectly through chemical actinometry using both O_2 photolysis leading to production of O_3, and N_2O photolysis leading to NO production.[31,40] This calibration method has been compared to other methods, including the decay of hydrocarbons in a chamber as well as the ozonolysis of alkenes, and good agreement was found between the different methods, giving confidence in the water–vapor photolysis method as an accurate calibration technique.[40,41] The accuracy of this calibration method is approximately $\pm 20\%(1\sigma)$ based on uncertainties associated with the calculation of the OH concentration from Eq. (7.28).[40]

7.3.2. *Chemical Ionization Mass Spectrometry (CIMS)*

Although several non-spectroscopic techniques to measure OH have been developed in the past,[1,15] only the Chemical Ionization Mass Spectrometry (CIMS) technique is currently used extensively.[4] This indirect chemical technique was first developed in 1989 and serves as an important complimentary technique to measurements by the LIF-FAGE technique.[1,4,42,43] Briefly, ambient air is sampled though a reactor where a small concentration of $^{34}SO_2$ is added to convert ambient OH radicals to isotopically labelled $H_2^{34}SO_4$ (Fig. 7.3):

$$OH + {}^{34}SO_2 \xrightarrow{M} H^{34}SO_3 \tag{7.29}$$

$$H^{34}SO_3 + O_2 \rightarrow HO_2 + {}^{34}SO_3 \tag{7.30}$$

$$^{34}SO_3 + H_2O \xrightarrow{M} H_2{}^{34}SO_4 \tag{7.31}$$

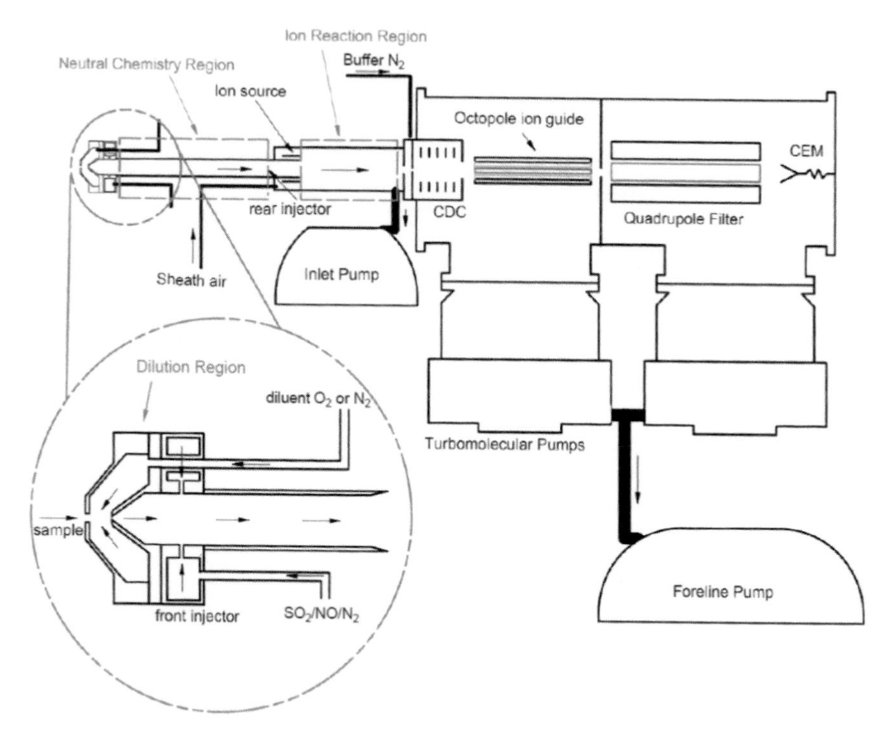

Fig. 7.3 Schematic diagram of the NCAR/University of Colorado PerCIMS instrument for the detection of OH, HO_2, and $HO_2 + RO_2$ radicals (reproduced from Ref. [51]).

The airstream then enters a chemical ionization region, where the $H_2^{34}SO_4$ is chemically ionized by the following charge transfer reaction:

$$H_2^{34}SO_4 + NO_3^- \cdot HNO_3 \rightarrow H^{34}SO_4^- \cdot HNO_3 + HNO_3 \qquad (7.32)$$

The $H^{34}SO_4^-/NO_3^-$ ion ratio is then measured using a quadrupole mass spectrometer and is used to determine the OH concentration. The use of isotopically labelled $^{34}SO_2$ allows the $H_2^{34}SO_4$ produced from reaction with ambient OH to be distinguished from ambient concentrations of H_2SO_4. To account for other chemical reactions that may produce $H_2^{34}SO_4$ in the reactor, propane is added to the reactor to remove ambient OH, and the resulting "background" $H_2^{34}SO_4$ signal is subtracted from the overall signal when the

OH scavenger is not present.[43] The ion signal is converted to a concentration using a calibration factor determined using the photolysis of water vapor to produce a known concentration of OH, similar to the calibration method for the LIF-FAGE instruments described above. In contrast to the different LIF-FAGE instrument designs, all CIMS instruments currently used to measure OH operate using a similar design (Georgia Institute of Technology,[44,45] NCAR/University of Colorado/University of Helsinki,[46,47] German Weather Service/National University of Ireland[48,49]). Limits of detection for OH are typically on the order of 1×10^5 molecules cm^{-3} for an averaging time of 5–10 min.[44,46,48]

The CIMS technique can also measure ambient concentrations of peroxy radicals through chemical conversion to OH by reaction with NO (PerCIMS).[44,45,50,51] Similar to the LIF-FAGE method, NO is added to the reactor in addition to $^{34}SO_2$ to convert ambient RO_2 and HO_2 radicals to OH, which are then converted to $H_2^{34}SO_4$ and detected as described above. Through variations in the ratio of $[NO]/[O_2]$ in the reactor, the conversion of RO_2 to HO_2 can be suppressed, allowing measurements of HO_2 in addition to ambient $RO_2 + HO_2$. Under conditions where the $[NO]/[O_2]$ is high (through addition of a high NO concentrations and dilution of air with added N_2), the reactor chemistry favors conversion of HO_2 to OH as the rate of reaction (7.8) is suppressed and reaction of RO with NO leading to the formation of RONO is enhanced. Under conditions where the $[NO]/[O_2]$ is low (through the use of lower added NO concentrations and additional O_2), the reactor chemistry favors conversion of both RO_2 and HO_2 to OH as reaction (7.8) is favored and reaction of RO with NO is suppressed.[51] Limits of detection using this technique are typically near 1×10^7 molecules cm^{-3} for an averaging time of 10–15 s.[50]

7.3.3. *OH Instrument Intercomparisons*

Intercomparisons of instruments using fundamentally different techniques can provide confidence in the ability of the instruments to accurately measure ambient OH concentrations, and several recent informal and formal intercomparisons have been reviewed previously.[1,4] One of the most recent intercomparisons involved measurements inside the Simulation of Atmospheric PHotochemistry In a Large Reaction (SAPHIR) chamber in addition to co-located ambient measurements as part of the HOxCOMP campaign in Jülich.[52] This formal blind intercomparison involved measurements with three LIF-FAGE instruments and one DOAS instrument inside the SAPHIR chamber, and ambient measurements from three LIF-FAGE instruments and one CIMS instrument. The chamber intercomparison involved measurements under a variety of conditions, including varying ozone, water vapor, NO_x, and sunlight. In addition, experiments were conducted where HO_x radicals were produced in the dark from the ozonolysis of pent-1-ene, and OH measurements were made during the photooxidation of different added alkanes, alkenes, and aromatic compounds, specifically benzene, 1-hexene, m-xylene, n-octane, n-pentane, and isoprene.

In general, the OH measurements conducted inside the SAPHIR chamber by the different LIF-FAGE instruments agreed well with the DOAS measurements and with each other, with pairwise linear regressions slopes between 1.01 and 1.13 that were well within the calibration accuracies of the instruments.[52] Similar results were found in a previous intercomparison of the Jülich LIF-FAGE instrument with the DOAS instrument in the SAPHIR chamber.[53] The agreement of the LIF-FAGE instruments with the DOAS measurements provides confidence in the calibration method for the LIF-FAGE instruments, as the DOAS technique is an absolute technique based on fundamental spectroscopic constants and does not require calibration. The agreement between the instruments during the daytime ambient measurements phase of the campaign was more varied, with pairwise linear regressions slopes between 1.06 and 1.69, and the greatest differences observed between the LIF-FAGE instruments and the CIMS instrument. The difference was often outside of the uncertainty of the instruments, suggesting possible systematic errors associated with the techniques, perhaps related to instrumental calibration.[52]

A subsequent intercomparison in the SAPHIR chamber of OH measurements using the Jülich DOAS and LIF-FAGE instruments was done to determine whether VOC oxidation products under low NO_x conditions could lead to interferences associated with the LIF-FAGE technique.[54] Similar to the results from the HOxCOMP campaign, they found that the LIF-FAGE measurements were generally in good agreement with the DOAS measurements and no instrument artifacts were observed during the oxidation of isoprene, methacrolein, and several aromatic compounds. However, the LIF-FAGE measurements of OH were approximately 30% larger than the DOAS measurements during the oxidation of up to 20 ppb of methyl vinyl ketone, and 40% greater during the oxidation of 40–100 ppb of toluene in the chamber.[54] Although this difference was only found to be significant within 1σ, it does suggest the possibility of an interference with the LIF-FAGE instrument under these conditions.

7.3.4. *OH Measurement Interferences*

Prior to HOxCOMP, the Max–Planck-Institute multi-pass instrument observed an artefact associated with their measurements at night,[52] and as a result measurements in the dark (either at night or with the SAPHIR louvers closed) were not included in the intercomparison. The source of the interference was unknown, but tests indicated that it was not laser generated.[52] High nighttime measurements of OH had also been observed during several field campaigns by the Penn State LIF-FAGE multipass instrument in both urban and forested environments.[55–61] Some of these nighttime measurements were found to correlate with ozone,[58] and although extensive laboratory tests did not reveal any interferences,[62] it was speculated that decomposition of weakly bound intermediates, such as Criegee intermediates produced from the ozonolysis of alkenes, may decompose inside the FAGE detection cell leading to an interference with the measurements.[56] Other LIF-FAGE instruments have also occasionally measured nighttime concentrations of OH above detection limit which could not be explained by model predictions,[63,64] while other nighttime measurements were below the detection limit of the instrument.[13,65,66]

As discussed above, a significant interference with measurements of OH using the LIF-FAGE technique is the photolysis of ozone by the laser leading to the production of OH through reactions (7.1) and (7.2). Although the use of low pressure sampling, excitation at 308 nm, and high repetition rate laser systems have reduced this interference,[1,4] some LIF-FAGE instruments must still take this interference into account through laboratory calibrations.[13] In an attempt to explain unusually high nighttime measurements of OH Ren *et al.*[62] performed several laboratory experiments that demonstrated that spectral interferences from naphthalene, sulfur dioxide, and formaldehyde were negligible. Addition of high concentrations of H_2O_2, HONO, SO_2, HNO_3, several alcohols and alkanes, propene, and isoprene to the detection cell also did not reveal any significant interferences, although small interferences were observed with addition of high amounts of ozone and acetone that would be insignificant under ambient conditions. Mixtures of ozone with ethene, propylene and isoprene did not result in any significant signal, suggesting that the ozonolysis of these compounds did not produce an interference in their instrument.[62] However, Hard *et al.* did observe an interference in their FAGE instrument during calibrations using the ozonolysis of trans-2-butene (T2B) when the concentration of T2B was greater than 3×10^{12} molecules cm^{-3}.[39] The interference was observed to disappear in air containing 1% of water vapor, and tests suggested that the interference was not laser generated. They suggested that the interference may be due to the dissociation of an intermediate in the ozonolysis mechanism that produces OH in the low-pressure cell of their FAGE instrument.[39]

To determine whether instrumental artefacts are interfering with measurements of OH using LIF-FAGE, several groups have adopted an external chemical titration technique during ambient measurements to routinely remove OH radicals prior to entering the FAGE detection cell.[12,13,30,67] These techniques have been motivated by both the high nighttime measurements observed by some groups described above and the high daytime measurements of OH by the LIF-FAGE technique in high VOC, low NO_x environments,[2] which will be discussed in the next section.

Traditionally, LIF-FAGE instruments have used spectral modulation to determine the background signal that must be subtracted from the overall signal to determine the signal attributable to OH fluorescence.[26,33] Typically, LIF-FAGE instruments use the strong $Q_1(2)$, $Q_1(3)$, or $P_1(1)$ transitions, and tuning the laser on and off on either side of these transitions provides a measure of the non-resonant background signals. This technique has the advantage of providing specificity to OH, as the tuning on and off the narrow OH spectral feature isolates the OH fluorescence from broadband fluorescence due to other species in ambient air as well as Raleigh and chamber scattering of the laser pulse.[68] However, spectral modulation does not discriminate ambient OH radicals from OH radicals produced in the detection cell from interferences, including laser-generated OH.

To measure OH produced from interfering processes, chemical modulation is often used. In this technique, a chemical scavenger that reacts rapidly with OH is added to the airstream to remove ambient OH, and spectral modulation is then used to measure any remaining OH.[23,26,68] Early chemical modulation techniques used n-butane and i-butane as the chemical scavenger,[23,38] while later techniques have incorporated perfluoropropylene, propane or propene as the chemical scavenger.[13,30,67] Perfluoropropylene (C_3F_6) as a scavenger has the advantage that it does not contain any hydrogens that could lead to the production of OH from laser-generated radical reactions.[26] Some early attempts to identify potential laser-generated interferences injected the chemical scavenger inside of the FAGE detection cell just below the inlet to scavenge OH. High concentrations of the scavenger had to be injected to remove OH as it entered the low pressure cell, but OH produced in the region illuminated by the laser would not have sufficient time to react and would be detected by spectral modulation.[26,68] Although this method can distinguish laser-generated OH from ambient OH, it cannot distinguish ambient OH radicals from OH radicals produced from non-photolytic processes, such as the rapid decomposition of ambient species inside the FAGE cell. To distinguish ambient OH from internally generated OH artefacts, a lower amount of the chemical scavenger is now added external to the FAGE detection cell to remove ambient OH before the air enters the inlet. Subsequent spectral modulation then provides a measure of internally generated OH.[13,30,67]

Using the chemical modulation technique, Mao *et al.* discovered a significant interference associated with the Penn State multi-pass instrument's measurements of OH during the BEARPEX 2009 campaign (Fig. 7.4).[67] They found that measurements using spectral modulation of the laser wavelength (OH_{wave}) were greater than the measurements using chemical modulation (OH_{chem}). The measurements using chemical modulation were also in better agreement with model predictions, both during the day and at night. The measurements using OH_{chem} were also consistent with measurements of glyoxal at this site,[69] suggesting that the chemistry is well represented by our current understanding of biogenic VOC oxidation.[67] The source of the interference was unclear, and several tests ruled out laser-generated interferences. However, the magnitude of the interference was observed to increase with the measured OH reactivity, indicating that the interference may be related to the oxidation of biogenic VOCs.

Similar results have been observed by the Max–Planck-Institute multi-pass instrument.[12,30] Using a similar external chemical modulation technique, Novelli *et al.* found that OH generated inside their detection cell comprised 30–80% of the daytime signal observed using spectral modulation, and 60–100% of the signal observed at night (Fig. 7.4).[30] The magnitude of the interference varied between different environments, with the highest interference observed during the Hyytiälä United Measurements of Photochemistry and Particles in Air–Comprehensive Organic Precursor Emission and Concentration (HUMPPA-COPEC) 2010 campaign in a boreal forest in Finland, and the smallest interference observed during the HOPE (Hohenpeißenberg Photochemistry Experiment) 2012 campaign in Bavaria. A major difference between the two sites was the magnitude of the observed OH reactivity, with relatively high levels of OH reactivity observed during the HUMPPA-COPEC 2010 campaign ($12\,s^{-1}$ on average with peaks over $40\,s^{-1}$), while OH reactivity measured during the HOPE 2012 campaign was often below the

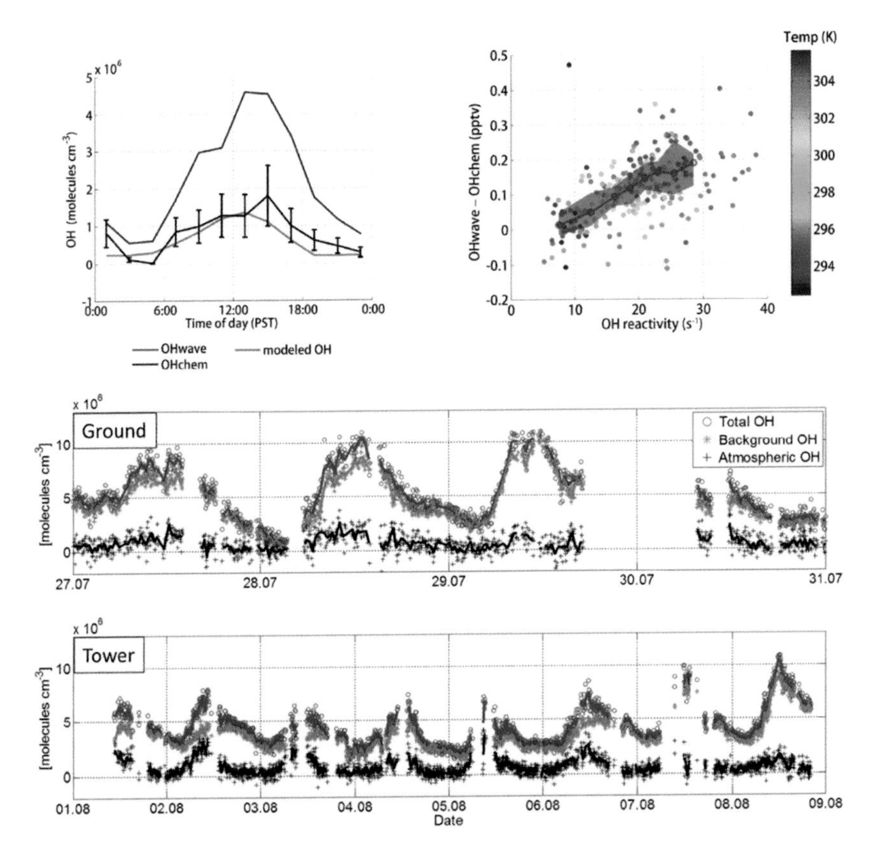

Fig. 7.4 Measurements of OH using both spectral modulation (OH_{wave}) and chemical modulation (OH_{chem}) during the BEARPEX 2009 campaign with model predictions (top left).[67] Measurements of the interference observed during BEARPEX 2009 as a function of OH reactivity and temperature (top right).[67] OH signals measured during the HUMPPA-COPEC 2010 campaign (bottom).[30] Blue circles represent the total signal measured using spectral modulation without the external OH scavenger, while red stars represent the signal measured with the addition of an external OH scavenger. Black crosses show the difference between the two measurements reflecting the ambient OH concentration. Lines are 30-min averages of the data (Reproduced from Refs. [30,67]).

detection limit of the reactivity instrument ($3.5 \pm 2\,s^{-1}$).[30] When the interference was subtracted from the overall signal, the resulting measured ambient OH concentrations were in good agreement with simultaneous measurements of OH by several CIMS instruments, although the agreement was better during HOPE 2012 as the measurements during HUMPPA-COPEC 2010 indicated a small systematic difference between the two techniques.[12,30]

The observation that the interference increases with increasing OH reactivity is consistent with that observed by Mao *et al.* during BEARPEX 2009 and suggests that the interference in these instruments is related to the level and composition of VOCs. Previous laboratory measurements had suggested that Criegee intermediates produced from the ozonolysis of various alkenes could decompose and produce OH radicals at low pressure,[70] and it was hypothesized that this could be the source of the interference.[67] Subsequent laboratory measurements confirmed the formation of OH radicals inside the FAGE cell from the unimolecular decomposition of stabilized Criegee intermediates produced from the ozonolysis of propene and (E)-2-butene, with the concentration of OH detected dependent on the concentration of the syn-CH_3COO intermediate at the inlet and rate of decomposition of the intermediate. The amount of OH detected in the FAGE detection cell showed a strong dependence on the reaction time inside the low pressure FAGE cell before detection by the laser, with the production of OH increasing rapidly with reaction time and then slowly decreasing due to the loss rate of both OH and the Criegee intermediate.[71]

These results suggested that the magnitude of this interference in LIF-FAGE instruments likely depends on the overall residence time in the detector, including the length of the inlet and the flow rate and velocity of the airstream. In addition, the level of this interference will depend on the rate coefficients for dissociation of the different stabilized Criegee intermediates in the ambient air, which have not been characterized.[71] As a result, the level of interference may differ depending on the ambient environment and the specific characteristics of each LIF-FAGE system. Griffith *et al.* conducted several measurements using a similar external chemical modulation technique in a northern Michigan forest, resulting in measured interferences that were generally near the detection limit of the instrument and consistent with the laser generated OH from reactions (7.1) and (7.2).[13] However, these measurements were done under uncharacteristically cold conditions, with mixing ratios of ozone and biogenic VOCs that were generally lower than in the forested areas where interferences have been previously observed.[13] Laboratory measurements of the OH radical yield produced from the ozonolysis of several alkenes suggest that under high ozone and alkene concentrations the Indiana University multi-pass instrument is sensitive to an interference associated with certain alkenes, including α- and β-pinene, ocimene, myrcene, and γ-terpinene.[72] The interference was found to increase with the length of the inlet and the pressure of the detection cell, suggesting that the interference increased with increasing reaction time. Addition of acetic acid to scavenge stabilized Criegee intermediates removed the interference, suggesting that the interference was due to decomposition of these intermediates in the detection cell.

Similarly, recent laboratory measurements using the Forschungszentrum Jülich single pass instrument using propane as the external OH chemical scavenger also revealed the presence of an interference under high concentrations of ozone and some alkenes, including α-pinene, limonene, and isoprene.[73] These authors found that the level of the interference increased as the turnover rate of the ozonolysis reaction increased and was only above the detection limit for their instrument for reactant concentrations that were orders of magnitude higher than in the atmosphere. They also found that the level of the interference increased with increasing residence time in the detection cell, as the interference increased with the length of the inlet and the pressure in the cell. However, the interference from the ozonolysis of α-pinene did not decrease with the addition of SO_2, which has been shown to react with some of the stabilized Criegee intermediates produced from α-pinene ozonolysis, suggesting that in these experiments the interference is not due to decomposition of Criegee intermediates.[73] In addition these authors also discovered an interference that scales with the mixing ratio of NO_3, corresponding to an equivalent OH concentration of $1.1 \times 10^5\,cm^{-3}$ for every 10 pptv of NO_3.[73] The mechanism for this interference could not be determined, but could involve surface reactions with water vapor on the chamber walls or from the formation of molecular clusters with water in the gas expansion.[73]

Given the uncertainty associated with these interferences, it has been recommended that all LIF-FAGE instruments employ an external chemical modulation technique during ambient measurements to quantify this and other potential interferences.[67,71] A similar chemical modulation technique has always been employed to measure background signals in CIMS instruments to account for the production of $H_2^{34}SO_4$ from the oxidation of $^{34}SO_2$ by oxidants other than OH.[43] In some areas it was observed that this background signal was larger than the actual OH signal, especially at night, suggesting that there was an oxidant besides OH that was converting SO_2 into H_2SO_4.[74,75] Field and laboratory tests suggested that the background signal was due to oxidation of SO_2 by stabilized Criegee intermediates.[75] Thus, the chemistry of stabilized Criegee intermediates may contribute to the background signals in some LIF-FAGE and CIMS instruments.

7.3.5. *HO_2 Instrument Intercomparisons*

There have been several recent intercomparisons of instruments measuring HO_2 and RO_2. Measurements of HO_2 and RO_2 radicals inside the SAPHIR chamber using the LIF-FAGE technique were in very good agreement with absolute measurements using matrix isolation electron spin resonance (MIESR).[76] These experiments included measurements of HO_2 and CH_3O_2 during methane photooxidation experiments as well as measurements of HO_2 and $C_2H_5O_2$ radicals during the ozonolysis of 1-butene. An earlier intercomparison of ambient HO_2 measurements between the Penn State LIF-FAGE instrument and the NCAR/University of Colorado PerCIMS instrument were in excellent agreement, with a regression slope close to unity.[77] However, subsequent corrections to the calibration of the LIF-FAGE instrument may require this result to be revisited.[78,79]

In contrast, during the HOxCOMP campaign three LIF-FAGE instruments measuring HO_2 through chemical conversion to OH conducted a formal blind intercomparison of HO_2 measurements both in ambient air and in the SAPHIR chamber.[79] Although the measurements between the instruments were highly correlated, the agreement between the different instruments was found to be variable. For the measurements inside the chamber, regression slopes of the measurements were between 0.69 and 1.26, while for the ambient measurements the regression slopes were between 0.59 and 1.46. Although the overall differences were within the range of the combined uncertainties of the measurements, there were several measurement periods when the differences were larger and cannot be explained by uncertainties associated with the sensitivities of the instruments. The agreement between the measurements inside the chamber was improved when the data was grouped into subsets of similar water vapor concentrations, suggesting that an unknown factor related to the concentration of water vapor in the chamber may have impacted the sensitivities of the instruments or may have caused an unknown interference.[79] However, these results suggested the possibility of potential interferences in the HO_2 calibration or measurement technique.[28]

7.3.6. *HO_2 Measurement Interferences*

It was previously believed that the detection of HO_2 radicals by chemical conversion to OH by reaction with added NO (reaction (7.6)) in LIF-FAGE instruments was free from interference from reactions of RO_2 radicals. Model simulations at the time suggested that the rate of conversion of RO_2 radicals to HO_2 and subsequently to OH through reactions (7.7), (7.8), and (7.6) was suppressed due to the slow rate of reaction (7.8) under the reduced oxygen concentration in the low pressure cell and the short reaction time between injection of NO and detection of OH[26,32,36,80] Consistent with these model predictions, measurements of the conversion efficiency of RO_2 radicals produced from the OH-initiated oxidation of methane, ethane, propane, and n-butane did not reveal a significant interference.[62]

However, recent measurements by Fuchs *et al.* have revealed that RO_2 radicals derived from the OH-initiated oxidation of alkenes and aromatics, including isoprene, can interfere with measurements of HO_2 through chemical conversion with NO by rapidly reacting with

NO and leading to the production of OH.[28] They found that in the configuration of their instrument during previous field campaigns, alkene- and aromatic-based peroxy radicals from the OH-initiated oxidation of isoprene, ethene, propene, and benzene were detected with greater than 79% efficiency, while peroxy radicals from the OH-initiated oxidation of small alkanes such as methane and ethane were detected with less than 10% efficiency.[28] The high conversion efficiency of alkene-based peroxy radicals is due to the ability of the β-hydroxyalkoxy radicals formed in reaction (7.7) to rapidly decompose forming a hydroxyalkyl radical which then reacts rapidly with O_2 leading to the production of a carbonyl compound and HO_2. The lifetime of the β-hydroxyalkoxy radicals is much shorter than the reaction time inside the LIF-FAGE detection cell compared to the longer lifetime of alkoxy radicals relative to reaction with O_2 which limits the conversion of small alkane-based peroxy radicals to HO_2.[28] However, similar to that found for the conversion of RO_2 radicals to HO_2 in the PerCIMS instrument, the conversion efficiency of alkane-based peroxy radicals increases with the size of the alkane due to increased rates of alkoxy radical isomerization in these compounds, leading to more rapid production of HO_2.[51]

This interference can be minimized by either reducing the reaction time in the FAGE detection cell, or by reducing the concentration of added NO.[28] Similar results have been reported for other LIF-FAGE instruments, and measurements of HO_2 where this interference has not been minimized are now often reported as $HO_2^* = HO_2 + \Sigma(\alpha_i RO_{2i})$ where α_i is the instrument's detection efficiency of peroxy radical RO_{2i}.[13,27,81,82] Although it is likely that this interference impacted the measurements of HO_2 by the different LIF groups during the ambient measurement portion of the HOxCOMP campaign, the RO_2 interference probably affected the measurements by the different instruments in a similar way and would be indistinguishable from calibration errors.[28] However, interferences associated with NO_3 radicals leading to internal OH generation could explain part of the systematic differences between instruments observed inside the chamber during dark conditions.[73,79]

In contrast to the interferences associated with OH by LIF instruments, interferences associated with the chemical conversion of RO_2 to HO_2 radicals may be less sensitive to the specific characteristics of each instrument, as the level of the interference depends primarily on the reaction time and concentration of NO. Maximizing the HO_2 to OH conversion efficiency through changes in reaction time and the concentration of added NO will also maximize the conversion efficiency of RO_2 to HO_2 conversion, especially for the conversion of alkene-based β-hydroxyalkoxy radicals to HO_2. This chemistry can also impact measurements of HO_2 by the PerCIMS instrument.[51,83] As a result of these interferences, past measurements of HO_2 and the conclusions drawn from these measurements will need to be re-evaluated, as it is likely that the actual HO_2 concentrations in these studies are lower than the reported measurements. This is especially true for past measurements of HO_2 radicals in environments influenced by BVOCs as the detection efficiency of RO_2 radicals from the oxidation isoprene and other biogenic alkenes is high.

7.4. Comparison of Measured Concentrations of OH and HO_2 with Models

Several previous reviews have discussed the measured and modeled OH and HO_2 concentrations for a number of measurement campaigns, including those in the remote marine boundary layer,[1,4] in polluted urban environments,[1,4] and in low NO_x forested environments.[1,3,4] In general for the remote marine boundary layer, the measured OH and HO_2 concentrations agree with modeled concentrations to within their combined uncertainties when a comprehensive suite of measurements are available to constrain the model.[4] In polluted urban environments, some measurements of OH radical concentrations were up to a factor of two lower than modeled concentrations in some studies,[84,85] while others have reported observed concentrations that were up to a factor of three greater than model predictions.[9,60,81,86] However, many studies have shown good agreement between observed and modeled OH radical concentrations to within their combined uncertainties.[6,8,56,58,87–91] Measurements of HO_2 concentrations in urban environments have been both greater than model predictions in some studies,[56,58,60,86,87,89,90] less than model predictions in others,[84,85] as well as in good agreement with model predictions in some cases.[6,8,9,81,88,91] Most of these measurements in polluted urban areas were performed using the LIF-FAGE technique, and given the potential interferences associated with measurements of OH and HO_2 described above it is not clear to what extent these interferences may have impacted the measurements in these environments.

In general measurements of OH in urban environments have been shown to be in good agreement with model predictions. To allow a simple comparison between different measurements with large differences in ambient conditions, Rohrer *et al.* used a coordinate system that normalizes the observed OH concentrations to the maximum modeled concentration as a function of nitrogen oxide concentrations (OH/OH_{max}, Fig. 7.5).[2] Using this coordinate system, Rohrer *et al.* demonstrated that for campaigns where total OH reactivity was greater than $10\,s^{-1}$ the measured OH concentrations were in good agreement with models in areas where the NO_2 concentrations were near or greater than the value that produces the maximum modeled OH concentration ($NO_2(OH_{max})$) (Fig. 7.5).[2] These high NO_x areas corresponded to measurements in polluted urban environments. However, measurements in areas where the NO_2 concentrations were much less than $NO_2(OH_{max})$,the measured OH concentrations were much greater than model predictions. Most of these measurements occurred near the surface in regions impacted by high mixing ratios of BVOCs such as isoprene, suggesting that OH concentrations are less suppressed by high concentrations of VOCs than predicted by current models.[2]

7.4.1. *OH Measurements in Low NO_x Forested Environments*

Table 7.1 summarizes measurement-model ratios for several campaigns in forest environments characterized by low mixing ratios

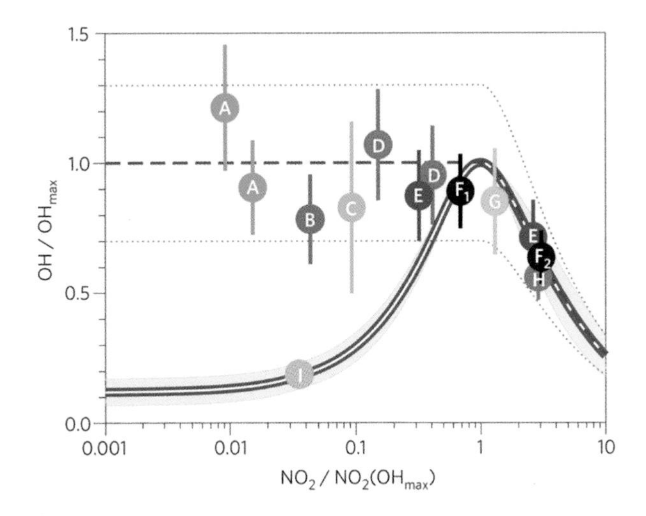

Fig. 7.5 The ratio of measured OH to the maximum modeled OH concentration (OH/OH_{max}) plotted as a function of the ratio of measured NO_2 to the concentration of NO_2 that produces the maximum modeled OH concentration ($NO_2/NO_2(OH_{max})$) for OH measurements in areas where OH reactivity was greater than $10\,s^{-1}$. (A) GABRIEL (Amazonia)[78,92]; (B) OP3 (Borneo)[93]; (C) PROPHET 1998 (N. Michigan)[94]; (D) PRIDE (Pearl River Delta)[64,81]; (E) CAREBeijing[9]; (F_1) MCMA-2003 (Mexico City)[61]; (F_2) MCMA-2006 (Mexico City)[8]; (G) IMPACT-L (Tokyo)[91]; (H) PMTACS-NY 2001 (New York)[57]; (I) BEARPEX 2009 (California)[67] (Reproduced from Ref. [2]).

of NO and high mixing ratios of BVOCs and in particular high mixing ratios of isoprene. For example, daytime measurements of OH radicals in a northern Michigan forest during the Program for Research on Oxidants: Photochemistry, Emissions and Transport (PROPHET) in 1998 by the Penn State LIF instrument were 2.7 times greater than that predicted by a constrained photochemical model (point C in Fig. 7.5).[94] Similar results were obtained by the Leeds LIF instrument in a forested environment in Greece during the Aerosol formation from Biogenic organic Carbon (AEROBIC) campaign in 1997 where the measured OH was approximately 2–6 times greater than model predictions.[95] Aircraft measurements by the Penn State LIF instrument during the Intercontinental Chemical Transport Experiment-A (INTEX-A) campaign found daytime boundary layer observed-to-modeled OH ratios of 1.5 at

Table 7.1 Measurement-model ratios for OH and HO_2 in low NO forested environments

Campaign	Method	Isoprene (ppbv)	NO (pptv)	Measurement/Model OH	Measurement/Model HO_2
AEROBIC (1997)[95]	LIF	2	60–100	1.7–6.2	0.1–1.7
PROPHET (1998)[94]	LIF	0–6.9	30–150	2.7	0.85
INTEX-A (2004)[78]	LIF	0.1–2.5	<20–2000	1–5	1.3
GABRIEL (2005)[78,92]	LIF	0.5–6	20	8–12	3–4
OP3 (2008)[93,96]	LIF	0–3	20–200	2–7	0.6–1.5
BEARPEX (2009)[67]	LIF	1.7; MBO 3.0	70	1.4	0.6
CABINEX (2009)[13]	LIF	1–2	30–120	0.9	0.64
BEACHON-ROCS (2010)[11]	CIMS	MBO 0.5–1.5	100	4	8
HUMPPA-COPEC (2010)[12]	LIF, CIMS	<0.2	50–100	0.96	3.3

isoprene mixing ratios of 500 pptv that increased to ratios greater than 4 with increasing isoprene mixing ratios.[78] Measurements of OH in the boundary layer over the Amazon during the Guyanas Atmosphere–Biosphere Exchange and Radicals Intensive Experiment with the Learjet (GABRIEL) campaign by the Max–Planck LIF instrument were 8–12 times greater than predicted by several photochemical models (point A in Fig. 7.5).[78,92] Similarly, measurements of OH radical concentrations by the Leeds LIF instrument over the tropics during the Oxidants and Particle Photochemical Processes (OP3) campaign in Borneo were approximately a factor of 2–7 times greater than model predictions (point B in Fig. 7.5).[93,96]

However, recent measurements of OH by the Penn State LIF instrument in a ponderosa pine forest during the Biosphere Effects on Aerosols and Photochemistry Experiment (BEARPEX) 2009 campaign were in good agreement with model predictions when interferences measured using the external chemical scavenging technique described above were subtracted from their ambient measurements (point I in Fig. 7.5).[67] Similar results were observed by the Max–Planck-Institute LIF instrument during the HUMPPA-COPEC 2010 study, where measurements of OH in a boreal forest using the external scrubbing technique were in good agreement with a model incorporating an expanded chemical mechanism for terpene oxidation.[12] The OH measurements by the Max–Planck LIF instrument in this study were in good agreement with measurements by the German Weather Service CIMS instrument, suggesting that the chemical scavenging technique was accurately accounting for interferences in the LIF instrument.[12,30] Similar results were reported during the Community Atmosphere–Biosphere Interactions Experiment (CABINEX) 2009, where measurements of OH using the Indiana University LIF instrument in a Michigan forest were in good agreement with model predictions.[13] Although a chemical scavenger technique was not used during this study, subsequent measurements at this site using the chemical scavenger technique did not reveal any unknown interference.

These results suggest that our understanding of atmospheric chemistry in low NO_x and high BVOC environments may be better than previously believed. These results may also suggest that some LIF measurements of OH using the traditional spectral modulation technique without the use of a chemical scavenger were overestimated, and bring into question previous measurements by LIF instruments.[67] For the BEARPEX 2009 measurements, the apparent OH concentration using spectral modulation was approximately a factor of 3 greater than the model predictions (OH_{wave} in Fig. 7.4),[67] which when normalized to OH_{max} appear to be consistent with measurements from the GABRIEL, OP3, and PROPHET campaigns (points A–C in Fig. 7.5), while the measurements using the chemical scavenger (OH_{chem} in Fig. 7.4) are in contrast to these measurements (point I in Fig. 7.5). Similar results would be seen from the HUMPPA-COPEC 2010 measurements, where most of the apparent OH measured using spectral modulation alone was due to an interference.[30]

It is not clear that all LIF instruments suffer from interferences. Although a chemical scavenger was not used, the high measured concentrations of OH measured by the Leeds LIF instrument during the OP3 campaign (point B in Fig. 7.5) were consistent with measurements of formaldehyde and glyoxal using long-path differential optical absorption spectroscopy, suggesting that the OH concentrations measured in this environment were not the result of an instrument artifact.[97] Measurements of the ratio of isoprene oxidation products methacrolein (MACR) and methyl vinyl ketone (MVK) to isoprene in the central Amazon Basin suggested a high oxidation capacity, consistent with OH concentrations an order of magnitude higher than model predictions.[98] And as discussed above measurements of OH using the chemical scavenger technique in a northern Michigan forest during and after the CABINEX campaign did not reveal any unknown interference, although these measurements were done under uncharacteristically cold conditions, with mixing ratios of ozone and biogenic VOCs that were generally lower than in the forested areas where interferences have been previously observed.[13]

In addition to the LIF measurements discussed above, measurements of OH by the NCAR CIMS instrument during the Bio-hydro-atmosphere interactions of Energy, Aerosols, Carbon, H_2O, Organics, and Nitrogen-Rocky Mountain Organic Carbon Study (BEACHON-ROCS) 2010 campaign in a Colorado Forest were approximately a factor of 4 greater than model predictions, suggesting that model underestimations of OH are not confined to LIF measurements.[11] However, the modeled OH was found to be in better agreement with the measurements when the model was constrained to the measured HO_2 concentrations, suggesting that the reason for the discrepancy with the OH measurements is an underestimation of the measured HO_2 concentrations by the model. Constraining the model to the measured OH did not significantly improve the model agreement for HO_2, implying that there are unknown sources of HO_2 in this environment missing from the model mechanism.

7.4.2. HO_2 Measurements in Low NO_x Forested Environments

In contrast to measurements of OH, measurements of peroxy radicals in forested environments have often shown better agreement with model predictions, although the agreement is highly variable (Table 7.1). For example, measurements of HO_2 radicals during PROPHET 1998 using chemical conversion to OH and subsequent detection of OH by LIF were in good agreement with model predictions,[94] while measurements during AEROBIC resulted in observed-to-modeled ratios in the range 0.1–1.7.[95] However, it is not clear whether these previous measurements were free from interferences from organic peroxy radicals. As discussed in Sec. 7.3.6, the hydroxyl alkyl peroxy radicals from the OH-initiated oxidation of biogenic alkenes can be efficiently detected by instruments using chemical conversion of HO_2 to OH through the addition of NO.[28] As a result it is likely that previous HO_2 measurements in forested environments reflected the concentration of both HO_2 and a fraction of RO_2 radicals derived from the oxidation of BVOCs such as isoprene (HO_2^*).

Taking these interferences into account, measurements of peroxy radicals during the BEARPEX and CABINEX campaigns were generally lower than model predictions.[13,67] The agreement with the measured HO_2 concentration by the Penn State LIF instrument during BEARPEX was improved when the model incorporated new mechanisms for the formation of isoprene nitrates and isoprene epoxides into the base model.[99,100] Including isomerization of isoprene peroxy radicals in the mechanism (see Sec. 7.4.3) led to an overestimation of the measured HO_2 concentrations similar to the base model, which could be attributed to the absence of reactive aerosol uptake of peroxy radicals in the model.[67] Similar peroxy radical loss mechanisms could also explain the model overestimation of HO_2^* during CABINEX.[13]

In contrast to these measurements in forest environments with high mixing ratios of isoprene, measurements of HO_2 radicals by the Max–Planck LIF instrument during the HUMPPA-COPEC campaign were 3 times greater than model predictions.[12] Isoprene was not the dominant BVOC in this boreal forest environment, contributing less than 10% to the total OH reactivity. The greatest discrepancy between the measured and modeled HO_2 concentrations occurred when the measured reactivity was the highest, but constraining the model to the observed OH reactivity did not improve the agreement with the measured HO_2 concentrations due to the rapid OH radical recycling in this environment. The authors concluded that a significant HO_2 source related the observed OH reactivity was missing from the model.[12]

Similar results were obtained in a forest with low mixing ratios of isoprene, and high mixing ratios of 2-methyl-3-butene-2-ol (MBO) and monoterpenes (MT) during BEACHON-ROCS, where measurements of HO_2^* and total peroxy radicals ($HO_2 + RO_2$) by the NCAR/University of Colorado CIMS instrument were underestimated by a model incorporating MBO and MT oxidation chemistry by as much as a factor of 3 when the model was constrained to the measured OH concentration.[83] When unconstrained to the measured OH, the model underpredicted the concentration of all radicals, with the underprediction of OH likely due to the underprediction of peroxy radicals as the model is able to reproduce the measured OH when constrained to the measured HO_2^*. The greatest discrepancy occurred in the model underestimation of HO_2^* at midday and displayed a strong dependence on radiation, suggesting that the model was missing a photolytic source of HO_2 radicals. The model tended to underestimate RO_2 radicals throughout the day, suggesting a missing non-photolytic source of RO_2 radicals, such as from the ozonolysis of reactive VOCs. Including the ozonolysis of an unmeasured very reactive VOC (similar in reactivity to β-caryophyllene)[101] in the model improved the agreement with measurements, although the nature of the missing photolytic source of HO_2 remained unidentified.[83]

7.4.3. *Sources of Radicals in Forested Environments*

The model underestimation of observed OH radical concentrations in forested areas has led to a reexamination of radical sources in these low NO_x environments. The discrepancy appears to depend on the mixing ratio of isoprene, suggesting that the reason for the model underestimation is related to our understanding of HO_x sources in low NO_x forested environments. As mentioned above, aircraft measurements during the INTEX-A campaign found daytime boundary layer observed-to-modeled OH ratios of approximately 1 at isoprene mixing ratios less than 100 pptv that increased to ratios greater than 4 at isoprene mixing ratios greater than 2 ppbv.[78] Similar results were observed during the GABRIEL campaign,[92] and this trend was found to be consistent with measurements from several studies, including measurements of OH by the Jülich LIF instrument during the Program of Regional Integrated Experiments of Air Quality over the Pearl River Delta (PRIDE-PRD2006) campaign for measurements when NO was less than 500 pptv, in addition to the results from the PROPHET, and OP3 campaigns (Fig. 7.6).[81] As discussed above, the measurements during both the BEARPEX campaign (average isoprene mixing ratios of 1.7 ppbv) where instrument interferences were measured and taken into account,[67] as well as the CABINEX campaign (average isoprene mixing ratios between 1 and 2 ppbv)[13] are in contrast to these previous measurements, as the measured OH concentrations were in good agreement with model predictions.

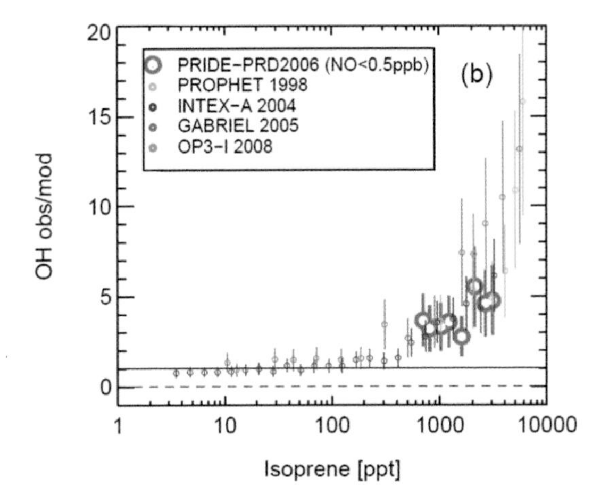

Fig. 7.6 The observed-to-modeled concentration ratio of OH as a function of isoprene for several campaigns in low NO_x forested environments (reproduced from Ref. [81]).

This again raises the question of whether instrument interferences are partly responsible for the disagreement.

For CABINEX, model results indicated that photolysis of ozone followed by reaction with water vapor (reactions (7.1) and (7.2)) was the main source of radicals in this environment, accounting for approximately 30% of the daytime average rate of radical initiation.[13] Alkene ozonolysis (reaction (7.21)) and photolysis of HONO (reaction (7.18)) were also predicted to be significant radical sources, accounting for 27 and 17% of the daytime average rate of radical initiation, respectively, while photolysis of formaldehyde and other aldehydes (reaction (7.20)) only contributed 5% to radical initiation during this campaign. However, the HCHO contribution to radical production at this site was found to be higher the previous year due to higher measured HCHO concentrations, perhaps the result of warmer temperatures and greater isoprene concentrations.[13] A similar radical budget for this site was predicted for the conditions during PROPHET 1998.[94] The high contribution of HONO to radical initiation in this low NO_x forest environment is due to the high observed concentrations of HONO at this site due to a surface canopy source of HONO that correlated with leaf-surface nitrate loading and the rate constant of nitrate photolysis.[102]

To explain the higher than expected measured concentrations of OH in forested environments, several new isoprene oxidation mechanisms were proposed to efficiently recycle HO_x radicals. Some of the proposed mechanisms involved the production of OH from reactions previously thought to terminate HO_x radicals, such as reactions of HO_2 with RO_2 radicals (reaction (7.16)),[78,103] the reformation of OH during the formation of isoprene-based dihydroxyepoxides,[100] an unknown species that efficiently recycles OH independent of NO,[64] or from the isomerization and decomposition of isoprene-based peroxy radicals leading to photolabile species that lead to additional HO_x.[104–107] Including these proposed recycling reactions could explain some of the observed discrepancies between measured and modeled OH concentrations.[93,108] If correct, these proposed mechanisms would imply that forest emissions, and in particular isoprene emissions, are not as significant a sink of OH radicals on the global scale as previously believed.[109] Several experimental and theoretical studies have recently addressed these potential HO_x radical recycling mechanisms.

7.5. Laboratory Studies of HO_x Radical Recycling Mechanisms

Regardless of whether the measurements of OH radicals in low NO_x forested environments described above are influenced by instrument artefacts, the observed discrepancies have stimulated a reexamination of BVOC oxidation under low NO_x conditions, and in particular the oxidation of isoprene. This section will include a summary of recent laboratory measurements of several HO_x radical recycling mechanisms in the oxidation of BVOCs. Whalley *et al.*[3] have recently provided an extensive review of advances in our understanding of the mechanism of isoprene oxidation and its impact on HO_x radical chemistry, which will only be briefly discussed here.

7.5.1. *Radical Cycling in $HO_2 + RO_2$ Reactions*

Based on evidence from laboratory and field studies, Leileveld *et al.* proposed that reactions of isoprene-based peroxy radicals with HO_2 may have a significant product channel that leads to HO_x radical propagation rather than termination.[78] For example, measurements of the product yields from the reaction of HO_2 radicals with $C_2F_5C(O)O_2$ suggested that OH radicals are produced with a yield of (0.76 ± 0.04).[110] Similarly, measurements of the product yields from the reaction of acetyl peroxy radicals ($CH_3C(O)O_2$) with HO_2 suggested that CH_3OOH is produced with significant yield through a channel that leads to the formation of $CH_3C(O)O$ and OH with a yield of (0.43 ± 0.10).[111,112] In addition, OH radicals were directly detected as a product of this reaction with a yield of (0.5 ± 0.2) using pulsed laser photolysis with laser-induced fluorescence detection of OH radicals.[113] Recent direct measurements of the OH radical yield from this reaction found a higher OH radical yield of (0.61 ± 0.09).[114] Similar OH radical yields have been obtained for the reaction of HO_2 with other carbonyl containing peroxy radicals.[113,115]

Including OH radical recycling in the HO_2 + isoprene RO_2 reaction could help explain the high measured OH concentration discussed above only if a large recycling efficiency was assumed. Several studies were able to demonstrate improved agreement with the measurements of OH during the GABRIEL campaign assuming an OH recycling efficiency of 40–80%, corresponding to an additional source of 2–3 OH radicals for every molecule of isoprene oxidized.[78,116] However, the improved agreement with OH results in unrealistically low mixing ratios of isoprene in the model.[116]

There is some evidence from field campaigns that suggest that the reaction of organic peroxy radicals, and in particular isoprene-based peroxy radicals, may efficiently recycle HO_x radicals. Thornton *et al.* used a chemical coordinate analysis of radical production and loss during the Southern Oxidant Study (SOS) to suggest that the rate constant for the reaction of organic peroxy radicals with HO_2 leading to the formation of organic peroxides was overestimated by a factor of 3–12.[103] Because isoprene-based peroxy radicals were estimated to account for approximately 40% of organic peroxy radicals at this site, the authors suggested that models were overestimating the rate constant of the radical terminating channel for the reaction of isoprene-based peroxy radicals with HO_2, which would be consistent with a channel from this reaction that significantly recycles OH radicals. It should be noted, however, that this analysis assumed that the measured HO_2 concentrations from the Penn State LIF instrument were free from interferences. Given that isoprene-based peroxy radicals were a significant fraction of peroxy radicals at this site and that the β-hydroxy peroxy radicals from isoprene can be efficiently converted to HO_2, it is likely that the measured HO_2 concentrations reflected $HO_2^*(= HO_2 + \alpha RO_2)$. Accounting for this interference would likely change the conclusion of the chemical coordinate analysis in this study.

In contrast to measurements of carbonyl containing peroxy radicals, OH recycling in reactions of HO_2 with peroxy radicals where the carbonyl functionality is absent was found to be minimal.[113] Theoretical studies of the reaction of HO_2 with $CH_3C(O)O_2$ radicals found that production of OH radicals was favorable through the formation of a hydrotetroxide complex, while this pathway was found to be endothermic for the reaction of HO_2 with $CH_3CH_2O_2$ and $CH_3C(O)CH_2O_2$ radicals.[117] Formation of OH radicals through the organic hydrotetroxide complex was found to be more favorable for the acetyl peroxy radical due to a lowering of the energy of the intermediates and transition states as a result of internal hydrogen bonding with the carbonyl group.[117] For the reaction of isoprene-based peroxy radicals with HO_2, Dillon and Crowley reported no direct production of OH,[113] consistent with the results of Paulot *et al.*, who found that formation of isoprene hydroxyhydroperoxide was the dominant product from this reaction.[100] However, experimental studies by Paulot *et al.* revealed that further oxidation of isoprene hydroxyhydroperoxide by OH radicals leads to the formation of isoprene dihydroxyepoxides accompanied by OH reformation.[100]

Measurement of the product yields in the Cl-initiated oxidation of MACR and MVK suggest that the yield of OH radicals from the $HO_2 + RO_2$ reaction for these carbonyl species can be large.[115,118] Given the structural similarity of the peroxy radicals produced in the Cl-initiated oxidation of MACR and MVK it is likely that the yield of OH radicals for the $HO_2 + RO_2$ reaction for peroxy radicals from the OH-initiated oxidation of MACR and MVK would be similar.[115] Experimental measurements of the product yields from the oxidation of MVK oxidation under conditions where the $HO_2 + RO_2$ reaction dominates the fate of peroxy radicals suggest that OH radicals are efficiently recycled through several channels.[119] The detection of large yields of glycolaldehyde with peracetic acid and acetic acid (produced from the reaction of $CH_3C(O)O_2$ radicals with HO_2) suggests that OH is efficiently recycled through one channel, while detection of a C4 α-diketone suggests an additional OH radical recycling channel through a tetroxide intermediate leading to the formation of OH and HO_2.[119] In addition, the hydroxy hydroperoxide product from the reaction was found to rapidly photolyze, potentially leading to additional OH radical recycling. Incorporating these HO_x radical recycling reactions into atmospheric models increased the modeled OH concentration by approximately 6% in localized areas of high isoprene concentrations.[119] Although these OH recycling reactions can contribute to higher levels of OH radicals in models, they alone cannot explain the high measured concentrations of OH in high isoprene, low NO_x environments.

7.5.2. *HO_x Production from Peroxy Radical Isomerization Reactions*

To explain the high observed OH concentrations in isoprene environments, Peeters *et al.* proposed that some peroxy radicals produced in the OH-initiated oxidation of isoprene could isomerize leading to the formation of hydroxyperoxyaldehydes and HO_2 radicals (Fig. 7.7). Specifically, a 1,6-H-shift isomerization of the Z-conformers of the δ-hydroxyperoxy radicals produced from OH addition to the 1

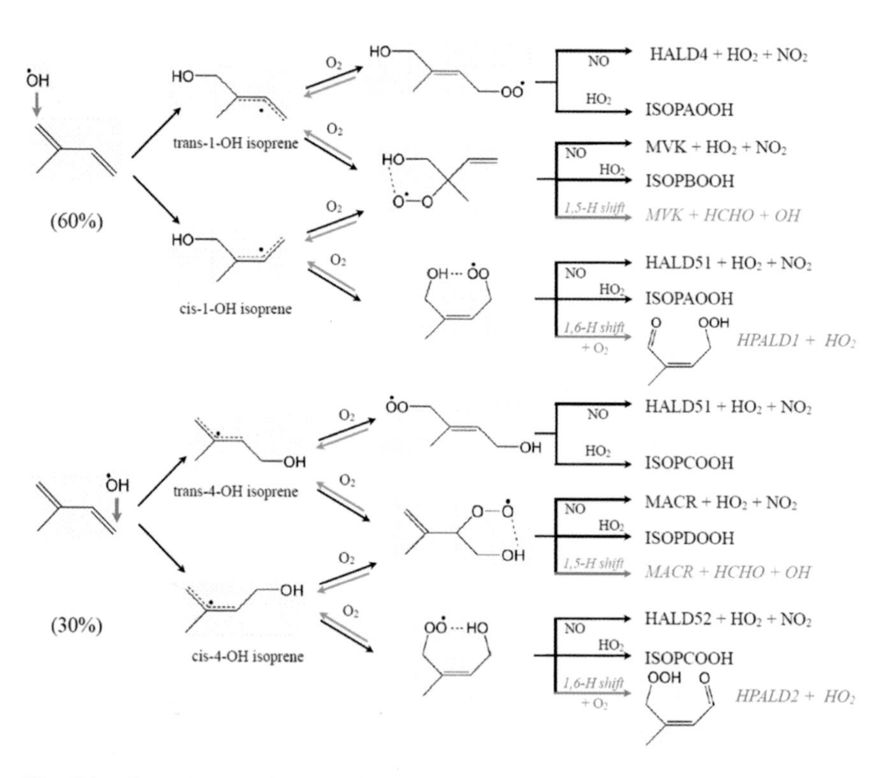

Fig. 7.7 Reaction mechanism for the OH addition to isoprene, including the 1,6 H shift and 1,5-H-shift isomerization channels (reproduced from Ref. [108]).

and 4 carbons of isoprene can lead to the production of HO_2 radicals and C5-hydroperoxy aldehydes (HPALDs).[104,105] Subsequent photolysis of the HPALDs produced from these isomerization reactions could lead to additional HO_x radical production, including the production of peroxyacid aldehydes (PACALDs), which upon photolysis could produce an additional 1–3 OH molecules (Leuven Isoprene Mechanism, LIM).[105] This isoprene oxidation mechanism could lead to an amplification of HO_x radicals and a significant increase in modeled OH concentrations.[104,108] A similar isomerization mechanism involving a 1,5-H shift of the β-hydroxyperoxy radicals leading to the formation OH radicals was also proposed (Fig. 7.7).[106]

Including these isomerization reactions using theoretical isomerization rates and assuming approximately 1–3 OH radicals produced from the photolysis of HPALDs resulted in modeled OH concentrations that were a factor of 4 times higher compared to the base mechanism and were in better agreement with the measured OH concentrations during the GABRIEL campaign.[108] The modeled HO_2 concentrations in this simulation were also found to be in better agreement with the measurements, although the modeled isoprene concentrations were significantly reduced.[108] Increasing the isoprene emission rate by 50% brought the modeled isoprene concentrations into better agreement with the measurements.[108] Including a similar recycling mechanism improved the modeled agreement with the measured OH during the OP3 campaign, although the improved agreement with OH came at the expense of the agreement with the HO_2 measurements.[93,120] It was concluded that none of the newly proposed mechanisms could simultaneously improve the modeled agreement with measurements of both OH and HO_2 when the model was constrained to the observed mixing ratios of isoprene and other measured species.[120]

Laboratory measurements of the formation of HPALDs from the isoprene oxidation mechanism by Crounse *et al.* confirmed the importance of the 1,6-H-shift isomerization pathway.[121] However, these measurements also found that the rate of production of HPALDs from two of the six peroxy radical isomers produced in the

OH-initiated oxidation isoprene was a factor of 2–50 times slower than theoretical predictions. Higher level theoretical calculations have improved the agreement between the theoretical 1,6-H shift isomerization rates and the experimental values to within their combined uncertainties.[122] In addition, experimental measurements of the OH radical yield from the photolysis of a proxy for isoprene-derived HPALDs suggests that the photolysis of these compounds can lead to significant OH radical recycling consistent with the theoretical estimates.[123] Including these rates of HPALD production and photolysis parameters in models increased calculated OH concentrations by 5–16%.[123]

Further experimental evidence that isoprene oxidation can efficiently recycle OH radicals was demonstrated by Fuchs *et al.* through measurements of OH radicals during the oxidation of isoprene in the SAPHIR chamber.[124] In these experiments, OH radicals were measured by the Jülich LIF-FAGE instrument and by the DOAS instrument inside the chamber. The measurements of OH by the two instruments were in good agreement, suggesting that interferences associated with the LIF-FAGE instrument were negligible. In the presence of CO the measured OH concentrations could be explained by model predictions, but when isoprene was added to the chamber the measured OH concentrations were under-predicted by a model employing the Master Chemical Mechanism (MCM v. 3.2).[124] The model could be brought into agreement with the measurements by including the 1,6-H shift isomerization of isoprene peroxy radicals leading to the formation of HO_2 and HPALDs, with subsequent photolysis of HPALDs and PACALDs leading to the formation of multiple OH radicals. The isomerization rate that provided the best agreement with both the measured OH and HO_2 radicals was consistent with the measurements of Crounse *et al.*,[121] although the best agreement was achieved with a rate at the upper end of the measurement uncertainty.[124] These results suggest that under low NO_x conditions typical of the atmosphere over the Amazonian rainforest the peroxy radical isomerization mechanism could increase the modeled OH concentrations by up to a factor of 2, which could explain part of the discrepancy between the measured and modeled OH concentrations in this environment.[124]

Additional experimental evidence has suggested that isomerization of peroxy radicals leading to recycling of OH radicals may be important in the atmospheric oxidation of MACR, a first generation product of isoprene oxidation. Crounse *et al.* measured ratio of methacrolein hydroxy nitrate to hydroxyacetone, two products produced from the reaction of NO with peroxy radicals formed by the addition of OH and O_2 to the olefinic carbon atoms of MACR. They found that the ratio decreases as the lifetime of the peroxy radical increases and attributed the change to a 1,4 H-shift of the aldehydic hydrogen atom to the peroxy group leading to the formation of a hydroxy hydroperoxy carbonyl radical. This radical decomposes rapidly, producing hydroxyacetone and re-forming OH.[125] Accounting for the yield of MACR from isoprene oxidation and the fraction of MACR oxidation that occurs through OH addition rather than H-atom abstraction, these authors estimate that this isomerization chemistry could regenerate approximately 4% OH for each isoprene molecule oxidized.[125] Similar experiments examining a proposed 1,5 H-shift isomerization mechanism for peroxy radicals produced from the oxidation of MVK[104,126] suggest that these isomerization reactions were not significant under the conditions of the experiment.[119]

Chamber studies of the oxidation of methacrolein provide additional evidence of OH radical recycling.[127] Similar to the results from the oxidation of isoprene, measurements of OH in the SAPHIR chamber during the oxidation of MACR were 50% greater than predicted by the MCM. In these experiments, measurements of OH by the Jülich LIF-FAGE instrument were in good agreement with measurements by the DOAS instrument, again suggesting that interferences with the LIF-FAGE instrument were negligible.[127] Including peroxy radical isomerization through the 1,4-H shift described above improves the modeled agreement with the measurements of OH. Additional isomerization mechanisms through proposed 1,5-H shift and decomposition of double-activated peroxy radicals[126,128] could not explain the observed OH concentrations in these experiments.[127]

Given the smaller reaction rate constant for the reaction of OH with MACR, this contribution of additional OH from MACR oxidation cannot explain the discrepancy between modeled and measured OH concentrations in low NO_x forested environments.[127] However, these results provide additional evidence that the atmospheric oxidation of isoprene is less oxidant consuming than expected.[125]

Although there is evidence that similar peroxy radical isomerization reactions in highly oxidized molecules may lead to the formation of extremely low volatile organic compounds (ELVOC) through autoxidation,[129] these isomerization reactions in the atmospheric oxidation of isoprene may not be able to account for the discrepancy in HO_x levels between observations and model simulations in high isoprene, low NO_x environments.[121] If the measured OH concentrations in these environments are correct they imply that either a significant source of radicals or an unknown radical cycling reaction is still missing from current models of atmospheric chemistry.[2] As illustrated in Eq. (7.25), measurements of total OH radical reactivity can be used to help identify these missing OH radical sources.

7.6. Measuring Total OH Reactivity in the Troposphere

This section includes a description of OH reactivity techniques, including intercomparisons of instruments, a review of their deployment during field campaigns, and a summary of recent laboratory measurements using OH reactivity measurements to better constrain the OH radical budget during kinetic experiments.

7.6.1. *Rationale Underlying OH Reactivity Measurements*

As described above, the total loss rate of OH, k_{OH}, is a sum of pseudo-first-order reaction rates of this radical with all the reactive compounds in ambient air, including VOCs, CO, NO_x, and other inorganic compounds (ICs) such as ozone and hydrogen peroxide. k_{OH} can be calculated as shown in Eq. (7.33), with the reactions included in the first three terms (VOCs, CO, ICs) leading to the propagation of OH to peroxy radicals and the reactions included in the last two terms (NO_x) leading to radical termination.

$$k_{OH} = \sum_i k_{OH+VOC_i}[VOC_i] + k_{OH+CO}[CO] + \sum_j k_{OH+IC_j}[IC_j] + k_{OH+NO}[NO] + k_{OH+NO_2}[NO_2] \quad (7.33)$$

The VOC reactivity (VOC_R) is defined as the first term on the right hand side of Eq. (7.33) and the NO_x reactivity ($NO_{x,R}$) as the sum of the last two terms. It is interesting to note that during field campaigns involving chromatographic and mass spectrometric techniques, the number of VOCs measured is usually on the order of several tens of compounds, which is at least one order of magnitude lower than expected.[130] In addition, the measured VOCs are usually C2-C12 compounds as higher molecular mass species evade detection. VOC_R will likely be incompletely characterized during field campaigns and it is important to determine whether the unidentified fraction is important for our understanding of atmospheric chemistry.

Direct *in situ* measurements of total OH reactivity provide information on the reactivity of these unidentified VOCs to assess their relevance for atmospheric chemistry. The difference between the k_{OH} measurements and values calculated from trace gas measurements using Eq. (7.33) is commonly called the "missing reactivity," and is used to quantify the unmeasured fraction of VOCs. When the missing reactivity is statistically significant this budget analysis indicates that important OH sinks have not been characterized by trace gas measurements, allowing the contribution of these unidentified sinks to k_{OH} to be assessed.

In addition to an exhaustive apportionment of OH sinks, k_{OH} measurements can also be used to perform a meaningful investigation of OH sources providing that collocated OH measurements are available. Due to the short lifetime of OH ($<1\,s$, $k_{OH} > 1\,s^{-1}$), production and loss rates of this radical are in balance and the total production rate of OH can be calculated from Eq. (7.34). Measurements of known OH precursors, including O_3, HONO, peroxide compounds (ROOH), unsaturated VOCs and HO_2, can also be used together with measurement of photolysis frequencies and NO to calculate the expected production rate of OH as shown in Eq. (7.35):

$$P_{\mathrm{OH}}^{\mathrm{meas}} = L_{\mathrm{OH}} = k_{\mathrm{OH}} \times [\mathrm{OH}] \tag{7.34}$$

$$\begin{aligned} P_{\mathrm{OH}}^{\mathrm{calc}} &= 2 \times f_{\mathrm{OH}} \times J_{\mathrm{O}(1D)}[\mathrm{O_3}] + J_{\mathrm{HONO}}[\mathrm{HONO}] + J_{\mathrm{H_2O_2}}[\mathrm{H_2O_2}] \\ &\quad + J_{\mathrm{ROOH}}[\mathrm{ROOH}] + \sum_i \alpha_i k_{\mathrm{VOC}_i+\mathrm{O_3}}[\mathrm{VOC}][\mathrm{O_3}] \\ &\quad + k_{\mathrm{HO_2+NO}}[H\mathrm{O_2}][\mathrm{NO}] + k_{\mathrm{HO_2+O_3}}[\mathrm{HO_2}][\mathrm{O_3}] \end{aligned} \tag{7.35}$$

Here f_{OH} is the fraction of $O(^1D)$ reacting with water vapor to produce OH and α_i is the OH yield from the ozonolysis reactions of unsaturated VOCs. Comparing $P_{\mathrm{OH}}^{\mathrm{calc}}$ and $P_{\mathrm{OH}}^{\mathrm{meas}}$ provides a critical test to assess whether known OH sources are sufficient to explain the measured production rate of OH. This type of analysis allows a more robust assessment of missing OH sources than the use of model simulations. Indeed, when measured OH concentrations are compared to model-calculated values, a disagreement could be either due to miscalculated OH loss rates, miscalculated production rates, or both. Measurements of k_{OH} provide additional constraints to investigate the atmospheric OH budget. Since collocated measurements of OH and OH reactivity are necessary, only a few laboratory[127,131] and field[9,61,93] studies have relied on this approach.

Collocated measurements of k_{OH} and NO_x can also provide meaningful information to assess whether ozone production is NO_x-limited or NO_x-saturated at the measurement site, which is of particular interest to develop strategies of ozone reduction from control measures on primary anthropogenic emissions. Indeed, Kirchner *et al.*[132] proposed the NO_x-to-VOC reactivity ratio ($\theta = \mathrm{NO}_{x,R}/\mathrm{VOC_R}$) as a new indicator, capable of distinguishing between these two regimes of ozone production, since θ gauges the competition between termination ($\mathrm{OH} + \mathrm{NO}_x$) and propagation ($\mathrm{OH} + \mathrm{VOCs}$) reactions, with the latter being the ozone production pathway. $NO_{x,R}$ and VOC_R are calculated from NO_x measurements and the subtraction between the measured k_{OH} and $NO_{x,R}$, respectively, with the contribution of CO included in VOC_R. The OH reactivity from other inorganic species is assumed to be negligible, which is a reasonable assumption for most areas, including urban, rural and forested environments. Box- and 3D-modeling performed by the authors showed that this indicator is robust to identify the ozone production regime ($\theta < 0.01$: NO_x-limited, $\theta > 0.2$: NO_x-saturated). This approach was successfully used by Sinha *et al.*[133] to identify ozone production regimes during the DOMINO (Diel Oxidant Mechanisms In relation to Nitrogen Oxides) field campaign.

7.6.2. *Description of OH Reactivity Techniques*

Ambient measurements of k_{OH} are currently performed using three different perturbation-based techniques as shown in Table 7.2 and illustrated in Fig. 7.8. Two of these techniques rely on monitoring OH decays in ambient air, i.e. the Total OH loss rate method (TOHLM) and the pump-probe method (LP-LIF), and a third technique is based on monitoring competitive OH-kinetics between a reference compound (Pyrrole) and ambient trace gases, i.e., the Comparative Reactivity Method (CRM).

7.6.2.1. *Total OH Loss Rate Method and the Pump-Probe Method*

TOHLM and LP-LIF instruments are equipped with a FAGE detection axis[29,33] to directly monitor the consumption of OH in a flow of ambient air, whose composition has been perturbed by the addition of a finite amount of OH radicals (see specifics below for each instrument). The initial concentration of OH is kept low enough (less than 10^{10} cm^{-3}) to ensure pseudo first-order kinetics and a negligible impact from OH-radical reactions, which in turn allows fitting the OH decay with a simple mono-exponential equation to retrieve k_{OH}:

$$[OH]_t = [OH]_0 \exp[-(k_{OH} + k_{loss})t] \tag{7.36}$$

Table 7.2 OH reactivity measurement techniques (adapted from Hansen *et al.*[134])

Technique	Institution	LOD (s^{-1})[a] Time res. (min)
Total OH loss rate Method (TOHLM)	Penn State Univ. (USA)[135]	2.4/4
	Univ. of Leeds (UK)[136]	2.0/5
	Indiana Univ. (USA)[137]	2.1/2.5
Pump-probe (LP-LIF)	Tokyo Metropolitan Univ. (Japan)[138]	—[b]/3[c]
	Forschungszentrum Jülich (Germany)[139]	0.9[d]/1–3
	Université Lille – PC2A (France)[140]	3.6–0.9/1–3
	Univ. of Leeds (UK)[141]	1–1.5/1–3
Comparative reactivity method (CRM)	Max Planck Institute Mainz (Germany)[142]	3.5–6[e]/1
	NCAR (USA)[143]	15/–[b]
	Max Planck Institute Mainz (Germany)[144]	3–6[f] / 1
	LSCE (France)[145]	3.0/2[g]
	IISER Mohali (India)[146]	4/1
	Mines Douai (France)[134]	3.6/5

[a]Limit of detection: 3σ unless otherwise stated; [b]value not reported; [c]value reported by Yoshino *et al.*[147]; [d]LOD determined from 3σ on zero-air decays reported by Lou *et al.*[139]; [e]LOD of $6\,s^{-1}$ reported by Sinha *et al.*,[142] LOD of $3.5\,s^{-1}$ reported by Sinha *et al.*[148]; [f]value reported for 2σ, relative to C2; [g]based on frequency measurements reported in Figure 8 of Ref. 145.

The observed pseudo-first-order decay rate of OH depends on the reaction of OH with ambient trace gases (k_{OH}) and an additional first-order loss rate (k_{loss}) due to the transfer of OH outside the probed volume (diffusion, turbulence) and wall losses. Time-resolved OH decays are recorded on a timescale of 20–300 ms as shown in Fig. 7.9 for both the LP-LIF and TOHLM techniques. These decays are fitted to Eq. (7.36) to extract the sum of k_{OH} and k_{loss}. To derive k_{OH}, k_{loss} has to be quantified during zeroing experiments when the instrument samples only zero air.

Three TOHLM instruments have been developed worldwide, two in the US at Penn State University.[135] and Indiana University,[137]

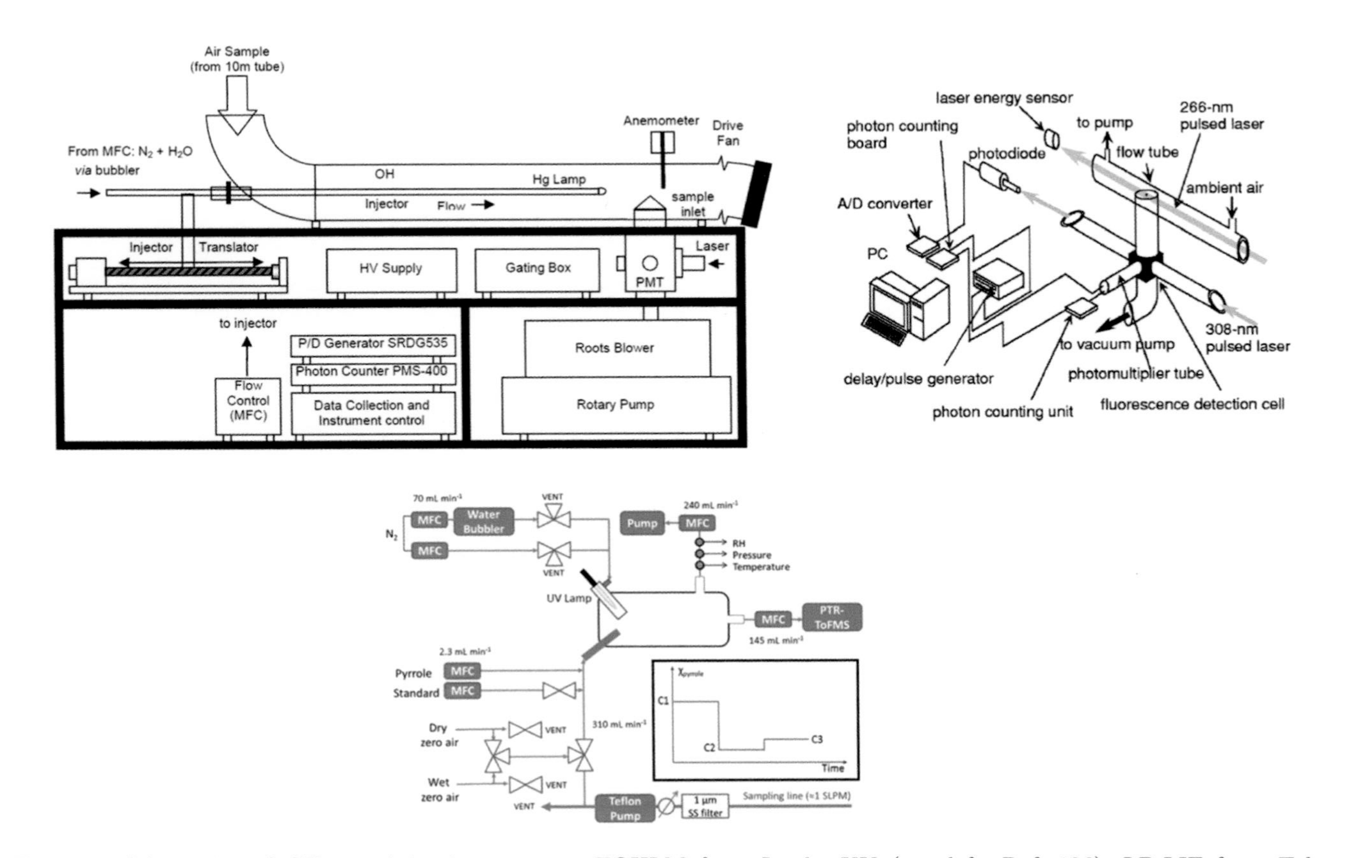

Fig. 7.8 Schematics of OH reactivity instruments: TOHLM from Leeds, UK (top left, Ref. 136), LP-LIF from Tokyo Metropolitan University, Japan (top right, Ref. [138]), CRM from Mines Douai, France (bottom, Ref. [149]).

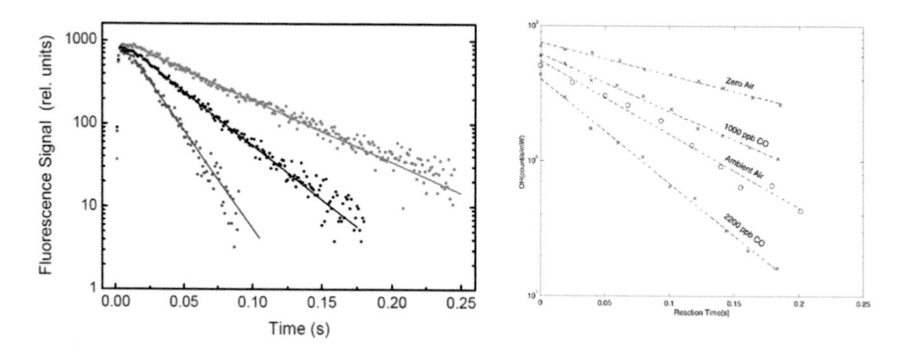

Fig. 7.9 Time-resolved decays of OH for the pump-probe method (left, Ref. [139]) and the total OH loss rate method (right, Ref. [135]). Left panel: first order loss rates of 17, 30, and 54 s^{-1} (top to bottom). A k_{loss} value of 1.4 s^{-1} has to be subtracted to derive k_{OH}. Right panel: $k_{loss} = 5.2\,s^{-1}$ in zero air (top). Additions of 1000 ppbv and 2200 ppbv of CO correspond to k_{OH} values of 5.7 s^{-1} and 12.5 s^{-1}, respectively.

and one in the UK at the University of Leeds.[136] This technique is based on sampling ambient air inside a flow tube (Fig. 7.8), which is operated under laminar[135] or turbulent[136,137] conditions. Sampling flow rates of 60–100 L min^{-1} and 100–400 L min^{-1} are required to develop the laminar and turbulent conditions, respectively.

In this technique, OH radicals, as well as unwanted hydrogen atoms, are produced within a central injector from the photolysis of water-vapor in nitrogen using the 185 nm emission line of a mercury pen-ray lamp (reactions (7.26)–(7.27)). Similar concentrations of OH and H atoms are injected in the main flow, where H atoms are quickly converted into HO_2 by reaction with molecular oxygen from ambient air. Time resolved OH decays are monitored by varying the distance between the tip of the OH injector and the FAGE detection, which is converted in reaction time based on the measured flow velocity. Keeping the distance between the injector and the detection constant allows measuring the OH concentration under steady conditions (constant reaction time) for several seconds to improve the measurement precision. The time needed to record a full OH decay ranges from 2.5 to 5 min depending on the instrument, and leads to limits of detection (3σ) of 1–2 s^{-1} (Table 7.2). Total OH reactivity values as high as $200\,s^{-1}$ have been measured with this technique in polluted areas such as Mexico City,[87] and an airborne version was recently used to measure the total OH reactivity in the troposphere up to an altitude of 12 km.[150]

A main drawback of this technique is the production of HO_2 radicals from water photolysis, which can propagate to OH in the flow tube when NO is present in ambient air. This secondary formation of OH disturbs OH decays and needs to be corrected for when measurements are performed in urban and suburban environments, where NO can be higher than 1 ppb. A correction procedure was proposed by Kovacs *et al.*[135] using simultaneous measurements of ambient NO and HO_2 within the reactor. This correction was tested during laboratory experiments[151] and was successfully used by Ren *et al.*[57] and Shirley *et al.*[87] during subsequent studies. Correction factors of 1.10, 1.48, and 1.97 have been reported for the Penn State instrument at NO mixing ratios of 1, 5 and 10 ppbv, respectively.[57]

Four pump-probe instruments (Fig. 7.8) have been developed at the University of Leeds in the UK,[141] at the Forschungszentrum Jülich in Germany,[139] at the Tokyo Metropolitan University in Japan,[138] and at the University of Lille in France.[140] The measurement principle was first introduced by Calpini *et al.*[152] who provided a proof of concept for local-point measurements, as well as space-resolved LIDAR measurements. Based on this pioneering work, several groups developed LP-LIF instruments designed for local-point measurements, in which ambient air is sampled within a reactor using flow rates of several liters per minute.[138–140] A small amount of ozone is usually added inside the reactor to produce a significant amount of OH from ozone-photolysis at 266 nm. The laser light is provided by a pulsed laser, called "pump laser", whose frequency is set at approximately 1 Hz, which is low enough to ensure that the reactor is sufficiently refreshed between each laser pulse. The FAGE instrument used to record OH decays employs the "probe laser."

In this technique, a full OH decay is recorded after each pump laser pulse, i.e., within a few hundred ms. However, the time resolution of OH reactivity measurements ranges from 10 s to several minutes, depending on the number of OH decays that are averaged together to reach a suitable signal-to-noise ratio. Detection limits (3σ) observed for Pump-probe instruments are in the range 1–4 s^{-1}, and ambient measurements of up to 120 s^{-1} have been reported.[139] There is no report of significant interferences in the literature for this technique. In contrast to TOHLM, HO_2 is not generated with OH and potential artifacts from the propagation of HO_2 into OH in the presence of ambient NO are greatly reduced. For instance, Sadanaga *et al.*[138] reports a systematic error lower than 5% at a NO mixing ratio of 20 ppbv. In addition, the photolysis of ambient species at 266 nm inside the instrument was found to be negligible.[138]

7.6.2.2. *The Comparative Reactivity Method*

Five instruments based on the Comparative Reactivity Method have been developed in Germany,[142] the US,[143] India,[146] and France.[134,145] The CRM was first proposed by Sinha *et al.* as an alternative technique to measure total OH reactivity without probing the OH radical.[142] This technique is based on monitoring competing OH kinetics between a reference molecule (Pyrrole) and ambient trace gases. Less than 400 mL min^{-1} of air are introduced together with a small flow of pyrrole into a sampling reactor (Fig. 7.8), which is equipped with a mercury lamp to produce OH from water photolysis at 185 nm (reactions (7.26)–(7.27)). The principle of the CRM measurement is based on monitoring changes in the pyrrole concentration when switching back and forth between zero air and ambient air sampling.

CRM instruments require three stages of measurements (C1, C2, and C3) of a few minutes each that are summarized in Fig. 7.10. During the C1 stage, only pyrrole, dry zero air and nitrogen are introduced into the reactor to quantify the initial amount of pyrrole available for reaction with OH. This measurement is usually made only once a day. During the second stage (C2), dry zero air and nitrogen are replaced by humid gases (same RH as in ambient air). These wet conditions lead to the formation of OH and HO_2 radicals through water photolysis as discussed above. The fast reaction of pyrrole with OH leads to a decrease of the pyrrole concentration to C2. The third measurement stage (C3) is achieved by replacing the

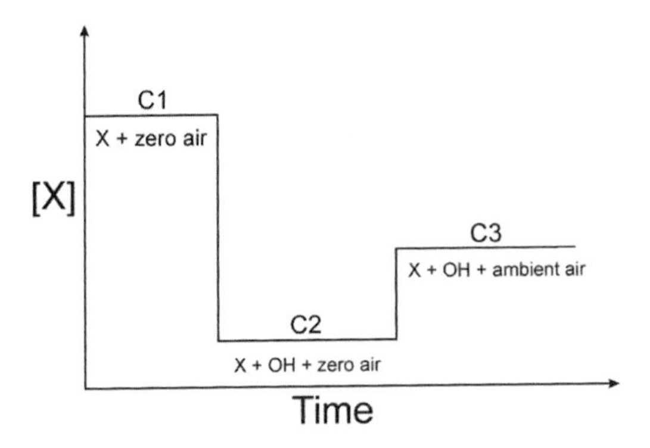

Fig. 7.10 Principle of the CRM. C1: initial concentration of pyrrole available for reaction with OH, C2: concentration of pyrrole after reaction with OH, C3: concentration of pyrrole observed during competing OH-kinetics (pyrrole vs. ambient trace gases, X) (reproduced from Ref. [42]).

humid zero air by ambient air. When switching from zero to ambient air, a fraction of OH that was initially reacting with pyrrole is instead consumed by ambient trace gases and an increase of the pyrrole concentration to C3 is observed. The amplitude of the modulation (C3−C2) depends on the reactivity of ambient trace gases and can be used as shown in Eq. (7.37) to derive the total OH reactivity assuming pseudo first-order kinetics in the reactor[142]:

$$k_{\mathrm{OH}} = \frac{(\mathrm{C3} - \mathrm{C2})}{(\mathrm{C1} - \mathrm{C3})} \times k_p \times \mathrm{C1}. \tag{7.37}$$

In this equation, k_p is the bimolecular rate constant for the reaction of OH with pyrrole ($1.28 \times 10^{-10}\,\mathrm{cm^3\,mol^{-1}\,s^{-1}}$).[153] Recent studies have shown that several measurement artifacts have to be corrected for with this technique.[142,145,149] It was found that a change in humidity between the C2 and C3 stages could lead to a change in OH concentrations within the reactor, which in turn could affect the C2–C3 modulation. Due to the large amount of OH produced inside the reactor, pseudo-first-order conditions are not met, and the use of Eq. (7.37) to calculate k_{OH} could lead to a measurement bias. In addition, similar to the TOHLM technique, HO_2 radicals are also concomitantly produced with OH from water-photolysis. CRM instruments are therefore impacted by the propagation of HO_2 into OH when ambient NO mixing ratios are higher than 1 ppbv. However, correction procedures were recently tested in the laboratory by Michoud *et al.*[149] and in the field by Hansen *et al.*[134] These authors showed that accurate corrections can be made for each of these artifacts. Michoud *et al.* also compared experimental characterizations to model simulations and showed that the chemistry occurring inside the CRM reactor is well understood, providing additional confidence in the CRM technique.[149]

While different types of detectors can be used to monitor pyrrole during field measurements, proton transfer reaction-mass spectrometry (PTR-MS) is commonly used on CRM instruments.[133,134,143,145,146] Only one study investigated the use of a gas chromatography photo-ionization detector (GC-PID) as a less expensive alternative.[144] This study demonstrated that the GC-PID could provide performance similar to a PTR-MS but that potential interferences from coelutions with pyrrole should be investigated. The potential for the use of a wide range of reference compounds and detection techniques is a main advantage of CRM instruments and this aspect should be further investigated to develop compact and low-cost OH reactivity instruments.

Due to the need to switch between zero and ambient air every few minutes, OH reactivity measurements are performed at a time resolution of approximately 2–5 min. Limits of detection (3σ) are currently 2–3 times higher than for the two other techniques discussed above and range from 3 to 4 s^{-1}. CRM instruments have been shown to measure total OH reactivity values as high as 160 s^{-1} in ambient air,[134,145] and up to 200 s^{-1} from branch enclosures.[143,154]

7.6.3. *Intercomparisons of OH Reactivity Techniques*

Only two intercomparison studies involving OH reactivity instruments have been published in the literature. Hansen *et al.* discussed collocated measurements made by a CRM and a LP-LIF instrument in an urban environment in France.[134] This intercomparison showed that the two techniques were in reasonable agreement, with a scatter plot (CRM vs. LP-LIF) exhibiting a regression slope of 0.78 and an intercept of $2.2\,s^{-1}$ ($R^2 = 0.82$), for OH reactivity measurements ranging from the detection limit of $1–3\,s^{-1}$ to approximately $60\,s^{-1}$, even at NO_x mixing ratios close to 100 ppbv (when appropriate corrections were applied to the CRM measurements).[134]

A second intercomparison took place between two CRM instruments at a remote location in Cap Corsica, France.[154] The two CRM instruments, constructed and operated by two different laboratories, were run under similar operating conditions to test the robustness of this technique. Zannoni *et al.* showed that the two instruments were in agreement for OH reactivity values ranging from the detection limit of $3\,s^{-1}$ to approximately $300\,s^{-1}$ (plant enclosure experiments), with a scatter plot exhibiting a unity slope and an intercept of $1.6\,s^{-1}$ ($R^2 = 71.2$), suggesting a good reproducibility of the CRM measurements.[154]

While these two studies provide confidence in the measurement of total OH reactivity, larger intercomparison exercises involving several instruments of each type under different atmospheric conditions are needed to test the accuracy of each technique.[155] In this context, an intercomparison study was conducted at the SAPHIR chamber in Germany during October 2015 including a TOHLM instrument, four LP-LIF instruments and four CRM instruments from five different countries. Results from this campaign will provide a better understanding of the measurement techniques.

7.6.4. *Field Measurements of Total OH Reactivity*

Field measurements of k_{OH}, together with concomitant measurements of trace gases, have been useful to evaluate our understanding of atmospheric composition in different types of environments (Table 7.3). This table indicates that investigations of the OH reactivity budget have been performed in marine, rural, forested, and urbanized areas over the last 15 years but only studies related to urban and forested areas are reviewed below.

Table 7.3 shows that the three OH reactivity techniques have been mainly used to perform ground-based measurements, with only

Table 7.3 Field measurements of OH reactivity(adapted from Refs. [134,139]).

Campaign	Technique	Collocated meas.[a]	Range of $k_{OH}(s^{-1})$[b]	MR[c]
Urban				
SOS[151] Nashville, TN, USA	TOHLM	HICOF	5–25 (8–15)	1.4 (1.2)
TexAQS 2000[61] Houston, TX, USA	TOHLM	HICOFB[d]	— (5–12)	~1
PMTACS-NY-s[57,58] New York City, USA	TOHLM	HICOF	4–50 (14–25)	~1
MCMA2003[87] Mexico City, Mexico	TOHLM	HICF[e]	5–200 (20–130)	—[e]
Tokyo, Japan[138,157]	LP-LIF	ICOFB	15–80	1.2–1.3
PMTACS-NY-w[59] New York City, USA	TOHLM	HICF	— (18–36)	1–1.4
Tokyo, Japan[147]	LP-LIF	ICOFB	10–90	1–1.3
Mainz, Germany[142]	CRM	—	— (6–18)	—
TRAMP2006[61] Houston, TX, USA	TOHLM	HICOFB	4–50 (9–22)	~ 1
Tokyo, Japan[158]	LP-LIF	ICOFB	— (15–55)	(1.1–1.4)
MEGAPOLI[145] Paris, France	CRM	ICO	3–130 —	1.1–4.0

(*Continued*)

Table 7.3 (*Continued*)

Campaign	Technique	Collocated meas.[a]	Range of $k_{OH}(s^{-1})^b$	MR[c]
Lille, France[134]	CRM LP-LIF	ICOF[g]	3–100 (4–80)	~ 1
Suburban and rural				
Pennsylvania, USA[159]	TOHLM	HI	2–15 (5–7)	—
PRIDE-PRD2006[139] Pearl River Delta, China	LP-LIF	HIC	8–80 (19–50)	~2 (1–1.4)
CalNex-SJV[160] California, USA	TOHLM	HICO	3–10	1–2
Forested				
PROPHET2000[161] Michigan, USA	TOHLM	HIC	— (1–12)	1–1.5
PMTACS-NY-wm[162] Whiteface Mountain, NY, USA	TOHLM	HICOF	2–12 (4–6)	~ 1
Brownsberg, Suriname[142]	CRM	CO	25–75 —	~ 3.5
OP-3[163] Borneo, Malaysia	TOHLM	HICOFB[f]	2–80 (7–30)	2.3–3
SMEAR-BFORM[148] Hyytiälä, Finland	CRM	ICOB	3.5–60 (6–12)	~1.6–2.8
BEACHON-SRM08[164] Colorado, USA	LP-LIF	ICOB	1–26 (4–11)	1–2

(*Continued*)

Table 7.3 (*Continued*)

Campaign	Technique	Collocated Meas.[a]	Range of $k_{OH}(s^{-1})$[b]	MR[c]
BEARPEX09[67] California, USA	TOHLM	HICOFB	5–40 —	1.3–2 (1–1.3)
CAB-INEX[137,143] Michigan, USA	CRM	COB[g]		~1
	TOHLM	HICOFB	2–30 (4–25)	1.3–3
HUMPPA-COPEC[165] Hyytiälä, Finland	CRM	ICOFB	1–70 —	1–10
ATTO[166] Manaus, Brazil	CRM	ICOB	(18–80)[h] (5–12)[i]	2.5–4 1.1-4
SOAS[167] Alabama, USA	TOHLM	HICOFB	3–55 (13–25)	(1.25)
CANOPEE[154] Haute Provence, France	CRM	COFB	<3–69 —	~1
Marine and continental				
TORCH-2[168] Weybourne, Norfolk, UK	TOHLM	HICOF	1–10	1.4–2.3
INTEX-B[150] Pacific Ocean	TOHLM	HICOF	<1–8	1–2.1
DOMINO[133] El Arenosillo, Spain	CRM	HIF	1–75 (15–26)	—

[a]H = HO_x, I = inorganics (including CO), C = anthropogenic NMHCs (including isoprene), O = OVOCs (excluding formaldehyde), F = formaldehyde, B = BVOCs; [b]Individual measurements and diurnal average or median values in parenthesis; [c]Missing OH reactivity fraction, expressed as a ratio of measured-to-calculated OH reactivity[139] OH reactivity values are calculated from trace gas measurements or model simulations when in parenthesis; [d]Based on description from Mao *et al.*;[61] [e]Limited coverage of VOC measurements; [f]Measurements of isoprene oxidation products not used in OH reactivity calculations; [g]No measurements of CO, limited coverage of measurements for OVOCs, formaldehyde; [h]Dry season; [i]Wet season.

one study reporting airborne measurements aboard the NASA DC-8 aircraft during INTEX-B.[150] Being the first developed technique, TOHLM was used for approximately half the field deployments but the two other techniques have also been widely used over the last 5 years. It is interesting to note that due to the wide use of PTR-MS instruments in the atmospheric scientific community, several groups have started to develop CRM instruments and an increasing use of this technique is expected in future field campaigns.

Campaigns performed in urban environments (Table 7.3), including cities from the US, Japan, and France, showed that OH reactivity values can range from a few s^{-1} to more than $100\,s^{-1}$ in polluted cities such as Mexico City.[87] These studies indicate that measured k_{OH} values are usually in agreement within 20–40% with values calculated from trace gas measurements. Urban field campaigns of at least one-week duration, and characterized by a suite of trace gas measurements, extensive enough to perform a meaningful investigation of the OH reactivity budget (Table 7.3), are reported in Fig. 7.11. This figure displays the range of measured OH reactivity values together with the missing OH reactivity (MR) defined as the calculated-to-measured k_{OH} ratio. However, to determine whether the missing OH reactivity is statistically significant, total uncertainties associated to both the measured and calculated values have to be estimated. While measurement uncertainties can be assessed for each measurement technique, it is not straightforward to estimate uncertainties associated with the calculated values, which should account for errors related to both OH rate constants and trace gas measurements, the latter including precision, accuracy, and differences in timescales for the different measurement techniques.[156] Taking the above mentioned sources of errors into account, calculated values are associated with 1σ uncertainties of at least 25–30% shown by the red arrows in Fig. 7.11. Only studies exhibiting missing OH reactivity higher than the red arrows may be significant, depending on the uncertainty of the k_{OH} measurements.

Considering the uncertainties mentioned above, the missing OH reactivity observed in urban areas is barely significant and suggests a relatively good understanding of OH sinks in environments

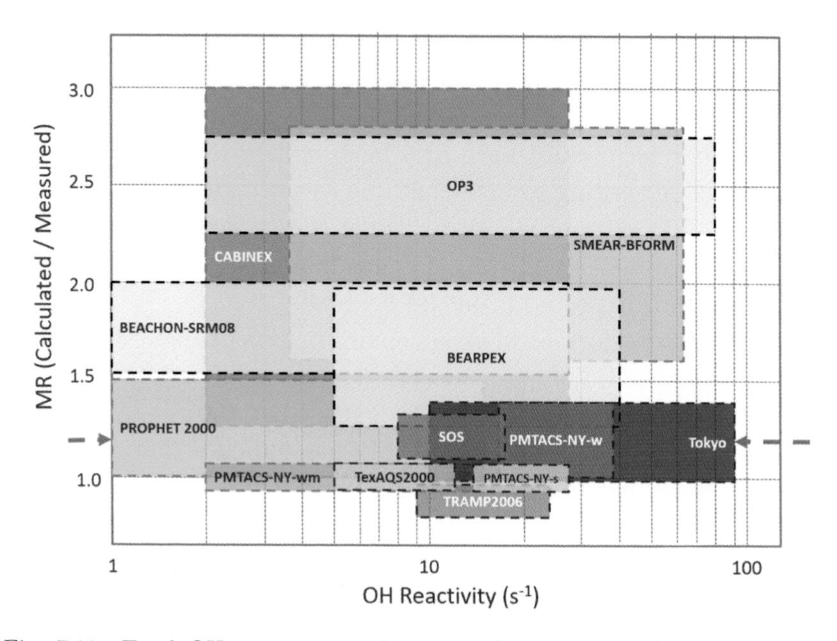

Fig. 7.11 Total OH reactivity and missing OH reactivity (MR) observed in urban and forested areas. The TRAMP2006 field campaign exhibits MR values of approximately unity but has been shifted down for clarity. Red arrows are shown as an approximate MR value above which the missing reactivity may be statistically significant at 1σ. Forested sites are shown in green and urban areas in brown.

impacted by anthropogenic emissions. These results also suggest that instrumental techniques used to monitor ambient trace gases are suitable to perform an exhaustive characterization of anthropogenic emissions. OH reactivity budgets reported for these urban areas are shown in Fig. 7.12 when sufficient details were available in the literature. This figure indicates contributions ranging from 15 to 50% for NO_x, 20–55% for hydrocarbons, and 10–25% for OVOCs. Interestingly, CO seems to exhibit a nearly constant contribution of 10–15% to k_{OH} and negligible contributions are observed for other inorganic species. However, in contrast to these studies, measurements performed in Paris highlighted a MR ratio of up to 4 (missing reactivity of 70–80 s^{-1}) for continental air masses.[145] On a campaign average, the missing reactivity accounts for approximately half the measured k_{OH}. The authors highlighted that air masses characterized by a large

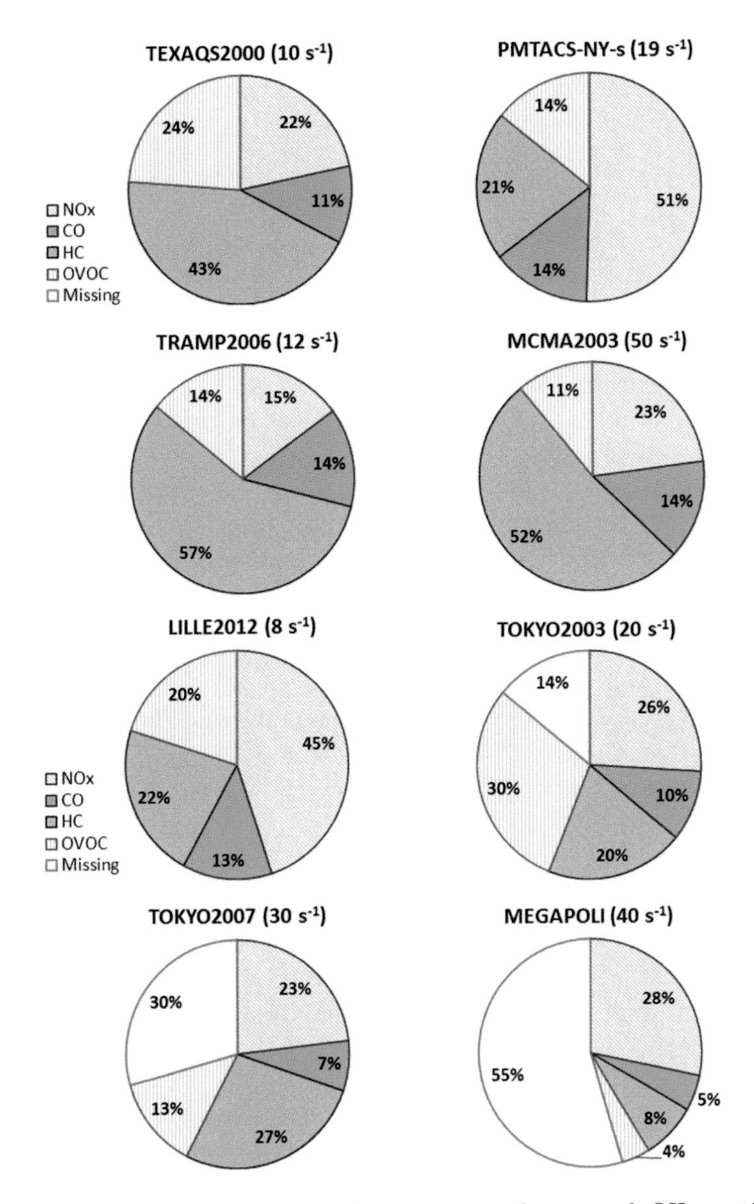

Fig. 7.12 Contributions of atmospheric compounds to total OH reactivity. TEXAQS2000, PMTACS-NY-s, MCMA2003, and TRAMP2006 from Ref. [61]; Lille2012 from Ref. [134]; MEGAPOLI from Ref. [145]; TOKYO2003 and 2007 from Ref. [147,158], respectively. Only the summer period is shown for TOKYO2003. k_{OH} measurements reported for TEXAQS2000 are only for ambient NO conditions <1 ppb.

extent of photochemical processing exhibit larger missing reactivity, likely due to unmeasured oxidation products of primary VOCs such as multifunctional oxygenated compounds that are not currently measured by conventional analytical techniques.[145] Measurements reported for Tokyo in 2007 may also highlight some missing OH reactivity.[158] While the recurrent 20–30% MR observed in Tokyo and most other urban field campaigns may not be statistically significant, these results put an upper limit on the contribution of unmeasured emissions and oxidation products on k_{OH}.

OH reactivity measurements performed in forested areas (Table 7.3) indicate that summer time biogenic emissions can compete with urban emissions in terms of VOC reactivity. As shown in Figs. 7.11 and 7.13, k_{OH} values observed in forests range from a few s^{-1} to approximately 80 s^{-1}, which is similar to the highest levels of OH reactivity observed in polluted urban areas. This behavior is partly due to the reactive nature of BVOCs, with isoprene being an interesting example because its reaction rate coefficient is close to the collision rate.

However, in contrast to urban areas, measurements performed in forests have highlighted a poor understanding of OH sinks, with most OH reactivity budgets showing a significant level of missing reactivity ($MR > 1.5$) (Figs. 7.11 and 7.13). Figure 7.11 illustrates that MR ratios of up to 3 (missing reactivity of 20–70 s^{-1}) have been observed in tropical[163] and Boreal[148] forests, but also in a Northern Michigan forest.[137] Even higher MR ratios of up to 10 have even been measured during the HUMPPA-COPEC campaign at the SMEAR station in Finland during heat stress conditions.[165]

OH reactivity budgets available in the literature for forested areas (Fig. 7.13) indicate that the contribution of the missing reactivity to k_{OH} is at least as high as the contribution of the measured BVOCs, with the exception of the CANOPEE campaign. It is interesting to note that high MR ratios are not specific to particular BVOC emissions since the different areas reported in Table 7.3 are characterized by either deciduous (OP3, CABINEX, CANOPEE) or coniferous (SMEAR-BFORM) trees, which are isoprene and monoterpene emitters, respectively. However, a common feature

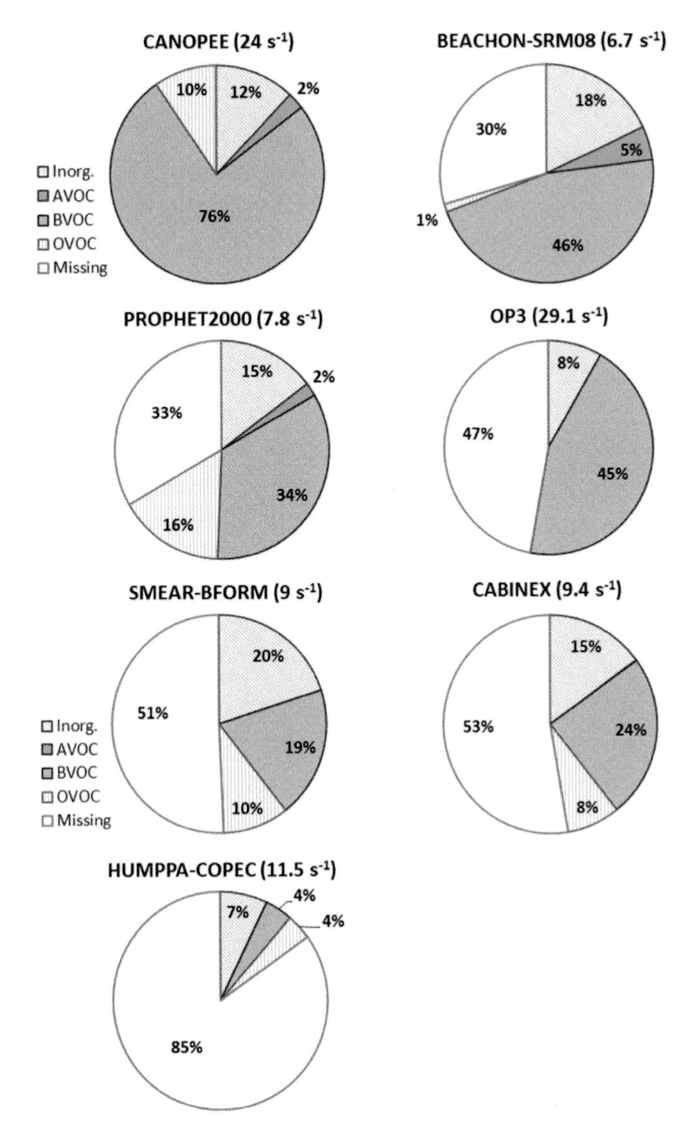

Fig. 7.13 Contributions of atmospheric compounds to k_{OH} in forested areas. CANOPEE from Zannoni *et al.*[154]; BEACHON-SRM08 from Ref. [164]; PROPHET 2000 from Ref. [161]; OP3 from Ref. [163]; SMEAR-BFORM from Ref. [148]; CABINEX from Ref. [137]; HUMPPA-COPEC from Ref. [165] 24-h average values of k_{OH} are shown in the titles, except for OP3 where noontime values are given. Inorg.: sum of NO_x, O_3, CO, and CH_4; AVOC: anthropogenic hydrocarbons; BVOC: sum of isoprene, monoterpenes, MBO and sesquiterpenes. Methanol and other primary emitted oxygenated VOCs are included in OVOC. OVOC measurements were not available for OP3.

of these different forests is low ambient mixing ratios of NO_x ($NO < 150$ pptv, $NO_2 < 2$ ppbv).

Reasons provided to explain the observed missing reactivity are either (i) emissions of unknown BVOCs.[165] or (ii) unmeasured BVOC oxidation products.[137,163] Hints about the origin of the missing reactivity were obtained from correlation analyses of the MR with environmental parameters such as temperature and light, as well as concentrations of primary and secondary species, and the use of box models to estimate the concentration of unmeasured oxidation products. Unfortunately, no compelling evidence was found in these studies to apportion the missing reactivity between primary emissions and secondary production. A few recent studies coupled branch enclosure experiments to OH reactivity instruments to investigate whether the most abundant trees in a forest can emit reactive unknown compounds. For instance, a study performed during the CABINEX field campaign demonstrated that the OH reactivity budget for red oaks (isoprene emitter) and white pines (monoterpene emitter) were closed when the branch enclosure measurements of total OH reactivity were compared to values calculated from concomitant BVOC measurements.[143] These results indicate that these two types of trees did not emit significant amounts of unknown reactive compounds under the temperature conditions of CABINEX. In contrast, a study performed in Germany highlighted measurements of significant missing reactivity from branch enclosure experiments performed on a Norway Spruce (monoterpene emitter) under heat stress conditions.[169]

In contrast to these studies, where high levels of ambient missing OH reactivity were observed, Zannoni *et al.* showed that the OH reactivity budget could be closed in a Mediterranean forest of downy oak trees in France (Table 7.3).[154] The measured daytime OH reactivity, ranging from a few s^{-1} to approximately $70\,s^{-1}$, could be accounted for by isoprene and its first oxidation products, i.e., formaldehyde, methacrolein, and methyl vinyl ketone. Similar results were reported for another isoprene-dominated environment by Kaiser *et al.*,[167] who showed that during the SOAS campaign modeled OH reactivity values were in agreement with the measurements to within their uncertainties (MR $\approx$ 1.25, 1σ combined model-measurement errors of 28%). Interestingly, these two environments appeared to exhibit a low level of photochemical processing, and as a consequence, levels of unmeasured multigenerational oxidation products are expected to be low. However, the nighttime budgets were not closed and missing OH reactivity ratios of up to 2 were observed, similar to other forested campaigns. Zannoni *et al.* speculated that unmeasured OVOCs, produced during the nighttime chemistry of reactive BVOCs, as well as the accumulation of multigenerational oxidation products of isoprene from the daytime chemistry, could cause this missing OH reactivity.[154]

Urban areas impacted by biogenic emissions have also been shown to exhibit large missing OH reactivity. For instance, missing reactivity ratios of up to 2 have been observed during the Pearl River Delta study conducted in China, where isoprene contributes to approximately 70% of the measured OH reactivity during daytime.[139] Interestingly, the authors showed that concentrations of unmeasured oxidation products of primary VOCs, calculated by a box model, could account for the missing OH reactivity. While concentrations of long-lived secondary species calculated by box models are very sensitive to the dilution approach implemented in the model (physical lifetime of 24 h in this study), the good agreement observed between modeled and measured values strongly suggests that the missing reactivity is mainly due to photochemically produced OVOCs in this environment.[139] Similar conclusions from the results of box models were reached in other studies.[67,163]

It is clear from the results discussed above that oxidation products of primary VOCs contribute significantly to the missing OH reactivity and more detailed investigations will require the development of techniques capable of measuring highly oxidized compounds in the gas-phase. Edwards *et al.*[163] discussed the different types of techniques (GC/FID, GC/MS, PTR-MS) that are usually used during large scale field campaigns and reached the conclusion that only a small subset of gas-phase compound classes would not be detected, i.e., multifunctional species. However, this pool of unmeasured compounds is likely composed of several tens to thousands of species at mixing ratios in the pptv range. For instance, a missing reactivity of $20\,s^{-1}$ could be the results of 250 unmeasured OVOCs ($k = 10^{-11}\,cm^3/mol\,s^{-1}$) present at a mixing ratios of approximately 300 pptv each. This represents a major analytical challenge and may require developing novel approaches, other than an exhaustive identification of all the OH co-reactants, to characterize the missing OH reactivity.

7.6.5. *Laboratory Measurements of Total OH Reactivity*

Measurements of total OH reactivity at emission sources (tunnel experiments, vehicular exhausts, gasoline evaporation) have been reviewed in Yang *et al.* and are not discussed here.[5] In this section, we discuss the coupling of OH reactivity instruments to atmospheric chambers, whose aim was to investigate (i) the extent to which oxidation products of VOCs contribute to the missing OH reactivity observed in field experiments, and (iii) OH recycling mechanisms.

For instance, photo-oxidation experiments of isoprene (Table 7.4) were performed in several atmospheric chambers to investigate the oxidative capacity and/or the OH reactivity budget of this chemical system under low (0.1 ppbv),[124] medium (up to 1.5 ppbv),[170] and high (≈200 ppbv)[171] NO mixing ratios. Fuchs *et al.*[124] performed experiments in the SAPHIR chamber under high isoprene reactivity and low NO mixing ratios (100 pptv) and low HO_2 concentrations ($<10^9\,cm^{-3}$), similar to that observed in forests and the Pearl River Delta area where both elevated OH concentrations and large ratios of missing OH reactivity have been observed. Total OH reactivity was measured using the pump-probe technique, and the modeled OH reactivity was found to be in good agreement with the observations.[124]

Nölscher *et al.*[170] performed experiments at the EUPHORE chamber in Valencia where isoprene, its first generation oxidation products (methacrolein, methyl vinyl ketone, formaldehyde), and multigenerational oxidation products (acetaldehyde, acetic acid, glycolaldehyde, methylglyoxal, hydroxyacetone) were measured by

Table 7.4 Atmospheric chamber experiments incorporating total OH reactivity measurements.

Chamber	Technique	VOC (ppb) NO (ppb)	Collocated Meas.[a]	MR[b]
NIES[171] Isoprene	LP-LIF	816 195	—	1.4
SAPHIR[54] Isoprene, MVK, MACR, Aromatics	LP-LIF	4–24[c] 0.1–0.3	HICOFB	—
SAPHIR[124] Isoprene	<LP-LIF	2–10 0.1	HICOFB	(~1)
SAPHIR[127] MACR	LP-LIF	7—14 0.06–0.09	HICOFB	(~1)
SAPHIR[131] Aromatics	LP-LIF	6–250 0.1–0.2 & 7–8	HICOF	—
EUPHORE[170] Isoprene	LP-LIF	200 <1.5	—	~1

[a]Key: H = HO_x, I = Inorganics (including CO), C = anthropogenic NMHCs (including isoprene), O = OVOCs (excluding formaldehyde), F = formaldehyde, B = biogenic VOCs (BVOCs); [b]Missing OH reactivity fraction, expressed as a ratio of measured-to-calculated OH reactivity [139] OH reactivity values are calculated from trace gas measurements or model simulations when in parenthesis; [c]Except for toluene and benzene experiments where 90 ppbv and 250 ppbv were used, respectively.

PTR-ToFMS. Total OH reactivity values measured using the CRM technique ranged from 500 s^{-1}, with an initial isoprene mixing ratio of 200 ppbv, to approximately 200 s^{-1} after 6 h of oxidation. Several mechanisms such as MCM 3.2, MIME (Mainz Isoprene Mechanism Extend), and modified versions of the later, were used to compare time profiles of VOCs and total OH reactivity to model simulations. Overall, the authors showed that MCM and the modified versions of MIME were able to reproduce the observations within uncertainty, including the total OH reactivity.[170] These results contrast with the large missing OH reactivity observed during field measurements. However, while the NO mixing ratio in the chamber was lower than the detection limit of a chemiluminescence instrument, an upper limit was estimated to 1.5 ppbv, which is more than an order of magnitude higher than the field conditions (<100 pptv). The nature of the oxidation products formed during the oxidation of VOCs is highly dependent on NO_x concentrations and the oxidation of isoprene in pristine areas, such as boreal and tropical forests, will likely lead to a different speciated pool of oxidation products. In addition, only 25% of the total OH reactivity was due to oxidation products in this experiment, with the rest still due to isoprene. If some missing reactivity is generated together with the production of the detected oxidation products, the missing reactivity may be masked due to the low contribution of these unmeasured oxidation products to the total OH reactivity in these experiments. In addition, as discussed by Kaiser *et al.*,[167] the behavior of ambient OH reactivity in forested areas may not be well reproduced in an atmospheric chamber due to the loss of multigenerational OVOCs on the wall of the chamber, and the contribution of heterogeneous chemistry of oxidants on leaf surfaces which could play a role in real forests.

Nakashima *et al.* performed isoprene oxidation experiments in a Teflon coated stainless steel chamber using approximately 820 ppbv and 200 ppbv of isoprene and NO, respectively.[171] The total OH reactivity was measured using the pump-probe technique and several species, including NO_x, O_3, formaldehyde, formic acid, methacrolein, and methyl vinyl ketone were monitored by FTIR. These authors showed that about 40% of missing OH reactivity was observed at the end of the experiments, when approximately 90% of isoprene had reacted away.[171]

As mentioned above, chamber experiments have also taken advantage of total OH reactivity measurements to investigate OH recycling mechanisms in the degradation chemistry of isoprene,[124] methacrolein,[127] and aromatic compounds.[131] The measured OH reactivity and OH concentrations were used to calculate the total production rate of OH ($P_{OH} = L_{OH} = k_{OH} \times [OH]$), which is balanced by its destruction rate due to the short lifetime of OH. The measured P_{OH} was compared to model simulations,[124,127] or values calculated from measured OH precursors (O_3, HONO, HO_2).[131] The level of agreement observed between the measured and calculated

P_{OH} values indicate whether all OH production pathways are accounted for in the calculations. Based on this approach, Nehr *et al.*.[131] showed that the OH budget in aromatic systems could be balanced within experimental uncertainties under low (0.1–0.2 ppbv) and high (7–8 ppbv) NO mixing ratios. The authors demonstrated that OH recycling from $RO_2 + HO_2$ reactions plays a minor role in the OH budget, even under low NO conditions.[131] In contrast, as discussed above, Fuchs *et al.*[124,127] reported that the oxidation of isoprene and methacrolein under low NO mixing ratios (0.09 ppbv) leads to OH recycling yields of 0.76–0.90 and 0.77 ± 0.31, respectively. The authors also demonstrated that approximate yields of 0.38–0.45 and 0.55 could be explained by unimolecular RO_2 isomerization reactions,[104,121,125] suggesting that these HO_x radical recycling reactions contribute significantly to the oxidizing capacity of the atmosphere.

7.7. Summary

Recent measurements of OH radicals have revealed serious discrepancies with model predictions both during the day and at night especially in low NO_x forested environments with high emissions of biogenic VOCs. These measurements have led to advances in our understanding of both the challenge associated with measuring these radicals in the atmosphere, as well as the complex radical chemistry associated with the atmospheric oxidation of VOCs. Improving our understanding of this important chemistry will improve the ability to project how changes in biogenic emissions as a result of climate change will affect the future composition and chemistry of the atmosphere.

Although several intercomparisons of HO_x radical measurements have been conducted that generally demonstrated good agreement between different instruments, especially in controlled chamber experiments, the recent discovery of potential interferences associated with measurements of OH radicals in low NO_x forested environments by some instruments brings into question whether past measurements of OH are free from interferences. Additional measurements of OH and HO_2 radicals as well as their sources and sinks in different forest environments are needed to resolve the discrepancies with current atmospheric chemistry models and to characterize and minimize unknown instrument interferences. Although it is not clear that all instruments suffer from interferences, future measurements should incorporate techniques to characterize potential instrumental artifacts, such as chemical scavenging, to help identify and minimize these artifacts as well as provide confidence that the measurements are free from interferences. The development of procedures to investigate and, if necessary, eliminate possible interferences is one of the main goals of a working group established to guide the community on the continued improvement of HO_x measurements.[172]

Even if the past measurements of HO_x radicals in these low NO_x forested environments prove to be influenced by instrument interferences, the perceived discrepancy has stimulated a reexamination of the complex chemistry of biogenic VOC oxidation, and in particular the atmospheric oxidation of isoprene. New insights into HO_x radical recycling have emerged, including new peroxy radical isomerization reactions and new mechanisms in the self-reaction of peroxy radicals. Although these new radical production mechanisms do lead to an increase in modeled HO_x radical concentrations, they have not been able to fully explain the higher than expected HO_x concentrations observed in low NO_x forested environments. Additional discrepancies associated with measurements of oxidation products in these environments suggest that our understanding of this complex chemistry and the budget of radicals in these environments is still incomplete.

New instruments capable of measuring total OH reactivity have been developed and used intensively in the field over the last 15 years to better constrain the OH radical budget. These instruments have also proved to be useful in laboratory experiments to help investigate HO_x radical chemistry. However, there have not been sufficient intercomparison studies to ensure that these new instruments provide reliable measurements under various atmospheric conditions typical of urban, suburban, forested and remote areas. This gap is being addressed by the recent intercomparison experiment performed at the SAPHIR chamber in October 2015 but a complete understanding of the measurement techniques will also require additional studies in a variety of real environments.

While the budget of OH sinks seems to be relatively well understood in urban areas, the picture is different for areas impacted by biogenic emissions. Indeed, significant missing OH reactivity fractions have been observed in forests characterized by either isoprene or monoterpene emissions. While unknown biogenic emissions may contribute to the observed missing OH reactivity, it is very likely that a large fraction of the missing reactivity is due to unidentified oxidation products of primary VOCs. A full characterization of this missing fraction represents a major technological challenge because the pool of unmeasured multifunctional species is likely composed of hundreds to thousands of species at mixing ratios in the pptv range. To tackle this issue, novel approaches, other than an exhaustive identification of all the OH co-reactants, may be necessary to characterize the missing OH reactivity. Combined improvements in both HO_x and k_{OH} measurements, together with their concomitant deployment in future field campaigns, will provide additional insights into the chemistry of OH and HO_2 radicals in the atmosphere, which in turn will address the important issues of air quality and climate change.

References

1. D. E. Heard and M. J. Pilling. Measurement of OH and HO_2 in the Troposphere. *Chem. Rev.* **103**, 5163–5198 (2003).
2. F. Rohrer, K. Lu, A. Hofzumahaus, B. Bohn, T. Brauers, C.-C. Chang, H. Fuchs, R. Haseler, F. Holland, M. Hu, K. Kita, Y. Kondo, X. Li, S. Lou, A. Oebel, M. Shao, L. Zeng, T. Zhu, Y. Zhang and A. Wahner. Maximum efficiency in the hydroxyl-radical-based self-cleansing of the troposphere. *Nature Geosci.* **7**, 559–563 (2014).
3. L. K. Whalley, D. Stone and D. E. Heard. New Insights into the tropospheric oxidation of isoprene: Combining field measurements, laboratory studies, chemical modelling and quantum theory. *Top. Curr. Chem.* **339**, 55–96 (2014).
4. D. Stone, L. K. Whalley and D. E. Heard. Tropospheric OH and HO_2 radicals: Field measurements and model comparisons. *Chem. Soc. Rev* **41**, 6348–6404 (2012).

5. Y. Yang, M. Shao, X. Wang, A. C. Nölscher, S. Kessel, A. Guenther and J. Williams. Towards a quantitative understanding of total OH reactivity: A review. *Atmos. Env.* **ASAP**, (2016).
6. X. Ren, D. van Duin, M. Cazorla, S. Chen, J. Mao, L. Zhang, W. H. Brune, J. H. Flynn, N. Grossberg, B. L. Lefer, B. Rappenglück, Wong, K. W. C. Tsai, J. Stutz, J. E. Dibb, B. Thomas Jobson, W. T. Luke and P. Kelley. Atmospheric oxidation chemistry and ozone production: Results from SHARP 2009 in Houston, Texas. *J. Geophys. Res.* **118**, 5770–5780 (2013).
7. R. Volkamer, P. Sheehy, L. T. Molina and M. J. Molina. Oxidative capacity of the Mexico City atmosphere — Part 1: A radical source perspective. *Atmos. Chem. Phys.* **10**, 6969–6991 (2010).
8. S. Dusanter, D. Vimal, P. S. Stevens, R. Volkamer, L. T. Molina, A. Baker, S. Meinardi, D. Blake, P. Sheehy, A. Merten, E. Fortner, J. Zheng, R. Zhang, W. Junkermann, M. Dubey, T. Rahn, B. Eichinger, P. Lewandowski, J. Prueger and H. Holder. Measurements of OH and HO_2 Concentrations during the MCMA-2006 Field Campaign: Part 2 — Model Comparison and Radical Budget. *Atmos. Chem. Phys.* **9**, 9823–9877 (2009).
9. K. D. Lu, A. Hofzumahaus, F. Holland, B. Bohn, T. Brauers, H. Fuchs, H. Hu, Häseler, Kita, K. Kondo, Y. Li, X. Lou, S. R. Oebel, A. Shao, M. R. L M. Zeng, A. Wahner, T. Zhu, Y. H. Zhang and F. Rohrer. Missing OH source in a suburban environment near Beijing: observed and modelled OH and HO_2 concentrations in summer 2006. *Atmos. Chem. Phys.* **13**, 1057–1080 (2013).
10. V. Michoud, A.Kukui, M. Camredon, A. Colomb, A. Borbon, K. Miet, B. Aumont, M. Beekmann, R. Durand-Jolibois, S. Perrier, P. Zapf, G. Siour, W. Ait-Helal, N. Locoge, S. Sauvage, C. Afif, V. Gros, M. Furger, G. Ancellet and J. F. Doussin. Radical budget analysis in a suburban European site during the MEGAPOLI summer field campaign. *Atmos. Chem. Phys.* **12**, 11951–11974 (2012).
11. S. Kim, G. M. Wolfe, L. Mauldin, C. Cantrell, A. Guenther, T. Karl, A. Turnipseed, J. Greenberg, S. R. Hall, K. Ullmann, E. Apel, R. Hornbrook, Y. Kajii, Y. Nakashima, F. N. Keutsch, J. P. DiGangi, S. B. Henry, L. Kaser, R. chnitzhofer, M. Graus, A. Hansel, W. Zheng, F. F. Flocke. Evaluation of HO_x sources and cycling using measurement-constrained model calculations in a 2-methyl-3-butene-2-ol (MBO) and monoterpene (MT) dominated ecosystem. *Atmos. Chem. Phys.* **13**, 2031–2044 (2013).
12. K. Hens, A. Novelli, M. Martinez, J. Auld, R. Axinte, B. Bohn, H. Fischer, P. Keronen, P. Kubistin, A. C. Nölscher, R. Oswald, P. Paasonen, T. Petäjä, E. Regelin, R. Sander, V.Sinha, M. Sipilä, D. Taraborrelli, C. Tatum Ernest, J. Williams, J. Lelieveld and H. Harder. Observation and modelling of HO_x radicals in a boreal forest. *Atmos. Chem. Phys.* **14**, 8723–8747 (2014).
13. S. M. Griffith, R. F. Hansen, S. Dusanter, P. S. Stevens, M. Alaghmand, S. B. Bertman, M. A. Carroll, M. Erickson, M. Galloway, N. Grossberg, J. Hottle, J. Hou, B. T. Jobson, A. Kammrath, F. N. Keutsch, B. L. Lefer, L. H. Mielke, A. O'Brien, P. B. Shepson, M. Thurlow, W. Wallace, N. Zhang

and X. L. Zhou. OH and HO_2 radical chemistry during PROPHET 2008 and CABINEX 2009 Part 1: Measurements and model comparison. *Atmos. Chem. Phys.* **13**, 5403–5423 (2013).
14. L. K. Whalley, K. L. Furneaux, A. Goddard, J. D. Lee, A. Mahajan, H. Oetjen, K. A. Read, N. Kaaden, L. J. Carpenter, A. C. Lewis, J. M. C. Plane, E. S. Saltzman, A. Wiedensohler and D. E. Heard. The chemistry of OH and HO_2 radicals in the boundary layer over the tropical Atlantic Ocean. *Atmos. Chem. Phys.* **10**, 1555–1576 (2010).
15. D. R. Crosley. The Measurement of OH and HO_2 in the Atmosphere, *J. Atmos. Sci.* **52**, 3299–3314 (1995).
16. D. E. Heard. Atmospheric field measurements of the hydroxyl radical using laser-induced fluorescence spectroscopy. *Annu. Rev. Phys. Chem.* **57**, 191–216 (2006).
17. E. L. Baardsen and R. W. Terhune. Detection of OH in the atmosphere using a dye laser. *Appl. Phys. Lett.* **21**, 209–211 (1972).
18. C. C. Wang and L. I. J. Davis. Measurement of Hydroxyl Concentrations in Air Using a Tunable uv Laser Beam. *Phys. Rev. Lett.* **32**, 349–352 (1974).
19. R. M. Stimpfle and J. G. Anderson. In-situ detection of OH in the lower stratosphere with a balloon borne high repetition rate laser system. *Geophys. Res. Lett.* **15**, 1503–1506 (1988).
20. P. O. Wennberg, R. C. Cohen, N. L. Hazen, L. B. Lapson, N.T. Allen, T. F. Hanisco, J. F. Oliver, N. W. Lanham, J. N.Demusz and J. G. Anderson. Aircraft-borne, laser-induced fluorescence instrument for the in situ detection of hydroxyl and hydroperoxyl radicals. *Rev. Sci. Instrum.* **65**, 1858–1876 (1994).
21. T. M. Hard, T. M. O'Brien, T. B. Cook and G. A. Tsongas. Interference suppression in HO fluorescence detection. *Applied Optics* **18**, 3216–3217 (1979).
22. T. M. Hard, R. J. O'Brien and T. B. Cook. Pressure dependence of fluorescent and photolytic interferences in HO detection by laser-excited fluorescence. *J. Appl. Phys.* **51**, 3459–3464 (1980).
23. T. M. Hard, R. J. O'Brien, C. Y. Chan and A. A. Mehrabzadeh. Tropospheric free radical determination by fluorescence assay with gas expansion. *Environ. Sci. Technol.* **18**, 768–777 (1984).
24. G. P. Smith and D. R. Crosley. A photochemical model of ozone interference effects in laser detection of tropospheric OH. *J. Geophys. Res.* **95**, 16427–16442 (1990).
25. C. Y. Chan, T. M. Hard, A. A. Mehrabzadeh, L. A. George and R. J. O'Brien. Third-generation FAGE instrument for tropospheric hydroxyl radical measurement. *J. Geophys. Res.* **95**, 18569–18576 (1990).
26. P. S. Stevens, J. H. Mather and W. H. Brune. Measurement of tropospheric OH and HO_2 by laser-induced fluorescence at low pressure. *J. Geophys. Res.* **99**, 3543–3557 (1994).
27. L. K. Whalley, M. A. Blitz, M. Desservettaz, P. W. Seakins and D. E. Heard. Reporting the sensitivity of laser-induced fluorescence instruments used for HO_2 detection to an interference from RO_2 radicals and introducing

a novel approach that enables HO_2 and certain RO_2 types to be selectively measured. *Atmos. Meas. Tech.* **6**, 3425–3440 (2013).

28. H. Fuchs, B. Bohn, A. Hofzumahaus, F. Holland, K. D. Lu, S. Nehr, F. Rohrer and A. Wahner. Detection of HO_2 by laser-induced fluorescence: Calibration and interferences from RO_2 radicals. *Atmos. Meas. Tech.* **4**, 1209–1225 (2011).
29. S. Dusanter, D. Vimal, P. S. Stevens, R. Volkamer and L. T. Molina. Measurements of OH and HO_2 concentrations during the MCMA-2006 field campaign — Part 1: Deployment of the Indiana University laser-induced fluorescence instrument. *Atmos. Chem. Phys.*, **9** 1665–1685 (2009).
30. A. Novelli, K. Hens, C. Tatum Ernest, C. Kubistin, E. Regelin, T. Elste, C. Plass-Dülmer, M. Martinez, J. Lelieveld and H. Harder. Characterisation of an inlet pre-injector laser-induced fluorescence instrument for the measurement of atmospheric hydroxyl radicals. *Atmos. Meas. Tech.* **7**, 3413–3430 (2014).
31. F. Holland, A. Hofzumahaus, J. Schäfer, A. Kraus and H.-W. Pätz. Measurements of OH and HO_2 radical concentrations and photolysis frequencies during BERLIOZ. *J. Geophys. Res.* **108**, 8246 (2003).
32. Y. Kanaya, Y. Sadanaga, J. Hirokawa, Y. Kajii and H. Akimoto. Development of a Ground-Based LIF Instrument for Measuring HO_x Radicals: Instrumentation and Calibrations. *J. Atmos. Chem.* **38**, 73–110, (2001).
33. I. C. Faloona, D. Tan, R. L. Lesher, N. L. Hazen, C. L. Frame, J. B. Simpas, H. Harder, M. Martinez, P. Di Carlo, X. Ren and W. H. Brune. A Laser-Induced Fluorescence Instrument for Detecting Tropospheric OH and HO_2: Characteristics and Calibration. *J. Atmos. Chem.* **47**, 139–167 (2004).
34. M. Martinez, H. Harder, D. Kubistin, M. Rudolf, H. Bozem, Eerdekens, G. H. Fischer, T. Klüpfel, C. Gurk, R. Königstedt, U. Parchatka, C. L. Schiller, A. Stickler, J. Williams and J. Lelieveld. Hydroxyl radicals in the tropical troposphere over the Suriname rainforest: Airborne measurements. *Atmos. Chem. Phys.* **10**, 3759–3773 (2010).
35. D. Amedro, K. Miyazaki, A. Parker, C. Schoemaecker and C. Fittschen. Atmospheric and kinetic studies of OH and HO_2 by the FAGE technique. *J. Environ. Sci.* **24**, 78–86 (2012).
36. J. H. Mather, P. S. Stevens and W. H. Brune. OH and HO_2 measurements using laser-induced fluorescence. *J. Geophys. Res.* **102**, 6427–6436, (1997).
37. H. Fuchs, F. Holland and A. Hofzumahaus. Measurement of tropospheric RO_2 and HO_2 radicals by a laser-induced fluorescence instrument. *Rev. Sci. Instr.* **79**, 084–104 (2008).
38. T. M. Hard, L. A. George and R. J. O'Brien. FAGE determination of tropospheric HO and HO_2. *J. Atmos. Sci.* **52**, 3354–3372 (1995).
39. T. M. Hard, L. A. George and R. J. O'Brien. An Absolute Calibration for Gas-Phase Hydroxyl Measurements. *Environ. Sci. Technol.* **36**, 1783–1790 (2002).
40. S. Dusanter, D. Vimal and P. S. Stevens. Technical note: Measuring tropospheric OH and HO_2 by laser-induced fluorescence at low pressure.

A comparison of calibration techniques. *Atmos. Chem. Phys.* **8**, 321–340 (2008).
41. W. J. Bloss, J. D. Lee, C. Bloss, D. E. Heard, M. J. Pilling, K. Wirtz, M. Martin-Reviejo and M. Siese. Validation of the calibration of a laser-induced fluorescence instrument for the measurement of OH radicals in the atmosphere. *Atmos. Chem. Phys.* **4**, 571–583 (2004).
42. F. L. Eisele and D. J. Tanner. Ion-assisted tropospheric OH measurements. *J. Geophys. Res.* **96**, 9295–9308 (1991).
43. D. J. Tanner, A. Jefferson and F. L. Eisele. Selected ion chemical ionization mass spectrometric measurement of OH. *J. Geophys. Res.* **102**, 6415–6425 (1997).
44. S. J. Sjostedt, L. G. Huey, D. J. Tanner, J. G. Peischl, J. E. Chen, B. Dibb, M. A. Lefer, A. J. Hutterli, N. J. Beyersdorf, D. R. Blake, D. Blake, T. Sueper, J. Ryerson, A. Burkhart and A. Stohl. Observations of hydroxyl and the sum of peroxy radicals at Summit, Greenland during summer 2003. *Atmos. Environ.* **41**, 5122–5137 (2007).
45. J. Liao, L. G. Huey, D. J. Tanner, N. Brough, S. Brooks, J. E. Dibb, J. Stutz, J. L. Thomas, B. Lefer, C. Haman and K. Gorham. Observations of hydroxyl and peroxy radicals and the impact of BrO at Summit, Greenland in 2007 and 2008. *Atmos. Chem. Phys.* **11**, 8577–8591 (2011).
46. T. R. Petäjä, L. Mauldin Iii, E. Kosciuch, J. McGrath, T. Nieminen, P. Paasonen, M. Boy, A. Adamov, T. Kotiaho and M. Kulmala. Sulfuric acid and OH concentrations in a boreal forest site. *Atmos. Chem. Phys.* **9**, 7435–7448 (2009).
47. R. Mauldin, E. Kosciuch, F. Eisele, G. Huey, D. Tanner, S. Sjostedt, D. Blake, G. Chen and D. Davis. South Pole Antarctica observations and modeling results: New insights on HOx radical and sulfur chemistry. *Atmos. Environ.* **44**, 572–581 (2010).
48. H. Berresheim, M. Adam, C. Monahan, C. O'Dowd, J. M. C. Plane, B. Bohn and F. Rohrer. Missing SO_2 oxidant in the coastal atmosphere? — Observations from high-resolution measurements of OH and atmospheric sulfur compounds. *Atmos. Chem. Phys.* **14**, 12209–12223 (2014).
49. H. Berresheim, T. Elste, C. Plass-Dülmer, F. L. Eisele and D. J. Tanner. Chemical ionization mass spectrometer for long-term measurements of atmospheric OH and H_2SO_4. *Int. J. Mass Spectrom.* **202**, 91–109 (2000).
50. G. D. Edwards, C. A. Cantrell, S. Stephens, B. Hill, O. Goyea, R. E. Shetter, R. L. Mauldin, E. Kosciuch, D. J. Tanner and F. L. Eisele. Chemical Ionization Mass Spectrometer Instrument for the Measurement of Tropospheric HO_2 and RO_2. *Anal. Chem.* **75**, 5317–5327 (2003).
51. R. S. Hornbrook, J. H. Crawford, G. D. Edwards, O. Goyea, R. L. Mauldin Iii, J. S. Olson and J. S. Cantrell. Measurements of tropospheric HO_2 and RO_2 by oxygen dilution modulation and chemical ionization mass spectrometry. *Atmos. Meas. Tech.* **4**, 735–756 (2011).
52. E. Schlosser, T. Brauers, H. P. Dorn, H. Fuchs, R. Häseler, A. Hofzumahaus, F. Holland, A. Wahner, Y. Kanaya, Y. Kajii, K. Miyamoto, S. Nishida, K. Watanabe, A. Yoshino, D. Kubistin,

M. Martinez, M. Rudolf, H. Harder, H. Berresheim, T. Elste, C. Plass-Dülmer, G. Stange and U. Schurath. Technical Note: Formal blind intercomparison of OH measurements: Results from the international campaign HOxComp. *Atmos. Chem. Phys.* **9**, 7923–7948 (2009).

53. E. Schlosser, B. Bohn, T. Brauers, H.-P. Dorn, H. Fuchs, R.H äseler, A. Hofzumahaus, F. Holland, F. Rohrer, L. Rupp, M. Siese, R. Tillmann and A. Wahner. Intercomparison of two hydroxyl radical measurement techniques at the atmosphere simulation chamber SAPHIR. *J. Atmos. Chem.* **56**, 187–205 (2007).
54. H. Fuchs, H. P. Dorn, M. Bachner, B. Bohn, T. Brauers, S. Gomm, A. Hofzumahaus, F. Holland, S. Nehr, F. Rohrer, R. Tillmann and A. Wahner. Comparison of OH concentration measurements by DOAS and LIF during SAPHIR chamber experiments at high OH reactivity and low NO concentration. *Atmos. Meas. Tech.* **5**, 1611–1626 (2012).
55. I. Faloona, D. Tan, W. Brune, J. Hurst, D. Barket, T. L. Couch, P. Shepson, E. Apel, D. Riemer, T. Thornberry, M. A. Carroll, S. Sillman, G. J. Keeler, J. Sagady, D. Hooper and K. Paterson. Nighttime observations of anomalously high levels of hydroxyl radicals above a deciduous forest canopy. *J. Geophys. Res.* **106**, 24315–24333 (2001).
56. M. Martinez, H. Harder, T. A. Kovacs, J. B. Simpas, J. Bassis, R. Lesher, W. H. Brune, G. J. Frost, E. J. Williams, C. A. Stroud, B. T. Jobson, J. M. Roberts, S. R. Hall, R. E. Shetter, B. Wert, A. Fried, B. Alicke, J. Stutz, V. L. Young, A. B. White and R. J. Zamora. OH and HO_2 concentrations, sources, and loss rates during the Southern oxidants Study in Nashville, Tennessee, summer 1999. *J. Geophys. Res.* **108**, 4617 (2003).
57. X. Ren, H. Harder, M. Martinez, R. L. Lesher, A. Oliger, T. Shirley, J. Adams, J. B. Simpas and W. H. Brune. HO_x concentrations and OH reactivity observations in New York City during PMTACS-NY2001. *Atmos. Env.* **37**, 3627-3637 (2003).
58. X. Ren, H. Harder, M. Martinez, R. L. Lesher, A. Oliger, J. B. Simpas, W. H. Brune, J. J. Schwab, K. L. Demerjian, Y. He, X. Zhoua and H. Gao. OH and HO_2 Chemistry in the urban atmosphere of New York City. *Atmos. Env.* **37**, 3639–3651 (2003).
59. X. Ren, W. H. Brune, J. Mao, M. J. Mitchell, R. L. Lesher, J. B. Simpas, A. R. Metcalf, J. J. Schwab, C. Cai, Y. Li, K. L. Demerjian, H. D. Felton, G. Boynton, A. Adams, J. Perry, Y. He, X. Zhou and J. Hou. Behavior of OH and HO_2 in the winter atmosphere in New York City. *Atmos. Env.* **40, Supplement 2**, 252–263 (2006).
60. S. Chen, X. Ren, J. Mao, Z. Chen, W. H. Brune, B. Lefer, B. Rappenglück, J. Flynn, J. Olson and J. H. Crawford. A comparison of chemical mechanisms based on TRAMP-2006 field data. *Atmos. Env.* **44**, 4116–4125 (2010).
61. J. Mao, X. Ren, S. Chen, W. H. Brune, Z. Chen, M. Martinez, H. Harder, B. Lefer, B. Rappenglück, J. Flynn and M. Leuchner. Atmospheric oxidation capacity in the summer of Houston 2006: Comparison with summer measurements in other metropolitan studies. *Atmos. Env.* **44**, 4107–4115 (2010).

62. X. Ren, H. Harder, M. Martinez, I. Faloona, D. Tan, R. Lesher, P. Di Carlo, J. Simpas and W. Brune. Interference testing for atmospheric HO_x measurements by laser-induced fluorescence. *J. Atmos. Chem.* **47**, 169–190 (2004).
63. K. M. Emmerson and N. Carslaw. Night-time radical chemistry during the TORCH campaign. *Atmos. Env.* **43**, 3220–3226. (2009).
64. A. Hofzumahaus, F. Rohrer, K. Lu, B. Bohn, T. Brauers, C.-C. Chang, H. Fuchs, F. Holland, K. Kita, Y. Kondo, X. Li, S. Lou, M. Shao, L. Zeng, A. Wahner and Y. Zhang, Amplified trace gas removal in the troposphere. *Science* **324**, 1702–1704 (2009).
65. F. Holland, U. Aschmutat, M. Heßling, A. Hofzumahaus and D. H. Ehhalt. Highly time resolved measurements of OH during POPCORN using laser-induced fluorescence spectroscopy. *J. Atmos. Chem.* **31**, 205–225 (1998).
66. A. Geyer, K. Bächmann, A. Hofzumahaus, F. Holland, S. Konrad, T. Klüpfel, H.-W. Pätz, D. Perner, D. Mihelcic, H.-J. Schäfer, A. Volz-Thomas and U. Platt. Nighttime formation of peroxy and hydroxyl radicals during the BERLIOZ campaign: Observations and modeling studies. *J. Geophys. Res.* **108**, 8249 (2003).
67. J. Mao, X. Ren, L. Zhang, D. M. Van Duin, R. C. Cohen, J. H. Park, A. H. Goldstein, F. Paulot, M. R. Beaver, J. D. Crounse, P. O. Wennberg, J. P. DiGangi, S. B. Henry, F. N. Keutsch, C. Park, G. W. Schade, G. M. Wolfe, J. A. Thornton and W. H. Brune. Insights into hydroxyl measurements and atmospheric oxidation in a California forest. *Atmos. Chem. Phys.* **12**, 8009–8020 (2012).
68. W. H. Brune, P. S. Stevens and J. H. Mather. Measuring OH and HO_2 in the Troposphere by Laser-Induced Fluorescence at Low Pressure. *J. Atmos. Sci.* **52**, 3328–3336 (1995).
69. A. J. Huisman, J. R. Hottle, M. M. Galloway, J. P. DiGangi, K. L. Coens, W. Choi, I. C. Faloona, J. B. Gilman, W. C. Kuster, J. de Gouw, N. C. Bouvier-Brown, A. H. Goldstein, B. W. LaFranchi, R. C. Cohen, G. M. Wolfe, J. A. Thornton, K. S. Docherty, D. K. Farmer, M. J. Cubison, J. L. Jimenez, J. Mao, W. H. Brune and F. N. Keutsch. Photochemical modeling of glyoxal at a rural site: Observations and analysis from BEARPEX 2007. *Atmos. Chem. Phys.* **11**, 8883–8897 (2011).
70. J. H. Kroll, S. R. Sahay, J. G. Anderson, K. L. Demerjian and N. M. Donahue. Mechanism of HO_x Formation in the gas-phase ozone-alkene reaction. 2. Prompt versus thermal dissociation of carbonyl oxides to Form OH. *J. Phys. Chem. A* **105**, 4446–4457 (2001).
71. A. Novelli, L. Vereecken, J. Lelieveld and H. Harder. Direct observation of OH formation from stabilised Criegee intermediates. *Phys. Chem. Chem. Phys.*, **16** 19941–19951 (2014).
72. P. Rickly, R. F. Hansen and P. S. Stevens *Laboratory Measurements of Potential Interferences with the Detection of OH Radicals using Laser-Induced Fluorescence at Low Pressure, Abstract A13D-0237 presented at 2013 Fall Meeting.* (AGU, San Francisco, Calif. 9–13 Dec. 2013).

73. H. Fuchs, Z. Tan, A. Hofzumahaus, S. Broch, H. P. Dorn, F. Holland, C. Künstler, S. Gomm, F. Rohrer, S. Schrade, R. Tillmann and A. Wahner. Investigation of potential interferences in the detection of atmospheric RO_x radicals by laser-induced fluorescence under dark conditions. *Atmos. Meas. Tech. Discuss.* **8**, 12475–12523 (2015).
74. D. Heard. Atmospheric chemistry: The X factor. *Nature* **488**, 164–165 (2012).
75. R. L. Mauldin III, T. Berndt, M. Sipila, P. Paasonen, T. Petaja, S. Kim, T. Kurten, F. Stratmann, V. M. Kerminen and M. Kulmala. A new atmospherically relevant oxidant of sulphur dioxide. *Nature* **488**, 193–196 (2012).
76. H. Fuchs, T. Brauers, R. Häseler, F. Holland, D. Mihelcic, P. Müsgen, F. Rohrer, R. Wegener and A. Hofzumahaus. Intercomparison of peroxy radical measurements obtained at atmospheric conditions by laser-induced fluorescence and electron spin resonance spectroscopy. *Atmos. Meas. Tech.* **2**, 55–64 (2009).
77. X. Ren, G. D. Edwards, C. A. Cantrell, R. L. Lesher, A. R. Metcalf, T. Shirley and W. H. Brune. Intercomparison of peroxy radical measurements at a rural site using laser-induced fluorescence and Peroxy Radical Chemical Ionization Mass Spectrometer (PerCIMS) techniques. *J. Geophys. Res.* **108**, (2003) doi:10.1029/2003jd003644.
78. J. Lelieveld, T. M. Butler, J. N. Crowley, T. J. Dillon, H. Fischer, L. Ganzeveld, H. Harder, M. G. Lawrence, M. Martinez, D. Taraborrelli and J. Williams. Atmospheric oxidation capacity sustained by a tropical forest. *Nature* **452**, 737–740 (2008).
79. H. Fuchs, T. Brauers, H. P. Dorn, H. Harder, R. Häseler, A. Hofzumahaus, F. Holland, Y. Kanaya, Y. Kajii, D. Kubistin, S. Lou, M. Martinez, K. Miyamoto, S. Nishida, M. Rudolf, E. Schlosser, A. Wahner, A. Yoshino and U. Schurath. Technical Note: Formal blind intercomparison of HO_2 measurements in the atmosphere simulation chamber SAPHIR during the HOxComp campaign. *Atmos. Chem. Phys.* **10**, 12233–12250 (2010).
80. D. J. Creasey, D. E. Heard and J. D. Lee. Eastern Atlantic Spring Experiment 1997 (EASE97) 1. Measurements of OH and HO2 concentrations at Mace Head, Ireland. *J. Geophys. Res.* **107**, ACH 3-1-ACH 3-15 (2002).
81. K. D. Lu, F. Rohrer, F. Holland, H. Fuchs, B. Bohn, T. Brauers, C. C. Chang, R. Häseler, M. Hu, K. Kita, Y. Kondo, X. Li, S. R. Lou, S. Nehr, M. Shao, L. M. Zeng, A. Wahner, Y. H. Zhang and A. Hofzumahaus. Observation and modelling of OH and HO_2 concentrations in the Pearl River Delta 2006: A missing OH source in a VOC rich atmosphere. *Atmos. Chem. Phys.* **12**, 1541–1569 (2012).
82. M. Lew, S. Dusanter, J. Liljegren, B. Bottorff and P. S. Stevens. *Measurement of Interferences Associated with the Detection of the Hydroperoxy Radical in the Atmosphere using Laser-Induced Fluorescence, Abstract A13D-0243 presented at 2013 Fall Meeting* (AGU, San Francisco, Calif. 9-13 Dec. 2013).

83. G. M. Wolfe, C. Cantrell, S. Kim, R. L. Mauldin Iii, T. Karl, P. Harley, A. Turnipseed, W. Zheng, F. Flocke, E. C. Apel, R. S. Hornbrook, S. R. Hall, K. Ullmann, S. B. Henry, J. P. DiGangi, E. S. Boyle, L. Kaser, R. Schnitzhofer, A. Hansel, M. Graus, Y. Nakashima, Y. Kajii, A. Guenther and F. N. Keutsch. Missing peroxy radical sources within a summertime ponderosa pine forest. *Atmos. Chem. Phys.* **14**, 4715–4732 (2014).
84. L. A. George, T. M. Hard and R. J. O'Brien. Measurement of free radicals OH and HO_2 in Los Angeles smog. *J. Geophys. Res.* **104**, 11643–11655 (1999).
85. S. Konrad, T. Schmitz, H. J.Buers, N. Houben, K. Mannschreck, D. Mihelcic, P. Müsgen, H.W. Pätz, F. Holland, A. Hofzumahaus, H. J. Schäfer, S.Schröder, A. Volz-Thomas, K. Bächmann, S. Schlomski, G. Moortgat and D. Großmann. Hydrocarbon measurements at Pabstthum during the BERLIOZ campaign and modeling of free radicals. *J. Geophys. Res.* **108**, 8251 (2003).
86. K. M. Emmerson, N. Carslaw, L. J. Carpenter, D. E. Heard, J. D. Lee and M. J. Pilling. Urban atmospheric chemistry during the PUMA campaign 1: Comparison of Modelled OH and HO_2 concentrations with measurements. *J. Atmos. Chem.* **52**, 143–164 (2005).
87. T. R. Shirley, W. H. Brune, X. Ren, J. Mao, R. Lesher, B. Cardenas, R. Volkamer, L. T. Molina, M. J. Molina, B. Lamb, E. Velasco, T. Jobson and M. Alexander. Atmospheric oxidation in the Mexico City Metropolitan Area (MCMA) during April 2003. *Atmos. Chem. Phys.* **6**, 2753–2765 (2006).
88. K. M. Emmerson, N. Carslaw, D. C. Carslaw, J. D. Lee, G. McFiggans, W. J. Bloss, T. Gravestock, D. E. Heard, J. Hopkins, T. Ingham, M. J. Pilling, S. C. Smith, M. Jacob and P. S. Monks. Free radical modelling studies during the UK TORCH Campaign in Summer 2003. *Atmos. Chem. Phys.* **7**, 167–181 (2007).
89. P. M. Sheehy, R. Volkamer, L. T. Molina, and M. J. Molina. Oxidative capacity of the Mexico City atmosphere — Part 2: A RO_x radical cycling perspective. *Atmos. Chem. Phys.* **10**, 6993–7008 (2010).
90. B. H. Czader, X. Li and B. Rappenglueck CMAQ modeling and analysis of radicals, radical precursors, and chemical transformations. *J. Geophys. Res.* **118**, 11376–311387 (2013).
91. Y. Kanaya, R. Cao, H. Akimoto, M. Fukuda, Y. Komazaki, Y. Yokouchi, M. Koike, H. Tanimoto, N. Takegawa and Y. Kondo. Urban photochemistry in central Tokyo: 1. Observed and modeled OH and HO_2 radical concentrations during the winter and summer of 2004. *J. Geophys. Res.* **112**, (2007), 10.1029/2007jd008670.
92. D. Kubistin, H. Harder, M. Martinez, M. Rudolf, R. Sander, H. Bozem, G. Eerdekens, H. Fischer, C. Gurk, T. Klüpfel, R. Königstedt, U. Parchatka, C. L. Schiller, A. Stickler, D. Taraborrelli, J. Williams and J. Lelieveld. Hydroxyl radicals in the tropical troposphere over the Suriname rainforest: comparison of measurements with the box model MECCA. *Atmos. Chem. Phys.* **10**, 9705–9728 (2010).

93. L. K. Whalley, P. M. Edwards, K. L. Furneaux, A. Goddard, T. Ingham, M. J. Evans, D. Stone, J. R. Hopkins, C. E. Jones, A. Karunaharan, J. D. Lee, A. C. Lewis, P. S. Monks, S. J. Moller and D. E. Heard. Quantifying the magnitude of a missing hydroxyl radical source in a tropical rainforest. *Atmos. Chem. Phys.* **11**, 7223–7233 (2011).
94. D. Tan, I. Faloona, J. B. Simpas, W. Brune, P. B. Shepson, T. L. Couch, A. L. Sumner, M. A. Carroll, T. Thornberry, E. Apel, D. Riemer and W. Stockwell. HO_x budgets in a deciduous forest: Results from the PROPHET summer 1998 campaign. *J. Geophys. Res.* **106**, 24407–24427 (2001).
95. N. Carslaw, D. J. Creasey, D. Harrison, D. E. Heard, M. C. Hunter, P. J. Jacobs, M. E. Jenkin, J. D. Lee, A. C. Lewis, M. J. Pilling and S. M. Saunders. OH and HO_2 radical chemistry in a forested region of north-western Greece. *Atmos. Environ.* **35**, 4725–4737 (2001).
96. T. A. M. Pugh, A. R. MacKenzie, C. N. Hewitt, B. Langford, P. M. Edwards, K. L. Furneaux, D. E. Heard, J. R. Hopkins, C. E. Jones, A. Karunaharan, J. Lee, G. Mills, P. Misztal, S. Moller, P. S. Monks, and L. K. Whalley. Simulating atmospheric composition over a South-East Asian tropical rainforest: Performance of a chemistry box model. *Atmos. Chem. Phys.* **10**, 279-298 (2010).
97. S. M. MacDonald, H. Oetjen, A. S. Mahajan, L. K. Whalley, P. M. Edwards, D. E. Heard, C. E. Jones and J. M. C. Plane. DOAS measurements of formaldehyde and glyoxal above a south-east Asian tropical rainforest. *Atmos. Chem. Phys.* **12**, 5949–5962 (2012).
98. U. Kuhn, M. O. Andreae, C. Ammann, A. C. Araújo, E. Brancaleoni, P. Ciccioli, T. Dindorf, M. Frattoni, L. V. Gatti, L. Ganzeveld, B. Kruijt, J. Lelieveld, J. Lloyd, F. X. Meixner, A. D. Nobre, U. Pöschl, C. Spirig, P. Stefani, A. Thielmann, R. Valentini and J. Kesselmeier. Isoprene and monoterpene fluxes from Central Amazonian rainforest inferred from tower-based and airborne measurements, and implications on the atmospheric chemistry and the local carbon budget. *Atmos. Chem. Phys.* **7**, 2855–2879 (2007).
99. F. Paulot, J. D. Crounse, H. G. Kjaergaard, J. H. Kroll, J. H. Seinfeld and P. O. Wennberg. Isoprene photooxidation: New insights into the production of acids and organic nitrates. *Atmos. Chem. Phys.* **9**, 1479–1501 (2009).
100. F. Paulot, J. D. Crounse, H. G. Kjaergaard, A. Kürten, J. M. St. Clair, J. H. Seinfeld and P. O. Wennberg. Unexpected Epoxide Formation in the Gas-Phase Photooxidation of Isoprene. *Science* **325**, 730–733 (2009).
101. G. M. Wolfe, J. A. Thornton, M. McKay and A. H. Goldstein. Forest-atmosphere exchange of ozone: Sensitivity to very reactive biogenic VOC emissions and implications for in-canopy photochemistry. *Atmos. Chem. Phys.* **11**, 7875–7891 (2011).
102. X. Zhou, N. Zhang, M. TerAvest, D. Tang, J. Hou, S. Bertman, M. Alaghmand, P. B. Shepson, M. A. Carroll, S. Griffith, S. Dusanter and P. S. Stevens. Nitric acid photolysis on forest canopy surface as a source for tropospheric nitrous acid. *Nature Geosci.* **4**, 440–443 (2011).

103. J. A. Thornton, P. J. Wooldridge, R. C. Cohen, M. Martinez, H. Harder, W. H. Brune, E. J. Williams, J. M. Roberts, F. C. Fehsenfeld, S. R. Hall, R. E. Shetter, B. P. Wert and A. Fried. Ozone production rates as a function of NO_x abundances and HO_x production rates in the Nashville urban plume. *J. Geophys. Res.* **107**, (2002) ACH 7-1-ACH 7-17.
104. J. Peeters, T. L. Nguyen and L. Vereecken. HO_x radical regeneration in the oxidation of isoprene. *Phys. Chem. Chem. Phys.* **11**, 5935–5939, (2009).
105. J. Peeters and J.-F. Muller. HO_x radical regeneration in isoprene oxidation via peroxy radical isomerisations. II: Experimental evidence and global impact. *Phys. Chem. Chem. Phys.* **12**, 14227–14235 (2010).
106. G. da Silva, C. Graham and Z.-F. Wang. Unimolecular β-hydroxyperoxy radical decomposition with OH recycling in the photochemical oxidation of isoprene. *Environ. Sci. Technol.* **44**, 250–256 (2010).
107. D. Taraborrelli, M. G. Lawrence, J. N. Crowley, T. J. Dillon, S. Gromov, C. B. M. Grosz, L. Vereecken and L. Lelieveld. Hydroxyl radical buffered by isoprene oxidation over tropical forests. *Nature Geosci.* **5**, 190–193 (2012).
108. T. Stavrakou, J. Peeters and J. F. Müller. Improved global modelling of HOx recycling in isoprene oxidation: Evaluation against the GABRIEL and INTEX-A aircraft campaign measurements. *Atmos. Chem. Phys.* **10**, 9863–9878 (2010).
109. A. Guenther. Atmospheric chemistry: Are plant emissions green?. *Nature* **452**, 701–702 (2008).
110. M. P. Sulbaek Andersen, M. D. Hurley, T. J. Wallington, J. C. Ball, J. W. Martin, D. A. Ellis and S. A. Mabury. Atmospheric chemistry of C_2F_5CHO: mechanism of the $C_2F_5C(O)O_2$ + HO_2 reaction. *Chem. Phys. Lett.* **381**, 14–21 (2003).
111. A. S. Hasson, G. S. Tyndall and J. J. Orlando. A product yield study of the reaction of HO_2 radicals with ethyl peroxy ($C_2H_5O_2$), acetyl peroxy ($CH_3C(O)O_2$), and acetonyl peroxy ($CH_3C(O)CH_2O_2$) radicals. *J. Phys. Chem. A* **108**, 5979–5989 (2004).
112. M. E. Jenkin, M. D. Hurley and T. J. Wallington. Investigation of the radical product channel of the $CH_3C(O)O_2$ + HO_2 reaction in the gas phase. *Phys. Chem. Chem. Phys.* **9**, 3149–3162 (2007).
113. T. J. Dillon and J. N. Crowley. Direct detection of OH formation in the reactions of HO_2 with $CH_3C(O)O_2$ and other substituted peroxy radicals. *Atmos. Chem. Phys.* **8**, 4877–4889 (2008).
114. C. B. M. Groß, T. J. Dillon, G. Schuster, J. Lelieveld and J. N. Crowley. Direct Kinetic Study of OH and O_3 Formation in the Reaction of $CH_3C(O)O_2$ with HO_2. *J. Phys. Chem. A* **118**, 974–985 (2014).
115. A. S. Hasson, G. S. Tyndall, J. J. Orlando, S. Singh, S. Q. Hernandez, S. Campbell and Y. Ibarra. Branching Ratios for the Reaction of Selected Carbonyl-Containing Peroxy Radicals with Hydroperoxy Radicals. *J. Phys. Chem. A* **116**, 6264–6281 (2012).
116. T. M. Butler, D. Taraborrelli, C. Brühl, H. Fischer, H. Harder, M. Martinez, J. Williams, M. G. Lawrence and J. Lelieveld. Improved simulation of isoprene oxidation chemistry with the ECHAM5/MESSy chemistry-climate

model: Lessons from the GABRIEL airborne field campaign. *Atmos. Chem. Phys.* **8**, 4529–4546 (2008).

117. A. S. Hasson, K. T. Kuwata, M. C. Arroyo and E. B. Peterson. Theoretical studies of the reaction of hydroperoxy radicals (HO_2) with ethyl peroxy ($CH_3CH_2O_2$), acetyl peroxy ($CH_3C(O)O_2$), and acetonyl peroxy ($CH_3C(O)CH_2O_2$) radicals. *J. Photochem. Photobiol. A* **176**, 218–230 (2005).
118. E. W. Kaiser, I. R. Pala and T. J. Wallington. Kinetics and mechanism of the reaction of methacrolein with chlorine atoms in 1-950 Torr of N_2 or N_2/O_2 diluent at 297 K. *J. Phys. Chem. A* **114**, 6850-6860, (2010).
119. E. Praske, J. D. Crounse, K. H. Bates, T. Kurtén and H. G. Kjaergaard, P. O. Wennberg. Atmospheric fate of methyl vinyl ketone: Peroxy radical reactions with NO and HO_2. *J. Phys. Chem. A* **119**, 4562–4572 (2015).
120. D. Stone, M. J. Evans, P. M. Edwards, R. Commane, T. Ingham, A. R. Rickard, D. M. Brookes, J. Hopkins, R. J. Leigh, A. C. Lewis, P. S. Monks, D. Oram, C. E. Reeves, D. Stewart and D. E. Heard, Isoprene oxidation mechanisms: Measurements and modelling of OH and HO_2 over a South-East Asian tropical rainforest during the OP3 field campaign. *Atmos. Chem. Phys.* **11**, 6749–6771 (2011).
121. J. D. Crounse, F. Paulot, H. G. Kjaergaard and P. O. Wennberg. Peroxy radical isomerization in the oxidation of isoprene. *Phys. Chem. Chem. Phys.* (2011).
122. J. Peeters, J.-F. Müller, T. Stavrakou and V. S. Nguyen. Hydroxyl radical recycling in isoprene oxidation driven by hydrogen bonding and hydrogen tunneling: The upgraded LIM1 mechanism. *J. Phys. Chem. A* **118**, 8625–8643 (2014).
123. G. M. Wolfe, J. D. Crounse, J. D. Parrish, J. M. St. Clair, M. R. Beaver, F. Paulot, T. P. Yoon, P. O. Wennberg and F. N. Keutsch. Photolysis. OH reactivity and ozone reactivity of a proxy for isoprene-derived hydroperoxyenals (HPALDs). *Phys. Chem. Chem. Phys.* **14**, 7276–7286 (2012).
124. H. Fuchs, A. Hofzumahaus, F. Rohrer, B. Bohn, T. Brauers, H. P. Dorn, R. Haseler, F. Holland, M. Kaminski, X. Li, K. Lu, S. Nehr, R. Tillmann, R. Wegener and A. Wahner. Experimental evidence for efficient hydroxyl radical regeneration in isoprene oxidation. *Nature Geosci* **6**, 1023–1026 (2013).
125. J. D. Crounse, H. C. Knap, K. B. Ørnsø, S. Jørgensen, F. Paulot, H. G. Kjaergaard and P. O. Wennberg. Atmospheric fate of methacrolein. 1. Peroxy radical isomerization following addition of OH and O_2. *J. Phys. Chem. A* **116**, 5756–5762 (2012).
126. R. Asatryan, G. d. Silva and J. W. Bozzelli. Quantum chemical study of the acrolein (CH_2CHCHO) + OH + O_2 reactions. *J. Phys. Chem. A* **114**, 8302–8311 (2010).
127. H. Fuchs, I. H. Acir, B. Bohn, T. Brauers, H. P. Dorn, R. Häseler, A. Hofzumahaus, F. Holland, M. Kaminski, X. Li, K. Lu, A. Lutz, S. Nehr, F. Rohrer, R. Tillmann, R. Wegener and A. Wahner. OH regeneration from

methacrolein oxidation investigated in the atmosphere simulation chamber SAPHIR. *Atmos. Chem. Phys.* **14**, 7895–7908 (2014).
128. G. da Silva. Reaction of Methacrolein with the Hydroxyl Radical in Air: Incorporation of Secondary O2 Addition into the MACR + OH Master Equation. *J. Phys. Chem. A* **116**, 5317–5324 (2012).
129. T. F. Mentel, M. Springer, M. Ehn, E. Kleist, I. Pullinen, T. Kurtén, M. Rissanen, A. Wahner and J. Wildt. Formation of highly oxidized multifunctional compounds: Autoxidation of peroxy radicals formed in the ozonolysis of alkenes–deduced from structure–product relationships. *Atmos. Chem. Phys.* **15**, 6745–6765 (2015).
130. A. H. Goldstein and I. E. Galbally. Known and unexplored organic constituents in the earth's atmosphere. *Environ. Sci. Technol.* **41**, 1514–1521 (2007).
131. S. Nehr, B. Bohn, H. P. Dorn, H. Fuchs, R. Häseler, A. Hofzumahaus, X. Li, F. Rohrer, R. Tillmann and A. Wahner. Atmospheric photochemistry of aromatic hydrocarbons: OH budgets during SAPHIR chamber experiments. *Atmos. Chem. Phys.* **14**, 6941–6952 (2014).
132. F. Kirchner, F. Jeanneret, A. Clappier, B. Kruger, H. Van den Bergh and B. Calpini. Total VOC reactivity in the planetary boundary layer 2. A new indicator for determining the sensitivity of the ozone production to VOC and NO_x. *J. Geophys. Res.* **106**, 3095–3110 (2001).
133. V. Sinha, J. Williams, J. M. Diesch, F. Drewnick, M. Martinez, H. Harder, E. Regelin, D. Kubistin, H. Bozem, Z. Hosaynali-Beygi, H. Fischer, M. D. Andrés-Hernández, D. Kartal, J. A. Adame and J. Lelieveld. Constraints on instantaneous ozone production rates and regimes during DOMINO derived using in-situ OH reactivity measurements. *Atmos. Chem. Phys.* **12**, 7269–7283 (2012).
134. R. F. Hansen, M. Blocquet, C. Schoemaecker, T. Léonardis, N. Locoge, C. Fittschen, B. Hanoune, P. S. Stevens, V. Sinha and S. Dusanter. Intercomparison of the comparative reactivity method (CRM) and pump–probe technique for measuring total OH reactivity in an urban environment. *Atmos. Meas. Tech.* **8**, 4243–4264 (2015).
135. T. A. Kovacs and W. H. Brune, Total OH Loss Rate Measurement. *J. Atmos. Chem.* **39**, 105–122 (2001).
136. T. Ingham, A. Goddard, L. K. Whalley, K. L. Furneaux, P. M. Edwards, C. P. Seal, D. E. Self, G. P. Johnson, K. A. Read, J. D. Lee and D. E. Heard. A flow-tube based laser-induced fluorescence instrument to measure OH reactivity in the troposphere. *Atmos. Meas. Tech.* **2**, 465–477 (2009).
137. R. F. Hansen, S. M. Griffith, S. Dusanter, P. S. Rickly, P. S. Stevens, S. B. Bertman, M. A. Carroll, M. H. Erickson, J. H. Flynn, N. Grossberg, B. T. Jobson, B. L. Lefer and H. W. Wallace. Measurements of total hydroxyl radical reactivity during CABINEX 2009 — Part 1: Field measurements. *Atmos. Chem. Phys.* **14**, 2923–2937 (2014).
138. Y. Sadanaga, A. Yoshino, K. Watanabe, A. Yoshioka, Y. Wakazono, Y. Kanaya and Y. Kajii. Development of a measurement system of OH

reactivity in the atmosphere by using a laser-induced pump and probe technique. *Rev. Sci. Instr.* **75**, 2648–2655 (2004).

139. S. Lou, F. Holland, F. Rohrer, K. Lu, B. Bohn, T. Brauers, C. C. Chang, H. Fuchs, R. Haseler, K. Kita, Y. Kondo, X. Li, M. Shao, L. Zeng, A. Wahner, Y. Zhang, W. Wang and A. Hofzumahaus. Atmospheric OH reactivities in the Pearl River Delta — China in summer 2006: Measurement and model results. *Atmos. Chem. Phys.* **10**, 11243–11260 (2010).
140. A. E. Parker, D. Amédro, C. Schoemaecker and C. Fittschen. OH Radical Reactivity Measurements by FAGE. *Environmental Engineering Management* **10**, 107–114 (2011).
141. D. Stone, L. K. Whalley, T. Ingham, P. M. Edwards, D. R. Cryer, C. A. Brumby, P. W. Seakins and D. E. Heard. Measurement of OH reactivity by laser flash photolysis coupled with laser-induced fluorescence spectroscopy. *Atmos. Meas. Tech.* **9**, 2827–2844 (2016).
142. V. Sinha, J. N. Crowley and J. Lelieveld. The comparative reactivity method–a new tool to measure total OH Reactivity in ambient air. *Atmos. Chem. Phys* **8**, 2213–2227 (2008).
143. S. Kim, A. Guenther, T. Karl and J. Greenberg. Contributions of primary and secondary biogenic VOC to total OH reactivity during the CABINEX (Community Atmosphere-Biosphere Interactions Experiments)-09 field campaign. *Atmos. Chem. Phys.* **11**, 8613–8623 (2011).
144. A. C. Nölscher, V. Sinha, S. Bockisch, T. Klüpfel and J. Williams. Total OH reactivity measurements using a new fast Gas Chromatographic Photo-Ionization Detector (GC-PID). *Atmos. Meas. Tech.* **5**, 2981–2992 (2012).
145. C. Dolgorouky, V. Gros, R. Sarda-Esteve, V. Sinha, J. Williams, N. Marchand, S. Sauvage, L. Poulain, J. Sciare and B. Bonsang. Total OH reactivity measurements in Paris during the 2010 MEGAPOLI winter campaign. *Atmos. Chem. Phys.* **12**, 9593–9612 (2012).
146. V. Kumar and V. Sinha. A new technique for rapid measurements of ambient total OH reactivity and volatile organic compounds using a single proton transfer reaction mass spectrometer. *Int. J. Mass Spectrom*, **374**, 55–63 (2014).
147. A. Yoshino, Y. Sadanaga, K. Watanabe, S. Kato, Y. Miyakawa, J. Matsumoto and Y. Kajii. Measurement of total OH reactivity by laser-induced pump and probe technique — comprehensive observations in the urban atmosphere of Tokyo. *Atmos. Env.* **40**, 7869–7881 (2006).
148. V. Sinha, J. Williams, J. Lelieveld, T. M. Ruuskanen, M. K. Kajos, J. Patokoski, H. Hellen, H. Hakola, D. Mogensen, M. Boy, J. Rinne and M. Kulmala. OH reactivity measurements within a boreal forest: Evidence for unknown reactive emissions. *Environ. Sci. Technol.* **44**, 6614–6620 (2010).
149. V. Michoud, R. F. Hansen, N. Locoge, P. S. Stevens and S. Dusanter. Detailed characterizations of the new Mines Douai comparative reactivity method instrument via laboratory experiments and modeling. *Atmos. Meas. Tech.* **8**, 3537–3553 (2015).

150. J. Mao, X. Ren, W. H. Brune, J. R. Olson, J. H. Crawford, A. Fried, G. Huey, R. C. Cohen, B. Heikes, H. B. Singh, D. R. Blake, G. W. Sachse, G. S. Diskin, S. R. Hall and R. E. Shetter. Airborne measurement of OH reactivity during INTEX-B. *Atmos. Chem. Phys.* **9**, 163–173 (2009).
151. T. A. Kovacs, W. H. Brune, H. Harder, M. Martinez, J. B. Simpas, G. J. Frost, E. Williams, T. Jobson, C. Stroud, V. Young, A. Fried and B. Wert. Direct measurements of urban OH reactivity during Nashville SOS in summer 1999. *J. Environ. Monit.* **5**, 48–74 (2003).
152. B. Calpini, F. Jeanneret, M. Bourqui, A. Clappier, R. Vajtai and H. van den Bergh. Direct measurement of the total reaction rate of OH in the atmosphere. *Analysis* **27**, 328–336 (1999).
153. T. J. Dillon, M. E. Tucceri, K. Dulitz, A. Horowitz, L. Vereecken and J. N. Crowley. Reaction of hydroxyl radicals with C_4H_5N (Pyrrole): Temperature and pressure dependent rate coefficients. *J. Phys. Chem. A* **116**, 6051–6058 (2012).
154. N. Zannoni, S. Dusanter, V. Gros, R. Sarda Esteve, V. Michoud, V. Sinha, N. Locoge and B. Bonsang. Intercomparison of two comparative reactivity method instruments in the Mediterranean basin during summer 2013. *Atmos. Meas. Tech.* **8**, 3851–3865 (2015).
155. J. Williams and W. Brune. A roadmap for OH reactivity research. *Atmos. Env.* **106**, 371–372 (2015).
156. H. Sonderfeld, I. R. White, I. C. A. Goodall, J. R. Hopkins, A. C. Lewis, R. Koppmann and P. S. Monks. What effect does VOC sampling time have on derived OH reactivity? *Atmos. Chem. Phys.* **16**, 6303–6318 (2016).
157. Y. Sadanaga, A. Yoshino, S. Kato, A. Yoshioka, K. Watanabe, Y. Miyakawa, I. Hayashi, M. Ichikawa, J. Matsumoto, A. Nishiyama, N. Akiyama, Y. Kanaya and Y. Kajii. The importance of NO_2 and volatile organic compounds in the urban air from the viewpoint of the OH reactivity. *Geophys. Res. Lett.* **31**, (2004).
158. S. Chatani, N. Shimo, S. Matsunaga, Y. Kajii, S. Kato, Y. Nakashima, K. Miyazaki, K. Ishii and H. Ueno. Sensitivity analyses of OH missing sinks over Tokyo metropolitan area in the summer of 2007. *Atmos. Chem. Phys.* **9**, 8975–8986 (2009).
159. X. Ren, W. H. Brune, C. A. Cantrell, G. D. Edwards, T. Shirley, A. R. Metcalf and R. L. Lesher, Hydroxyl and peroxy radical chemistry in a rural area of central Pennsylvania: Observations and model comparisons. *J. Atmos. Chem.* **52**, 231–257 (2005).
160. S. E. Pusede, D. R. Gentner, P. J. Wooldridge, E. C. Browne, A. W. Rollins, K. E. Min, A. R. Russell, J. Thomas, L. Zhang, W. H. Brune, S. B. Henry, J. P. DiGangi, F. N, Keutsch, S. A. Harrold, J. A. Thornton, M. R. Beaver, J. M. St. Clair, P. O. Wennberg, J. Sanders, X. Ren, T. C. VandenBoer, M. Z. Markovic, A. Guha, R. Weber, A. H. Goldstein and R. C. Cohen. On the temperature dependence of organic reactivity, nitrogen oxides, ozone production, and the impact of emission controls in San Joaquin Valley. California. *Atmos. Chem. Phys.* **14**, 3373–3395 (2014).

161. P. Di Carlo, W. H. Brune, M. Martinez, H. Harder, R. Lesher, X. Ren, T. Thornberry, M. A. Carroll, V. Young, P. B. Shepson, D. Riemer, E. Apel and C. Campbell. Missing OH Reactivity in a Forest: Evidence for Unknown Reactive Biogenic VOCs. *Science* **304**, 722–725 (2004).
162. X. Ren, W. H. Brune, A. Oliger, A. R. Metcalf, J. B. Simpas, T. Shirley, J. J. Schwab, C. Bai, U. Roychowdhury, Y. Li, C. Cai, K. L. Demerjian, Y. He, X. Zhou, H. Gao and J. Hou. OH, HO_2, and OH reactivity during the PMTACS-NY Whiteface Mountain 2002 campaign: Observations and model comparison. *J. Geophys. Res.* **111**, D10S03, (2006), doi:1029/2005JD 006126.
163. P. M. Edwards, M. J. Evans, K. L. Furneaux, J. Hopkins, T. Ingham, C. Jones, J. D. Lee, A. C. Lewis, S. J. Moller, D. Stone, L. K. Whalley and D. E. Heard. OH reactivity in a South East Asian tropical rainforest during the Oxidant and Particle Photochemical Processes (OP3) project. *Atmos. Chem. Phys.* **13**, 9497–9514 (2013).
164. Y. Nakashima, S. Kato, J. Greenberg, P. Harley, T. Karl, A. Turnipseed, E. Apel, A. Guenther, J. D. Smith and Y. Kajii. Total OH reactivity measurements in ambient air in a southern Rocky mountain ponderosa pine forest during BEACHON-SRM08 summer campaign. *Atmos. Env.* **85**, 1–8 (2014).
165. A. C. Nölscher, J. Williams, V. Sinha, T. Custer, W. Song, A. M. Johnson, R. Axinte, H. Bozem, H. Fischer, N. Pouvesle, G. Phillips, J. N. Crowley, P. Rantala, J. Rinne, M. Kulmala, D. Gonzales, J. Valverde-Canossa, A. Vogel, T. Hoffmann, H. G. Ouwersloot, J. Vilà-Guerau de Arellano and J. Lelieveld. Summertime total OH reactivity measurements from boreal forest during HUMPPA-COPEC 2010. *Atmos. Chem. Phys.* **12**, 8257–8270 (2012).
166. A. C. Nölscher, A. M. Yanez-Serrano, S. Wolff, A. C. de Araujo, J. V. Lavric, J. Kesselmeier and J. Williams. Unexpected seasonality in quantity and composition of Amazon rainforest air reactivity *Nat. Commun.* **7**, (2016).
167. J. Kaiser, K. M. Skog, K. Baumann, S. B. Bertman, S. B. Brown, W. H. Brune, J. D. Crounse, J. A. de Gouw, E. S. Edgerton, P. A. Feiner, A. H. Goldstein, A. Koss, P. K. Misztal, T. B. Nguyen, K. F. Olson, J. M. St. Clair, A. P. Teng, S. Toma, P. O. Wennberg, R. J. Wild, L. Zhang and F. N. Keutsch. Speciation of OH reactivity above the canopy of an isoprene-dominated forest. *Atmos. Chem. Phys.* **16**, 9349–9359 (2016).
168. J. D. Lee, J. C. Young, K. A. Read, J. F. Hamilton, J. R. Hopkins, A. C. Lewis, B. J. Bandy, J. Davey, P. Edwards, T. Ingham, D. E. Self, S. C. Smith, M. J. Pilling and D. E. Heard. Measurement and calculation of OH reactivity at a United Kingdom coastal site. *J. Atmos. Chem.* **64**, 53–76 (2010).
169. A. C. Nölscher, E. Bourtsoukidis, B. Bonn, J. Kesselmeier, J. Lelieveld and J. Williams. Seasonal measurements of total OH reactivity emission rates from Norway spruce in 2011. *Biogeosciences* **10**, 4241–4257 (2013).
170. A. C. Nölscher, T. Butler, J. Auld, P. Veres, A. Muñoz, D. Taraborrelli, L. Vereecken, J. Lelieveld and J. Williams. Using total OH reactivity to

assess isoprene photooxidation via measurement and model. *Atmos. Env.* **89**, 453–463 (2014).

171. Y. Nakashima, H. Tsurumaru, T. Imamura, I. Bejan, J. C. Wenger and Y. Kajii. Total OH reactivity measurements in laboratory studies of the photooxidation of isoprene. *Atmos. Env.* **62**, 243–247 (2012).
172. A. Hofzumahaus and D. Heard. Assessment of local HO_x and RO_x measurement techniques: Achievements, challenges, and future directions. *IGAC news* **55**, 20–21 (2015).

Index